英汉地学专业术语集注系列
丛书主编 董元兴 张红燕

英汉岩石学专业术语集注

许峰 唐晓云 张莉 冯迪 编著

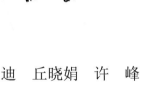

丛书编委会

主　编　董元兴　张红燕
编　委（按姓氏笔画排序）
　　　　王　伟　江　敏　冯　迪　丘晓娟　许　峰
　　　　李　慷　张红燕　张　莉　杨红燕　周　艳
　　　　赵　妍　唐晓云　董元兴

武汉大学出版社

图书在版编目(CIP)数据

英汉岩石学专业术语集注/许峰等编著. —武汉：武汉大学出版社，2018.9
英汉地学专业术语集注系列/董元兴，张红燕主编
ISBN 978-7-307-19912-5

Ⅰ.英… Ⅱ.许… Ⅲ.岩石学—术语—英、汉 Ⅳ.P58-61

中国版本图书馆 CIP 数据核字(2017)第 309195 号

责任编辑：谢群英 李 玚　　责任校对：汪欣怡　　版式设计：韩闻锦

出版发行：武汉大学出版社　　(430072　武昌　珞珈山)
　　　　　(电子邮件：cbs22@whu.edu.cn　网址：www.wdp.com.cn)
印刷：北京虎彩文化传播有限公司
开本：787×1092　1/16　印张：42.5　字数：1031 千字　插页：1
版次：2018 年 9 月第 1 版　　2018 年 9 月第 1 次印刷
ISBN 978-7-307-19912-5　　定价：88.00 元

版权所有，不得翻印；凡购我社的图书，如有质量问题，请与当地图书销售部门联系调换。

序　　言

《英汉地学专业术语集注》以中国地质大学重点培育学科建设项目《基于语料库的跨学科科技英语翻译研究》为依托，由专门从事英汉古生物学、岩石学和矿物学领域的专家精心选择专业术语编撰而成。《英汉地学专业术语集注》词条选择科学，力图涵盖古生物学、岩石学及矿物学主要核心术语。词条配有中英文对照释义、词条来源并辅助学术期刊及专著例子，有利于读者全面了解学习该术语词条。词条汉语拼音的标注利于国际学生学习并掌握地学领域专业术语词汇。《英汉地学专业术语集注》主要包括三个分册词典，即《英汉矿物学专业术语集注》、《英汉古生物学专业术语集注》及《英汉岩石学专业术语集注》。

三个分册的内容主要由中英文释义、汉语拼音、词条来源、例子及延伸词条等组成，例如：

Sedimentary

Definition：（Geology）（of rock）that has formed from sediment deposited by water or air. （地质）（岩石）沉积形成的（yán shí）chén jī xíng chéng de

Origin：Mid-16th century：from French sédiment or Latin sedimentum "settling"，from sedere "sit".

Example：

1. The determination of semidentary characteristics of the Yanqi basin in Mesozoic Era and the boundaries of paleobasin is a very important scientific research task for the exploration and development of oil and gas in this region. 焉耆中生代古盆地沉积特征与古盆地边界的确定对于该区域油气勘探与开发是一项非常重要的科研任务。

2. Ordos Basin is the second largest sedimentary basin on land in China that possesses abundant oil and gas resources and has bright prospects in oil and gas exploration. 鄂尔多斯盆地是我国陆上第二大沉积盆地，油气资源非常丰富，油气勘探潜力很大。

Extended Terms：

sedimentary-environment 沉积环境

sedimentary-exhalation 沉积喷流

sedimentary-metamorphic 沉积变质

sedimentary-reformation 沉积改造

sedimentary structural evolution 沉积构造演化

sedimentary-structural episode 沉积-构造幕

sedimentary-tectonic 沉积-构造

sedimentary-transformation 沉积-改造

exhalative-sedimentary 喷流-沉积

hydrothermal-sedimentary 热水沉积
tectonic-sedimentary 构造沉积
volcano-sedimentary 火山沉积

2009年10月中国地质大学重点培育学科建设项目《基于语料库的跨学科科技英语翻译研究》获批启动,《英汉地学专业术语集注》于2009年10月开始编撰,2011年10月完成初稿,2014年10月《英汉地学专业术语集注》获得中国地质大学地球生物学基金资助。这里还要说明的是,本书的词条释义,例句的汉语译文并不与原文的专业术语词条和英文例句完全对应,但都是对上述词条与例句的最主要的表述。编写及校订数易其稿,由于经验所限,本书在选词和例句选用等方面难免有不足或错误之处,我们诚恳地希望广大读者提出批评和建议,以利我们进一步改正和修订。

董元兴　张红燕
2016年10月

前　言

　　作为地质学一门分支学科，古生物学是生命科学和地球科学的交叉科学。该学科既是生命科学中唯一具有历史科学性质的一个独特分支，又是地球科学的一个分支。古生物学是历史生物学的一个重要基础和组成部分，不仅研究生命起源、发展历史、生物宏观进化模型、节奏与作用机制等，而且也研究保存在地层中的生物遗体、遗迹、化石，用以确定地层的顺序、时代，了解地壳发展的历史，推断地质史上水陆分布、气候变迁和沉积矿产形成与分布的规律。因此，古生物研究既具有理论意义，又对地质矿产的开发和应用有重要的实际意义。近些年来，中外地质矿产合作日益紧密，地质学以及古生物学研究日益趋向国际化，地质工作者和学生迫切需要有一本古生物学方面的专业术语英汉工具书，为此我们编写了这本专业术语集注。

　　《英汉古生物学专业术语集注》根据古生物的学科特点，共分为六个大部分，分别是古生物学总论、古无脊椎动物学、古脊椎动物学、古植物学、古生态学和地球生物学以及地球生物学和分子生物学。词条选择严格参照古生物学专业术语库。在体例上，以英语音标、英语释义、对应汉语名、汉语拼音注音、词源、例句(含汉语译文)和拓展词组为主要组成部分。在词汇释义和例句选择上紧扣专业主题，对部分较为复杂的复合词术语中的单词也进行了单独释义。词源部分除了提供部分术语在该专业领域的缘起，还包含许多普通词汇的由来和演变，有助于广大师生、研究员、从业人员了解该词汇的来龙去脉。本书兼顾专业和语言，融合了英汉科技专业词典和英语语言词典的双重特色。

　　本书的例句不少来自国内外有关著作和文献，限于体例，未予以注明出处；本词典的编写得到了中国地质大学(武汉)地球科学学院和地球科学学院相关专家的支持和帮助；外语学院赵妍老师的编撰以及2012届研究生许岚、张涵、张馨引、郑红红、辛华等做了大量资料收集整理工作，在此一并表示诚挚的谢意。

　　这里还要说明的是，本书的专业术语词条释义，例句的汉语译文并不与原文的专业术语词条和英文例句完全对应，但都是对上述词条、例句等的最主要的表述。由于时间和经验所限，本书在选词和例句选用等方面还存在许多不足的地方，我们诚恳地希望广大读者提出批评和建议，以便我们今后做进一步的改正和修订。

作者
2016年10月

目　　录

序言　…………………………………………………………………………… 1

前言　…………………………………………………………………………… 1

词汇正文　……………………………………………………………………… 1

A-type granite [ei taip ˈgrænit]

Definition: A 型花岗岩 A xíng huā gāng yán

granite：a very hard, granular, crystalline, igneous rock consisting mainly of quartz, mica, and feldspar and often used as a building stone. 花岗岩 huā gāng yán

Example:

Qianshan granite has character of A-type granite, and it is the product of partial fusion.
潜山花岗岩有着 A 型花岗岩的特征,它是部分熔融的产物。

aa lava [ei ei ˈlɑːvə]

Definition: blocky basalt lava. 块熔岩 kuài róng yán

Example:

Aa lava, on the other hand, moves faster and doesn't have time to develop a skin, resulting in a cooler flow with a more angular texture.
另一方面,块状熔岩移动更快,没有时间形成外皮,这就导致了岩浆以较冷却的流动而形成了更有棱角的质地。

abyssal facies [əˈbisəl ˈfeiʃiiːz]

Definition: 深海相 shēn hǎi xiāng

abyssal：relating to or denoting the depths or bed of the ocean, especially between about 3,000 and 6,000 metres down. 深海的 shēn hǎi de

facies：(Geology) the character of a rock expressed by its formation, composition, and fossil content. (地质) 相 xiàng

Example:

The bathyal-abyssal facies, littoral facies, neritic facies and delta facies are beneficial to oil and gas exploration.
半深海-深海相区、滨浅海相区和三角洲相区是较为有利的油气分布区。

Extended Terms:

open abyssal belvedere facies 开阔海台地相
bathyal-abyssal facies 次深海-深海相

accessory mineral [əkˈsesəri ˈminərəl]

Definition: any mineral in an igneous rock not essential to the naming of the rock. When it is present in small amounts, as is common, it is called a minor accessory. 副矿物 fù kuàng wù

> Example:

According to its main rock forming mineral, accessory mineral formation, petrochemistry and geochemistry character, it shows that I type is the main genetic type for intrusive, only a few is A Type or S type.
根据主要造岩矿物及副矿物组合、岩石化学和地球化学特征判别结果，岩体成因类型以 I 型为主，少数为 A 型和 S 型。

> Extended Terms:

accessory mineral assemblages 副矿物组合
accessory mineral chrome spinel 副矿物铬尖晶石

accessory rock [əkˈsesəri rɒk]

> Definition: 附生岩 fù shēng yán

accessory: a thing which can be added to something else in order to make it more useful, versatile, or attractive. 附件 fù jiàn
rock: the solid mineral material forming part of the surface of the earth and other similar planets, exposed on the surface or underlying the soil. 岩石 yán shí, 岩 yán

accreting plate [æˈkriːtɪŋ pleit]

> Definition: 增生板块 zēng shēng bǎn kuài

accreting: growing by accumulation or coalescence. 增生 zēng shēng
plate: a thin, flat sheet or strip of metal or other material, typically one used to join or strengthen things or forming part of a machine. 板 bǎn

> Extended Term:

accreting convergent plate boundary
增生聚敛板块边界

accretionary wedge [əˈkriːʃənəri wedʒ]

> Definition: 增生楔 zēng shēng xiē; 加积楔 jiā jī xiē; 增积岩体 zēng jī yán tǐ

accretionary: a process by which the size of a continent increases as a result of the moving together and deforming of tectonic plates. 增生 zēng shēng, 增大 zēng dà
wedge: a piece of wood, metal, or some other material having one thick end and tapering to a thin edge, that is driven between two objects or parts of an object to secure or separate them. 楔子 xiē zi

achondrite [eiˈkɔndrait]

Definition: a rare stony meteorite that consists mainly of silicate minerals and has the texture of igneous rock but contains no chondrules. 无球粒陨石 wú qiú lì yǔn shí

Origin: 1900-1905; a- +chondrite.

Example:
GR99027 meteorite from Blue Ice Area in Antarctica is an achondrite.
来自南极格罗夫山蓝冰地区的 GR99027 陨石为一无球粒陨石。

Extended Terms:
calcium-poor achondrite 贫钙无球粒陨石
calcium-rich achondrite 富钙无球粒陨石

achondritic meteorite [ˌeikɔnˈdritik ˈmiːtiərait]

Definition: 无球粒陨石 wú qiú lì yǔn shí

achondritic: relating to a stony meteorite containing no small mineral granules (chondrules). 无球粒陨星的 wú qiú lì yǔn xīng de

meteorite: a piece of rock or metal that has fallen to the earth's surface from outer space as a meteor. Over 90 percent of meteorites are of rock while the remainder consist wholly or partly of iron and nickel. 陨石 yǔn shí

acid rock [ˈæsid rɔk]

Definition: a form of psychedelic rock, which is characterized with long instrumental solos, few (if any) lyrics and musical improvisation. 酸性岩 suān xìng yán

Examples:
①The Carboniferous volcanic rocks in eastern Junggar belong to basic and intermediate-acid rock assemblages and are characterized by early island-arc compression and late intraplate extension.
东准噶尔石炭系火山岩为一套基性、中酸性岩石组合,具早期岛弧挤压、晚期板内伸展环境特征。
②I Siliceous soil—Soil composed of acid rock that are crystalline in nature.
I 硅酸土壤——由透明的酸性岩石组成的土壤。

acmite [ˈækmait]

Definition: a rare pyroxene mineral, sodium-ferric iron silicate, $NaFe(Si_2O_6)$, found in feldspathoid rocks 锥辉石 zhuī huī shí

Origin: 1830-1940; from Greek akm(ē) sharp point + -ite.

Extended Terms:

sahlite acmite 次透锥辉石
acmite-augite 霓辉石
jadeite-acmite 玉质锥辉

active continental margin ['æktiv ˌkɔnti'nentəl 'mɑːdʒin]

Definition: 活动大陆边缘 huó dòng dà lù biān yuán

active: moving or tending to move about vigorously or frequently. 活跃的 huó yuè de
continental: forming or belonging to a continent. 大陆的 dà lù de
margin: the edge or border of something. 边 biān, 边沿 biān yán

Example:

Seventeen petroliferous basins have been found in circum-Pacific belt now and have been classified and analyzed in basinal origin, for the basins are all in the active continental margins. 环太平洋已发现17个含油气盆地，它们都处于活动大陆边缘，在盆地成因方面有类似之处，可以归类进行分析。

active continental margin basin ['æktiv ˌkɔnti'nentəl 'mɑːdʒin 'beisən]

Definition: 活动大陆边缘盆地 huó dòng dà lù biān yuán pén dì

active: moving or tending to move about vigorously or frequently. 活跃的 huó yuè de
continental: forming or belonging to a continent. 大陆的 dà lù de
margin: the edge or border of something. 边 biān, 边沿 biān yán
basin: a place where the earth's surface is lower than in other areas of the world. 盆地 pén dì; 凹地 āo dì

Example:

During the Neoproterozoic to the Early Paleozoic, the South China was a marginal basin captured by the Yangtze active continental margin, filled with very thick flysch, sandstone, and shale sediments.
华南新元古代到早古生代是扬子活动大陆边缘构造"捕获"的一个边缘海盆地，其间充填了巨厚的复理层、砂岩、页岩沉积。

active margin ['æktiv 'mɑːdʒin]

Definition: 活动边缘 huó dòng biān yuán

active: moving or tending to move about vigorously or frequently. 活跃的 huó yuè de
margin: the edge or border of something. 边 biān, 边沿 biān yán

Example:

Volcanic rifted margins geodynamic evolution is as complex as that of active margins.
火山型被动陆缘演化过程与活动陆缘一样复杂多变。

Extended Terms:

active plate margin 活动板块边缘
active type continental margin 主动型大陆边缘

adakite [əˈdeikait]

Definition: (Geology) rocks formed from lavas that melted from subducting slabs associated with either volcanic arcs or arc/continent collision zones; they were first described from Adak Island in the Aleutians. (地质) 埃达克岩 āi dá kè yán

Examples:

①The adakite in Heishishan of Ganshu Province is of O-type composed of trondhjemite.
甘肃黑石山埃达克岩属 O 型埃达克岩，其岩石类型为奥长花岗岩。
②The Bayan Bold gold deposit is of porphyry-type related to adakite.
矿床类型属于与埃达克岩有关的斑岩型金矿床。

Extended Term:

Adakite-like rocks 埃达克质岩

adiabat [ˈædiəbæt]

Definition: a line on a thermodynamic chart relating the pressure and temperature of a substance undergoing an adiabatic change. 绝热线 jué rè xiàn

Origin: back formation from adiabatic.

Example:

Although Scheme B considers the condensation latent heat, the curvature of wet adiabat is not considered in the computational method. Scheme B is superior to Scheme A, but the errors are still large.
方案 B 虽然考虑水汽凝结潜热，但计算方法没有考虑湿绝热线的曲率等问题，计算结果虽优于 A 方案，但误差仍较大。

Extended Terms:

condensation adiabat 凝结绝热线
dry adiabat 干绝热线

pseudo adiabat 假绝热线
saturation adiabat 饱和绝热线
wet adiabat 湿绝热线

adiabatic decompression [ˌædiəˈbætik ˌdiːkəmˈpreʃən]

Definition: In thermodynamics, an adiabatic process or an isocaloric process is a thermodynamic process in which no heat is transferred to or from the working fluid. 绝热减压 jué rè jiǎn yā

adiabatic gradient [ˌædiəˈbætik ˈgreidiənt]

Definition: 绝热梯度 jué rè tī dù

adiabatic: relating to or denoting a process or condition in which heat does not enter or leave the system concerned. 绝热的 jué rè de
gradient: a measure of change in a physical quantity such as temperature or pressure over a particular distance. 梯度 tī dù, 倾斜度 qīng xié dù, 坡度 pō dù

Extended Term:

adiabatic temperature gradient 绝热温度梯度

advection [ædˈvekʃən]

Definition: the transference of heat energy in a horizontal stream of gas, especially of air. 水平对流 shuǐ píng duì liú

Origin: C20: from Latin advectiō conveyance, from advehere, from ad- to + vehere to carry.

Example:

We must pay attention to the southern warm thermal advection as well as the northern cold thermal advection in the cold front shear type rainstorm process.
在该暴雨的发生过程中，北部冷平流具有十分重要的作用，但南部暖平流的作用也不可忽视。

Extended Terms:

advection effluent 平流流出
advection scale 平流尺度
geostrophic advection 地转平流
warm advection 暖平流

thermal advection 温度平流

aegirine ['eigəˌriːn]

Definition: a mineral, mainly sodium-ferric iron silicate, $NaFe \cdot (Si_2O_6)$, occurring in feldspathoid rocks in slender prismatic crystals 霓石 ní shí

Origin: 1830-1840; Aegir + -ite$_2$.

Example: There exist quite a few deep xenoliths in silicified aegirine syenite porphyry within Jinhe intrusive body of Jianchuan, western Yunnan Province.
云南剑川金河岩体中的硅化霓辉正长斑岩含有较多的镁铁、超镁铁质深源岩石包体。

Extended Terms:
aegirine aplite 霓细晶岩
aegirine-augite 霓辉石
aegirine carbonatite 霓碳酸岩
aegirine granite aplite 霓岗细晶岩
aegirine grorudite 多霓细岗岩

aenigmatite [iːˈnigməˌtait]

Definition: an inosilicate mineral of sodium, iron, and titanium, forming brown to black triclinic lamellar crystals. 三斜闪石 sān xié shǎn shí; 钠铁闪石 nà tiě shǎn shí

aeolian cross bedding [iːˈəuliən krɔs ˈbediŋ]

Definition: 风成交错层理 fēng chéng jiāo cuò céng lǐ

aeolian: relating to or arising from the action of the wind. 风成的 fēng chéng de
cross: pass in an opposite or different direction; intersect. 相交 xiāng jiāo
bedding: (Geology) the stratification or layering of rocks or other geological materials. (地质)层理 céng lǐ

agglomerate [əˈglɔmərət]

Definition: a coarse-grained volcanic rock composed of sharp or sub-angular fragments of lava, set in a fine matrix 集块岩 jí kuài yán

Origin: 17 century: from Latin agglomerāre, from glomerāre to wind into a ball, from glomus ball, mass.

Extended Terms:

agglomerate lava 集块熔岩
slaggy agglomerate 多孔集块角砾岩；渣状集块岩
vent agglomerate 喷溢道集块岩
welded agglomerate 熔结集块岩

agglomeratic texture [əˌglɔməˈrætik ˈtekstʃə]

Definition: 集块结构 jí kuài jié gòu

agglomeratic：of or related to agglomerate. 集块岩的 jí kuài yán de

texture：the arrangement of the particles or constituent parts of any material, as wood, metal, etc., as it affects the appearance or feel of the surface; structure, composition, grain, etc. 结构 jié gòu

Al-spinel lherzolites [ɑːl-spiˈnel lˈheəzəlaits]

Definition: 铝尖晶石二辉橄榄岩 lǚ jiān jīng shí èr huī gǎn lǎn yán

Al：the chemical element aluminium.（化学元素）铝 lǚ

spinel：a hard glassy mineral occurring as octahedral crystals of variable colour and consisting chiefly of magnesium and aluminium oxides. 尖晶石 jiān jīng shí

lherzolite：a coarse-grained rock containing minerals high in iron and magnesium that is believed to originate in the Earth's mantle. 二辉橄榄岩 èr huīgǎn lǎn yán

alaskite [əˈlæskait]

Definition: (Petrology) a granitic rock composed mainly of quartz and alkali feldspar, with few dark mineral components.（岩石学）白岗岩 bái gǎng yán

Example:

Rossing uranium mine is the only operating uranium mine in the world where the uranium occurs in intrusive alaskite.
罗辛铀矿山是目前世界上唯一尚在运营中的侵入体白岗岩型铀矿。

Extended Terms:

alaskite aplite 白岗细晶岩
alaskite porphyry 白岗斑岩
soda alaskite 钠白岗岩

albite [ˈælbait]

Definition: a colourless, milky-white, yellow, pink, green, or black mineral of the feldspar group and plagioclase series, found in igneous sedimentary and metamorphic rocks. It is used in the manufacture of glass and ceramics. Composition: sodium aluminium silicate. 钠长石 nà cháng shí

Formula: $NaALSi_3O_8$. Crystal structure: triclinic.

Origin: 19 century: from Latin albus white.

Example:

As a common secondary mineral, feldspar minerals, especially albite, have shared a close relationship with jadeite minerals.

长石矿物特别是钠长石作为翡翠的常见次要矿物，与硬玉矿物之间存在着密切关系。

Extended Terms:

albite granite 钠长花岗岩
albite monzonite 钠长二长岩
albite moonstone 钠长月光石
albite twin 钠长石双晶；钠长双晶
secondary albite 次生钠长石

albite-anorthite system [ˈælbait æˈnɔːθait ˈsistəm]

Definition: 钠长石-钙长石系 nà cháng shí-gài cháng shí xì

albite: a sodium-rich mineral of the feldspar group, typically white, occurring widely in silicate rocks. 钠长石 nà cháng shí

anorthite: a calcium-rich mineral of the feldspar group, typically white, occurring in many basic igneous rocks. 钙长石 gài cháng shí

system: (Geology)(in chronostratigraphy) a major range of strata that corresponds to a period in time, subdivided into series. （地质）（在年代地层学中）地层的系 dì céng de xì

albite-orthoclase-quartz system [ˈælbait- ˈɔːθəukleis-kwɔːts ˈsistəm]

Definition: 钠长石-正长石-石英系 nà cháng shí-zhèng cháng shí-shí yīng xì

albite: a sodium-rich mineral of the feldspar group, typically white, occurring widely in silicate rocks. 钠长石 nà cháng shí

orthoclase: a common rock-forming mineral occurring typically as white or pink crystals. It is a potassium-rich alkali feldspar and is used in ceramics and glass-making. 正长石 zhèng cháng shí

quartz: a hard mineral consisting of silica, found widely in igneous and metamorphic rocks and typically occurring as colourless or white hexagonal prisms. It is often coloured by impurities (as in amethyst, citrine, and cairngorm). 石英 shí yīng

system: (Geology) (in chronostratigraphy) a major range of strata that corresponds to a period in time, subdivided into series. (地质) (在年代地层学中) 地层的系 dì céng de xì

albite epidote hornfels facies [ˈælbait ˈepidəut ˈhɔːnfelz ˈfeiʃiːz]

Definition: 钠长石-绿帘石-角岩相 nà zhǎng shí-lǜ lián shí-jiǎo yán xiāng

albite: a sodium-rich mineral of the feldspar group, typically white, occurring widely in silicate rocks. 钠长石 nà cháng shí

epidote: a lustrous yellow-green crystalline mineral, common in metamorphic rocks. It consists of a basic, hydrated silicate of calcium, aluminium, and iron. 绿帘石 lǜ lián shí

hornfels: a dark, fine-grained metamorphic rock consisting largely of quartz, mica, and particular feldspars. 角页岩 jiǎo yè yán

facies: the character of a rock expressed by its formation, composition, and fossil content. 相 xiàng

albitization [ˈælbitiˈzeiʃən]

Definition: (Petrology) the formation of albite in a rock as a secondary mineral. (岩石学) 钠长石化 nà zhǎng shí huà

Example:
Therefore, the albitization of plagioclase is shown in the early stage of diagenesis, and albitization of potash feldspar appears only in the late stage of diagenesis.
因此，斜长石的钠长石化可见于成岩早期，而钾长石的钠长石化只发生于成岩晚期。

Extended Term:
glaucophane-albitization 蓝闪钠长石化

algal dolomite [ˈælgəl ˈdɔləmait]

Definition: 藻白云岩 zǎo bái yún yán

algal: relating to a simple, non-flowering, and typically aquatic plant of a large assemblage that includes the seaweeds and many single-celled forms. Algae contains chlorophyll but lacks true stems, roots, leaves, and vascular tissue. 藻类的 zǎo lèi de, 海藻的 hǎi zǎo de

dolomite: a translucent mineral consisting of a carbonate of calcium and magnesium, usuallyalso containing iron. 白云石 bái yún shí

Example:

The main reservoirs of Carboniferous are allochemical dolomite and algal dolomite, and the void space is mainly constituted by secondary solutional pores.

石炭系的主要储集岩是粒屑白云岩、藻白云岩,孔隙以次生溶孔为主。

algal limestone ['ælgəl 'laimstəun]

Definition: 藻灰岩 zǎo huī yán

algal: relating to a simple, non-flowering, and typically aquatic plant of a large assemblage that includes the seaweeds and many single-celled forms. Algae contains chlorophyll but lacks true stems, roots, leaves, and vascular tissue. 藻类的 zǎo lèi de, 海藻的 hǎi zǎo de

limestone: a hard sedimentary rock, composed mainly of calcium carbonate or dolomite, used as building material and in the making of cement. 石灰岩 shí huī yán

Example:

The best lithology is (silty) algal limestone, and the second are silty micrite limestone and limy (muddy) siltstone.

储集层物性最好的为藻灰岩(含粉砂),其次为砂质泥晶灰岩和灰质(泥质)粉砂岩。

Extended Term:

algal structure 藻类构造

alkali ['ælkəlai]

Definition: a carbonate or hydroxide of an alkali metal, the aqueous solution of which is bitter, slippery, caustic, and characteristically basic in reactions. 碱性 jiǎn xìng

Origin: 14 century: from Medieval Latin, from Arabic al-qili the ashes (of the plant saltwort).

Example:

The soil transformed from acid to alkali. 土壤由弱碱性向弱酸性演替。

Extended Terms:

alkali antimonide 碱金属锑化物
alkali granite 碱性花岗岩
alkali liquor 碱液
alkali meter 碱量计;碳酸定量计
alkali soil 碱土

alkali basalt [ˈælkəlai ˈbæsɔːlt]

Definition: 碱性玄武岩 jiǎn xìng xuán wǔ yán

alkali: a carbonate or hydroxide of an alkali metal, the aqueous solution of which is bitter, slippery, caustic, and characteristically basic in reactions. 碱性 jiǎn xìng

basalt: a dark fine-grained volcanic rock that sometimes displays a columnar structure. It is typically composed largely of plagioclase with pyroxene and olivine. 玄武岩 xuán wǔ yán

Example:

The deposit is located in alkali basalt which is controlled by the secondary structural zones of the Dunha-Mishan fault.
矿床为碱性玄武岩，以位于大断裂次级构造上为特征。

Extended Terms:

alkali olivine basalt 碱性橄榄石玄武岩
alkali olivine basalt magma 碱性橄榄玄武岩岩浆
calc alkali basalt 钙碱性玄武岩

alkali dolerite [ˈælkəlai ˈdɔlərait]

Definition: 碱性粗玄岩 jiǎn xìng cū xuán yán

alkali: a carbonate or hydroxide of an alkali metal, the aqueous solution of which is bitter, slippery, caustic, and characteristically basic in reactions. 碱性 jiǎn xìng

dolerite: (Geology) a dark, medium-grained igneous rock, typically with ophitic texture, containing plagioclase, pyroxene, and olivine. It typically occurs in dykes and sills. (地质)粒玄岩 lì xuán yán, 粗玄岩 cū xuán yán

alkali feldspar [ˈælkəlai ˈfeldspɑː]

Definition: The alkali feldspar groups are those feldspar minerals rich in the alkali elements like potassium. The alkali feldspars include: anorthoclase, microcline, orthoclase and sanidine. 碱性长石 jiǎn xìng cháng shí

Example:

The Penshan granitic body is a calc-alkaline pluton composed of muscovite alkali granite, two-mica alkali feldspar granite and biotite adamellite.
彭山岩体是由白云母碱长花岗岩、云母碱长花岗岩和黑云母二长花岗岩组成的钙碱性花岗复式岩体。

Extended Term:
alkali-feldspar series 碱性长石系

alkali feldspar gneiss ['ælkəlai 'feldspɑː nais]

Definition: 碱性长石片麻岩 jiǎn xìng cháng shí piàn má yán

alkali: a carbonate or hydroxide of an alkali metal, the aqueous solution of which is bitter, slippery, caustic, and characteristically basic in reactions. 碱性 jiǎn xìng

feldspar: an abundant rock-forming mineral typically occurring as colourless or pale-coloured crystals and consisting of aluminosilicates of potassium, sodium, and calcium. 长石 cháng shí

gneiss: a metamorphic rock with a banded or foliated structure, typically coarse-grained and consisting mainly of feldspar, quartz, and mica. 片麻岩 piàn má yán

alkali feldspar granite ['ælkəlai 'feldspɑː 'grænit]

Definition: 碱性长石花岗岩 jiǎn xìng cháng shí huā gāng yán

alkali: a carbonate or hydroxide of an alkali metal, the aqueous solution of which is bitter, slippery, caustic, and characteristically basic in reactions. 碱性 jiǎn xìng

feldspar: an abundant rock-forming mineral typically occurring as colourless or pale-coloured crystals and consisting of aluminosilicates of potassium, sodium, and calcium. 长石 cháng shí

granite: a very hard, granular, crystalline, igneous rock consisting mainly of quartz, mica, and feldspar and often used as a building stone. 花岗岩 huā gāng yán

Example:

The deposit is an alkali-feldspar granite type Ta deposit. 该矿床为碱性长石花岗岩型钽矿床。

alkali feldspar syenite ['ælkəlai'feldspɑː 'sainait]

Definition: 碱性长石正长岩 jiǎn xìng cháng shí zhèng cháng yán

alkali: a carbonate or hydroxide of an alkali metal, the aqueous solution of which is bitter, slippery, caustic, and characteristically basic in reactions. 碱性 jiǎn xìng

feldspar: an abundant rock-forming mineral typically occurring as colourless or pale-coloured crystals and consisting of aluminosilicates of potassium, sodium, and calcium. 长石 cháng shí

syenite: a coarse-grained grey igneous rock composed mainly of alkali feldspar and ferromagnesian minerals such as hornblende. 正长岩 zhèng cháng yán

alkali gabbro [ˈælkəlai ˈɡæbrəu]

Definition: 碱性辉长岩 jiǎn xìng huī cháng yán

alkali: a carbonate or hydroxide of an alkali metal, the aqueous solution of which is bitter, slippery, caustic, and characteristically basic in reactions. 碱性 jiǎn xìng

gabbro: a dark, coarse-grained plutonic rock of crystalline texture, consisting mainly of pyroxene, plagioclase feldspar, and often olivine. 辉长岩 huī cháng yán

Example:

The intrusive body is consisted mainly of dioritic rocks showing the geochemical characters of calc-alkalic andesite and accompany with minor of alkali gabbro.
岩体主要由钙碱性闪长质岩组成，具有钙碱性安山岩的地球化学特点，伴有少量碱性辉长岩。

Extended Term:

calc-alkali gabbro 钙碱辉长岩

alkali granite [ˈælkəlai ˈɡrænit]

Definition: 碱性花岗岩 jiǎn xìng huā gāng yán

alkali: a carbonate or hydroxide of an alkali metal, the aqueous solution of which is bitter, slippery, caustic, and characteristically basic in reactions. 碱性 jiǎn xìng

granite: a very hard, granular, crystalline, igneous rock consisting mainly of quartz, mica, and feldspar and often used as a building stone. 花岗岩 huā gāng yán

Extended Term:

calc-alkali granite 钙碱花岗岩

alkali lime index [ˈælkəlai laim ˈindeks]

Definition: 碱钙指数 jiǎn gài zhǐ shù

alkali: a carbonate or hydroxide of an alkali metal, the aqueous solution of which is bitter, slippery, caustic, and characteristically basic in reactions. 碱性 jiǎn xìng

lime: a white caustic alkaline substance consisting of calcium oxide, which is obtained by heating limestone and combines with water with the production of much heat. 生石灰 shēng shí huī，氧化钙 yǎng huà gài

index: a figure in a system or scale representing the average value of specified prices, shares, or other items as compared with some reference figure. 指数 zhǐ shù

alkali index [ˈælkəlai ˈindeks]

Definition: 碱指数 jiǎn zhǐ shù

alkali: a carbonate or hydroxide of an alkali metal, the aqueous solution of which is bitter, slippery, caustic, and characteristically basic in reactions. 碱性 jiǎn xìng

index: a figure in a system or scale representing the average value of specified prices, shares, or other items as compared with some reference figure. 指数 zhǐ shù

Example:

By calculating the calc-alkali index (CA) and alumina saturation index (ASI), the magmatypes of the granites can be roughly determined.
通过钙碱指数(CA)和铝饱和指数(ASI)的计算, 可以大体获知花岗岩的岩浆类型。

Extended Terms:

calc-alkali index 钙碱指数
silic-alkali index 硅碱指数

alkali olivine basalt [ˈælkəlai ˌɒliˈviːn ˈbæsɔːlt]

Definition: 碱性橄榄玄武岩 jiǎn xìng gǎn lǎn xuán wǔ yán

alkali: a carbonate or hydroxide of an alkali metal, the aqueous solution of which is bitter, slippery, caustic, and characteristically basic in reactions. 碱性 jiǎn xìng

olivine: an olive-green, grey-green, or brown mineral occurring widely in basalt, peridotite, and other basic igneous rocks. It is a silicate containing varying proportions of magnesium, iron, and other elements. 橄榄石 gǎn lǎn shí

basalt: a dark fine-grained volcanic rock that sometimes displays a columnar structure. It is typically composed largely of plagioclase with pyroxene and olivine. 玄武岩 xuán wǔ yán

Extended Term:

alkali olivine basalt magma 碱性橄榄玄武岩岩浆

alkali rock [ˈælkəlai rɒk]

Definition: any igneous rock with a marked preponderance of alkali and a low percentage of silica. 碱性岩 jiǎn xìng yán

Example:

The paper concludes genetic relationship between the ore-bearing dolomitite, K-rich slate and Na-rich slate and believes that they are one alkali rock series of comagmatic origin. 文章总结含矿白云岩、富钾板岩和富钠岩石三者之间在成因方面的关系, 认为其属于同一岩浆起源的

碱性岩系。

> Extended Terms:

alkali-carbonate rock reaction 碱-碳酸盐岩反应
alkali rock series 碱性岩系
rich-alkali magmatic rock 富碱岩体

alkali syenite [ˈælkəlai ˈsainait]

> Definition: 碱性正长岩 jiǎn xìng zhèng zhǎng yán

alkali: a carbonate or hydroxide of an alkali metal, the aqueous solution of which is bitter, slippery, caustic, and characteristically basic in reactions. 碱性 jiǎn xìng

syenite: a coarse-grained grey igneous rock composed mainly of alkali feldspar and ferromagnesian minerals such as hornblende. 正长岩 zhèng chǎng yán

> Extended Term:

calci-alkali syenite 钙碱正长岩

alkalic olivine basalt series [ˈælkəlik ˌɔliˈviːn ˈbæsɔːlt ˈsiəriːz]

> Definition: 碱性橄榄玄武岩系列 jiǎn xìng gǎn lǎn xuán wǔ yán xì liè

alkalic: (Geology) (of a rock or mineral) richer in sodium and/or potassium than is usual for its type. (地质) (岩石或矿物) 碱性的 (yán shí huò kuàng wù) jiǎn xìng de

olivine: an olive-green, grey-green, or brown mineral occurring widely in basalt, peridotite, and other basic igneous rocks. It is a silicate containing varying proportions of magnesium, iron, and other elements. 橄榄石 gǎn lǎn shí

basalt: a dark fine-grained volcanic rock that sometimes displays a columnar structure. It is typically composed largely of plagioclase with pyroxene and olivine. 玄武岩 xuán wǔ yán

series: a number of things, events, or people of a similar kind or related nature coming one after another. 系列 xì liè

alkaline igneous rock [ˈælkəlain ˈigniəs rɒk]

> Definition: 碱性火成岩 jiǎn xìng huǒ chéng yán

alkaline: having the properties of an alkali, or containing alkali; having a pH greater than 7. 碱性的 jiǎn xìng de

igneous: (Geology) (of rock) having solidified from lava or magma. (地质) (岩石) 火成的 huǒ chéng de

rock: the solid mineral material forming part of the surface of the earth and other similar planets, exposed on the surface or underlying the soil. 岩石 yán shí, 岩 yán

alkaline series [ˈælkəlain ˈsiəriːz]

Definition: 碱性系列 jiǎn xìng xì liè

alkaline: having the properties of an alkali, or containing alkali; having a pH greater than 7. 碱性的 jiǎn xìng de

series: a number of things, events, or people of a similar kind or related nature coming one after another. 系列 xì liè

Example:
The chemical differences between these two series resulted from the higher melting degree of the tholeiite series than that of the calc-alkaline series.
两个系列间的化学差异起因于拉斑系列的熔融程度高于钙碱系列。

Extended Term:
alkaline basalt series 碱性玄武岩系列

alkaline volcanic rock [ˈælkəlain vɔlˈkænik rɒk]

Definition: 碱性火山岩 jiǎn xìng huǒ shān yán

alkaline: having the properties of an alkali, or containing alkali; having a pH greater than 7. 碱性的 jiǎn xìng de

volcanic: of, resembling, or caused by a volcano or volcanoes. 火山的 huǒ shān de

rock: the solid mineral material forming part of the surface of the earth and other similar planets, exposed on the surface or underlying the soil. 岩石 yán shí; 岩 yán

alkalinity ratio (AR) [ˈælkəˈliniti ˈreiʃiəu]

Definition: 碱度率 jiǎn dù lǜ

alkalinity: the alkali concentration or alkaline quality of an alkali-containing substance. 碱度 jiǎn dù; 碱性 jiǎn xìng; 含碱量 hán jiǎn liàng

ratio: the quantitative relation between two amounts showing the number of times, one value contains or is contained within the other. 比 bǐ; 比率 bǐ lǜ; 比例 bǐ lì

Example:
The paper investigates the peroxide bleaching properties of chemithemomechanical pulp from Acaciamangium, including optimum alkalinity ratio for H_2O_2 bleaching, the adding method of DTPA and the two-stage peroxide bleaching technology. 本文对马占相思 CTMP 浆料性能、

H_2O_2 漂白适宜碱比、DTPA 添加方式及两段 H_2O_2 漂白工艺进行了研究。结果表明，适宜碱比与 H_2O_2 用量之间有较强的相关关系。

allanite [ˈælənait]

Definition: a rare black or brown mineral consisting of the hydrated silicate of calcium, aluminium, iron, cerium, lanthanum, and other rare earth minerals. It occurs in granites and other igneous rocks. 褐帘石 hè lián shí

Origin: 19 century: named after T. Allan (1777-1833), English mineralogist.

Example:
Mineralogically, zoisite is a member of the epidote group, the other members of which are clinozoisite, epidote perse, piemontite, and allanite.
从矿物学上讲，黝帘石是绿帘石家族的一员，其他成员是绿帘石还有褐帘石等。

Extended Terms:
allanite-(Ce) 铈褐帘石
allanite-(Y) 钇褐帘石

allochem [ˈæləkem]

Definition: Allochem is a term introduced by Folk (1959) to describe the recognisable "grains" in carbonate rocks. Any fragment from around 1/2mm upwards in size may be considered an allochem, including ooids, peloids, oncolites, fossil or pre-existing carbonate fragments. 异化颗粒 yì huà kē lì, 又称粒屑 (also called grained clast) lì xiè

Example:
According to the result of analysis, the basin and basin margin facies are the favourable zones for the genesis of petroleum, while the tidal beach, allochem beach are the favourable zones to reservoir the oil and gas.
根据化验结果分析，盆地相、盆地边缘相是有利生油相带，潮汐砂滩、粒屑滩是有利油气储集相带。

allotriomorphic granular texture [əˌlɔtriəˈmɔːfik ˈɡrænjulə ˈtekstʃə]

Definition: 他形粒状结构 tā xíng lì zhuàng jié gòu

allotriomorphic: of minerals in igneous rock not bounded by their own crystal faces but having their outlines impressed on them by the adjacent minerals. Also known as anhedral; xenomorphic. 他形的 tā xíng de

granular: composed or appearing to be composed of granules or grains. 颗粒的 kē lì de
texture: the arrangement of the particles or constituent parts of any material, as wood, metal, etc., as it affects the appearance or feel of the surface; structure, composition, grain, etc. 结构 jié gòu

alluvial conglomerate [əˈljuːvɪəl kənˈɡlɔməreit]

Definition: 河成砾岩 hé chéng lì yán

alluvial: of, relating to, or derived from alluvium. 冲积层的 chōng jī céng de，来自冲积层的 lái zì chōng jī céng de

conglomerate: (Geology) a coarse-grained sedimentary rock composed of rounded fragments embedded in a matrix of cementing material such as silica. (地质)砾岩 lì yán

Example:
There are few researches focus on the fracture characteristics and distribution rules of alluvial fan conglomerate reservoirs. 有关冲积扇砾岩储层中裂缝特征及分布规律预测的研究很少。

alluvial facies [əˈljuːvɪəl ˈfeiʃiːz]

Definition: 冲积相 chōng jī xiāng

alluvial: of, relating to, or derived from alluvium. 冲积层的 chōng jī céng de，来自冲积层的 lái zì chōng jī céng de

facies: the character of a rock expressed by its formation, composition, and fossil content. 相 xiàng

Example:
A new concept tsunami is put forward to explain the formation of the alluvial facies in the upper part of the second member of the Shahejie Formation throughout the Bohai Bay Basin and the reason for maximum lake flooding in the first member of the Shahejie Formation from the Tertiary onwards.
这指出了遍布于整个渤海湾盆地的沙二段上部冲积相是由强海啸形成的，它也是造成沙一段时期发生第三纪以来最大湖淹的主要原因。

Extended Term:
alluvial flat facies 河漫滩相

almandine [ˈælməndiːn]

Definition: a deep violet-red garnet that consists of iron aluminium silicate and is used as a gemstone. 铁铝榴石 tiě lǚ liú shí

Origin: 17 century: from French, from Medieval Latin alabandīna, from Alabanda, ancient city of Asia Minor where these stones were cut.

Example:

The variety of garnet includes almandine, pyrope, magnesium-iron garnet, manganese-aluminum garnet, andradite and calcium-chromium garnet.
石榴石的种类包括铁铝榴石、镁铝榴石、镁铁榴石、锰铝榴石、钙铁榴石、钙铬榴石。

Extended Terms:

almandine garnet 贵榴石
almandine sapphire 贵榴石蓝宝石
almandine zone 铁铝榴石带
pyrope-almandine 镁铝-铁铝榴石
star almandine sapphire 紫红色星彩蓝宝石

alnoite [ˈælnəˌait]

Definition: (Petrology) a variety of biotite lamprophyres characterized by lepidomelane phenocrysts; it is feldspar-free but contains melitite, perovskite, olivine, and carbonate in the matrix. (岩石学) 黄长煌斑岩 huáng cháng huáng bān yán

Example:

The kamafugite is a rare type of ultra-potassic rocks, and a rock series contained generally that rock types of alnoite, olivine-melilitite, mafurite, katungite and Wugantite and so on.
钾霞橄黄长岩是一种非常稀少的超钾质火山岩,也是一个岩石系列,包括黄长煌斑岩、橄榄石黄长岩、白橄黄长岩、橄辉钾霞岩和乌干达岩等多种岩石类型。

Extended Term:

biotite alnoite 黑云黄长煌斑岩

alteration [ˌɔːltəˈreiʃən]

Definition: the act of altering or state of being altered 蚀变 shí biàn

Origin: 1350-1400; Middle English alteracioun. See alter, -ation.

Example:

Alteration is very extensive in some areas and serpentinites are common in Greece.
在一些地区蚀变作用非常广泛,常见在希腊蛇纹岩上。

Extended Terms:

alteration halo 蚀变晕

alteration mineral 蚀变矿物
alteration zone 蚀变晕；蚀变带
biogenic alteration 生物蚀变
deuteric alteration 岩浆后期蚀变

alumina rich gneiss [əˈljuːmɪnə rɪtʃ naɪs]

Definition: 富铝片麻岩 fù lǚ piàn má yán

alumina: a white solid that is a major constituent of many rocks, especially clays, and is found crystallized as corundum, sapphire, and other minerals. 矾土 fán tǔ

rich: (of a mine or mineral deposit) yielding a large quantity or proportion of precious metal. (矿，矿藏)富的 fù de，含量丰富的 hán liàng fēng fù de

gneiss: a metamorphic rock with a banded or foliated structure, typically coarse-grained and consisting mainly of feldspar, quartz, and mica. 片麻岩 piàn má yán

alumina saturation [əˈljuːmɪnə ˌsætʃəˈreɪʃən]

Definition: 铝饱和 lǚ bǎo hé

alumina: a white solid that is a major constituent of many rocks, especially clays, and is found crystallized as corundum, sapphire, and other minerals. 矾土 fán tǔ

saturation: the state or process that occurs when no more of something can be absorbed, combined with, or added. 饱和 bǎo hé

Example:

By calculating the calc-alkali index (CA) and alumina saturation index (ASI), the magma types of the granites can be roughly determined.
通过钙碱指数(CA)和铝饱和指数(ASI)的计算，可以大体获知花岗岩的岩浆类型。

aluminum silicate system [əˈljuːmɪnəm ˈsɪlɪkɪt ˈsɪstəm]

Definition: 铝硅酸盐体系 lǚ guī suān yán tǐ xì

aluminum: a chemical element in the boron group with symbol Al and atomic number 13; a silvery white, soft, nonmagnetic, ductile metal. 铝 lǚ

silicate: a salt in which the anion contains both silicon and oxygen, especially one of the anion SiO_{42}. (化)硅酸盐 guī suān yán

system: a set of things working together as parts of a mechanism or an interconnecting network. 系统 xì tǒng

amorphous [əˈmɔːfəs]

Definition: (Petrography, Mineralogy) occurring in a mass, as without stratification or crystalline structure.(岩相学，矿物学)非晶质 fēi jīng zhì

Origin: 18 century: from New Latin, from Greek amorphos shapeless.

Example:

The research results show the force sensors can be constructed using amorphous alloy materials with high stress sensitivity.
研究结果表明具有大的应力敏感性的非晶体材料可用来制作力传感器。

Extended Terms:

amorphous cellulose 非晶态纤维素
amorphous film 非晶体膜
amorphous magnetism 非晶体磁性
amorphous polyolefins 无结晶聚烯烃；非晶性聚烯烃
amorphous polymer 非晶态聚合物

amorphous rock [əˈmɔːfəs rɔk]

Definition: 非晶质岩 fēi jīng zhì yán

amorphous: (Petrography, Mineralogy) occurring in a mass, as without stratification or crystalline structure. (岩相学，矿物学)非晶质 fēi jīng zhì

rock: the solid mineral material forming part of the surface of the earth and other similar planets, exposed on the surface or underlying the soil. 岩石 yán shí；岩 yán

amphibole eclogite [ˈæmfibəul ˈeklədʒait]

Definition: 闪榴辉岩 shǎn liú huī yán

amphibole: any of a class of rock-forming silicate or aluminosilicate minerals typically occurring as fibrous or columnar crystals. 闪石 shǎn shí

eclogite: a metamorphic rock containing granular minerals, typically garnet and pyroxene. 榴辉岩 liú huī yán

Example:

The paragenetic sequence in eclogite indicates that eclogite facies rocks experienced progressive metamorphism from epidote amphibolite facies to coesite eclogite facies and retrograde metamorphism from amphibole eclogite facies through epidote amphibolite facies to greenschist facies.

矿物共生序列研究表明，榴辉岩相岩石经历了从绿帘角闪岩相、柯石英榴辉岩相、角闪榴辉岩相、绿帘角闪岩相到绿片岩相的演化过程。

amphibolite facies [æmˈfibəlait ˈfeiʃiːz]

Definition: one of the major divisions of the mineral facies classification of metamorphic rocks, encompassing rocks that formed under conditions of moderate to high temperatures (950°F, or 500°C, maximum) and pressures. Less-intense temperatures and pressures form rocks of the epidote-amphibolite facies, and more-intense temperatures and pressures form rocks of the granulite facies. 角闪岩相 jiǎo shǎn yán xiàng

Example:
Granulite—the metamorphic gneiss district level are mostly granulite facies to senior amphibolite facies.
麻粒岩—片麻岩区的变质程度多属麻粒岩相到高级角闪岩相。

Extended Terms:
albite-epidote-amphibolite facies 钠长帘闪岩相
epidote-amphibolite facies 绿帘石角闪岩相

amygdale [əˈmigdeil]

Definition: also called amygdule, a vesicle in a volcanic rock, formed from a bubble of escaping gas, that has become filled with light-coloured minerals, such as quartz and calcite. 杏仁孔 xìng rén kǒng

Origin: 19 century: from Greek almond.

Example:
Voices, emotional content either positive or negative, induce a very strong activity, in this amygdale.
人类声音，无论其内容是积极还是消极的，都在这个杏仁孔中包含了一种强烈的活性。

Extended Terms:
amygdale complex 杏仁复合体
basolateral amygdale 基底外侧杏仁核
basolateral nucleus of amygdale 外侧杏仁核
basomedial nucleus of the amygdale 杏仁基底内侧核

amygdaloidal structure [əˌmigdəˈlɔidəl ˈstrʌktʃə]

Definition: 杏仁状构造 xìng rén zhuàng gòu zào

amygdaloidal: a volcanic rock containing many amygdules. 杏仁岩的 xìng rén yán de
structure: the arrangement of and relations between the parts or elements of something complex.
结构 jié gòu;构造 gòu zào

analcime [əˈnælsiːm]

Definition: a white, grey, or colourless zeolite mineral consisting of hydrated sodium aluminium silicate in cubic crystalline form. 方沸石 fāng fèi shí

Formula: $NaAlSi_2O_6 \cdot H_2O$

Origin: 19 century: from Greek analkimos weak (from an- + alkimos strong, from alkē strength) + -ite.

Example:
The discovery of a huge analcime deposit in Nei Monggo fills up a gap in minerals in China.
内蒙古特大型方沸石矿藏的发现，填补了我国矿产品种的又一空白。

Extended Terms:
analcime tephrite 方沸岩碱玄岩
petalite-analcime 透锂长石-方沸石

anatexis [ˌænəˈteksis]

Definition: a high-temperature process of metamorphosis by which plutonic rock in the lowest levels of the crust is melted and regenerated as a magma 重熔作用 chóng róng zuò yòng

Example:
Anatexis is pervasive in high grade metamorphic rocks and has received much attention from geo-scientists.
重熔作用在高级变质岩中非常普遍并受到广泛关注。

Extended Terms:
differential anatexis 分异深熔酌
partial anatexis 余深溶酌

andalusite [ˌændəˈluːsait]

Definition: a grey, pink, or brown hard mineral consisting of aluminium silicate in orthorhombic crystalline form. It occurs in metamorphic rocks and is used as a refractory and as a gemstone. 红柱石 hóng zhù shí

Formula: Al_2SiO_5

Origin: 1830-1840; named after Andalusia, where it was first found.

Example:

Based on the experiment of decomposition dynamics of andalusite, the dimensions of the decomposed grains have been measured by means of the perimeter-area method.
在对红柱石的热分解进行实验研究的基础上,运用周长-面积法测定了分解颗粒的分形维数,并利用分形维数与反应时间的关系探讨了红柱石分解反应的历程。

Extended Terms:

andalusite-hornstone 红柱角岩
andalusite-mica schist 红柱云片岩

---

andalusite zone [ˌændəˈluːsait ˈzəun]

Definition: 红柱石带 hóng zhù shí dài

andalusite: a grey, green, brown, or pink aluminosilicate mineral occurring mainly in metamorphic rocks as elongated rhombic prisms, sometimes of gem quality. 红柱石 hóng zhù shí

zone: an area or stretch of land having a particular characteristic, purpose, or use, or subject to particular restrictions. 地带 dì dài, 区域 qū yù

---

andesine anorthosite [ˈændiziːn ænˈɔːθəsait]

Definition: 中长斜长岩 zhōng cháng xié cháng yán

andesine: a plagioclase feldspar with a composition ranging from Ab70An30 to Ab50An50, where Ab = $NaAlSi_3O_8$ and An = $CaAl_2Si_2O_8$; it is a primary constituent of intermediate igneous rocks, such as andesites. 中长石 zhōng cháng shí

anorthosite: a variety of diorite consisting chiefly of feldspar. 钙长石 gài cháng shí

---

andesite [ˈændizait]

Definition: a fine-grained tan or grey volcanic rock consisting of plagioclase feldspar, especially andesine, amphibole, and pyroxene. 安山岩 ān shān yán

Origin: 19 century: from Andes + -ite.

Example:

The pyroclastic rock fragments comprise quartzite, schist, gneiss, basalt, andesite, rhyolite.
火成岩的碎块成分主要为石英岩、岩、麻岩、武岩、山岩和流纹岩。

andesite line

> Extended Terms:

andesite line 安山岩线
basaltic andesite 玄武安山岩；安山岩
biotite andesite 黑云安山岩
circumoceanic andesite 环洋安山岩
hornblende andesite 角闪安山岩

andesite line [ˈændizait lain]

> Definition: 安山岩线 ān shān yán xiàn

andesite：a fine-grained tan or grey volcanic rock consisting of plagioclase feldspar, especially andesine, amphibole, and pyroxene. 安山岩 ān shān yán
line：a long, narrow mark or band. 线 xiàn

andradite [ˈændrədait]

> Definition: a yellow, green, or brownish-black garnet that consists of calcium iron silicate and is used as a gemstone. 钙铁榴石 gài tiě liú shí

> Formula: $Ca_3Fe_2(SiO_4)_3$

> Origin: 19 century：named after J. B. d'Andrada e Silva (1763-1838), Brazilian mineralogist.

> Example:

The end member of garnet is dominated by andradite, with minor grossular and spessartine. 石榴子石端员组分以钙铁榴石为主，伴以少量钙铝榴石和锰铝榴石。

> Extended Terms:

andradite demantoid(gamet) 钙铁榴石
andradite syenite 黑榴正长岩
black andradite garnet 黑色钙铁榴石

angularity [ˌæŋgjuˈlærəti]

> Definition: angularities, sharp corners; angular outlines：the angularities of the coastline. 棱角度 léng jiǎo dù；角状 jiǎo zhuàng

> Origin: 1635-1945；angular + -ity.

Example:

In order to ensure the use safety of pressure container, we must test the welded joint angularity and stagger joint.

为了保证压力容器的使用安全,必须检测焊接棱角度和错边量。

Extended Terms:

angularity chart 菱角对比图
angularity correction 角度校正
angularity factor 菱角因数
gravel angularity 砾石棱角度

anhedral [æn'hiːdrəl]

Definition: the downward inclination of an aircraft wing in relation to the lateral axis. 他形的 tā xíng de

Origin: 1895-1900; an- + -hedral.

Example:

Middle-Upper Cambrian dolomites in Shimen, Hunan Province, are mainly composed of anhedral micritic dolomite, euhedral and subhedral granular dolomites, grain dolomite.

湖南石门中上寒武统白云岩极其发育,类型多样,不同类型的白云岩其成因机理和白云石化模式各不相同。

Extended Terms:

anhedral crystal 他形晶
anhedral grain 他形晶粒
anhedral texture 他形结构

anhedral crystal [æn'hiːdrəl 'krɪstəl]

Definition: 他形晶 tā xíng jīng

anhedral: the downward inclination of an aircraft wing in relation to the lateral axis. 他形的 tā xíng de

crystal: a piece of a homogeneous solid substance having a natural geometrically regular form with symmetrically arranged plane faces. 晶体 jīng tǐ

anhydrite [æn'haɪdraɪt]

Definition: a colourless or greyish-white mineral, found in sedimentary rocks. It is used in the manufacture of cement, fertilizers, and chemicals. Composition: anhydrous calcium sulphate. 硬石膏 yìng shí gāo

Formula: $CaSO_4$. Crystal structure: orthorhombic.

Origin: 19 century: from anhydr (ous) + -ite.

Example:

The types of ore are gypsum and anhydrite, with the high quality, simple mining conditions, the convenient traffic and the abundant energy.
矿石类型为石膏及硬石膏，矿石质量好，矿床开采条件简单，交通方便，能源充足，具有广阔的开发利用前景。

Extended Terms:

anhydrite formation 硬石膏层
anhydrite gypsum 硬石膏质石膏
anhydrite rock 硬石膏岩
chemical anhydrite 化学硬石膏
soluble anhydrite 可溶硬石膏

ankaramite [ˌæŋkəˈrɑːmait]

Definition: a dark porphyritic basanite which has abundant pyroxene and olivine phenocrysts. It contains minor amounts of plagioclase and accessory biotite, apatite, and iron oxides. 橄榄辉玄岩 gǎn lǎn huī xuán yán

Example:

Three different kinds of clinopyroxenes were found in ankaramite and basalts: big phenocryst, small phenocrysts and microcrystals in groundmass respectively.
北塔山组富辉橄玄岩，玄武岩中存在着三种类型的单斜辉石：大斑晶辉石、小斑晶辉石和基质辉石。

annealing [əˈniːliŋ]

Definition: to heat and then cool (double-stranded nucleic acid) in order to separate strands and induce combination at lower temperatures especially with complementary strands 退火 tuì huǒ

Example:

At present, hot-air circulation annealing lehrs is the most advanced glass annealing device.
热风循环退火炉是目前最先进的玻璃制品退火设备。

Extended Terms:

annealing activation 退火杂质活化
annealing furnace 退火炉；退火窑
annealing oven 退火炉；热处理炉
annealing point 退火点；钝化点
annealing temperature 退火温度；回温温度

---

anorthite [æˈnɔːθait]

Definition: a white to greyish-white or reddish-white mineral of the feldspar group and plagioclase series, found chiefly in igneous rocks and more rarely in metamorphic rocks. It is used in the manufacture of glass and ceramics. Composition: calcium aluminium silicate. 钙长石 gài cháng shí

Formula: $CaAl_2Si_2O_8$. Crystal structure: triclinic

Origin: 19 century: from an- + ortho- + -ite.

Example:

Light Anorthite heat-isolated material is a new type of refractory material. 钙长石质轻质隔热材料是一种新型耐火材料。

Extended Terms:

anorthite andesite 钙长安山岩
anorthite basalt 钙长玄武岩
anorthite syenite 钙长正长岩
anorthite peridotite 钙长橄榄岩

---

anorthoclase [æˈnɔːθəuˌkleiz]

Definition: a white to colorless triclinic mineral of the alkali feldspar group. It is a common constituent in the matrices of slightly alkaline volcanic rocks. 歪长石 wāi cháng shí
Chemical formula: $(Na, K)AlSi_3O_8$.

Origin: an-1 + orthoclase

Example:

The basalts in this region also contain abundant megacrysts of augite and anorthoclase and rare megacrysts garnet.
本区玄武岩还含有丰富的普通辉石巨晶和歪长石巨晶，而石榴石巨晶罕见。

anorthosite

> Extended Terms:

anorthoclase-aplite granite 长细晶花岗岩
anorthoclase basalt 歪长玄武岩

anorthosite [æˈnɔːθəsait]

> Definition: a coarse-grained plutonic igneous rock consisting almost entirely of plagioclase feldspar. 斜长岩 xié chǎng yán

> Origin: 19 century: from French anorthose (see an-, ortho-) + -ite.

> Example:

Illite jade, in vein, was discovered in some copper mine, located in Daye County, Hubei Province, and is a rind hydrothermal alteration rock from anorthosite.
伊利石玉产于湖北省大冶市某铜矿区内，玉石矿体呈脉状，系由斜长岩脉经热液蚀变而成。

> Extended Terms:

bytownite anorthosite 倍长斜长岩；钠钙斜长岩
corundum anorthosite 刚玉斜长石
labradorite-anorthosite 喇长斜长岩
plagioclasite anorthosite 斜长岩
quartz anorthosite 石英斜长岩

anorthositic gabbro [æˌnɔːθəˈsitik ˈgæbrəu]

> Definition: 斜长辉长石 xié cháng huī cháng shí

anorthositic: relating to a coarse-grained igneous rock composed of at least 90 percent feldspar. 斜长石的 xié cháng shí de
gabbro: a dark, coarse-grained plutonic rock of crystalline texture, consisting mainly of pyroxene, plagioclase feldspar, and often olivine. 辉长岩 huī cháng yán

apatite [ˈæpətait]

> Definition: a pale green to purple mineral found in igneous rocks and metamorphosed limestones. It is used in the manufacture of phosphorus, phosphates, and fertilizers. Composition: calcium fluorophosphate or calcium chlorophosphate. General formula: $Ca_5(PO_4, CO_3)_3(F, OH, Cl)$. Crystal structure: hexagonal. 磷灰石 lín huī shí

> Origin: 19 century: from German Apatit, from Greek apatē deceit; from its misleading

similarity to other minerals.

Example:

The apatite plays an important role in Xiangshan rich uranium mineralization.
相山铀矿田中磷灰石对生成特富铀矿有着十分重要的作用。

Extended Terms:

apatite calculus 磷灰石结石
apatite crystal 磷灰石结晶
apatite pyroxenite 磷灰辉岩
talc apatite 滑石磷灰石
yttrium apatite 钇磷灰石

aphanitic igneous rock [ˌæfəˈnitik ˈigniəs rɒk]

Definition: 隐晶火成岩 yǐn jīng huǒ chéng yán

aphanitic: relating to an igneous rock with mineral components that are too fine to be seen by the naked eye. 隐晶岩的 yǐn jīng yán de

igneous: (of rock) having solidified from lava or magma(岩石)火成的. huǒ chéng de

rock: the solid mineral material forming part of the surface of the earth and other similar planets, exposed on the surface or underlying the soil. 岩石 yán shí；岩 yán

aplite [ˈæplait]

Definition: a light-coloured fine-grained acid igneous rock with a sugary texture, consisting of quartz and feldspars. 细晶岩 xì jīng yán

Origin: 19 century: from German Aplit, from Greek haploos simple + -ite.

Example:

Alkaline rocks in the area are mainly composed of gray-black trachyte, and some breccia lava, syenite porphyry and syenite aplite.
碱性岩的岩性组合主要以灰黑色粗面岩为主，并含有角砾熔岩，正长斑岩、正长细晶岩等。

Extended Terms:

aegirine aplite 霓细晶岩
diabase aplite 辉绿细晶岩
euritic aplite 霏细细晶岩
kersantite aplite 云斜细晶岩
nepheline aplite 霞石细晶岩

apophysis [əˈpɔfisis]

Definition: a branch from a dike or vein. 岩支 yán zhī；隆起 lóng qǐ

Example:

After forming of ancient apophysis, it will have different format with different affects of movement of structure in the long evolvement of history of geology.

岩脉隆起后，在漫长的地史演化过程中，受不同时期构造运动的影响而具有不同的表现形式。

Extended Terms:

apophysis ossium 松果体
annular apophysis 环状突
posterior apophysis 后内突
ring apophysis 环状骨突

aragonite [əˈrægənait]

Definition: a generally white or grey mineral, found in sedimentary rocks and as deposits from hot springs. Composition: calcium carbonate. 文石 wén shí，霰石 xiàn shí

Formula: $CaCO_3$. Crystal structure: orthorhombic.

Origin: 19 century: from Aragon + -ite.

Example:

Polar surface waters most likely will become undersaturated for aragonite before the end of this century.

在本世纪结束前，极地海水表层很有可能变得对霰石不饱和。

Extended Terms:

aragonite sinter 霰石华
botryoid aragonite 葡萄状霰石
micritic aragonite 微晶文石
ruin aragonite 废墟霰石
zinc-aragonite 锌霰石

Archean [ɑːˈkiːən]

Definition: noting or pertaining to rocks of the Archeozoic portion of the Precambrian Era. 太古庙 tài gǔ miào，太古代的 tài gǔ dài de

Origin: 1870-1875；from Greek archaî (os), ancient (see archaeo-) + -an.

> Example:

Conglomerate of the Middle Proterozoic Tietonggou Formation is in unconformity relation with the plagioclase schist of the Archean Taihua Group.
中元古界铁铜沟组砾岩，砾石形态多样，与下面的太古界太华群斜长片岩呈角度不整合接触。

> Extended Terms:

Archean eon 太古代；太古宙
Archean Eonothem 太古宇
Archean erathem 太古界
Archean protaxis 太古陆块
Archean stratum 太古代层

arenite [ˈærəˌnait]

> Definition:

Arenite (Latin Arena, sand) is a sedimentary clastic rock with sand grain size between 0.0625 mm (0.00246 in) and 2 mm (0.08 in) and contains less than 15% matrix. 砂屑岩 shā xiè yán，又称纯砂岩(also called clean sandstone) chún shā yán，砂粒碎屑岩 shā lì suì xiè yán

> Example:

They mainly consist of sandstone, in which quartz arenite and litharenite are dominant and arkose is lack.
碎屑岩以砂岩类为主，其中又以石英砂岩类和岩屑砂岩类为主，缺少长石砂岩类。

> Extended Terms:

feldspathic arenite 长石质砂岩
volcanic arenite 火山砂屑岩；火山质砂岩
subfeldspathic lithic arenite 亚长石质岩屑砂屑岩

argillite [ˈɑːdʒilait] =argillyte

> Definition:

a fine-grained sedimentary rock composed predominantly of indurated clay particles. Argillites are basically lithified muds and oozes. They contain variable amounts of silt-sized particles. The argillites grade into shale when the fissile layering typical of shale is developed. 泥板岩 ní bǎn yán

> Origin:

late 18th century：from Latin argilla "clay" + -ite.

> Example:

The Cardium sandstone is lithic, being composed of grains of chert, quartz, quartzite, silicified

argillite, and other rock fragments.
卡迪砂岩是岩屑砂岩，由燧石、石英、石英岩、硅化泥板岩和其他岩石碎屑颗粒所组成。

Extended Terms:
argillite argillolite 泥质层凝灰岩
argillite fissure 泥岩裂缝
argillite formation 泥质岩建造
argillic horizon 黏化层
argillite rock 黏土岩
argillite series 泥质岩系
argillite shale 泥质页岩
detrital rock and argillite 碎屑岩和泥质岩
ferruginous argillite 铁质硬页岩

argon [ˈɑːgɔn]

Definition: the chemical element of atomic number 18, an inert gaseous element of the noble gas group. Argon is the commonest noble gas, making up nearly one percent of the earth's atmosphere (Symbol: Ar). （化学元素）氩 yà（符号：Ar）

Origin: late 19 century: from Greek, neuter of argos "idle", from a- "without" +ergon "work"

Example:
The effect of nitrogen can be overcome by sheathing the flame in argon.
氮的影响则可以通过用氩气包覆火焰而予以克服。

Extended Terms:
argon method 氩法
liquid argon 液态氩；液氩

arkose [ˈɑːkəus]

Definition: a detrital sedimentary rock, specifically a type of sandstone containing at least 25% feldspar. Arkosic sand is sand that is similarly rich in feldspar, and thus the potential precursor of arkose. 长石砂岩 cháng shí shā yán

Origin: mid-19th century. From French, probably, and from Greek arkhaios "ancient".

Example:
The main types of clastic sandstones are feldspathic litharenite and lithic arkose in the Cretaceous of the Songliao basin.

松辽盆地白垩系砂岩以长石岩屑砂岩和岩屑长石砂岩为特征。

Extended Terms:

basal arkose 底长石砂岩
meta-arkose 变长石砂岩
plagioclase arkose 斜长石砂岩
quartzose arkose 石英长石砂岩
residual arkose 残积长石砂岩

---

ash [æʃ]

Definition: also called volcanic ash. (Geology) finely pulverized lava thrown out by a volcano in eruption. (地质) 火山灰 huǒ shān huī

Example:

Gas bubbles in the ash left holes that store large amounts of water for plant use.
火山灰内的气泡会留下空洞，它可以贮藏大量水分供植物利用。

Extended Terms:

ash gun 吹灰枪；吹灰器
ash handling 除灰；沉渣池
coal ash 煤灰；碎煤机；粉煤灰综合利用；煤炭

---

ash flow [æʃ fləu]

Definition: 火山灰流 huǒ shān huī liú

ash：(Geology) finely pulverized lava thrown out by a volcano in eruption. (地质) 火山灰 huǒ shān huī

flow：the steady and continuous movement of something in one direction. 流 liú；流动 liú dòng

Example:

Beyond causing immediate destruction from scalding ash flows, active super volcanoes spew gases that severely disrupt global climate for years afterward.
除了灼热火山灰流造成的立即性毁灭之外，活跃的超级火山也会喷出气体，在喷发后的数年仍严重影响着全球的气候。

---

ash flow tuff [æʃ fləu tʌf]

Definition: 火山灰凝流灰岩 huǒ shān huī níng liú huī yán

ash：(Geology) finely pulverized lava thrown out by a volcano in eruption. (地质) 火山灰 huǒ shān huī

flow: the steady and continuous movement of something in one direction 流 liú；流动 liú dòng

tuff: a light, porous rock formed by consolidation of volcanic ash 凝灰岩 níng huī yán

assimilation [əˌsimiˈleiʃən]

Definition: the process of becoming part of or more like something greater. 同化 tóng huà

Origin: 1595-1605; from Latin assimilātiōn-(s. of assimilātiō).

Example:

Individual topics include photosynthesis, respiration, nitrogen assimilation, lipid metabolism, cell wall formation, and biosynthesis of plant hormones and secondary metabolites.
个别的主题包括光合作用、呼吸作用、氮同化作用、脂质代谢、细胞壁形成、植物荷尔蒙和次及代谢物的合成。

Extended Terms:

assimilation ratio 同化率
juxtapositional assimilation 邻近同化；邻接同化
magmatic assimilation 岩浆同化
reciprocal assimilation 交互同化；相互同化(作用)

assimilation-fractional crystallization

[əˌsimiˈleiʃən-ˈfrækʃənl ˌkristəlaiˈzeiʃən]

Definition: 同化混染与分离结晶作用 tong huà hún rǎn yǔ fēn lí jié jīng zuò yòng

assimilation: the process of becoming part of or more like something greater. 同化 tóng huà

fractional: relating to or denoting the separation of components of a mixture by making use of their differing physical properties. 分级的 fēn jí de

crystallization: the process of forming or causing to form crystals. 结晶 jié jīng；晶化 jīng huà

asthenosphere [æsˈθenəsfiə]

Definition: (Geology) the region below the lithosphere, variously estimated as being from fifty to several hundred miles (eighty-five to several hundred kilometers) thick, in which the rock is less rigid than that above and below but rigid enough to transmit transverse seismic waves. (地质)软流圈 ruǎn liú quān

Origin: 1910-1915; from Greek asthen (ḗs) frail (see asthenia) + -o- + -sphere.

Example:

The rise of the asthenosphere, the position of Moho and the mantle plume are the indication of the mantle convection.
软流圈上涌高度、莫霍面或地幔羽的位置是地幔对流的具体表现。

Extended Terms:

underlying asthenosphere 下伏软流圈
upper asthenosphere 重软流圈上部

augen structure [ˈɔːdʒən ˈstrʌktʃə]

Definition: (Petrology) a structure found in some gneisses and granites in which certain of the constituents are squeezed into elliptic or lens-shaped forms and, especially if surrounded by parallel flakes of mica, resemble eyes. (岩石学)眼球状构造 yǎn qiú zhuàng gòu zào

augite [ˈɔːdʒait]

Definition: a black or greenish-black mineral of the pyroxene group, found in igneous rocks. Composition: calcium magnesium iron aluminium silicate. General formula: $(Ca, Mg, Fe, Al)(Si, Al)_2O_6$. Crystal structure: monoclinic. 普通辉石 pǔ tōng huī shí

Origin: 19 century: from Latin augītēs, from Greek, from augē brightness.

Example:

Thyon volcanic rocks are classified as alkali basalts, with pyroxene megacrysts, and deep-seated mantle xenoliths such as augite and peridotite have been observed in some basaltic layers.
火山岩主要为碱性玄武岩,部分玄武岩中可见橄榄岩类等深源岩石包体和普通辉石等矿物巨晶。

Extended Terms:

augite akerite 辉石英辉正长岩
augite andesite 辉石安山岩
basaltic augite 玄武辉石
egirine augite 霓辉石
vanadian augite 钒辉石

authigenic [ˌɔːθiˈdʒenik]

Definition: (of minerals) having crystallized in a sediment during or after deposition. 自生的 zì shēng de

Origin: 19 century: from German authigene, from Greek authigenēs native + -ic.

Example:

The formation of authigenic clay mineral is related to clastic components of clastic rocks. 自生黏土矿物的形成与碎屑岩的碎屑组分有着密切的关系。

Extended Terms:

authigenic element 自生元素
authigenic kaolinite 自生高岭石
authigenic mineralizations 自生矿化作用
authigenic minerals 自生矿物
authigenic sediment 自生沉积

authigenic mineral [ˌɔːθiˈdʒenik ˈminərəl]

Definition: an authigenic mineral or sedimentary rock deposit is one that was generated where it is found or observed. 自生矿物 zì shēng kuàng wù

Example:

Authigenic mineral assemblage formed in different ambients is distinct to each other.
两种环境下形成的自生矿物组合截然不同。

autochthonous granite [ɔːˈtɔkθənəs ˈɡrænit]

Definition: 原地花岗岩 yuán dì huā gǎng yán

autochthonous: (of a deposit or formation) formed in its present position. (沉积或形成) 原地 (生成) 的 yuán dì de

granite: a very hard, granular, crystalline, igneous rock consisting mainly of quartz, mica, and feldspar and often used as a building stone. 花岗岩 huā gǎng yán

autoclastic texture [ˌɔːtəˈklæstik ˈtekstʃə]

Definition: 自碎结构 zì suì jié gòu

autoclastic: of rock, fragmented in place by folding due to orogenic forces when the rock is not so heavily loaded as to render it plastic. 自碎 zì suì

texture: the arrangement of the particles or constituent parts of any material, as wood, metal, etc., as it affects the appearance or feel of the surface; structure, composition, grain, etc. 结构 jié gòu

axial magma chamber (AMC) [ˈæksiəl ˈmæɡmə ˈtʃeimbə]

Definition: 轴向岩浆房 zhóu xiàng yán jiāng fáng

axial: of, forming, or relating to an axis. 与轴有关的 yǔ zhóu yǒu guān de

magma: hot fluid or semi-fluid material below or within the earth's crust from which lava and other igneous rock are formed by cooling. 岩浆 yán jiāng

chamber: a large room used for formal or public events. 大厅 dà tīng

back-arc spreading [bæk ɑːk sprediŋ]

Definition: 弧后扩张 hú hòu kuò zhāng

back-arc: relating to or denoting the area behind an island arc. (岛)弧后部的 hú hòu bù de, 后弧的 hòu hú de

spreading: the act of gradually reaching or causing to reach a larger and larger area or more and more people. 传开 chuán kāi; 散布 sàn bù; 蔓延 màn yán

backarc basin [ˈbækɑːk ˈbeisən]

Definition: (Geology) the region (small ocean basin) between an island arc and the continental mainland formed during oceanic plate subduction, containing sediment eroded from both. (地质)弧后盆地 hú hòu pén dì

Example:
These study refers to the petrologic, geochemical and isotopic variations, along and across mid-ocean ridges and backarc basin spreading centres, and of the Hawaiian and other oceanic islands and seamounts.
这些研究涉及岩石学、地球化学、同位素变异、中尺度洋脊、大洋盆地扩散，以及夏威夷岛屿与其他海洋岛屿及海山等方面。

backshore facies [ˈbækʃɔː ˈfeiʃiːz]

Definition: 后滨相 hòu bīn xiāng

backshore: the area of a shore that is above the high-water mark except in very severe weather. 后滨 hòu bīn

facies: the character of a rock expressed by its formation, composition, and fossil content. 相 xiàng

baffle stone [ˈbæfl stəun]

Definition: 障积灰岩 zhàng jī huī yán

baffle: a flat plate that controls or directs the flow of fluid or energy 挡板 dǎng bǎn
stone: a lump or mass of hard consolidated mineral matter 岩石 yán shí

Example:

The reef core can be divided into three microfacies from bottom to top: 1. bind-stone, 2. baffle-bindstone, 3. bind-bafflestone. 礁核从底到顶可分出三个微相：1）黏结岩，2）障积-黏结岩，3）黏结-障积岩。

banakite [ˈbænəkait]

Definition: （Petrology）an alkalic basalt made up of plagioclase, sanidine, and biotite, with small quantities of analcime, augite, and olivine; quartz or leucite may be present.（岩石学）橄云安粗岩 gǎn yún ān cū yán

Extended Term:

quartz-banakite 石英粗面粒玄岩

banding bedding [ˈbændiŋ ˈbediŋ]

Definition: 带状层理 dài zhuàng céng lǐ

banding: the presence or formation of visible stripes of contrasting colour. 条纹 tiáo wén
bedding:（Geology）the stratification or layering of rocks or other geological materials.（地质）层理 céng lǐ

banded iron formation（BIF）[ˈbændid ˈaiən fɔːˈmeiʃən]

Definition: 条带状含铁建造 tiáo dài zhuàng hán tiě jiàn zào

banded: marked with bands of different or contrasting colors. 有带的 yǒu dài de, 有条子的 yǒu tiáo zǐ de
iron: a strong, hard magnetic silvery-grey metal, the chemical element of atomic number 26, much used as a material for construction and manufacturing, especially in the form of steel (symbol: Fe).（化学元素）铁 tiě（符号：Fe）
formation: a structure or arrangement of something. 结构 jié gòu, 排列 pái liè

Example:

The Earth: An Intimate History mentions a type of rock found in crust known as banded iron formations.

《地球:亲密的历史》提到一种在地壳中发现的岩石被称为带状铁生成物的石头。

banded structure ['bændid 'strʌktʃə]

Definition: (Petrology) an outcrop feature in igneous and metamorphic rocks due to alternation of layers, stripes, flat lenses, or streaks that obviously differ in mineral composition or texture. (岩石学) 条带状构造 tiáo dài zhuàng gòu zào

Extended Terms:

concentric banded structure 同心条带状结构
parallel banded structure 平行带状构造
symmetrical banded structure 对称带状构造

barite ['bɛərait]

Definition: also called heavy spar, a colourless or white mineral consisting of barium sulphate in orthorhombic crystalline form, occurring in sedimentary rocks and with sulphide ores; a source of barium. 重晶石 zhòng jīng shí

Formula: $BaSO_4$

Origin: 18 century; from bar (ium) + -ite.

Example:
The principal minerals of lead-zinc deposit in Zhenxun are sphalerite, galena, pyrite, quartz and barite.
闪锌矿、方铅矿、黄铁矿、石英及重晶石是镇旬铅锌矿田的主要矿物。

Extended Terms:

barite plug 重晶石塞
barite powder 重晶石粉
barite slurry 重晶石浆
insoluble barite 不溶性重晶石

barometry [bəˈrɔmitri]

Definition: the process of measuring atmospheric pressure. 气压测定法 qì yā cè dìng fǎ

Origin: 1705-1715; baro- + -metry.

> Example:

Specialised programmes include *The Weather Channel*, 24 hours a day of barometry and precipitation forecasts.
专门的节目包括《天气频道》，一天 24 小时预报气压和降雨量。

barrier island deposit [ˈbæriə ˈailənd diˈpɔzit]

> Definition: 障壁岛沉积 zhàng bì dǎo chén jī

barrier：a fence or other obstacle that prevents movement or access. 屏障 píng zhàng；障碍物 zhàng ài wù
island：a piece of land surrounded by water. 岛 dǎo，岛屿 dǎo yǔ
deposit：a layer or body of accumulated matter. 沉积物 chén jī wù，沉淀 chén diàn

basal cementation [ˈbeisəl ˌsiːmenˈteiʃən]

> Definition: 基底胶结 jī dǐ jiāo jié

basal：forming or belonging to a bottom layer or base. 基底的 jī dǐ de；基层的 jī céng de
cementation：the binding together of particles or other things by cement. 黏结 nián jié，胶接作用 jiāo jiē zuò yòng

basal conglomerate [ˈbeisəl kənˈglɔməreit]

> Definition: 底砾岩 dǐ lì yán

basal：forming or belonging to a bottom layer or base. 基底的 jī dǐ de；基层的 jī céng de
conglomerate：a coarse-grained sedimentary rock composed of rounded fragments embedded in a matrix of cementing material such as silica. 砾岩 lì yán

> Example:

Vertically, rocks near the unconformity appear layers which are in turn semi-weathering rock, weathering clay bed and basal conglomerate.
纵向上不整合附近的岩层呈层分布：从下至上依次为半风化岩石、风化黏土层和底砾岩。

basalt [ˈbæsɔːlt]

> Definition: the dark, dense igneous rock of a lava flow or minor intrusion, composed essentially of labradorite and pyroxene and often displaying a columnar structure. 玄武岩 xuán wǔ yán

> Origin: 18 century：from Late Latin basaltēs, variant of basanitēs, from Greek basanitēs

touchstone, from basanos, of Egyptian origin.

> Example:

The blast furnace slag by basalt fiber is made, and the price is low, friction stability is good substitution of asbestos.

该纤维由玄武岩和高炉炉渣制成，价格低、摩擦性能稳定是石棉的良好代用纤维。

> Extended Terms:

decayed basalt 风化玄武岩
gallaston basalt 辉绿基橄玄岩
melilite basalt 黄长玄武岩
ocean basalt 大洋玄武岩
tholeiitic basalt 拉斑玄武岩；硅质玄武岩

basalt-andesite-rhyolite association

[ˈbæsɔːlt ˈændizait ˈraiəlait əˌsəusiˈeiʃən]

> Definition: 玄武岩-安山岩-流纹岩组合 xuán wǔ yán-ān shān yán-liú wén yán zǔ hé

basalt: the dark, dense igneous rock of a lava flow or minor intrusion, composed essentially of labradorite and pyroxene and often displaying a columnar structure. 玄武岩 xuán wǔ yán

andesite: a dark, fine-grained, brown or greyish intermediate volcanic rock which is a common constituent of lavas in some areas. 安山岩 ān shān yán

rhyolite: a pale fine-grained volcanic rock of granitic composition, typically porphyritic in texture. 流纹岩 liú wén yán

association: a linking or joining of people or things. 联合 lián hé

basalt-rhyolite association [ˈbæsɔːlt ˈraiəlaitə əˌsəusiˈeiʃən]

> Definition: 玄武岩-流纹岩组合 xuán wǔ yán-liú wén yán zǔ hé

basalt: the dark, dense igneous rock of a lava flow or minor intrusion, composed essentially of labradorite and pyroxene and often displaying a columnar structure. 玄武岩 xuán wǔ yán

rhyolite: a pale fine-grained volcanic rock of granitic composition, typically porphyritic in texture. 流纹岩 liú wén yán

association: a linking or joining of people or things. 联合 lián hé

basalt tetrahedron [ˈbæsɔːlt ˌtetrəˈhiːdrən]

> Definition: 玄武岩四面体 xuán wǔ yán sì miàn tǐ

basalt: the dark, dense igneous rock of a lava flow or minor intrusion, composed essentially of

labradorite and pyroxene and often displaying a columnar structure. 玄武岩 xuán wǔ yán
tetrahedron: a solid having four plane triangular faces; a triangular pyramid. 四面体 sì miàn tǐ

basaltic andesite [bəˈsɔːltik ˈændizait]

Definition: 玄武安山岩 xuán wǔ ān shān yán

basaltic: relating to the dark, dense igneous rock of a lava flow or minor intrusion, composed essentially of labradorite and pyroxene and often displaying a columnar structure. 玄武岩的 xuán wǔ yán de

andesite: a dark, fine-grained, brown or greyish intermediate volcanic rock which is a common constituent of lavas in some areas. 安山岩 ān shān yán

Example:

The protolith of the metavolcanic rocks is a volcanic-subvolcanic rock series composed chiefly of low-K tholeiite with subordinate low-K basaltic andesite formed in an oceanic extensional environment.
变质火山岩的原岩为大洋拉张环境下形成的以低钾拉斑玄武岩为主，低钾玄武安山岩次之的火山-次火山岩系，其物源为亏损地幔。

basaltic glass [bəˈsɔːltik glɑːs]

Definition: 玄武玻璃 xuán wǔ bō lí

basaltic: relating to the dark, dense igneous rock of a lava flow or minor intrusion, composed essentially of labradorite and pyroxene and often displaying a columnar structure. 玄武岩的 xuán wǔ yán de

glass: a hard, brittle substance, typically transparent or translucent, made by fusing sand with soda, lime, and sometimes other ingredients and cooling rapidly. It is used to make windows, drinking containers, and other articles. 玻璃 bō lí

basanite [ˈbæsənait]

Definition: a basaltic extrusive rock closely allied to chert, jasper, or flint; also known as Lydian stone; lydite. 碧玄岩 bì xuán yán

Example:

The native minerals originating from Tertiary basanite of Yaoshan formation, the placer is in water systems around Fangshan.
原生矿产于第三纪的尧山组玄武质碧玄岩中，砂矿则分布于方山周围的水系中。

Extended Terms:

biotite leucite basanite 云白碧玄岩
hauyne-basanite 蓝方碧玄岩
leucite-basanite 白榴石碧玄岩
noselite basanite 黝方碧玄岩

basic intrusion ['beisik in'truːʒən]

Definition: 基性侵入岩 jī xìng qīn rù yán

basic: forming an essential foundation or starting point; fundamental. 基本的 jī běn de; 基础的 jī chǔ de
intrusion: the action or process of forcing a body of igneous rock between or through existing formations, without reaching the surface. 侵入 qīn rù

basic rock ['beisik rɒk]

Definition: an igneous rock with a relatively low silica content, and rich in iron, magnesium, or calcium. 基性岩 jī xìng yán

Example:
The primary gold deposits in western Hunan are included mainly in three types as quartz vein type, fracture zone altered rock type and altered basic rock type.
湘西地区的岩金矿，主要有石英脉型、破碎带蚀变岩型和基性岩蚀变型三类。

Extended Terms:

basic rock strata 基础岩层
basic rock zone 基性岩带
basic igneous rock 基性火成岩
hard basic rock 坚硬基岩

basin plain deposit ['beisən plein di'pɔzit]

Definition: 深海盆地平原沉积 shēn hǎi pén dì píng yuán chén jī

basin: a place where the earth's surface is lower than that in other areas of the world. 盆地 pén dì; 凹地 āo dì
plain: a large expanse of fairly flat dry land, usually with few trees. 平原 píng yuán
deposit: a layer or body of accumulated matter. 沉积物 chén jī wù, 沉淀 chén diàn

batch melting [bætʃ ˈmeltɪŋ]

Definition: 间歇熔融 jiàn xiē róng róng

batch: a quantity of people or things treated or regarded as a group, especially when subdivided from a larger group. 整批的 zhěng pī de; 间歇式的 jiàn xiē shì de

melting: the process whereby heat changes something from a solid to a liquid. 熔化 róng huà

batholith [ˈbæθəlɪθ]

Definition: (Geology) a large body of intrusive igneous rock believed to have crystallized at a considerable depth below the earth's surface; pluton. (地质) 岩基 yán jī

Origin: 1900-1905; batho- + -lith.

Example:
A batholith has been defined as a huge intrusive mass of granitic rock.
岩基的定义是巨大的花岗石侵入岩体。

Extended Terms:
anatectic batholith 深源岩基
concordant batholith 整合岩基
homogeneous batholith 均匀岩基
simple batholith 单岩基
subautochthonous batholith 半原地岩基

bathyal facies [ˈbæθiːəl ˈfeɪʃiːz]

Definition: 半深海相 bàn shēn hǎi xiāng

bathyal: of or relating to the zone of the sea between the continental shelf and the abyssal zone. 半深海的 bàn shēn hǎi de

facies: the character of a rock expressed by its formation, composition, and fossil content 相 xiàng

bauxite [ˈbɔːksaɪt]

Definition: a white, red, yellow, or brown amorphous claylike substance comprising aluminium oxides and hydroxides, often with such impurities as iron oxides. It is the chief ore of aluminium. General formula: $Al_2O_3 \cdot nH_2O$ 铝土矿 lǚ tǔ kuàng

Origin: 19 century: from French, from (Les) Baux in southern France, where it was originally found.

Example:

By analyzing its grindability, the study bases on the digestion performance of black bauxite to find out the optimum condition.

从分析黑铝土矿的可磨性出发,针对其溶出性能研究,旨在找出黑铝土矿的最佳溶出条件。

Extended Terms:

bauxite cement 矾土水泥
bauxite chamotte 高铝矾土熟料
bauxite fumes 铝土矿烟尘
activated bauxite 活化铝矾土
cellular bauxite 多孔状铝土矿

beach facies [biːtʃ ˈfeiʃiːz]

Definition: 海滩相 hǎi tān xiàng

beach: a pebbly or sandy shore, especially by the sea between high- and low-water marks. 海滩 hǎi tān

facies: the character of a rock expressed by its formation, composition, and fossil content. 相 xiàng

Example:

Beach facies is only the basic condition for reservoir development. The forming of effective reservoir in the beach body depends on the later diagenesis.

由此可见,滩相沉积体仅是储层发育的基本条件,滩体最终能否形成有效的储层体,还要取决于后期成岩作用的改造。

Extended Term:

reef beach facies 礁滩相

beach ridge facies [biːtʃ ridʒ ˈfeiʃiːz]

Definition: 海滩脊相 hǎi tān jǐ xiàng

beach: a pebbly or sandy shore, especially by the sea between high- and low-water marks 海滩 hǎi tān

ridge: a long narrow hilltop, mountain range, or watershed. 山脊 shān jǐ;分水岭 fēn shuǐ lǐng

facies: the character of a rock expressed by its formation, composition, and fossil content 相 xiàng

bed [bed]

Definition: the bottom of the sea or a lake or river. 河床 hé chuáng

Origin: Old English bed, bedd (noun), beddian (verb), of Germanic origin; related to Dutch bed and German Bett.

Example:
It is a research on the effect of bed erosion and fall of water at the lower reach of Danjiangkou Reservoir on the waterway.
这是一个关于丹江口水库下游河床冲刷与水位降落对航道影响的研究。

Extended Terms:
bed form 河床形体
ditch bed 沟床
flowing bed 流化床
gully bed 沟床
planar bed 二维床
rubble bed 抛石基床

bedding ['bedɪŋ]

Definition: (Geology) the stratification or layering of rocks or other geological materials. (地质) 层理 céng lǐ

Origin: later Old English beddinge "bedding, bed covering," from bed. Meaning "bottom layer of anything" is from c. 1400.

Example:
Bedding is the most important structural feature of sedimentary rock.
层理是沉积岩最主要的构造特征。

Extended Terms:
convolute bedding 旋绕层理
current bedding 流水层理
discordant bedding 不整合层理
dune bedding 沙丘层理
false bedding 假层理

bedding plane ['bedɪŋ pleɪn]

Definition: the surface that separates one stratum, layer, or bed of stratified rock from another 层面 céng miàn

Origin: 1895-1900.

Example:

The mathematical model with finite solution for bedding plane air-leakage is established, in which the air leakage intensity is regarded as attenuating follows proximately negative exponential law.
建立了层面漏风的定解数学模型,近似地把地表漏风强度看作按指数规律衰减。

Extended Terms:

bedding plane cleavage 顺面劈理
bedding plane fault 层面断层
bedding-plane power spectrum 层面功率谱
bedding-plane shear surface 层面剪切滑面
faulted bedding plane 错动层面

bedding plane structure ['bediŋ plein 'strʌktʃə]

Definition: 层面构造 céng miàn gòu zào

bedding plane: the surface that separates one stratum, layer, or bed of stratified rock from another. 层面 céng miàn

structure: the arrangement of and relations between the parts or elements of something complex. 结构 jié gòu;构造 gòu zào

bentonite ['bentənait]

Definition: a light-colored clay that expands in water used in oil drilling, paper, pharmaceutical industries. 膨润土 péng rùn tǔ

Origin: late 19th century. After Fort Benton, Montana.

Example:

If more bentonite clay is used, the lead temperature must be decreased.
如果多用膨润土,铅的温度必须降低。

Extended Terms:

bentonite cement 胶质水泥;膨润土水泥
bentonite liner 膨润土衬垫

beresite [be'resaɪt]

Definition: a granite in which the orthoclase has been hydrothermally altered to muscovite. It was first described from Beresovsk by Rose in 1837. 黄铁绢英岩 huáng tiě juān yīng yán

Example:
The beresite from the Shuangshanzi gold mine contains a number of coarse-grained pyrite crystals with regular zonal structure.
胶东乳山金矿双山子黄铁石英岩中含有粗粒黄铁矿发育规则环带结构。

Extended Terms:
beresite porphyry 黄铁石英斑岩
carbonate-beresite 碳酸黄铁石英岩

bimodal igneous tectonic assemblage
[baiˈməudəl ˈigniəs tekˈtɔnik əˈsemblidʒ]

Definition: 双峰式火成岩组合 shuāng fēng shì huǒ chéng yán zǔ hé

bimodal：having two modes. 双峰的 shuāng fēng de
igneous：formed by solidification from a molten state. 火成的 huǒ chéng de
tectonic：relating to, causing, or resulting from structural deformation of the earth's crust. 构造的 gòu zào de
assemblage：a form of art involving the assembly and arrangement of unrelated objects, parts, and materials in a kind of sculptured collage. 集合 jí hé

bimodal magmatism [baiˈməudəl ˈmægmətizm]

Definition: 双峰式岩浆活动 shuāng fēng shì yán jiāng huó dòng

bimodal：having two modes. 双峰的 shuāng fēng de
magmatism：the motion or activity of magma. 岩浆活动

binary system [ˈbainəri ˈsistəm]

Definition: a system consisting of two parts. 二元系统 èr yuán xì tǒng, 二元体制 èr yuán tǐ zhì

bindstone [ˈbaindstəun]

Definition: 黏结灰岩 nián jié huī yán

bind：stick to firmly. 黏结 nián jié
stone：a lump or mass of hard consolidated mineral matter. 岩石 yán shí

Examples:

①The reef rocks thus formed are mainly sponge (or hydrozoa) bafflestone, grading upwards into laminated algae and Archaeolithoporella bindstone, and eventually covered by stromatolitic lime mudstone of tidal flat factes.
其主要的礁岩类型为海绵(或水螅)障积岩,在上部出现纹层藻和古石孔藻黏结岩,最后为潮坪相叠层石灰泥岩所覆盖。

②The shallow ramp which is between the wave base and the storm wave base, has type A, B tempestites, pattern A, B nodular limestone, reef limestone and bindstone deposits.
浅缓坡位于浪基面至风暴浪基面之间,岩性组合为A、B类风暴岩、A、B类瘤状灰岩、生物礁灰岩及黏结岩。

biochemical sedimentary rock [ˌbaiəuˈkemikəl ˌsediˈmentəri rɔk]

Definition: 生物化学沉积岩 shēng wù huà xué chén jī yán

biochemical: relating to the chemical processes and substances which occur within living organisms. 生物化学的 shēng wù huà xué de

sedimentary: (of rock) that has formed from sediment deposited by water or air. (岩石)沉积形成的 chén jī xíng chéng de

rock: the solid mineral material forming part of the surface of the earth and other similar planets, exposed on the surface or underlying the soil. 岩石 yán shí; 岩 yán

bioclastic limestone [ˈbaiəuˈklæstik ˈlaimstəun]

Definition: 生物碎屑灰岩 shēng wù suì xiè huī yán, 生物贝屑灰岩 shēng wù bèi xiè huī yán

bioclastic: a fragment of a shell or fossil forming part of a sedimentary rock. 生物碎屑 shēng wù suì xiè

limestone: a hard sedimentary rock, composed mainly of calcium carbonate or dolomite, used as building material and in the making of cement. 灰岩 huī yán, 石灰岩 shí huī yán

Example:

Permian bioclastic limestone and bioherm are the most favorable reservoirs.
二叠系生屑灰岩和生物礁为最有利的储层。

biogenetic texture [ˌbaiəudʒiˈnetik ˈtekstʃə]

Definition: 生物结构 shēng wù jié gòu

biogenetic: the synthesis of substances by living organisms. 生物合成 shēng wù hé chéng

texture: the arrangement of the particles or constituent parts of any material, as wood, metal, etc., as it affects the appearance or feel of the surface; structure, composition, grain, etc. 结构 jié gòu

Example:

SME and polarizing microscope investigations reveal abundant cyanoalgal fossils and evident biogenetic texture in the speleothems.
经扫描电子显微镜和偏光显微镜鉴定，其中含有大量的蓝藻化石，具有明显的生物结构。

biogenic imprint [ˌbaiəuˈdʒenik ˈimprint]

Definition: 生物痕迹 shēng wù hén jì

biogenic: produced or brought about by living organisms. 由生物作用所产生的 yóu shēng wù zuò yòng suǒ chǎn shēng de

imprint: a mark made by pressing something on to a softer substance so that its outline is reproduced. 印记 yìn jì, 印痕 yìn hén

biogenic sedimentary structure [ˌbaiəuˈdʒenik ˌsediˈmentəri ˈstrʌktʃə]

Definition: 生物成因构造 shēng wù chéng yīn gòu zào

biogenic: produced or brought about by living organisms. 由生物作用所产生的 yóu shēng wù zuò yòng suǒ chǎn shēng de

sedimentary: (of rock) that has formed from sediment deposited by water or air. (岩石)沉积形成的 chén jī xíng chéng de

structure: the arrangement of and relations between the parts or elements of something complex. 结构 jié gòu; 构造 gòu zào

biotite [ˈbaiətait]

Definition: a black or dark green mineral of the mica group, found in igneous and metamorphic rocks. Composition: hydrous magnesium iron potassium aluminium silicate. 黑云母 hēi yún mǔ

Formula: $K(Mg, Fe)_3(Al, Fe)Si_3O_{10}(OH)_2$. Crystal structure: monoclinic.

Origin: 1860-1865; named after J. B. Biot (1774-1862), French mineralogist and mathematician.

Example:

The talc-schist, talcmarble and biotite-schist are the most favourable hosting rocks to mineralization.

最有利于矿化的围岩是富镁的滑石片岩,含滑石片岩、含滑石大理岩和黑云母片岩等。

Extended Terms:

biotite camptonite 黑云康煌岩

biotite granite 黑云花岗岩

biotite kersantite 多云云斜煌岩

biotite klausenite 黑云苏斜岩

biotite isograd 黑云母变度

biotite zone [ˈbaɪətaɪt zəun]

Definition: 黑云母带 hēi yún mǔ dài

biotite: a black or dark green mineral of the mica group, found in igneous and metamorphic rocks. Composition: hydrous magnesium iron potassium aluminium silicate. 黑云母 hēi yún mǔ

Formula: $K(Mg, Fe)_3(Al, Fe)Si_3O_{10}(OH)_2$. Crystal structure: monoclinic.

zone: an area or stretch of land having a particular characteristic, purpose, or use, or subject to particular restrictions. 地带 dì dài, 区域 qū yù

bioturbation [ˌbaɪəutəːˈbeɪʃən]

Definition: the alteration and disturbance of a site by living organisms; the turning and mixing of sediments by organisms, as rodents. 生物扰动作用 shēng wù rǎo dòng zuò yòng

Example:

X-radiography has sometimes been used with success to detect bedding structures and bioturbation in apparently uniform sandstone.

有时射线照片已成功地被用来发现在外表均一的砂岩中的层次构造和生物扰动。

Extended Terms:

bioturbation heaving 生物扰动隆起

bioturbation structure 生物扰动构造

bird eye structure [bəːd aɪ ˈstrʌktʃə]

Definition: 鸟眼构造 niǎo yǎn gòu zào, 窗孔构造 chuāng kǒng gòu zào

bird: a warm-blooded egg-laying vertebrate animal distinguished by the possession of feathers, wings, a beak, and typically by being able to fly. 鸟 niǎo, 禽 qín

eye: each of a pair of globular organs in the head through which people and vertebrate animals see, the visible part typically appearing almond-shaped in animals with eyelids. 眼睛 yǎn jīng
structure: the arrangement of and relations between the parts or elements of something complex. 结构 jié gòu；构造 gòu zào

black shale [blæk ʃeil]

Definition: a dark, muddy rock, rich in sulfides and organic material. 黑色页岩 hēi sè yè yán

Extended Term:
black alum shale 黑色明矾页岩

black smoker [blæk ˈsməukə]

Definition: a geothermal vent on the seabed which ejects superheated water containing much suspended matter, typically black sulphide minerals. 海底地热排放口 hǎi dǐ dì rè pái fang kǒu

Example:
Large amounts of microfossil records have been discovered in the seafloor black smoker chimney from the Okinawa Trough.
报道了冲绳海槽伊平屋海洼海底黑烟囱样品中保留着完好的矿化微生物群，它们的形态完好。

blastic [ˈblæstik]

Definition: a combining form meaning "having a given type or number of buds, cells, or cell layers," or "undergoing a given type of development," as specified by the initial element: holoblastic. 变晶的 biàn jīng de

Origin: Greek, form of blastós a bud, sprout.

Example:
The raw jade is in multi-colors with scaly blastic and massive textures under microscope.
玉石原石色彩多样，岩石以显微鳞片变晶结构和块状构造为主。

Extended Terms:
blastic deformation 爆炸变形；冲击变形；重结晶变形
acicular blastic texture 针状变晶结构

blocking temperature [ˈblɒkɪŋ ˈtempərɪtʃə]

Definition: 阻挡温度 zǔ dǎng wēn dù

blocking: the action or process of obstructing movement, progress, or activity, in particular. 妨碍 fáng ài，阻塞 zǔ sè，阻挡 zǔ dǎng

temperature: the degree or intensity of heat present in a substance or object, especially as expressed according to a comparative scale and shown by a thermometer or perceived by touch. 温度 wēn dù，气温 qì wēn

---

block lava [blɒk ˈlɑːvə]

Definition: 块状熔岩 kuài zhuàng róng yán

block: a large solid piece of hard material, especially rock, stone, or wood, typically with flat surfaces on each side. 大块 dà kuài；大块石板 dà kuài shí bǎn

lava: hot molten or semi-fluid rock erupted from a volcano or fissure, or solid rock resulting from cooling of this. 熔岩 róng yán；火山岩 huǒ shān yán

---

blueschist facies [ˈbluːʃɪst ˈfeɪʃiːz]

Definition: 蓝片岩相 lán piàn yán xiàng

blueschist: a metamorphic rock with a blue colour, formed under conditions of high pressure and low temperature. 蓝片岩 lán piàn yán

facies: the character of a rock expressed by its formation, composition, and fossil content. 相 xiàng

---

boehmite [ˈbəːmaɪt]

Definition: a grey, red, or brown mineral that consists of alumina in rhombic crystalline form and occurs in bauxite. 水软铝石 shuǐ ruǎn lǚ shí；勃姆石 bó mǔ shí

Formula: AlO(OH)

Origin: C20：from German Böhmit, after J. Böhm, 20th-century German scientist.

Example:
The low bulk density and large porous pseudo-boehmite was obtained by the addition of crystal seeds.
采用加入晶种的方法制备了低堆比和含有大孔的假水软铝石。

Extended Term:

pseudo-boehmite 拟薄水铝石

bomb [bɔm]

Definition: (Geology) also called volcanic bomb, a rough spherical or ellipsoidal mass of lava, ejected from a volcano and hardened while falling. (地质) 火山弹 huǒ shān dàn

Origin: C17: from French bombe, from Italian bomba, probably from Latin bombus a booming sound, from Greek bombos, of imitative origin; compare Old Norse bumba drum.

Example:

Two volcanic structures, hornitos and accretionary lava ball, were wrongly named as "fumaroles" and "volcanic bomb" respectively in the Geopark.
景点中有些火山构造的命名却存在着明显的错误，如：熔岩丘和增生熔岩球分别被错误地命名为"喷气锥"和"火山弹"。

Extended Terms:

incendiary bomb 燃烧弹；烧夷炸弹
penetrating bomb 钻地弹

boninite [ˈbɔninait]

Definition: a mafic extrusive rock high in both magnesium and silica, formed in fore-arc environments, typically during the early stages of subduction. 玻古安山岩 bō gǔ ān shān yán

Origin: The rock is named for its occurrence in the Izu-Bonin arc south of Japan.

Example:

The extrusive rock unit includes basalt, andesite, boninite and albitite as well as breccia and tuff.
喷出岩单元由玄武岩、安山岩、玻古安山岩、钠长岩以及角砾岩、凝灰岩等组成。

Extended Term:

garnetiferous boninite 石榴玻紫安山岩

borax [ˈbɔːræks]

Definition: also called tincal, a soluble readily fusible white mineral consisting of impure hydrated disodium tetraborate in monoclinic crystalline form, occurring in alkaline soils and salt deposits. 硼砂 péng shā

Formula: $Na_2B_4O_7 \cdot 10H_2O$

Origin: late 14 century, from Anglo-French boras, from Middle English baurach, from Arabic buraq.

Example:

The 4 methods used to disinfect endotracheal tubes are all effective, especially the methods of boiling and soaking with glutaral, borax phenol and formaldehyde solution.
4种方法用于气管内套管消毒均有效，其中尤以煮沸法、戊二醛浸泡法、硼砂酚醛浸泡法为佳（双氧水浸泡法稍差）。

Extended Terms:

borax bead 硼砂珠
borax bead test 硼砂珠试验
borax lake 硼砂湖

braid index [breid ˈindeks]

Definition: 辫状指数 biàn zhuàng zhǐ shù

braid: something that is made of three or more interwoven strands. 辫状 biàn zhuàng
index: a figure in a system or scale representing the average value of specified prices, shares, or other items as compared with some reference figure. 指数 zhǐ shù

braided river deposit [ˈbreidid ˈrivə diˈpɔzit]

Definition: 辫状河沉积 biàn zhuàng hé chén jī

braided: composed of interconnected tracks or channels that divide and reunite. 辫状 biàn zhuàng
river: a large natural stream of water flowing in a channel to the sea, a lake, or another such stream. 河 hé；水道 shuǐ dào
deposit: a layer or body of accumulated matter. 沉积物 chén jī wù，沉淀 chén diàn

Example:

The Zao 0 oil group in the Kong 1 Member of Duanliubo oilfield, Dagang area is braided river deposit.
大港段六拨油田孔一段枣 0 油组为网状河沉积。

breccia [ˈbretʃiə]

Definition: a coarse-grained sedimentary rock made of sharp fragments of rock and stone cemented together by finer material. Breccia is produced by volcanic activity or erosion, including frost shattering. 角砾岩 jiǎo lì yán

Origin: late 18th century. From Italian, "gravel".

Example:
Geologists involved in the 2005 study suspect that the impact upturned bedrock, pushing it outward and breaking much of the rock into angular fragments, or breccia.
参与2005年那次研究的地质学家们怀疑，撞击翻转了下层的岩石，将它们朝外推出，并且大多破裂成棱角分明的碎片或者角砾岩。

Extended Terms:
breccia dike 角砾岩脉
dislocation breccia 断层角砾岩
rubble breccia 破碎角砾岩；碎石角砾岩
shatter breccia 震裂角砾岩
solution breccia 溶塌角砾岩

brush cast [brʌʃ kɑːst]

Definition: 刷模 shuā mó，刷铸型 shuā zhù xíng

brush: an implement with a handle, consisting of bristles, hair, or wire set into a block, used for cleaning or scrubbing, applying a liquid or powder to a surface, arranging the hair, or other purposes. 刷子 shuā zi，毛刷 máo shuā

cast: shape (metal or other material) by pouring it into a mould while molten. 浇铸 jiāo zhù

buffer reaction [ˈbʌfə riˈækʃən]

Definition: 缓冲反应 huǎn chōng fǎn yìng

buffer: a solution which resists changes in pH when acid or alkali is added to it. Buffers typically involve a weak acid or alkali together with one of its salts. (化)缓冲溶液 huǎn chōng róng yè，缓冲剂 huǎn chōng jì

reaction: a chemical process in which two or more substances act mutually on each other and are changed into different substances, or one substance changes into two or more other substances. 反应 fǎn yìng

Example:
The experiment results show that using the reaction buffer and stop reagent is the key of the synthesis of spherical calcium carbonate.
研究结果表明：加入反应缓冲剂及结晶生长停止剂是合成球状碳酸钙的关键。

bulk assimilation [bʌlk əˌsimiˈleiʃən]

Definition: 全岩同化作用 quán yán tóng huà zuò yòng

bulk: a large mass or shape, such as a building or a heavy body. 大块 dà kuài
assimilation: the process of becoming part of or more like something greater. 同化 tóng huà

burial metamorphism [ˈberiəl ˌmetəˈmɔːfizəm]

Definition: a form of regional metamorphism that acts on rocks covered by 5 to 10 kilometers of rock or sediment, caused by heat from the Earth's interior and lithostatic pressure. 埋藏变质作用 mái cáng biàn zhì zuò yòng

Example:
Medium-rank bituminites are formed by pneumato-hydrothermal metamorphism on the basis of burial metamorphism, and it is partial.
(深大断裂带及其附近局部发育的)中级烟煤是在埋藏变质作用的基础上,叠加了高温气液热变质作用而形成。

Extended Term:
tectonic burial metamorphism 构造埋藏变质作用

buried dolomitization [ˈberid ˌdɔləmitaiˈzeiʃən]

Definition: 埋藏白云石化作用 mái cáng bái yún shí huà zuò yòng

buried: relating to sinking something deeply into something else so that it is difficult to see or retrieve. 埋藏的 mái cáng de
dolomitization: a geological process by which the carbonate mineral dolomite is formed when magnesium ions replace calcium ions in another carbonate mineral, calcite. 白云石化作用 bái yún shí huà zuò yòng

Example:
The research on dolomite genesis is in postulation stage at present, which includes five models of vaporization, infiltration and circumfluence, sea water, mixing water and buried dolomitization.
白云岩成因的研究目前还处于假说阶段,主要包括蒸发、渗透回流、海水、混合水以及埋藏白云岩化作用五种模式。

burnt metamorphism [bəːnt ˌmetəˈmɔːfizəm]

Definition: 燃烧变质作用 rán shāo biàn zhì zuò yòng

burnt: having been burned. 烧过的 shāo guò de

metamorphism: alteration of the composition or structure of a rock by heat, pressure, or other natural agency. (岩石)变质作用 biàn zhì zuò yòng

bytownite ['baitaunait]

Definition: a bluish to dark gray triclinic mineral of the plagioclase feldspar group. Bytownite occurs in alkaline igneous rocks. Chemical formula: $(Ca, Na)(Si, Al)_4O_8$. 倍长石 bèi cháng shí

Origin: 1865-1870; named after Bytown (old name of Ottawa, Canada), where first found; see-ite1.

Example:
The plagioclase contains albite, oligoclase, andesine, laboratories, bytownite and anorthite. 钠长石、更长石、中长石、拉长石、倍长石和钙长石属于斜长石大类。

Extended Term:
bytownite sakalavite 倍长玻英玄武岩

calc alkalic volcanic rock [kælk 'ælkəlain vɔl'kænik rɒk]

Definition: 钙碱性火山岩 gài jiǎn xìng huǒ shān yán

calc: of lime or calcium. 石灰的 shí huī de, 钙的 gài de

alkaline: of, relating to, or containing an alkali. 碱性的 jiǎn xìng de

volcanic: of, resembling, or caused by a volcano or volcanoes. 火山的 huǒ shān de

rock: the solid mineral material forming part of the surface of the earth and other similar planets, exposed on the surface or underlying the soil. 岩石 yán shí；岩 yán

calc-alkaline trend [kælk-'ælkəlain trend]

Definition: 钙碱性趋势 gài jiǎn xìng qū shì

calc: of lime or calcium. 石灰的 shí huī de, 钙的 gài de

alkaline: having the properties of an alkali, or containing alkali. 碱性的 jiǎn xìng de

trend: a general direction in which something is developing or changing. 趋向 qū xiàng, 趋势 qū shì

calc gneiss [kælk nais]

Definition: 钙质片麻岩 gài zhì piàn má yán

calc: of lime or calcium. 石灰的 shí huī de, 钙的 gài de

gneiss: a metamorphic rock with a banded or foliated structure, typically coarse-grained and consisting mainly of feldspar, quartz, and mica. 片麻岩 piàn má yán

calc-pelite rock [kælk-'piːlait rɒk]

Definition: 钙泥质岩岩石 gài ní zhì yán yán shí

calc: of lime or calcium. 石灰的 shí huī de, 钙的 gài de

pelite: a sediment or sedimentary rock composed of very fine clay or mud particles. 泥质岩 ní zhì yán

rock: the solid mineral material forming part of the surface of the earth and other similar planets, exposed on the surface or underlying the soil. 岩石 yán shí; 岩 yán

calc schist [kælk ʃist]

Definition: 钙质片岩 gài zhì piàn yán

calc: of lime or calcium. 石灰的 shí huī de, 钙的 gài de

schist: a coarse-grained metamorphic rock which consists of layers of different minerals and can be split into thin irregular plates. 片岩 piàn yán

Extended Terms:

calc-mica schist 钙质云母片岩

calc silicate [kælk 'silikit]

Definition: 钙硅酸盐 gài guī suān yán

calc: of lime or calcium. 石灰的 shí huī de, 钙的 gài de

silicate: a salt in which the anion contains both silicon and oxygen, especially one of the anion SiO_{42}-. (化)硅酸盐 guī suān yán

calc silicate rock [kælk 'silikit rɒk]

Definition: 钙硅酸盐岩 gài guī suān yán yán

calc: of lime or calcium. 石灰的 shí huī de, 钙的 gài de

silicate: a salt in which the anion contains both silicon and oxygen, especially one of the anion SiO_{42}-. (化)硅酸盐 guī suān yán

rock: the solid mineral material forming part of the surface of the earth and other similar planets, exposed on the surface or underlying the soil. 岩石 yán shí; 岩 yán

calcareous shale [kælˈkɛəriəs ʃeil]

Definition: 钙质页岩 gài zhì yè yán

calcareous: containing calcium carbonate; chalky. 含碳酸钙的 hán tàn suān gài de; 白垩的 bái è de

shale: soft finely stratified sedimentary rock that formed from consolidated mud or clay and can be split easily into fragile plates 页岩 yè yán

calcite [ˈkælsait]

Definition: a colourless or white mineral (occasionally tinged with impurities), found insedimentary and metamorphic rocks, in veins, in limestone, and in stalagmites and stalactites. It is used in the manufacture of cement, plaster, paint, glass, and fertilizer. Composition: calcium carbonate. 方解石 fāng jiě shí

Formula: $CaCO_3$. Crystal structure: hexagonal (rhombohedral)

Origin: 1849, from Germany. Calcit, coined by Austrian mineralogist Wilhelm Karl von Hardinger (1795-1871), from Latin. Calx (generally calcis) "lime" + mineral, suffix-ite (Germany -it).

Example:
Many carbonate masses older than Middle Palaeozoic are of dolomite rather than calcite.
中古生代以前的许多碳酸盐类矿物是白云石而不是方解石。

Extended Terms:
ferroan calcite 铁方解石
poikilotopic calcite 嵌晶方解石

caldera [kælˈdɛərə]

Definition: a large basin-shaped crater at the top of a volcano, formed by the collapse or explosion of the cone. 破火山口 pò huǒ shān kǒu

Origin: 1865, "cavity on the summit of a volcano," from Spanish; caldera "cauldron, kettle," from Latin; caldarium, caldarius "pertaining to warming," from calidus "warm, hot".

Example:
The mountain is encircled by the doughnut-shaped depression of the caldera.

山为环形的破火山口洼地环抱着。

Extended Terms:

caldera complex 火口组合体
collapse caldera 塌陷破火山口

caliche [kæˈliːtʃi]

Definition: calcium-rich duricrust, a hardened layer in or on a soil. It is formed on calcareous materials as a result of climatic fluctuations in arid and semiarid regions. Calcite is dissolved in groundwater and, under drying conditions, is precipitated as the water evaporates at the surface. Rainwater saturated with carbon dioxide acts as an acid and also dissolves calcite and then redeposits it as a precipitate on the surfaces of the soil particles; as the interstitial soil spaces are filled, an impermeable crust is formed. 钙质层 gài zhì céng

Origin: 20 century: from American Spanish, from Latin calx lime.

Example:

Caliche crusts are secondary products formed by the dissolution of meteoric water on former lithified sediments and rocks during their exposure into subaerial vadose environments.
钙结壳是土壤层之下的岩石在出露地表渗流带环境内受大气雨水的溶解,并通过上覆土壤层的淋滤而成的产物。

Extended Term:

dolomitic caliche 白云石质钙结岩

Canadian Shield [kəˈneidiən ʃiːld]

Definition: a large plateau which occupies over two fifths of the land area of Canada and is drained by rivers flowing into Hudson Bay. 加拿大地盾 jiā ná dà dì dùn(占据加拿大2/5土地面积的大型高原,由注入哈得孙湾的河流流灌)

capillary concentration [kəˈpiləri ˌkɔnsənˈtreiʃən]

Definition: 毛细管浓缩作用 máo xì guǎn nóng suō zuò yòng

capillary: a tube which has an internal diameter of hair-like thinness. 毛细管 máo xì guǎn
concentration: the action of strengthening a solution by the removal of water or other diluting agent or by the selective accumulation of atoms or molecules. 浓缩 nóng suō

Example:

Through selecting typical sections, two kinds of genetic models of dolomite in the study area are

put forward, i. e., capillary concentration model and mixed water model.
通过选取典型剖面，归纳出研究区白云岩的两种成因模式：毛细管浓缩作用模式和混合水作用模式。

carbon isotope dating ['kɑːbən 'aisəutəup 'deitiŋ]

Definition: 碳同位素测年 tàn tóng wèi sù cè nián

carbon: the chemical element of atomic number 6, a non-metal which has two main forms (diamond and graphite) and which also occurs in impure form in charcoal, soot, and coal (Symbol: C). (化学元素)碳 tàn (符号: C)

isotope: each of two or more forms of the same element that contain equal numbers of protons but different numbers of neutrons in their nuclei, and hence differ in relative atomic mass but not in chemical properties; in particular, a radioactive form of an element. (化)同位素 tóng wèi sù

dating: the activity of establishing how old an object or substance is, often with the use of sophisticated scientific techniques. 测定年龄 cè dìng nián líng

carbonaceous chondrite [ˌkɑːbəu'neiʃəs 'kɔndrait]

Definition: 碳质球粒陨石 tàn zhì qiú lì yǔn shí

carbonaceous: (chiefly of rocks or sediments) consisting of or containing carbon or its compounds. 由碳组成的 yóu tàn zǔ chéng de;含碳的 hán tàn de

chondrite: a stony meteorite containing small mineral granules (chondrules). 球粒陨石 qiú lì yǔn shí

carbonaceous shale [ˌkɑːbəu'neiʃəs ʃeil]

Definition: 炭质页岩 tàn zhì yè yán

carbonaceous: (chiefly of rocks or sediments) consisting of or containing carbon or its compounds. 由碳组成的 yóu tàn zǔ chéng de;含碳的 hán tàn de

shale: soft finely stratified sedimentary rock that formed from consolidated mud or clay and can be split easily into fragile plates. 页岩 yè yán

carbonate buildup ['kɑːbəneit 'bildʌp]

Definition: 碳酸盐岩隆 tàn suān yán yán lóng

carbonate: a salt of the anion $CO_{32}-$, typically formed by reaction of carbon dioxide with bases. 碳酸盐.tàn suān yán

buildup: a gradual accumulation or increase, typically of something negative and typically leading to a problem or crisis. 累积 lěi jī

Example:

Favorable reservoirs are paleo-karst related to regional unconformity, carbonate buildup and bioclastic limestone in TST and HST, turbidite in LST and dolostone or grainstone in lower LST and higher HST.
良好的储层与不整合相关的古喀斯特、TST 和 HST 的碳酸盐岩隆、生物碎屑滩和 LST 的浊积岩、LST 底部和 HST 顶部相关。

carbonate cement ['kɑːbəneit siˈment]

Definition: 碳酸盐胶结物 tàn suān yán jiāo jié wù

carbonate: a salt of the anion CO_3^{2-}, typically formed by reaction of carbon dioxide with bases. 碳酸盐. tàn suān yán

cement: the material which binds particles together in sedimentary rock. 胶结物 jiāo jié wù

Example:

The block of Yanmuxi in Tu Ha oil-field belongs to loose sandstone reservoir, which has features of loose consolidation, high content of shale and carbonate cement. 吐哈油田雁木西疏松砂岩油藏，储层胶结疏松，泥质含量和碳酸盐胶结物含量较高，(常规酸化易造成岩石骨架松散，地层出砂，酸后无效)。

Extended Term:

eogenetic carbonate cement 成岩初期碳酸盐胶结物

carbonate compensation depth (CCD)

[ˈkɑːbəneit ˌkɔmpenˈseiʃən depθ]

Definition: 碳酸盐补偿深度 tàn suān yán bǔ cháng shēn dù

carbonate: a salt of the anion CO_3^{2-}, typically formed by reaction of carbon dioxide with bases. 碳酸盐. tàn suān yán

compensation: something that counterbalances or makes up for an undesirable or unwelcome state of affairs. 弥补 mí bǔ, 抵消 dǐ xiāo

depth: the distance from the top or surface to the bottom of something. 深度 shēn dù

carbonate platform facies [ˈkɑːbəneit ˈplætfɔːm ˈfeiʃiːz]

Definition: 碳酸盐台地相 tàn suān yán tái dì xiàng

carbonate: a salt of the anion CO_3^{2-}, typically formed by reaction of carbon dioxide with bases.

碳酸盐. tàn suān yán

platform: a flat raised area of ground. 台地 tái dì

facies: the character of a rock expressed by its formation, composition, and fossil content. 相 xiàng

> Example:

Based on the study of drilling and outcrop data, from which deduced the depositional framework, and main types and characteristics of sedimentary facies in Feixianguan Formation. It can recognize there basic facies belt including carbonate platform, continental shelf (slope) and trough, etc., and carbonate platform facies can further classified into five subfacies such as tidal flats, open platform, beach, stricted platform (lagoon) and marginal platform, etc.
在对大量钻井和野外剖面资料研究的基础上，阐述了研究区飞仙关组的沉积格局以及沉积相类型与特征，识别出了碳酸盐台地、陆棚(斜坡)和海槽等三个基本相带，碳酸盐台地相可进一步分为潮坪、开阔台地、台地内滩、局限台地(潟湖)和台地边缘等五个亚相。

> Extended Term:

carbonate platform marginal facies 碳酸盐台地边缘相

carbonate ramp facies ['kɑːbəneit ræmp 'feiʃiːz]

> Definition: 碳酸盐缓坡相 tàn suān yán huǎn pō xiàng

carbonate: a salt of the anion $CO_3{}^{2-}$, typically formed by reaction of carbon dioxide with bases. 碳酸盐 tàn suān yán

ramp: a sloping surface that allows access from one level to a higher or lower level, or raises something up above floor or ground level. 斜面 xié miàn

facies: the character of a rock expressed by its formation, composition, and fossil content. 相 xiàng

carbonate rock ['kɑːbəneit rɔk]

> Definition: 碳酸盐岩 tàn suān yán yán

carbonate: a salt of the anion $CO_3{}^{2-}$, typically formed by reaction of carbon dioxide with bases. 碳酸盐. tàn suān yán

rock: the solid mineral material forming part of the surface of the earth and other similar planets, exposed on the surface or underlying the soil. 岩石 yán shí; 岩 yán

> Example:

According to the study of source of minerals the ore-bearing minerals are mainly derived from the laterite weathering of carbonate rock.
根据成矿物质来源方面的研究，认为含矿岩系的物源主要来自碳酸盐岩的红土风化壳。

carbonate sediment [ˈkɑːbəneit ˈsedimənt]

Definition: 碳酸盐沉积物 tàn suān yán chén jī wù

carbonate: a salt of the anion CO_3^{2-}, typically formed by reaction of carbon dioxide with bases. 碳酸盐. tàn suān yán

sediment: particulate matter that is carried by water or wind and deposited on the surface of the land or the seabed, and may in time become consolidated into rock. 沉积物 chén jī wù

Example:

Carbonate sediments in all environments are basically products of an ancient biochemical system.
一切环境的碳酸盐沉积物，基本上都是古代生物化学体系的产物。

carbonate solution level [ˈkɑːbəneit səˈljuːʃən ˈlevəl]

Definition: 碳酸盐溶解层 tàn suān yán róng jiě céng

carbonate: a salt of the anion CO_3^{2-}, typically formed by reaction of carbon dioxide with bases 碳酸盐 tàn suān yán

solution: a liquid mixture in which the minor component (the solute) is uniformly distributed within the major component (the solvent). 溶液 róng yè; 溶体 róng tǐ

level: a position on a real or imaginary scale of amount, quantity, extent, or quality. 等级 děng jí, 水平 shuǐ píng

carbonatite [ˈkɑːbənətait]

Definition: a calcitic or dolomitic carbonate rock emplaced as an igneous intrusion. 碳酸岩 tàn suān yán

Origin: carbonate + -ite.

Example:

The basic principle of forward modeling technique shows that it is essential to do forward modeling well when predicating carbonatite reservoir.
模型正演的基本原理表明，在预测碳酸盐岩储层时做好模型正演工作是十分必要的。

Extended Terms:

calcite carbonatite 方解石碳酸岩
dolomite carbonatite 白雪碳酸岩

carpet form structure [ˈkɑːpit fɔːm ˈstrʌktʃə] = felty structure

Definition: 毡状构造 zhān zhuàng gòu zào

carpet: a floor or stair covering made from thick woven fabric, typically shaped to fit a particular room. 地毯 dì tǎn

form: the visible shape or configuration of something. 形状 xíng zhuàng, 轮廓 lún kuò

structure: the arrangement of, and relations between the parts or elements of something complex. 结构 jié gòu; 构造 gòu zào

Cascade Range [kæsˈkeid reindʒ]

Definition: a range of volcanic mountains in western area of North America, extending from southern British Columbia through Washington and Oregon to northern California. 喀斯喀特山脉 kā sī kā tè shān mài(北美西部的火山山脉，从不列颠哥伦比亚省南部，经华盛顿州和俄勒冈州，延伸到加利福尼亚州北部)

cataclasite [ˈkætəklæˌsait]

Definition: a type of cataclastic rock that is formed by fracturing and comminution during faulting. It is normally cohesive and non-foliated, consisting of angular clasts in a finer-grained matrix. 碎裂岩 suì liè yán

Origin: 19 century: New Latin, from Greek, from cata- +klasis, a breaking.

Example:

The fault rock is characterized as a low temperature deformation of about 200℃ ± and it is mainly a type of cataclasite.

断层岩的变形特征为低温变形，断层岩主要为碎裂岩类，变形温度在200℃左右。

Extended Term:

cataclasite series 碎裂岩系列

cataclastic texture [ˌkætəˈklæstik ˈtekstʃə]

Definition: 碎裂结构 suì liè jié gòu

cataclastic: the fracture and breaking up of rock by natural processes. 岩石碎裂 yán shí suì liè

texture: the arrangement of the particles or constituent parts of any material, as wood, metal, etc., as it affects the appearance; structure, composition, grain, etc. 结构 jié gòu

Cathodoluminescence [ˌkæˌθəʊdəˌluːmɪˈnesəns]

Definition: (Physics) luminescence caused by irradiation with electrons (cathode rays). 阴极发光 yīn jí fā guāng

Origin: 1905-1910; cathode + -o- + luminescence.

Examples:

①Zircons from ultra-high pressure (UHP) metamorphic metabasite and metafelsic rock in Dabie orogen were investigated by cathodoluminescence and Laser Raman spectrometry.
利用阴极发光和激光拉曼研究了大别山俯冲带超高压基性和长英质变质岩中的锆石。

②In the Cathodoluminescence(CL) images the zircons extracted from the dioritic gneiss show good crystal morphology and clear oscillatory zoning, as is the case of magmatic zircon.
在阴极发光图像中，片麻岩中的锆石具有较好的晶形，并显示出清晰的岩浆锆石韵律环带结构。

Extended Terms:

cathodoluminescence microscope 阴极发光显微镜
cathodoluminescence spectrum 阴极射线发光谱
extrinsic cathodoluminescence 非固有阴极发光
intrinsic cathodoluminescence 本质阴极发光
state cathodoluminescence 固态阴极射线发光

---

cement [sɪˈment]

Definition: the material which binds particles together in sedimentary rock. 胶结物 jiāo jié wù

Origin: Middle English: from Old French ciment (noun), cimenter (verb), from Latin caementum "quarry stone", from caedere "hew".

Example:

Lithologically, the content quartz is lower than those of feldspar and detritus, and the cement is dominated by clay, which gives the feature of low compositional maturity and low textural maturity.
馆上段储层岩性以岩屑质长石细砂岩为主，其次为中细砂岩和粉砂岩，岩石中石英含量低，而长石、岩屑含量高，胶结物以泥质为主，表现出低成分成熟度和低结构成熟度的特点。

Extended Terms:

abrasive cement 磨料黏结剂
acryloid cement 丙烯酸黏合剂

cementation

air vulcanizing cement 常温硫化胶浆
casein cement 酪素黏合剂
contact cement 接触黏合法

cementation [ˌsiːmenˈteiʃən]

(Definition:) the heating of two substances in contact in order to effect some change in one of them, esp., the formation of steel by heating iron in powdered charcoal. 胶结作用 jiāo jié zuò yòng

(Origin:) 1585-1995; cement + -ation.

(Example:)
Clay coats can be particularly important as inhibitors of later quartz cementation.
黏土包壳作为石英晚期胶结作用的抑制剂可能是特别重要的。

(Extended Terms:)
cementation factor 胶结系数
gas cementation 气体渗碳
granular cementation 粒状胶结
metallic cementation 渗金属法；喷镀金属
superficial cementation 表面渗碳

Central Atlantic Magmatic Province (CAMP)
[ˈsentrəl ətˈlæntik ˈmægmətik ˈprɔvins]

(Definition:) 中大西洋岩浆区 zhōng dà xī yáng yán jiāng qū

central: at the point or in the area that is in the middle of something. 中心的，中央的
Atlantic: of or adjoining the Atlantic Ocean. 大西洋的 dà xī yang de；大西洋沿岸的 dà xī yang yán àn de
magmatic: relating to the hot fluid or semi-fluid material below or within the earth's crust from which lava, and other igneous rock is formed by cooling. 岩浆的 yán jiāng de
province: a biogeographical area within a region that is defined by the plants and animals that inhabits it. (生物地理)区域 qū yù

central volcano [ˈsentrəl vɔlˈkeinəu]

(Definition:) 中央火山 zhōng yāng huǒ shān

central: at the point or in the area that is in the middle of something. 中心的，中央的
volcano: a mountain or hill, typically conical, having a crater or vent through which lava, rock

fragments, hot vapour, and gas are or have been erupted from the earth's crust. 火山 huǒ shān

ceramic clay [siˈræmik klei]

Definition: 陶瓷黏土 táo cí nián tǔ

ceramic: made of clay and permanently hardened by heat. 陶瓷的 táo cí de

clay: a stiff, sticky fine-grained earth, typically yellow, red, or bluish-grey in colour and often forming an impermeable layer in the soil. It can be moulded when wet, and is dried and baked to make bricks, pottery, and ceramics. 黏土 nián tǔ, 泥土 ní tǔ, 陶土 táo tǔ

Example:

It is an effective way to reduce the energy consumption and improve the quality of ceramic products that ceramic clay is treated by means of germ.

对陶瓷泥料进行细菌处理,是一项降低产品能耗、提高产品质量的一个有效途径。

chain silicate [tʃein ˈsilikit]

Definition: 链硅酸盐 liàn guī suān yán

chain: line of rings joined together. 链子 liàn zi

silicate: a salt in which the anion contains both silicon and oxygen, especially one of the anion SiO_{42}-(化)硅酸盐 guī suān yán

Example:

An experimental method was present for determining the origins of infrared absorption spectra of chain silicate minerals.

提供了一种研究链状硅酸盐矿物红外吸收谱归属的实验方法。

chalcedony [kælˈsedəni]

Definition: a microcrystalline often greyish form of quartz with crystals arranged in parallel fibres: a gemstone. 玉髓 yù suǐ

Formula: SiO_2

Origin: 15 century: from Late Latin chalcēdōnius, from Greek khalkēdōn, a precious stone (Revelation 21:19), perhaps named after Khalkēdōn Chalcedon, town in Asia Minor.

Example:

Aventurine is a chalcedony that contains small inclusions of one of several shiny minerals which give the gem a glistening effect.

沙金石是玉髓的一种,由多种发光矿物质微小内含物构成,从而使这种宝石具有了闪耀发光的效果。

Extended Terms:
red chalcedony 红玉髓
occidental chalcedony 西方玉髓
purple chalcedony 紫玉髓

chalcophile element ['kælkəˌfail 'elimənt]

Definition: Chalcophile elements are those metals and heavier nonmetals that have a low affinity for oxygen and prefer to bond with sulfur as highly insoluble sulfides. 亲铜元素 qīn tóng yuán sù

chalcopyrite [ˌkælkə'paiərait]

Definition: also called copper pyrites, a widely distributed yellow mineral consisting of a sulphide of copper and iron in tetragonal crystalline form; the principal ore of copper. 黄铜矿 huáng tong kuàng

Formula: $CuFeS_2$

Origin: 1825-1835; chalco- + pyrite.

Example:
The minerals are likely to be from the suite pyrite chalcopyrite, sphalerite, galena, fluorite and barite, and to be deposited at temperatures of between 50℃ and 250℃.
这些矿物可能由黄铁矿、黄铜矿、闪锌矿、方铅矿、萤石和重晶石组成，并且可能是在50℃～250℃之间的温度下沉积出来的。

Extended Terms:
chalcopyrite concentrate 黄铜矿精矿
chalcopyrite disease 黄铜矿病毒
chalcopyrite type 黄铜矿式

chalk [tʃɔːk]

Definition: a soft white or gray fine-grained sedimentary rock consisting of nearly pure calcium carbonate originally formed under the sea and containing minute fossil fragments of marine organisms. 白垩 bái è

Origin: Old English cealc "lime(stone), chalk," via Germanic, from Latin calc-"lime(stone)", from Greek khalix "pebble".

> Example:

Carr is the first person to put chalk on his cue tip.
卡尔是首位将白垩用于磨削球杆尖端的人。

> Extended Term:

chalk soil 白垩土
chalk test 白垩试验
precipitated chalk 沉淀白垩

chamosite ['kæməsait]

> Definition: a greenish grey or black silicate of iron and aluminum. 鲕绿泥石 ér lǜ ní shí

> Origin: 1825-1835; named after Chamoison in the Valais, Switzerland, where found; see-ite1.

> Example:

In tropical areas chamosite develops in the remains of organisms and in faecal pellets in the marine environment shallower than about 60 m, but occasionally down to about 150 m.
在热带地区，鲕绿泥石在海洋环境下浅于 60 米的生物遗骸和粪粒中发育，但偶尔也会下降至约 150 米处。

> Extended Term:

Mg-chamosite 镁鲕绿泥石

channel bar deposit ['tʃænəl bɑː di'pɔzit]

> Definition: 滨河床沙坝沉积 bīn hé chuáng shā bà chén jī

channel: a hollow bed for a natural or artificial waterway. 河床 hé chuáng, 河槽 hé cáo
bar: a sandbank or shoal at the mouth of a harbour or an estuary. 沙洲 shā zhōu, 浅滩 qiǎn tān
deposit: a layer or body of accumulated matter. 沉积物 chén jī wù, 沉淀 chén diàn

channel structure ['tʃænəl 'strʌktʃə]

> Definition: 渠道构造 qú dào gòu zào

channel: a hollow bed for a natural or artificial waterway. 河床 hé chuáng, 河槽 hé cáo
structure: the arrangement of and relations between the parts or elements of something complex. 结构 jié gòu; 构造 gòu zào

> Example:

This article analyzes design issue, such as channel structure, geometry size and form length,

cooling of model and so on.
本文就模具流道结构、流道几何尺寸、定形模长度、定形模冷却等设计问题进行了分析。

charnockite [ˈtʃɑːnəkait]

Definition: Charnockite is applied to any orthopyroxene-bearing granite, composed mainly of quartz, perthite or antiperthite and orthopyroxene (usually hypersthene), as an end-member of the charnockite series. 紫苏花岗岩 zǐ sū huā gāng yán

Example:
Studies on the Petrology of Charnockite indicate the charnockite was produced by anatexis under water-unsaturated conditions. The crystaline temperature fugacity of Oxyyen (fo_2,) and fugacity of water(fH_2O) determined for Charnockite 800℃, 1.013 × 108.6 Pa and 45.4 MPa, respectively. Isotopic dating of Single grain Zircon suggests that charnockite emplaced at about 450 Ma±10 Ma.
这种紫苏花岗岩是麻粒岩在水不饱和的条件下发生深熔作用的结果,结晶温度在800℃以上,氧逸度(fo_2)和水逸度(fH_2O)分别为1.013×108.6 Pa 和45.4 MPa,单颗粒锆石U-Ph同位素年龄为450±10Ma。

Extended Terms:
charnockite facies 紫苏花岗岩相
charnockite series 紫苏花岗岩系

chelate [ˈkiːleit]

Definition: of or noting a heterocyclic compound having a central metallic ion attached by covalent bonds to two or more nonmetallic atoms in the same molecule. 螯合物 áo hé wù

Origin: 1820-1830; chel(a) + -ate.

Examples:
①Chlorophyll is also a chelate, the porphyrin part of the molecule being bound to iron.
叶绿素也是一种螯合物,分子中的卟啉部分可以和离子结合。
②One of the important factors to form chelate is that there should be enough atoms to coordinate with metal ion.
而要形成螯合物其中有一个因素非常重要,就是金属离子周围必须有足够的原子与它产生配位。

Extended Terms:
chelate action 螯合作用
chelate complex 螯形复合物

chelate stabilizer 螯合稳定剂
chelate substance 螯形物；钳合物

chelation [kiːˈleiʃən]

Definition: the chemical removal of metallic ions in a mineral or rock by weathering. 螯合作用 áo hé zuò yòng

Origin: 1930-1935; chelate + -ion.

Example:
It is difficult to define the detection limits obtained by chelation-solvent extraction.
确定螯合-溶剂萃取法所得到的检测限颇为困难。

Extended Terms:
chelation agents 螯合剂
chelation reagent 螯合剂(金属离子的)
sequestration chelation 多价螯合作用
iron chelation 铁螯合剂

chemical limestone [ˈkemikəl ˈlaimstəun]

Definition: 化学石灰岩 huà xué shí huī yán

chemical: of or relating to chemistry, or the interactions of substances as studied in chemistry. 化学的 huà xué de；化学反应的 huà xué fǎn yìng de

limestone: a hard sedimentary rock, composed mainly of calcium carbonate or dolomite, used as building material and in the making of cement. 灰岩 huī yán，石灰岩 shí huī yán

Example:
There are three types of section of lithological combination: (1) chemical limestone—reef limestone—mudstone; (2) terrigenous clastic rock—reef limestone—mudstone; and (3) anhydrite-bearing mudstone—reef limestone—marl.
剖面类型有三种：①化学石灰岩—礁灰岩—泥岩型，②陆源碎屑岩—礁灰岩—泥岩型，③膏泥岩—礁灰岩—泥灰岩型。

chemical plume [ˈkemikəl pluːm]

Definition: 化学羽流 huà xué yǔ liú

chemical: of or relating to chemistry, or the interactions of substances as studied in chemistry. 化学的 huà xué de；化学反应的 huà xué fǎn yìng de

plume: a long cloud of smoke or vapour resembling a feather as it spreads from its point of origin. 羽状烟柱(或气流) yǔ zhuàng yān zhù(huò qì liú)

chemical potential ['kemikəl pəu'tenʃəl]

Definition: 化学势 huà xué shì

chemical: of or relating to chemistry, or the interactions of substances as studied in chemistry. 化学的 huà xué de; 化学反应的 huà xué fǎn yìng de

potential: (Physics) the quantity determining the energy of mass in a gravitational field or of charge in an electric field. (物理)势 shì, 位 wèi; 电势 diàn shì, 电位 diàn wèi

chemical potential gradient ['kemikəl pəu'tenʃəl 'greidiənt]

Definition: 化学势梯度 huà xué shì tī dù

chemical: of or relating to chemistry, or the interactions of substances as studied in chemistry. 化学的 huà xué de; 化学反应的 huà xué fǎn yìng de

potential: (Physics) the quantity determining the energy of mass in a gravitational field or of charge in an electric field. (物理)势 shì, 位 wèi; 电势 diàn shì, 电位 diàn wèi

gradient: a measure of change in a physical quantity such as temperature or pressure over a particular distance. 梯度 tī dù, 倾斜度 qīng xié dù, 坡度 pō dù

chemical sedimentary differentiation
['kemikəl ˌsedi'mentəri ˌdifərenʃi'eiʃən]

Definition: 化学沉积分异作用 huà xué chén jī fēn yì zuò yòng

chemical: of or relating to chemistry, or the interactions of substances as studied in chemistry. 化学的 huà xué de; 化学反应的 huà xué fǎn yìng de

sedimentary: (of rock) that has formed from sediment deposited by water or air. (岩石)沉积形成的 chén jī xíng chéng de

differentiation: the act of making or becoming different in the process of growth or development. 变异 biàn yì, 变化 biàn huà

chemical sedimentary rock ['kemikəl ˌsedi'mentəri rɒk]

Definition: a rock composed of material precipitated directly from solution. 化学沉积岩 huà xué chén jī yán

chemical: of or relating to chemistry, or the interactions of substances as studied in chemistry.

化学的 huà xué de；化学反应的 huà xué fǎn yìng de

sedimentary：(of rock) that has formed from sediment deposited by water or air. (岩石)沉积形成的 chén jī xíng chéng de

rock：the solid mineral material forming part of the surface of the earth and other similar planets, exposed on the surface or underlying the soil. 岩石 yán shí；岩 yán

Example:

As we need energy continuously to grow and as oil and gas are less able to satisfy the demand for fuel, coal is becoming an extremely important chemical sedimentary rock once again.
当我们对能源的需求不断增长，当石油和天然气还不能满足我们对燃料的需求时，煤便成了一种极其重要的化学沉积岩石。

Extended Term:

organic chemical sedimentary rock 有机化学沉积岩

chenier [ˈʃiniə]

Definition: a hummock in a marshy region, with stands of evergreen oaks. 沼泽沙丘 zhǎo zé shā qiū(贝壳堤)

Example:

Chenier is an important geological phenomenon occurring during the continent-building process since almost 5,000~6,000 years ago on the west and northwest of the Bohai Bay.
贝壳堤是渤海湾西岸、西北岸近五六千年以来成陆过程中的一种重要地质现象。

Extended Term:

chenier plain 贝壳堤平原

chert [tʃəːt]

Definition: a hard, dark, opaque rock composed of silica (chalcedony) with an amorphous or microscopically fine-grained texture. It occurs as nodules (flint) or, less often, in massive beds. 燧石 suì shí

Origin: late 17th century. (originally dialect): of unknown origin.

Example:

The cardium sandstone is lithic, being composed of grains of chert, quartz, quartzite, silicified argillite, and other rock fragments.
卡迪砂岩是岩屑砂岩，由燧石、石英、石英岩、硅化泥板岩和其他岩石碎屑颗粒所组成。

Extended Terms:

chert clause 燧石条款

calcareous chert 钙质燧石
chalcedonic chert 玉髓状燧石
diatomaceous chert 硅藻燧石
radiolarian chert 放射虫燧石

chert nodule [tʃəːt ˈnɔdjuːl]

Definition: 燧石结核 suì shí jié hé

chert: a hard, dark, opaque rock composed of silica (chalcedony) with an amorphous or microscopically fine-grained texture. It occurs as nodules (flint) or, less often, in massive beds. 燧石 suì shí

nodule: all rounded lump of matter distinct from its surroundings, e. g. of flint in chalk, carbon in cast iron, or a mineral on the seabed. (矿物)结核 jié hé, 小圆块 xiǎo yuán kuài

chiastolite [kaiˈæstəlait]

Definition: also called macle, a variety of andalusite containing carbon impurities. 空晶石 kōng jīng shí

Origin: 19 century: from German Chiastolith, from Greek khiastos crossed, marked with a chi + lithos stone.

Example:
This variety is called "Chiastolite" (named after the Greek word for cross) and sometimes referred to in ancient texts as "Lapis Crucifer", meaning "Cross Stone" or "Macle".
这类晶体被称之为"空晶石"(来源于希腊语"十字"),在古籍中也称之为"十字石",意为有十字图案的石头或有斑点的石头。

Extended Terms:
chiastolite-schist 空晶石片岩
chiastolite-slate 空晶板岩

chilled margin [tʃild ˈmɑːdʒin]

Definition: a shallow intrusive or volcanic rock texture characterised by a glassy or fine grained zone along the margin where the magma or lava has contacted air, water, or particularly much cooler rock. 冷凝边 lěng níng biān

chlorite ['klɔːrait]

Definition: any of a group of green soft secondary minerals consisting of the hydrated silicates of aluminium, iron, and magnesium in monoclinic crystalline form; common in metamorphic rocks. 绿泥石 lǜ ní shí

Origin: 18 century: from Latin chlōrītis precious stone of a green colour, from Greek khlōritis, from khlōros greenish yellow

Examples:

①The relationship between crystallinity and crystal size of illite or chlorite can reflect the behavior and mechanism of deformed rock.
伊利石和绿泥石的结晶大小与其结晶度之间的关系可以反映岩石变形的机制和行为。

②It has been found that the distribution and occurrence of chlorite plays an important role in controlling and affecting the pore permeability in the reservoir.
经长期研究发现，绿泥石黏土矿物在储层中的分布和赋存状态对储层孔渗性有重要的控制和影响作用。

Extended Terms:

chlorite bleach 亚氯酸盐漂白
barium chlorite 亚氯酸钡
hydrated chlorite 多水绿泥石
mangan chlorite 锰绿泥石
nickel chlorite 氯化镍
silver chlorite 亚氯酸银
sodium chlorite 亚氯酸钠
swelling chlorite 膨胀绿泥石

chlorite schist ['klɔːrait ʃist]

Definition: (Petrology) a metamorphic rock whose composition is dominated by members of the chlorite group. (岩石学)绿泥片岩 lǜ ní piàn yán

Example:

Chlorite schist is a black mica schist or a carbonate schist of chlorite mica of metamorphic group.
绿泥片岩为变质岩类黑云母片岩或绿泥石化云母碳酸盐片岩。

Extended Terms:

chlorite green schist 绿泥石绿色片岩
chlorite-sericite schist 绿泥绢云母片岩

epidote-chlorite schist 绿帘绿泥片岩
serpentine-chlorite schist 蛇纹绿泥片岩
talc-chlorite schist 滑石绿泥片岩

chlorite zone [ˈklɔːraɪt zəun]

Definition: 绿泥石带 lǜ ní shí dài

chlorite: any of a group of green soft secondary minerals consisting of the hydrated silicates of aluminium, iron, and magnesium in monoclinic crystalline form; common in metamorphicrocks 绿泥石 lǜ ní shí

zone: an area or stretch of land having a particular characteristic, purpose, or use, or subject to particular restrictions 地带 dì dài, 区域 qū yù

chloritoid [ˈklɔːrɪtɔɪd]

Definition: an iron magnesium manganese alumino-silicate hydroxide with formula: $(Fe, Mg, Mn)_2$ 硬绿泥石 yìng lǜ ní shí

Extended Terms:

chloritoid-micaschist 硬绿云片岩
chloritoid-phyllite 硬绿千枚岩
chloritoid schist 硬绿泥片岩

chondrite [ˈkɔndraɪt]

Definition: a stony meteorite consisting mainly of silicate minerals in the form of chondrules. 球粒陨石 qiú lì yǔn shí

Origin: 1880-1885; chondr- + -ite.

Examples:

①The transitional element curves standardized by chondrite were all W type curves inclining gently to the right.
呈球粒陨石标准化的过渡元素曲线，均是向右倾斜的 W 形曲线。
②The large number of meteorites discovered from Antarctica and the desert areas of Australia, Africa and North America changed the previous chondrite classification in recent years.
近年来在南极及澳大利亚、非洲和北美的沙漠地区发现和回收了大量的陨石，改变了以前的球粒陨石分类。

Extended Terms:

carbonaceous chondrite 碳粒陨石
crystalline chondrite 晶质球粒
enstatite chondrite 顽辉石球粒陨石
H chondrite H 球粒陨石
ordinary chondrite 普通球粒陨石

chondrite meteorite ['kɔndrait 'miːtiərait]

Definition: 球粒陨石 qiú lì yǔn shí

chondrite: a stony meteorite containing small mineral granules (chondrules). 球粒陨石 qiú lì yǔn shí

meteorite: a piece of rock or metal that has fallen to the earth's surface from outer space as a meteor. Over 90 percent of meteorites are of rock while the remainders consist wholly or partly of iron and nickel. 陨石 yǔn shí

chondritic [kɔn'dritik]

Definition: of chondrite 球粒陨石的 qiú lì yǔn shí de

Examples:

①Because the chondritic asteroids formed after the chondrules did, the initial 26Al to 27Al ratio in chondrules places an upper limit on the amount of 26Al that was available to heat the rocky bodies.
因为球粒陨石小行星是在球粒之后形成，球粒中铝 26 对铝 27 的初始比值，决定了能够用来加热石质天体的铝 26 含量的上限。

②If you look over the element ratios we measured, you might conclude that our comet particles look more like chondritic porous IDPs than CI chondrite meteorites.
如果你查看我们测定的元素比率，你将得出结论——我们的彗星颗粒看上去更接近多孔球粒陨石 IDPs 而不是 CI 球粒陨石。

Extended Terms:

chondritic meteorite 球状陨石
chondritic model 球粒陨石型
non-chondritic material 非球粒物质

chondrule ['kɔndruːl]

Definition: one of the small spherical masses of mainly silicate minerals present in chondrites. 球粒 qiú lì

Origin: 1885-1890; chondr- + -ule; from Ancient Greek χόνδρος chondros, grain.

Example:

The difference between the chondrule and the matrix indicates that aqueous alteration in the chondrule occurred in the nebula, but the aqueous alteration occurring widely in the matrix may be occurred in the parent process.
球粒与基质的含水蚀变差异说明，球粒中的含水蚀变可能发生在星云阶段，而基质的广泛含水蚀变则主要发生在母体过程中。

Extended Terms:

meteoritic chondrule 陨石球粒
porphyritic chondrule 斑状球粒

chromite [ˈkrəumait]

Definition: brownish-black mineral consisting of a ferrous chromic oxide in cubic crystalline form, occurring principally in basic igneous rocks; the only commercial source of chromium and its compounds. 铬铁矿 gè tiě kuàng

Formula: $FeCr_2O_4$

Origin: 1830-1840; chrom(ium) or chrom(ate) + -ite.

Example:

Knowledge of the occurrence of chromite deposits is likewise incomplete.
对于铬铁矿矿层的产量状况的了解同样是不完整的。

Extended Terms:

chromite brick 铬砖
chromite deposit 铬铁矿床
chromite dunite 铬铁纯橄岩
magnesium chromite 铬镁尖晶石
Zinc chromite 氧化铬锌；铬酸锌

chute deposit [ʃuːt diˈpɔzit]

Definition: 串沟沉积 chuàn gōu chén jī

chute: a waterfall, rapids, or steep descent in a river or stream. 水流 shuǐ liú
deposite: a layer or body of accumulated matter. 沉积物 chén jī wù, 沉淀 chén diàn

cinder cone [ˈsɪndə kəun]

Definition: a cone formed round a volcanic vent by fragments of lava thrown out during eruptions. 火山渣锥 huǒ shān zhā zhuī

Example:
Cinder cone volcanoes are formed when magmas with high gas contents and high viscosity are blown high into the air during an eruption.
火山渣锥火山是在爆发时，随着高气体含量和高黏度的熔岩被喷发到空气中而形成的。

classification of igneous rock
[ˌklæsɪfɪˈkeɪʃən əv ˈɪgnɪəs rɒk]

Definition: 火成岩分类 huǒ chéng yán fēn lèi

classification: the action or process of classifying something according to shared qualities or characteristics. 分类 fēn lèi

igneous rock: rock formed by the cooling and solidifying of molten materials. Igneous rocks can form beneath the Earth's surface, or at its surface, as lava. 火成岩 huǒ chéng yán

classification of pyroclastic rock
[ˌklæsɪfɪˈkeɪʃən əv ˌpaɪərəʊˈklæstɪk rɒk]

Definition: 火山碎屑岩分类 huǒ shān suì xiè yán fēn lèi

classification: the action or process of classifying something according to shared qualities or characteristics. 分类 fēn lèi

pyroclastic: relating to, consisting of, or denoting fragments of rock erupted by a volcano. 火山碎屑的 huǒ shān suì xiè de

rock: the solid mineral material forming part of the surface of the earth and other similar planets, exposed on the surface or underlying the soil. 岩石 yán shí；岩 yán

clast [klæst]

Definition: a constituent fragment of a clastic rock 碎屑 suì xiè

Origin: mid-20th century: back-formation from clastic.

Example:
The tempestite and non-tempestite occur alternately in vertical section. The development of tempestite is related to stratum, ancient landform and water depth, tectonic location and clast supply.

在剖面上，风暴沉积与非风暴沉积交互出现，发育程度与层位、古地形、水深、构造位置及碎屑供给特征等条件有关。

Extended Terms:

clast-supported conglomerate 碎屑支撑砾岩
muddy rip-up clast 泥浆撕裂碎屑
resedimented clast 再沉积碎屑

clastic dyke ['klæstik daik] =clastic dike

Definition: 碎屑岩脉 suì xiè yán mài，又称沉积岩脉(sedimentary dike) chén jī yán mài

clastic: denoting rocks composed of broken pieces of older rocks. 碎屑的 suì xiè de，由碎屑构成的 yóu suì xiè gòu chéng de

dyke: an intrusion of igneous rock cutting across existing strata. 岩墙 yán qiáng，岩脉 yán mài

clastic texture ['klæstik 'tekstʃə]

Definition: 碎屑结构 suì xiè jié gòu

clastic: denoting rocks composed of broken pieces of older rocks. 碎屑的 suì xiè de，由碎屑构成的 yóu suì xiè gòu chéng de

texture: the arrangement of the particles or constituent parts of any material, as wood, metal, etc., as it affects the appearance or feel of the surface; structure, composition, grain, etc. 结构 jié gòu

Example:

In Akekule area, the maturity of component and texture of the clastic rock reservoir in Kalashayi formation are relatively low.
阿克库勒地区卡拉沙依组碎屑岩储层中的成分成熟度和结构成熟度均较低。

Extended Terms:

proto clastic texture 原生碎屑结构
crypto clastic texture 隐屑结构

clastic wedge ['klæstik wedʒ]

Definition: (地质学)碎屑楔体 suì xiè xiē tǐ

clastic: denoting rocks composed of broken pieces of older rocks. 碎屑的 suì xiè de，由碎屑构成的 yóu suì xiè gòu chéng de

wedge: an object or piece of something having such a shape. 楔形物 xiē xíng wù

> Example:

Analyses of seismic sequences, seismic facies and velocity suggest that the Pleistocene and Pliocene sequences developed into a clastic wedge.

由震波序列、震测相及速度分析结果认为，上新世及更新世两个沉积序列发展成一个楔形碎屑沉积岩体。

clay [klei]

> Definition: a stiff, sticky fine-grained earth, typically yellow, red, or bluish-grey in colour and often forming an impermeable layer in the soil. It can be moulded when wet, and is dried and baked to make bricks, pottery, and ceramics. 黏土 nián tǔ, 泥土 ní tǔ, 陶土 táo tǔ

> Example:

Clay soil retains water.

黏土能保持水分。

> Extended Terms:

clay minerals 黏土矿物
clay pit 采黏土场
clay puddle 黏土混合浆
ball clay 块状黏土；球黏土
fire clay 耐火黏土
granulated clay 颗粒状黏土
marine clay 海成黏

clay rock [klei rɒk]

> Definition: 黏土岩 nián tǔ yán, 又称泥质岩（also called argillite, argillaceous rock, pelite) ní zhì yán

clay: a stiff, sticky fine-grained earth, typically yellow, red, or bluish-grey in colour and often forming an impermeable layer in the soil. It can be moulded when wet, and is dried and baked to make bricks, pottery, and ceramics. 黏土 nián tǔ, 泥土 ní tǔ, 陶土 táo tǔ

rock: the solid mineral material forming part of the surface of the earth and other similar planets, exposed on the surface or underlying the soil. 岩石 yán shí；岩 yán

> Example:

The physic and chemical soft rock is a kind of soft rock in engineering, which mainly stems from clay rock. Under the effect of water and weathering, the rock mass quality of clay rock, becomes bad.

物化型软岩主要是由于黏土类岩石在水或风化作用下，岩体质量持续恶化而形成的一类工程软岩。

climate [ˈklaimit]

Definition: the long-term prevalent weather conditions of an area, determined by latitude, position relative to oceans or continents, altitude, etc. 气候 qì hòu, 风土 fēng tǔ

Origin: 14 century: from Late Latin clima, from Greek klima inclination, region; related to Greek klinein to lean

Example:
With the changes in the world's climate, dinosaurs died, but many smaller animals lived on. It was the survival of the fittest.
随着世界气候的变迁，恐龙绝迹了，但许多较小的动物却继续活了下来。这就是适者生存。

Extended Terms:
Climate Group 气候组织；气候集团
arctic climate 北极气候；高寒性气候
continental climate 大陆气候
oceanic climate 海洋性气候
urban climate 城市气候

climatic arkose [klaiˈmætik ˈɑːkəus]

Definition: 气候长石砂岩 qì hòu cháng shí shā yán

climatic: the weather conditions prevailing in an area in general or over a long period. 气候 qì hòu

arkose: a coarse-grained sandstone which is at least 25 percent feldspar. 长石砂岩 cháng shí shā yán

climbing ripple lamination [ˈklaimiŋ ˈripl ˌlæmiˈneiʃən]

Definition: 爬升沙纹层理 pá shēng shā wén céng lǐ

climbing: the sport or activity of ascending mountains or cliffs. 爬山 pá shān；攀崖 pān yá
ripple: a small wave or series of waves on the surface of water. 微波 wēi bō, 细浪 xì làng
lamination: a layered structure. 叠片结构 dié piàn jié gòu

clinopyroxene [klainəˈpirəziːn]

Definition: any variety of the mineral pyroxene that crystallizes in the monoclinic system.

Diopside and augite are clinopyroxenes. 单斜辉石 dān xié huī shí

Example:

All metamorphic rocks exhibit an equilibrium texture, but exsolution lamellae of orthopyroxene (pigeonite) occur in all clinopyroxenes in mafic granulites.
这些岩石一般都展示了平衡的矿物共生结构，但在镁铁质麻粒岩的单斜辉石中却发育着斜方辉石(变辉石)溶片晶。

clotted limestone ['klɔtid 'laimstəun]

Definition: 凝块石 níng kuài shí

clotted: thickened or coalesced in soft thick lumps. 凝结的 níng jié de
limestone: a hard sedimentary rock, composed mainly of calcium carbonate or dolomite, used as building material and in the making of cement. 灰岩 huī yán, 石灰岩 shí huī yán

coal [kəul]

Definition: a combustible black or brownish-black sedimentary rock usually occurring in rock strata in layers or veins called coal beds or coal seams. 煤 méi

Origin: from the Old English term col, which has meant "mineral of fossilized carbon" since the 13th century.

Example:

Some locals know of no opportunities other than coal, know nothing of wind as an energy source and do not have the resources to concern themselves with global climate change.
除了煤炭之外，一些当地人对别的一无所知，他们不知道风也能产生能源，也没有渠道去了解他们和全球气候变化的关系。

Extended Terms:

coal bunker 原煤斗；煤仓
Coal Shop 煤店
bituminous coal 烟煤；沥青煤
coking coal 炼焦煤；焦煤
pulverized coal 粉煤；煤粉；末煤

coated grain ['kəutid grein]

Definition: 包粒 bāo lì, 包壳颗粒 bāo ké kē lì

coated: having a coating; covered with an outer layer or film; often used in combination. 上涂

料的 shàng tú liào de；上胶的 shàng jiāo de

grain：a small hard piece of particular substances. 颗粒 kē lì；细粒 xì lì

Example:

It's classification and nomenclature of the Biogenetic Coated Grain.

这是生物包粒的分类与命名。

Extended Term:

metal coated abrasive grain 金属镀覆磨粒

coated grain dolomite [ˈkəutid grein ˈdɔləmait]

Definition: 包粒白云岩 bāo lì bái yún yán

coated：having a coating；covered with an outer layer or film；often used in combination. 上涂料的 shàng tú liào de；上胶的 shàng jiāo de

grain：a small hard piece of particular substances. 颗粒 kē lì；细粒 xì lì

dolomite：a sedimentary rock formed chiefly of this mineral. 白云岩 bái yún yán

coesite [ˈkəusait]

Definition: a high-pressure polymorph of silica found in extreme conditions such as the impact craters of meteorites, with the chemical composition of silicon dioxide, SiO_2 柯石英 kē shí yīng

Origin: 1950-1955；named after Loring Coes, Jr., 20th-century American who synthesized it

Example:

Pure N_2 inclusions have been observed only in fresh coesite-bearing eclogites, whereas highly brine inclusions exist almost throughout the ultrahigh pressure (UHP) metamorphic rocks.

N_2 包裹体仅见于含柯石英榴辉岩，而高盐度流体包裹体则几乎存在于所有的榴辉岩和硬玉石英岩中。

Extended Terms:

coesite-bearing 含柯石英

coesite-bearing gneiss 含柯石英片麻岩

coesite eclogite 柯石英榴辉岩

coesite pseudomorph 柯石英假象

quartz-coesite 石英-柯石英

Colima Volcano ['kəulima vɔl'keinəu]

Definition: Colima's Volcano is currently one of the most active volcanos in Mexico and in North America. It has erupted more than 40 times since 1576. 科利马火山 kē lì mǎ huǒ shān

collision zone igneous tectonic assemblage
[kə'liʒn zəun 'igniəs tek'tɔnik ə'semblidʒ]

Definition: 碰撞造山带火成岩构造组合 pèng zhuàng zào shān dài huǒ chéng yán gòu zào zǔ hé

collision: the act of colliding, or coming together with sudden, violent force. 碰撞 pèng zhuàng

zone: an area or stretch of land having a particular characteristic, purpose, or use, or subject to particular restrictions. 地带 dì dài, 区域 qū yù

igneous: formed by solidification from a molten state. 火成的 huǒ chéng de

tectonic: relating to, causing, or resulting from structural deformation of the Earth's crust. 构造的 gòu zào de

assemblage: a form of art involving the assembly and arrangement of unrelated objects, parts, and materials in a kind of sculptured collage. 集合 jí hé

colloid ageing ['kɔlɔid 'eidʒiŋ]

Definition: 胶体陈化作用 jiāo tǐ chén huà zuò yòng

colloid: a mixture with properties between those of a solution and fine suspension. 胶质 jiāo zhì, 胶体 jiāo tǐ

ageing: acquiring desirable qualities by being left undisturbed for some time. 老化 lǎo huà, 陈化 chén huà

colloidal solution [kə'lɔidəl sə'ljuːʃən]

Definition: a colloid that has a continuous liquid phase in which a solid is suspended in a liquid. 胶体溶液 jiāo tǐ róng yè

colloidal: a substance of gelatinous consistency. 胶质 jiāo zhì

solution: a liquid mixture in which the minor component (the solute) is uniformly distributed within the major component (the solvent). 溶液 róng yè; 溶体 róng tǐ

Example:

$Zn(OH)_2$ colloidal solution can be changed to non-crystalline ZnO superfine powder (about $0.2\mu m$) under microwave irradiation.

Zn(OH)$_2$胶体溶液在微波照射下，能够生成均匀的超细 ZnO 非晶粉末(直径约 0.2μm)。

Extended Terms:
lyophobic colloidal solution 疏液胶体溶液
nano-sized colloidal solution 纳米胶体溶液

collophane [kəˈlɔfein]

Definition: Apatite is a group of phosphate minerals, usually referring to hydroxyapatite, fluorapatite, chlorapatite and bromapatite, named for high concentrations of OH-, F-, Cl- or Br-ions 胶磷矿 jiāo lín kuàng

Example:
Glauconite, detrital carbonate skeletal debris, marine fossils, and collophane are commonly present.
海绿石、碳酸盐骨骸碎屑、海相化石和胶磷矿普遍存在。

Extended Term:
uraniferous collophane 含铀胶磷矿

colonnade jointing [ˌkɔləˈneid ˈdʒɔintiŋ]

Definition: 柱状节理下段 zhù zhuàng jié lǐ xià duàn

colonnade: a row of columns supporting a roof, an entablature, or arches. 柱廊 zhù láng
jointing: the action of providing with, connecting by, or preparing for a joint. 接合 jiē hé，连接 lián jiē

color index [ˈkʌlə ˈindeks]

Definition: the difference between the apparent photographic magnitude and the apparent visual magnitude of a star. 色率 sè lǜ

color: the property possessed by an object of producing different sensations on the eye as a result of the way it reflects or emits light. 颜色 yán sè
index: a figure in a system or scale representing the average value of specified prices, shares, or other items as compared with some reference figure. 指数 zhǐ shù

Example:
The energy saving rate and the color index all reached 80%.
节能率、显色指数均达到 80%以上。

> **Extended Terms:**

color index error 色彩指数误差
infrared color index 红外色指数
integrated color index 累积色指数
sea color index 海色指数
special color-index 特殊色指数

Columbia River basalt [kəˈlʌmbiə ˈrivə ˈbæsɔːlt]

> **Definition:** 哥伦比亚河玄武岩 gē lún bǐ yà hé xuán wǔ yán

Columbia River: a river in NW North America which rises in the Rocky Mountains of SE British Columbia, Canada, and flows 1,953 km (1,230 miles) generally southwards into the US, where it turns westwards to enter the Pacific south of Seattle. 哥伦比亚河 gē lún bǐ yà hé

basalt: a dark fine-grained volcanic rock that sometimes displays a columnar structure. It is typically composed largely of plagioclase with pyroxene and olivine. 玄武岩 xuán wǔ yán

Columbia River flood basalt [kəˈlʌmbiə ˈrivə flʌd ˈbæsɔːlt]

> **Definition:** 哥伦比亚河溢流玄武岩 gē lún bǐ yà hé yì liú xuán wǔ yán

Columbia River: a river in NW North America which rises in the Rocky Mountains of SE British Columbia, Canada, and flows 1,953 km (1,230 miles) generally southwards into the US, where it turns westwards to enter the Pacific south of Seattle. 哥伦比亚河 gē lún bǐ yà hé

flood: a large amount of water covering an area that is usually dry. 洪水 hóng shuǐ; 水灾 shuǐ zāi

basalt: a dark fine-grained volcanic rock that sometimes displays a columnar structure. It is typically composed largely of plagioclase with pyroxene and olivine. 玄武岩 xuán wǔ yán

Columbia River Plateau basalt [kəˈlʌmbiə ˈrivə ˈplætəu ˈbæsɔːlt]

> **Definition:** 哥伦比亚河高原玄武岩 gē lún bǐyà hé gāo yuán xuán wǔ yán

Columbia River: a river in NW North America which rises in the Rocky Mountains of SE British Columbia, Canada, and flows 1,953 km (1,230 miles) generally southwards into the US, where it turns westwards to enter the Pacific south of Seattle. 哥伦比亚河 gē lún bǐ yà hé

Plateau: an area of fairly level high ground. 高原 gāo yuán

basalt: a dark fine-grained volcanic rock that sometimes displays a columnar structure. It is typically composed largely of plagioclase with pyroxene and olivine. 玄武岩 xuán wǔ yán

columnar jointing [kəˈlʌmnə ˈdʒɔintiŋ]

Definition: parallel, prismatic columns that are formed as a result of contraction during cooling in basaltic flow and other extrusive and intrusive rocks. 柱状节理 zhù zhuàng jié lǐ

Example:
The columnar jointing in the Black Fish Gorge displays a most spectacular sight.
黑鱼河峡谷的柱状节理雄奇壮观。

comendite [ˈkɔməndait]

Definition: a hard, peralkaline igneous rock, a type of light blue grey rhyolite. 钠闪碱流岩 nà shǎn jiǎn liú yán

Example:
Baitoushan volcanic rocks (trachyte and comendite) may have a crystallization relationship with the basalts.
粗面岩和钠闪碱流岩与玄武岩有成因联系，可能是玄武岩浆通过分离结晶作用而形成的。

common pyroclastic rock [ˈkɔmən ˌpaiərəuˈklæstik rɔk]

Definition: 普通火山碎屑岩 pǔ tōng huǒ shān suì xiè yán

common: occurring, found, or done often; prevalent. 一般的 yì bān de; 常见的 cháng jiàn de
pyroclastic: relating to, consisting of, or denoting fragments of rock erupted by a volcano. 火山碎屑的 huǒ shān suì xiè de
rock: the solid mineral material forming part of the surface of the earth and other similar planets, exposed on the surface or underlying the soil. 岩石 yán shí; 岩 yán

compaction [kəmˈpækʃən]

Definition: the consolidation of sediments resulting from the weight of overlying deposits 压实作用 yā shí zuò yòng

Origin: 1350-1400.

Example:
Although warped by compaction, there is only slight folding of the sandstone lenses.
虽然由于压实作用而弯曲，但砂岩体褶皱很轻微。

Extended Terms:
differential compaction 分异致密

incremental compaction 增量压缩
isostatic compaction 等(静)压成型
loose compaction 疏松压制
curve fitting compaction 曲线拟合(法)压缩
curve-pattern compaction 曲线型(法)压缩

compatible [kəmˈpætəbl]

Definition: capable of forming a homogeneous mixture that neither separates nor is altered by chemical interaction. 相容的 xiāng róng de，兼容的 jiān róng de

Origin: late Middle English：from French, from medieval Latin compatibilis, from compati "suffer with"

Example:
During magmatic stage, platinum group elements are compatible and the compatibility of IPGE exceeds that of PPGE.
铂族元素在岩浆演化过程中表现为相容元素的性质，且铱族铂族的相容性较钯族铂族强。

Extended Terms:
downward compatible 向下兼容
compatible degree 兼容度
compatible interlaminate 层间协调
compatible topology 相容拓扑

compatible element [kəmˈpætəbl ˈelimənt]

Definition: 相容元素 xiāng róng yuán sù

compatible：capable of forming a homogeneous mixture that neither separates nor is altered by chemical interaction. 相容的 xiāng róng de，兼容的 jiān róng de

element：each of more than one hundred substances that cannot be chemically interconverted or broken down into simpler substances and are primary constituents of matter. Each element is distinguished by its atomic number, i. e., the number of protons in the nuclei of its atoms. (化学)元素 yuán sù

Example:
All reported lunar granites contain elevated K_2O and Ba abundances, and have lower compatible element contents (Cr, Sc, Co and V) than other lunar rocks. Generally, the REE profile is flat or "concave upward" with a strongly negative Eu anomaly.
月球花岗岩富 K_2O，富 Ba，相容元素(Cr, Sc, Co, V)含量比其他月岩低，具有平坦或 V 型的 REE 轮廓，负 Eu 异常明显。

> Extended Terms:

compatible isoparametric element 协调等参元
compatible finite element 谐调元
compatible multi-element 多元兼容
quasi-compatible element 准谐调元

complexation [kɔmplek'seiʃ(ə)n]

> Definition: the formation of complex chemical species by the coordination of groups of atoms termed ligands to a central ion, commonly a metal ion. 络合作用 luò hé zuò yòng

> Origin: mid-17th century. (in the sense "group of related elements"): from Latin complexus, past participle (used as a noun) of complectere "embrace, comprise", later associated with complexus "plaited"; the adjective is partly via French complexe.

> Example:

The acid base characteristics of different natural illites were investigated by potentiometric titrations, and interpreted using the constant capacitance surface complexation model.
利用电位滴定技术和恒定容量表面络合模式查明不同天然伊利石的表面酸碱性质。

> Extended Terms:

complexation chromatography 络合层析
complexometry complexometric titration 络合滴定法
host-guest complexation 主体-客体配位(作用)
inner sphere complexation 内层配位(作用)
polynuclear complex 多核络合物

component [kəm'pəunənt]

> Definition: each of two or more forces, velocities, or other vectors acting in different directions which are together equivalent to a given vector. 成分 chéng fèn, 组成 zǔ chéng

> Origin: mid-17th century: from Latin component- "putting together", from the verb componere, from com- "together" +ponere "put". Compare with compound.

> Example:

The absolute content of crustal elements in dust-fall of different zones were all higher, being the main component, but heavy metal elements such as Cu, Zn, Pb, Ni and As were relatively much higher in industrial zone and the mixed industrial and residential zone.
不同功能区降尘中地壳元素绝对含量均较高,为主要成分,而一些污染元素(Cu, Zn, Pb, Ni, As)的含量在工业混合区和钢铁工业区则相对较高。

> Extended Terms:

component identification 成分鉴定
acidic component 酸性成[组]分
alternating component 交变部分
aperiodic component 非周期部分
assignment component 设定组件

composite volcano ['kɔmpəzit vɔl'keinəu]

> Definition: a large, steep volcano built up of alternating layers of lava and ash or cinders. 复式火山 fù shì huǒ shān

composite: made up of recognizable constituents. 复合的 fù hé de
volcano: a mountain or hill, typically conical, having a crater or vent through which lava, rock fragments, hot vapour, and gas are or have been erupted from the earth's crust. 火山 huǒ shān

> Example:

shield volcano, cinder cones and composite volcanoes. 盾状火山、火山渣锥和复合火山

compositional gradient [ˌkɔmpə'zɪʃənl 'greidiənt]

> Definition: 成分梯度 chéng fèn tī dù

compositional: the nature of something's ingredients or constituents; the way in which a whole or mixture is made up. 组成方式的 zǔ chéng fāng shì de
gradient: a measure of change in a physical quantity such as temperature or pressure over a particular distance. 梯度 tī dù, 倾斜度 qīng xié dù, 坡度 pō dù

> Example:

At solder joints, different materials are brought into contact, and there are compositional gradients at the interfaces.
不同的材料在焊点相互接触，因此在界面处会组成浓度梯度。

compositional maturity [ˌkɔmpə'zɪʃənl mə'tjuəriti]

> Definition: 成分成熟度 chéng fèn chéng shú dù

compositional: the nature of something's ingredients or constituents; the way in which a whole or mixture is made up. 组成方式的 zǔ chéng fāng shì de
maturity: the state, fact, or period of being mature. 成熟 chéng shú

compound grain

> Example:

It shows that the reservoir rock consists of silty and fine-grained arkose, and has low compositional maturity and medium textural maturity.
结果表明，该储集岩为粉细、细粒岩屑质长石砂岩，成分成熟度普遍较低，结构成熟度中等较好。

compound grain [ˈkɔmpaund grein]

> Definition: 复合颗粒 fù hé kē lì

compound: made up or consisting of several parts or elements, in particular. 复合的 fù hé de
grain: a small hard piece of particular substances. 颗粒 kē lì；细粒 xì lì

> Example:

A compound grain plugging and profile control agent of GQ type is developed based on the geological characteristics of sandstone reservoirs with high permeability, high temperature and high salinity.
针对高渗、高温、高矿化度砂岩油藏的地质特点，研制了 GQ 复合颗粒堵调剂。

> Extended Terms:

compound abrasive grain 复合磨粒
compound grain plugging agent 复合颗粒堵剂
compound grain structure 复粒结构
half-compound grain 半复粒

compressibility [kəmˌpresəˈbiləti]

> Definition: the degree to which something is compressible. 压缩性 yā suō xìng

> Origin: 1685-1695; compressible + -ity.

> Example:

These two parameters are nowadays called the modulus of compressibility.
这两种参数现被称为压缩模量。

> Extended Terms:

compressibility effect 压缩性效应
adiabatic compressibility 绝热压缩系数
coefficient of compressibility/factor 压缩系数
isothermal compressibility 等温压缩率

concentration ratio diagram [ˌkɔnsənˈtreiʃən ˈreiʃiəu ˈdaiəɡræm]

Definition: 集中率图 jí zhōng lǜ tú

concentration: the action of strengthening a solution by the removal of water or other diluting agent or by the selective accumulation of atoms or molecules. 浓缩 nóng suō

ratio: the quantitative relation between two amounts showing the number of times one value contains or is contained within the other. 比 bǐ; 比率 bǐ lǜ; 比例 bǐ lì

diagram: a simplified drawing showing the appearance, structure, or workings of something; a schematic representation. 图解 tú jiě

concordance intrusive body [kənˈkɔːdəns inˈtruːsiv ˈbɔdi]

Definition: 整合侵入体 zhěng hé qīn rù tǐ

concordance: agreement; harmony. 调和 tiáo hé

intrusive: of or relating to igneous rock that is forced while molten into cracks or between other layers of rock. 侵入的 qīn rù de

body: the whole physical structure and substance of a human being, animal, or plant. 体 tǐ

concordant intrusion [kənˈkɔːdənt inˈtruːʒən]

Definition: 整合侵入 zhěng hé qīn rù

concordant: in agreement; consistent. 协调的 xié tiáo de, 和谐的 hé xié de

intrusion: the action or process of forcing a body of igneous rock between or through existing formations, without reaching the surface. 侵入 qīn rù

concordant diagrams [kənˈkɔːdənt ˈdaiəɡræms]

Definition: 谐和图 xié hé tú

concordia: in agreement; consistent. 协调的 xié tiáo de, 和谐的 hé xié de

diagram: a simplified drawing showing the appearance, structure, or workings of something; a schematic representation. 图解 tú jiě

concordia plot [ˈkɔŋkɔːdiə plɔt]

Definition: 谐和图(图表) xié hé tú

concordia: in agreement; consistent. 协调的 xié tiáo de, 和谐的 hé xié de

plot: a graph showing the relation between two variables. 图表 tú biǎo

concretion [kənˈkriːʃən]

Definition: a rounded mass of compact concentric layers within a sediment, built up around a nucleus such as a fossil. 结核 jié hé

Origin: mid-16th century, concrescere (see concrete).

Example:
For the interpretation of δ 18 O variation, it is assumed that δ 18 O depleted meteoric water was probably involved at the time when the concretion started to precipitate, and that the oxygen isotope composition of the pore water approached gradually the value of normal ocean water (0‰) with the increase of the burial depth.
在解释 δ18O 的变化情况时，推测在结核开始生长时可能有 δ18O 亏损的大气降水掺入，然后随着埋藏深度的逐渐加大，孔隙水的 O 同位素逐渐趋近正常海水的值(0‰)。

Extended Term:
siliceous concretion 硅质结核

cone in cone [kəun in kəun]

Definition: 叠锥 dié zhuī

cone: a solid or hollow object which tapers from a circular or roughly circular base to a point. 圆锥体 yuán zhuī tǐ

Example:
It is expounded the structural types of cone in cone, cone and pillar, cone in cone and pillar compound, cone in cone and pillar intercalated with each other, and the constitution characteristics of the bedding plane subtypes: arch convex, arch concave, horizontal, horizontal concave and so on.
阐明叠锥、叠柱、叠锥柱复合和叠锥柱彼此间夹复合等构造类型及类型中的拱凸、拱凹、水平、平凹等纹层（组）层面构造亚类的组成特征。

Extended Terms:
cone in cone coal 套锥煤
cone in cone structure 叠锥构造
cone in cone volcano 叠锥火山

cone sheet [kəun ʃiːt]

Definition: 锥状岩席 zhuī zhuàng yán xí

cone: a solid or hollow object which tapers from a circular or roughly circular base to a point. 圆锥体 yuán zhuī tǐ

sheet: a thin wide layer over the surface of something else. 大层（覆盖物）dà céng（fù gài wù）

conglomerate [kən'glɔməreit]

Definition: coarse-grained sedimentary rocks containing fragments of other rocks larger than 2 mm (0.08 in.) in diameter, held together with another material such as clay. 砾岩 lì yán

Origin: late Middle English (as an adjective describing something gathered up into a rounded mass): from Latin conglomeratus, past participle of conglomerare, from con-"together" + glomus, glomer-"ball". The geological sense dates from the early 19th century; the other noun senses are later.

Example:
Sandstone and conglomerate are the consolidated equivalent of sand and gravel.
砂岩和砾岩分别是砂和砾石固结的产物。

Extended Terms:
coal conglomerate 煤砾岩
crystal conglomerate 丛晶体
dolomitic conglomerate 白云石砾岩
glacial conglomerate 冰砾岩
orthoquartzitic conglomerate 正石英砾岩

congruent melting ['kɔŋgruənt 'meltiŋ]

Definition: Congruent melting occurs as a compound melts, when the composition of the liquid forms, it is the same as the composition of the solid. It contrasts with incongruent melting. 一致熔融 yí zhì róng róng

congruent: in agreement; consistent. 协调的 xié tiáo de, 和谐的 hé xié de

melting: the process whereby heat changes something from a solid to a liquid. 熔化 róng huà

Example:
Preparation methods, i.e., adsorption, congruent melting, micro-encapsulation and pressing-sintering, are reviewed. Application fields of shape stabilized phase change materials are presented.
总结了定形相变材料的制备方法，包括物理吸附法、熔融共混法、微胶囊化及压制烧结法4种。

> Extended Terms:

congruent melting point 共熔点
solid-liquid congruent melting 液固共融

consolidation [kənˌsɔli'deiʃən]

> Definition: lithification. 固结作用 gù jié zuò yòng

> Origin: early 16th century. (in the sense "combine into a single whole"): from Latin consolidare, from con-"together" +solidare"make firm" (from solidus"solid").

> Example:

During the late consolidation stage of sedimentation, mineralized solution are squeezed into fractures.
在沉积物固结的晚期，矿化溶液渗入到裂隙中。

> Extended Terms:

consolidation by vibration 振动固结
industrial consolidation 工业合并
one-line consolidation 单线并合
three-dimensional consolidation 三向固结
vertical consolidation 纵向合并

constructional delta deposit [kən'strʌkʃənəl 'deltə di'pɔzit]

> Definition: 建设性三角洲沉积 jiàn shè xìng sān jiǎo zhōu chén jī

constructional: the building of something, typically a large structure. 建造 jiàn zào；构筑 gòu zhù；建设 jiàn shè

delta: a triangular tract of sediment deposited at the mouth of a river, typically where it diverges into several outlets. 三角洲 sān jiǎo zhōu

deposit: a layer or body of accumulated matter. 沉积物 chén jī wù，沉淀 chén diàn

constructive margin [kən'strʌktiv 'mɑːdʒin]

> Definition: 增生性边缘 zēng shēng xìng biān yuán

constructive: constructing or tending to construct or improve or promote development. 建设性的 jiàn shè xìng de

margin: the edge or border of something. 边 biān，边沿 biān yán

> Example:

At constructive plate margins, the plates diverge. Submarine volcanoes, lava flows, volcanic islands are found, e. g. along Mid-Atlantic Ridge.
因为地壳碰撞,大洋板块向地幔俯冲,形成俯冲带。因为岩浆受压沿地壳的裂缝喷出,形成海面下的火山、熔岩流或火山岛。

> Extended Term:

constructive plate margins 增生性板块边缘

contact aureole ['kɔntækt 'ɔːriəul]

> Definition: 接触带 jiē chù dài

contact: the state or condition of physical touching. 接触 jiē chù
aureole: the zone of metamorphosed rock surrounding an igneous intrusion. 接触圈(晕) jiē chù quān (yūn)

> Example:

The lead-zinc deposit of Fenglishan occurs in the thermal contact metamorphic aureole of intermediate-acid magmatic rocks with marine carbonates of Triassic.
凤梨山铅锌矿床产于中酸性岩浆岩与三叠系海相碳酸盐地层的热接触变质晕内。

> Extended Terms:

aureole of contact metamorphism 接触变质圈
contact metamorphic aureole 接触变质圈

contact cementation ['kɔntækt ˌsiːmen'teiʃən]

> Definition: 接触胶结 jiē chù jiāo jié

contact: the state or condition of physical touching. 接触 jiē chù
cementation: the binding together of particles or other things by cement. 黏结 nián jié, 胶接作用 jiāo jiē zuò yòng

contact metamorphic aureole ['kɔntækt metə'mɔfik 'ɔːriəul]

> Definition: 接触变质圈 jiē chù biàn zhì quān

contact: the state or condition of physical touching. 接触 jiē chù
metamorphic: denoting rock that has undergone transformation by heat, pressure, or other natural agencies, e. g. in the folding of strata or the nearby intrusion of igneous rocks. (岩石)变质的(yán shí) biàn zhì de

aureole: the zone of metamorphosed rock surrounding an igneous intrusion. 接触圈(晕)jiē chù quān (yūn)

contact metamorphism [ˈkɔntækt ˌmetəˈmɔːfizəm]

Definition: a type of metamorphism in which the mineralogy and texture of a body of rock are changed by exposure to the pressure and extreme temperature associated with a body of intruding magma. Contact metamorphism often results in the formation of valuable minerals, such as garnet and emery, through the interaction of the hot magma with adjacent rock. 接触变质作用 jiē chù biàn zhì zuò yòng

Example:
On the condition of thermal contact metamorphism, clay minerals in coal are changed into andalusite.
无烟煤中红柱石是煤中黏土质矿物经热接触变质作用而形成的。

Extended Terms:
contact metasomatic metamorphism 接触交代变质作用
thermal contact metamorphism 热触点变质酌;热力变质酌

contact metasomatic metamorphism
[ˈkɔntækt ˌmetəsəuˈmætik ˌmetəˈmɔːfizəm]

Definition: 接触交代变质作用 jiē chù jiāo dài biàn zhì zuò yòng

contact: the state or condition of physical touching. 接触 jiē chù

metasomatic: change in the composition of a rock as a result of the introduction or removal of chemical constituents. 交代作用 jiāo dài zuò yòng

metamorphism: alteration of the composition or structure of a rock by heat, pressure, or other natural agency. (岩石)变质作用(yán shí) biàn zhì zuò yòng

contamination [ˌkɔntæmiˈneiʃən]

Definition: make (something) impure by exposure to or addition of a poisonous or polluting substance. 使不纯 shǐ bù chún;污染 wū rǎn

Example:
Others have said that all of Mars should be protected as a wildlife preserve where we do everything possible to avoid contamination; on the assumption Martian microbes could exist.
假设火星上可能存在微生物,其他人就认为火星应该像我们做任何事以免污染到野生动物保护区那样来被保护着。

Extended Terms:

interplanetary contamination 星际污染
radioactive contamination 放射性污染

contamination of magma [kənˌtæmiˈneiʃən ɔv ˈmægmə]

Definition: 岩浆混染 yán jiāng hún rǎn

contamination: make (something) impure by exposure to or addition of a poisonous or polluting substance. 使不纯 shǐ bù chún; 污染 wū rǎn

magma: hot fluid or semi-fluid material below or within the earth's crust from which lava and other igneous rock is formed by cooling. 岩浆 yán jiāng

Example:

On this basis, the paper points out that the calc-alkaline magmatic rock is the product formed by contamination of mantle derived magma with continental crust consolidated 1600 milion years ago.

在此基础上提出;本区钙碱性岩浆岩系由地幔岩浆与1600万年前固结的大陆地壳混熔作用而形成的岩浆产物。

contemporaneous deformation structure

[kənˌtempəˈreiniəs ˌdiːfɔːˈmeiʃən ˈstrʌktʃə]

Definition: 同生变形构造 tóng shēng biàn xíng gòu zào

contemporaneous: existing or occurring in the same period of time. 同时期存在(或发生)的 tóng shí qī cún zài (huò fā shēng) de

deformation: the action or process of changing in shape or distorting, especially through the application of pressure. 变形 biàn xíng

structure: the arrangement of and relations between the parts or elements of something complex. 结构 jié gòu; 构造 gòu zào

continental [ˌkɔntiˈnentəl]

Definition: forming or belonging to a continent 大陆的 dà lù de, 大陆性的 dà lù xìng de

Origin: 1750-1760; continent + -al.

Example:

Based on 191 bulk rock compositions of basic-intermediate volcanic rocks and selecting three efficient elements, the triangular plot of Al_2O_3-$TiO_2 \times 10$-$K_2O \times 10$ (wB, %, i. e. ATK diagram) are presented in this paper. The ATK diagram appeared to show that the plotted points concentrate at three areas: oceanic, continental and island arc or orogenic belt.

基于191个采自不同构造环境的中基性火山岩的全岩化学成分,选择 Al_2O_3-$TiO_2\times10$-$K_2O\times10$(氧化物质量百分数)做成 ATK 三角图解,可以发现大洋的、大陆的、岛弧或造山带的中基性火山岩的成分点分别集中在三个区。

Extended Terms:

continental climate 大陆性气候
continental deposit 陆相沉积
continental drift 大陆漂移
continental facies 陆相
continental island 陆边岛

continental collision [ˌkɔntiˈnentəl kəˈluːʒən]

Definition: 大陆碰撞 dà lù pèng zhuàng

continental: forming or belonging to a continent. 大陆的 dà lù de, 大陆性的 dà lù xìng de
collision: an instance of one moving object or person striking violently against another. (猛烈) 碰撞 (měng liè) pèng zhuàng

Example:

Mountains are built by the continental collisions between plates that force land upwards.
山脉是由于大陆板块之间的碰撞,迫使陆地上翘而形成的。

continental crust [ˌkɔntiˈnentəl krʌst]

Definition: Continental crust is composed of silica rich minerals and has a largely granitic composition (quartz, feldspars, micas). It also includes surface sediments and metamorphic rocks. 大陆地壳 dà lù dì qiào

continental: forming or belonging to a continent. 大陆的 dà lù de, 大陆性的 dà lù xìng de
crust: the outermost layer of rock of which a planet consists, especially the part of the earth above the mantle. 外壳 wài ké

Example:

As the leading edge of one plate is consumed beneath the other, some of the magma finds its way up between cracks and crevices in the continental crust.
随着一个板块的前边缘被淹没在另一个之下,有些岩浆会顺着大陆地壳中的裂缝和缝隙向上涌出。

Extended Terms:

continental crust underthrust 陆壳俯冲
archean continental crust 太古宙陆壳
lower continental crust 大陆下地壳
paleozoic continental crust 古生代陆块
upper continental crust 大陆上地壳

continental facies [ˌkɒntiˈnentəl ˈfeiʃiːz]

Definition: 陆相 lù xiàng

continental: forming or belonging to a continent. 大陆的 dà lù de, 大陆性的 dà lù xìng de
facies: the character of a rock expressed by its formation, composition, and fossil content. 相 xiàng

Example:

The depositional environments have the basic characteristics of progressing upward and northwestward from marine facies to continental facies.
沉积环境具有自下而上、自南东向北西方向从海相逐步过渡为陆相的基本特点。

Extended Terms:

continental facies basin 陆相盆地
continental facies strata 陆相层
continent facies phosphorite 陆相磷块岩
continental facies volcano 陆相火山

continental flood basalt (CFB) [ˌkɒntiˈnentəl flʌd ˈbæsɔːlt]

Definition: 大陆溢流玄武岩 dà lù yì liú xuán wǔ yán

continental: forming or belonging to a continent. 大陆的 dà lù de, 大陆性的 dà lù xìng de
flood: a large amount of water covering an area that is usually dry. 洪水 hóng shuǐ; 水灾 shuǐ zāi

basalt: a dark fine-grained volcanic rock that sometimes displays a columnar structure. It is typically composed largely of plagioclase with pyroxene and olivine. 玄武岩 xuán wǔ yán

Example:

The mafic rocks are geologically and chemically similar to the Permian Emeishan continental flood basalts and have experienced no mineralization, with normal content of ore-forming elements.
镁铁质岩在地质特征和地球化学特征上与峨眉山玄武岩相似，成矿元素含量正常。

continental flood tholeiite [ˌkɔntiˈnentəl flʌd ˈθəuliːait]

Definition: 大陆溢流拉斑玄武岩 dà lù yì liú lā bān xuán wǔ yán

continental: forming or belonging to a continent. 大陆的 dà lù de, 大陆性的 dà lù xìng de

flood: a large amount of water covering an area that is usually dry. 洪水 hóng shuǐ; 水灾 shuǐ zāi

tholeiite: a basaltic rock containing augite and a calcium-poor pyroxene (pigeonite or hypersthene), and with a higher silica content than an alkali basalt. 拉斑玄武岩 lā bān xuán wǔ yán

Example:
The picrite field, located in the southwest of Emeishan continental flood basalts (ECFB) and with the center in Lijiang—Dali—Panzhihua triangular area, is about 5×10^4 km^2 and regarded as the axial area of the Emei mantle plume.
位于峨眉山大陆溢流玄武岩(ECFB)的西南部以丽江、大理和攀枝花三角区为中心的苦橄岩分布区，面积约 5×10^4km^2，为峨眉地幔柱的轴部区。

continental interior [ˌkɔntiˈnentəl inˈtiəriə]

Definition: 内陆 nèi lù

continental: forming or belonging to a continent. 大陆的 dà lù de, 大陆性的 dà lù xìng de

interior: situated within or inside; relating to the inside; inner. 里面的 lǐ miàn de, 内部的 nèi bù de

Example:
Junggar basin is a large continental interior basin in western China, which has been developing since Early Permian.
准噶尔盆地是中国西部一大块内陆盆地，自早二叠世就一直发育着。

continental margin igneous tectonic assemblage
[ˌkɔntiˈnentəl ˈmɑːdʒin ˈigniəs tekˈtɔnik əˈsemblidʒ]

Definition: 大陆边缘弧火成岩组合 dà lù biān yuán hú huǒ chéng yán zǔ hé

continental: forming or belonging to a continent. 大陆的 dà lù de, 大陆性的 dà lù xìng de

margin: the edge or border of something. 边 biān, 边沿 biān yán

igneous: formed by solidification from a molten state. 火成的 huǒ chéng de

tectonic: relating to, causing, or resulting from structural deformation of the earth's crust. 构造的 gòu zào de

assemblage: a form of art involving assembly and arrangement of unrelated objects, parts, and materials in a kind of sculptured collage. 集合 jí hé

continental rift zone [ˌkɔntiˈnentəl rift zəun]

Definition: 大陆裂谷带 dà lù liè gǔ dài

continental: forming or belonging to a continent. 大陆的 dà lù de, 大陆性的 dà lù xìng de
rift: a crack, split, or break in something. 裂缝 liè fèng, 裂口 liè kǒu
zone: an area or stretch of land having a particular characteristic, purpose, or use, or subject to particular restrictions. 地带 dì dài, 区域 qū yù

Example:

Environments include subduction zones, continental rift zones, mid-oceanic ridges, and hotspots, some of which are interpreted as mantle plumes.
环境包括俯冲带、大陆裂谷带、大洋中脊和热点，其中一些可以解释为地幔柱。

continental slope facies [ˌkɔntiˈnentəl sləup ˈfeiʃiːz]

Definition: 大陆坡相 dà lù pō xiàng

continental: forming or belonging to a continent. 大陆的 dà lù de, 大陆性的 dà lù xìng de
slope: a surface of which one end or side is at a higher level than another; a rising or falling surface. 斜坡 xié pō
facies: the character of a rock expressed by its formation, composition, and fossil content. 相 xiàng

continental volcanic [ˌkɔntiˈnentəl vɔlˈkænik]

Definition: 大陆火山岩 dà lù huǒ shān yán

continental: forming or belonging to a continent. 大陆的 dà lù de, 大陆性的 dà lù xìng de
volcanic: of, relating to, or produced by a volcano or volcanoes. 火山的 huǒ shān de

Example:

Jinyun County of Zhejiang Province is a major production area of Mesozoic continental volcanic zeolite deposits in eastern China.
浙江省缙云县是我国东部中生代陆相火山岩型沸石矿床的重要产区。

continuous reaction [kənˈtinjuəs riˈækʃən]

Definition: 连续反应 lián xù fǎn yìng

continuous: forming an unbroken whole; without interruption. 连续的 lián xù de; 连贯的 lián guàn de

reaction: a chemical process in which two or more substances act mutually on each other and are changed into different substances, or one substance changes into two or more other substances. 反应 fǎn yìng

Example:

The production of zinc chloride from zinc ash and exhausted hydrochloric acid with continuous reaction device has advantages of low cost and no hydrogen chloride leaking out.
以锌灰和废盐酸为原料，用连续反应装置生产氯化锌，不仅成本低，而且无气体逸出。

Extended Terms:

continuous conversion reaction 连续转化反应
continuous reaction series 连续反应系列

contour current ['kɔntuə 'kʌrənt]

Definition: Undercurrent typical of the continental rise which flows along the western boundaries of ocean basins. Such currents occur particularly in regions in which density stratification is strong because of the supply of cold waters originating near the poles. A well-known example is the Western Boundary Undercurrent which hugs the continental rise of eastern America. Contour currents are persistent, slow-moving (velocity 5-30 cm/s) flows capable of transporting mud, silt, and sand. 等深流 děng shēn liú

Example:

Abyssal allochthonous deposits consisting of density current, contour current, internal tide and wave deposits are relatively coarser sediments transported and accumulated by density and contour current in abyssal environments.
深水异地沉积是指在深水环境中经重力流或牵引流搬运、改造、沉积而形成的相对粗粒的沉积物，是深水重力流、等深流、内潮汐和内波沉积的总称。

Extended Term:

contour current deposit 等深流沉积

contourite deposit ['kɔntuərait di'pɔzit]

Definition: a sedimentary deposit produced by deepwater bottom currents, which result from thermohaline, wind or tidal forces. 等深流沉积物 děng shēn liú chén jī wù

convection model [kən'vekʃən 'mɔdəl]

Definition: 对流模式 duì liú mó shì

convection: the movement caused within a fluid by the tendency of hotter and therefore less dense material to rise, and colder, denser material to sink under the influence of gravity, which consequently results in transfer of heat. 对流 duì liú

model: a system or thing used as an example to follow or imitate. 模式 mó shì

Example:

In mantle convection models, the mantle viscosity is generally assumed constant or dependent on depth.

黏度为常数或仅随深度变化是地幔对流模型中常用的假设。

Extended Terms:

convection-diffusion models 对流扩散模型
mantle convection models 地幔对流模式

convection of heat [kənˈvekʃən ɔv hiːt]

Definition: 热对流 rè duì liú

convection: the movement caused within a fluid by the tendency of hotter and therefore less dense material to rise, and colder, denser material to sink under the influence of gravity, which consequently results in transfer of heat. 对流 duì liú

heat: the quality of being hot; high temperature. 热 rè, 高温 gāo wēn

Example:

The coefficient of convection of heat transfer increases with the increasing of the particle diameters.

对流换热系数随着颗粒直径的增大而增大。

Extended Term:

coefficient of heat convection 热对流系数

convective fractionation [kənˈvektiv ˌfrækʃəˈneiʃən]

Definition: 对流分馏 duì liú fēn liú

convective: the movement caused within a fluid by the tendency of hotter and therefore less dense material to rise, and colder, denser material to sink under the influence of gravity, which consequently results in transfer of heat. 对流 duì liú

fractionation: separate (a mixture) by fractional distillation. 分馏 fēn liú; 分级 fēn jí

convergent margin [kənˈvəːdʒənt ˈmɑːdʒin]

Definition: 会聚边缘 huì jù biān yuán

convergent: coming closer together, especially in characteristics or ideas. 相似的 xiāng sì de, 相近的 xiāng jìn de

margin: the edge or border of something. 边 biān, 边沿 biān yán

Example:

They mainly occur in orogenic belts of various geological times and are formed during extension periods against the background of convergent margins.
它们主要产于各个时期的造山带中，形成于以会聚板块边界为大背景的局部引张阶段。

convergent plate boundary [kənˈvɜːʤənt pleit ˈbaundəri]

Definition: A convergent plate boundary is where crust is being destroyed. In the case of oceanic and continental crust meeting it is the denser oceanic crust which is subducted. 聚敛板块边界 jù liǎn bǎn kuài biān jiè

Example:

Pengxian copper deposit and the Masongling Formation acting as host formation are products of island arc volcanism in convergent plate boundary.
证据表明，马松岭组及其所赋存的铜矿床均为会聚板块边界古火山岛弧的产物。

convolution structure [ˌkɔnvəˈluːʃən ˈstrʌktʃə]

Definition: 包卷构造 bāo juǎn gòu zào

convolution: the state of being or process of becoming coiled or twisted. 盘绕 pán rào; 弯曲 wān qū

structure: the arrangement of and relations between the parts or elements of something complex. 结构 jié gòu; 构造 gòu zào

Example:

Finally, the paper has calculated the reliability of composite convolution structure by use of Montecarlo direct comparison method.
然后采用蒙特卡罗直接比较法，计算一复合材料回旋结构的可靠性。

coprecipitation [ˈkəupriˌsipiˈteiʃən]

Definition: Coprecipitation (CPT) or co-precipitation is the carrying down by a precipitate of substances normally soluble under the conditions employed. 共(同)沉淀 gòng (tóng) chén diàn

Origin: 1930-1935; co- + precipitate.

Example:

In this paper, the fundamental principle of coprecipitation preparation for nano-ceramics powder is given in detail.
本文介绍了共同沉淀法制备纳米级陶瓷粉体原理(分析了微粒子晶核的形成与长大过程)。

Extended Terms:

coprecipitation phenomena 共沉淀现象
adsorption coprecipitation 吸附共沉淀
carrier coprecipitation 载体共沉淀
chemical coprecipitation process 化学共沉淀工艺

coquina [kəuˈkiːnə]

Definition: an incompletely consolidated sedimentary rock. Coquina was formed in association with marine reefs and is a variety of "coral rag", technically a subset of limestone. 介壳灰岩 jiè qiào huī yán

Origin: mid-19th century: from Spanish, literally cockle, based on Latin concha.

Example:

Coquina of shoal facies, micrite of neritic facies and sandstone of tidal flat facies may constitute the reservoir rocks in which the fractures serve as reservoir space.
可能的储集岩是浅滩相介壳灰岩、浅海相泥晶灰岩及潮坪相砂岩,裂隙是其主要的储集空间。

Extended Terms:

coquina clam 斧蛤
coquina coq 贝壳灰岩

cordierite [ˈkɔːdiərait]

Definition: a dark blue mineral occurring chiefly in metamorphic rocks. It consists of an aluminosilicate of magnesium and iron, and also occurs as a dichroic gem variety. 堇青石 jǐn qīng shí

Origin: early 19th century: named after Pierre L. A. Cordier (1777-1861), French geologist, + -ite.

Example:

The article detailed the composition, structure, properties and application of cordierite material.
概述了堇青石材料的组成、结构、性能特点及应用领域。

Extended Terms:

cordierite-based glass fibre 堇青石基玻璃纤维
cordierite norite 堇青苏长岩
cordierite porcelain 堇青石瓷
cordierite stoneware 堇青石炻器
cordierite whiteware 堇青石白坯陶瓷制品
cordierite zone 堇青石带

core [kɔː]

Definition: a piece of flint from which flakes or blades have been removed 岩心 yán xīn

Origin: 1275-1325; 1945-1950.

Example:
In simulated conditions of Zhongyuan Wenming village oil field 70℃, salinity 7.15×10^4 mg/L and serious layer heterosphere, 0.8 PV of this system was injected into the two groups of parallel core with the permeability level difference 14.1 and 38.2, oil recovery rose by 44% and 28.6% respectively.
在模拟中原文明寨油田油藏温度70℃、矿化度为7.15×104mg/L和非均质严重的地层条件下,该体系在渗透率级差为14.1和38.2的两组并联岩心中注入0.8PV,分别提高采收率44%和28.6%。

Extended Terms:

baked core 干(砂)型心
black core 炭斑,黑心(斑)
body core 主芯
carbon core 碳芯
cover core 盖芯

corona texture [kəˈrəunə ˈtekstʃə]

Definition: 冠状边结构 guàn zhuàng biān jié gòu

corona: a crown or crown-like structure. 冠 guàn;冠状部位 guàn zhuàng bù wèi
texture: the arrangement of the particles or constituent parts of any material, as wood, metal, etc., as it affects the appearance or feel of the surface; structure, composition, grain, etc. 结构 jié gòu

cotectic [kəuˈtektik]

Definition: Describing the conditions of pressure, temperature and composition at which multiple solid phases crystallize at the same time from a single liquid when cooled. 共结的 gòng jié de

Example:
From high-temperature molten, a series of silicate melts near cotectic-line were quenched or cooled with different rate.
通过高温熔融，对同结线附近一系列硅酸盐熔体进行淬冷或不同速度冷却。

Extended Terms:
cotectic surface 共析面
cotectic line(cotectic-line) 共结线
cotectic point 共析点
cotectic temperature 共析温度

Cotopaxi volcano [ˌkəutəˈpæksi vɔlˈkeinəu]

Definition: Cotopaxi is a stratovolcano in the Andes Mountains, located about 45 kilometres (28mi) south of Quito, Ecuador, South America. 科托帕希火山 kē tuō pà xī huǒ shān

coulee [ˈkuːli]

Definition: a stream of lava 粘熔岩流 zhān róng yán liú

Origin: 1800-1810.

Example:
The volcanic hazards in Longgang region are descendent scoria, coulee and base-surge deposits.
龙冈区内火山灾害类型主要为降落火山渣、熔岩流和基浪堆积物。

Extended Terms:
coulee lake 熔岩湖
Grand Coulee Hydroelectric Power Station 大古力水电站

country rock [ˈkʌntri rɒk]

Definition: the rock which encloses a mineral deposit, igneous intrusion, or other feature. 围岩 wéi yán

Example:

The ore bodies of Mouping Gold Mine are steep extreme-thin to narrow veins with unstable country rock.
牟平金矿的矿体为极薄的急倾斜矿脉，且围岩不稳固。

Extended Terms:

country rock classification 围岩分类
country rock displacement 围岩位移
country rock mass 围岩岩体
country rock support 围岩支护
weak country rock 软岩

craton [ˈkreitɔn]

Definition: a large stable block of the earth's crust forming the nucleus of a continent. 稳定地块 wěn dìng dì kuài, 克拉通 kè lā tōng

Origin: 1930s；alteration of kratogen in the same sense, from Greek kratos "strength".

Example:

The early tectonic evolution of craton has been the target topic for continental geology.
大陆早期构造演化的研究一直是大陆地质学研究的焦点问题。

Extended Terms:

craton basin 克拉通盆地
cratonic block 克拉通地块
cratonic nuclei 克拉通核
crater rim 火山口边缘
epeirocratic craton 陆地克拉通
preorogenic craton 造山运动前稳定地块

crenulation [ˌkrenjuˈleiʃən]

Definition: crinkled, having an irregular wavy or serrate outline. 细褶皱 xì zhě zhòu

Origin: 1840-1850；crenulate + -ion.

Example:

Therefore, the crenulation cleavage is the basic character of the brittle-ductile shear zones as well as the prospecting criteria of meso-low temperature structural-hydrothermal deposit.
因此，区内细褶皱劈理构造既可作为脆-韧性剪切带的基本特征，又是寻找中低温构造热液型矿床的标志。

Extended Terms:

crenulation cleavage 细褶皱劈理
crenulation lineation 皱纹线理

crevasse splay deposit [kri'væs splei di'pɔzit]

Definition: 决口扇沉积 jué kǒu shàn chén jī

crevasse: a breach in the embankment of a river or canal. 裂隙 liè xi; 决口 jué kǒu
splay: a widening or outward tapering of something, in particular. 八字形展开 bā zì xíng zhǎn kāi
deposit: a layer or body of accumulated matter. 沉积物 chén jī wù, 沉淀 chén diàn

cristobalite [kris'təu,bəlait]

Definition: a form of silica which is the main component of opal and also occurs as small octahedral crystals. 方石英 fāng shí yīng

Origin: late 19th century: named after Cerro San Cristóbal in Mexico, where it was discovered, + -ite.

Example:
It is found that it is very difficult to separate the cristobalite and quartz with similar size with montmorillonte from montmorillonte.
研究证实，与蒙脱石粒度相近的方石英和石英微粒等杂质很难去除。

Extended Terms:

cristobalite content 方石英含量
cristobalite crystallite 方石英晶体
cristobalite phase 方石英相
sericite-tridymite-cristobalite 绢英粉

critical melting ['kritikəl 'meltiŋ]

Definition: 临界熔融 lín jiè róng róng

critical: maintaining a self-sustaining chain reaction. 临界的 lín jiè de
melting: the process whereby heat changes something from a solid to a liquid. 熔化 róng huà

Extended Terms:

critical melting point 临界熔点

critical point [ˈkritikəl pɔint]

Definition: a point on a phase diagram at which both the liquid and gas phases of a substance have the same density. 临界点 lín jiè diǎn

Example:
Then we combine with example—BTW model to simulate and analyze its characteristic at critical point.
接着我们以具体实例，即 BTW 模型来模拟分析其在临界点所具有的特性。

Extended Terms:
critical bearing point 临界承载量
infinity critical point 无穷远奇点
virtual critical point 虚拟临界点

critical radius [ˈkritikəl ˈreidiəs]

Definition: Critical radius is the minimum size that must be formed by atoms or molecules clustering together (in a gas, liquid or solid matrix) before a new-phase inclusion (a bubble, a droplet, or a solid particle) is stable and begins to grow. 临界半径 lín jiè bàn jìng

Example:
Moreover, in accordance with the micro-fracture compared with the indentation loading curve, dimensionless critical radius decrease in evidence with the wedge angle increase.
另外，再分析粒间微裂位置对应于贯彻加载历程之关系，发现随楔形刃口角度之渐增（即由尖变钝），其无因次化之塑性区半径则明显减小。

cross bedding [krɔːs ˈbediŋ]

Definition: a sedimentary rock texture characterised by overlapping and cross-cutting bedding at an angle to the main layers of bedding. Cross bedding is typical of aeolian sands, beach and deltaic deposits. 交错层理 jiāo cuò céng lǐ

Example:
Structure of sedimentary rocks: bedding, cross bedding, mud cracks and ripples.
沉积岩的构造：层理、交错层理、泥裂和波痕。

Extended Terms:
angular cross bedding 倾斜交错层
concave cross-bedding 凹面交错层理
deltaic cross bedding 三角洲交错层

large scale cross bedding 大型板状交错层理

crust [krʌst]

Definition: the outermost layer of rock, of which a planet consists, especially the part of the earth above the mantle. 地壳 dì qiào

Origin: 1275-1325.

Example:

The south-north crust in Qiangtang basin has been shortening successively since Indosinian movement, and respectively its shortening ratios are 38% at Trias, 24%~26.3% at Jurassic, 17.47%~19.2% at Tertiary.
盆地自印支运动以来 SN 向地壳缩短具递减性，地壳缩短率分别为上三叠统为 38%、侏罗系为 24%~26.3%，第三系为 17.47%~19.2%。

Extended Terms:

breakable crust 易碎雪层
calcareous crust 钙积层
continental crust 陆壳
cork crust 木栓壳
desert crust 沙漠壳
flinty crust 硬壳

crustal [ˈkrʌstəl]

Definition: of or pertaining to a crust, as of the earth. 地壳的 dì qiào de

Origin: 1855-1860; from Latin crūst shell, crust + -al.

Example:

Crustal deformation observation includes observations of horizontal deformation, vertical deformation, and ground tilting.
地壳形变观测应包括水平形变、垂直形变和倾斜形变观测。

Extended Terms:

crustal deformation 地壳形变作用
crustal dynamics 地壳动力学
crustal disturbance 地壳变动
crustal downbuckling 地壳下弯

crustal contamination [ˈkrʌstəl kənˌtæmiˈneiʃən]

Definition: 地壳混染作用 dì qiào hún rǎn zuò yòng

crustal: of or pertaining to a crust, as of the earth. 地壳的 dì qiào de
contamination: make (something) impure by exposure to or addition of a poisonous or polluting substance. 使不纯 shǐ bù chún; 污染 wū rǎn

Example:

In addition, the more evolved andesitic basalts and quartz tholeiites are often rich in radiogenic Sr, indicating that to a certain extent crustal contamination was involved in the genesis of these rocks.
比较演化的石英拉斑玄武岩、安山玄武岩往往富放射成因锶，反映其成因还涉及一定程度的地壳混染。

Extended Term:

crustal-mantle contamination 壳幔混染

crustal melting [ˈkrʌstəl ˈmeltiŋ]

Definition: 地壳熔化 dì qiào róng huà

crustal: of or pertaining to a crust, as of the earth. 地壳的 dì qiào de
melting: the process whereby heat changes something from a solid to a liquid. 熔化 róng huà

Example:

A stage of laboratory experiment modelling the natural crustal melting of rocks into granite has been initiated.
以地壳岩石为源岩的热力学实验的开展，是花岗岩热力学研究的开始。

Extended Terms:

crustal partial melting 壳内部分熔融
lower crustal melting 下地壳熔融

crustified cementation [ˈkrʌstifaid ˌsiːmenˈteiʃən]

Definition: 丛生胶结 cóng shēng jiāo jié

crustified: 皮壳状脉. pí qiào zhuàng mài
cementation: the binding together of particles or other things by cement. 黏结 nián jié, 胶接作用 jiāo jiē zuò yòng

crust source [krʌst sɔːs]

Definition: 壳源 qiào yuán

crust: the outermost layer of rock of which a planet consists, especially the part of the earth above the mantle. 地壳 dì qiào

source: a place, person, or thing from which something comes or can be obtained. 来源 lái yuán

Example:

The mineralized pedigree can be indicated as mantle derived-inner basin and deep source-crust source ore-forming materials.
在成矿物质上由幔源→盆内及深源→壳源的成矿谱系表示。

cryoscopic equation [kraiəsˈkɔpik iˈkweiʒən, -ʃən]

Definition: 冰点测定法方程 bīng diǎn cè dìng fǎ fāng chéng

cryoscopic: of or related to any instrument used to determine the freezing point of a substance. 冰点测定法的 bīng diǎn cè dìng fǎ de

equation: a statement that values of two mathematical expressions are equal. 等式 děng shì;方程 fāng chéng

cryptic [ˈkriptik]

Definition: similar to the background; camouflaged 隐造礁灰岩 yǐn zào jiāo huī yán

Origin: 1595-1605.

Example:

The cryptic layering and rhythmic layering are typical feature of layered intrusions.
层状侵入体的一个典型的特征是具有隐层理和韵律层理。

Extended Terms:

cryptic layering 隐蔽层状构造
cryptic plasmid 隐蔽性质粒
cryptic suture 隐蔽缝合

cryptocrystalline [ˌkriptəuˈkristəlain]

Definition: having a microscopic crystalline structure. 隐晶质 yǐn jīng zhì

Example:

Chalcedonies include many types of cryptocrystalline quartz gems and feature a number of

different colors.
玉髓包括很多类型的隐晶的石英宝石，并具有不同的颜色。

[Extended Terms:]

cryptocrystalline granophyric 隐晶文像的，隐晶斑状的
cryptocrystalline quartz 隐晶石英
cryptocrystalline texture 隐晶质结构
cryptocrystalline variety 隐晶质类

cryptocrystalline texture [ˌkriptəuˈkristəlain ˈtekstʃə]

[Definition:] 隐晶质结构 yǐn jīng zhì jié gòu

cryptocrystalline：having a microscopic crystalline structure. 隐晶质 yǐn jīng zhì
texture：the arrangement of the particles or constituent parts of any material, as wood, metal, etc., as it affects the appearance or feel of the surface; structure, composition, grain, etc. 结构 jié gòu

cryptovolcanic rock [ˈkriptəuvɔlˈkænik rɔk]

[Definition:] rocks of a small, nearly circular area of highly disturbed strata in which there is no evidence of volcanic materials to confirm the origin as being volcanic. 潜火山岩 qián huǒ shān yán

[Example:]

Two mineralization magmatic formation of the continental volcanism: continental volcanic rocks and cryptovolcanic complex are similar in geochemistry character of trace elements and REE.
陆相火山作用两类成矿岩浆建造——陆相火山岩与潜火山杂岩具有基本一致的微量元素和稀土元素地球化学特征。

crystal [ˈkristəl]

[Definition:] a piece of a homogeneous solid substance having a natural geometrically regular form with symmetrically arranged plane faces. 晶体 jīng tǐ

[Origin:] late Old English (denoting ice or a mineral resembling it), from Old French cristal, from Latin crystallum, from Greek krustallos "ice, crystal". The chemistry sense dates from the early 17th century.

[Example:]

The heat and pressure of this effect solidifies the ash into a hard, dark crystal.

此效果的高热和高压将使灰烬聚合成一块坚硬的黑色晶体。

Extended Terms:

crystal-controlled 晶体控制的，石英稳频的
crystal detector 晶体检波器
acicular crystal 针状晶体
acousto-optic crystal 声光晶体
aeolotropic crystal 各向异性晶体

crystal fragment ['krɪstəl 'frægmənt]

Definition: 晶屑 jīng xiè

crystal：a piece of a homogeneous solid substance having a natural geometrically regular form with symmetrically arranged plane faces. 晶体 jīng tǐ

fragment：a small part broken or separated off something. 碎片 suì piàn

crystal imprint ['krɪstəl 'ɪmprɪnt]

Definition: 晶体印痕 jīng tǐ yìn hén

crystal：a piece of a homogeneous solid substance having a natural geometrically regular form with symmetrically arranged plane faces. 晶体 jīng tǐ

imprint：a mark made by pressing something on to a softer substance so that its outline is reproduced. 印记 yìn jì，印痕 yìn hén

crystalline dolomite ['krɪstəlaɪn 'dɒləmaɪt]

Definition: 结晶白云岩 jié jīng bái yún yán

crystalline：having the structure and form of a crystal；composed of crystals. 水晶的 shuǐ jīng de；晶状的 jīng zhuàng de

dolomite：a sedimentary rock formed chiefly of this mineral. 白云岩 bái yún yán

Example:

Fine to coarse crystalline dolomite and coarse-grained dolomite cements in the gradular dolomite were derived from dolomitization in the deeply-burial diagenetic environment.
细至粗晶白云岩及颗粒白云岩中的粗粒白云石胶结物是深埋藏成岩环境的产物。

crystalline granular texture ['krɪstəlaɪn 'grænjulə 'tekstʃə]

Definition: 结晶粒状结构 jié jīng lì zhuàng jié gòu

crystalline：having the structure and form of a crystal；composed of crystals. 水晶的 shuǐ jīng

de；晶状的 jīng zhuàng de
granular：composed or appearing to be composed of granules or grains. 颗粒的 kē lì de
texture：the arrangement of the particles or constituent parts of any material, as wood, metal, etc., as it affects the appearance or feel of the surface; structure, composition, grain, etc. 结构 jié gòu

crystalline limestone [ˈkristəlain ˈlaimstəun]

Definition: 结晶灰岩 jié jīng huī yán，又称晶粒灰岩（also called crystal grain limestone）jīng lì huī yán

crystalline：having the structure and form of a crystal; composed of crystals. 水晶的 shuǐ jīng de；晶状的 jīng zhuàng de

limestone：a hard sedimentary rock, composed mainly of calcium carbonate or dolomite, used as building material and in the making of cement. 灰岩 huī yán，石灰岩 shí huī yán

Example:
It occurs in the upper part of Lower Permian strata. It is composed of crystalline limestone with white-lighter gray-gray marble.
矿床主矿体赋存于下二叠统上部地层，主要岩性为结晶灰岩夹白色-浅灰-灰色大理岩。

crystalline schist [ˈkristəlain ʃist]

Definition: 结晶片岩 jié jīng piàn yán

crystalline：having the structure and form of a crystal; composed of crystals. 水晶的 shuǐ jīng de；晶状的 jīng zhuàng de

schist：a coarse-grained metamorphic rock which consists of layers of different minerals and can be split into thin irregular plates. 片岩 piàn yán

Example:
The middle Proterozoic crystalline schist series in the Panzhihua-Xichang region has undergone progressive regional metamorphism.
攀西中元古结晶片岩系遭受了前进区域变质作用的影响。

Extended Terms:
crystalline schist system 结晶片岩系
Zaluskar crystalline schist zone 札洛斯喀结晶片岩带

crystallinity [ˌkristəˈlinəti]

Definition: the degree of structural order in a solid. In a crystal, the atoms or molecules are arranged in a regular, periodic manner. 结晶度 jié jīng dù

Origin: Middle English: from Old French cristallin, via Latin from Greek krustallinos, from krustallos (see crystal).

Example:

The results showed that the crystallinity of cotton fiber increased after being slightly oxidised, but decreased after being severely oxidised.
结果表明轻度氧化棉纤维的结晶度略有提高,而深度氧化棉纤维的结晶度降低。

Extended Terms:

crystallinity index 结晶度指数
liquid crystallinity 液晶度
quartz crystallinity 石英结晶度
relative crystallinity 相对结晶度
surface crystallinity 表面结晶度

crystallite [ˈkrɪstəˌlaɪt]

Definition: an individual perfect crystal or region of regular crystalline structure in the substance of a material, typically of a metal or a partly crystalline polymer. 雏晶 chú jīng

Origin: 1795-1805; crystall- + -ite.

Example:

The results show that with the increase of milling time, particle size and crystallite size are reduced.
结果表明,随着球磨时间的增加,粉末的颗粒度及晶粒度均不断减少。

Extended Terms:

crystallite size 雏晶大小
crystallite theory 微晶说
needle shaped crystallite 针状晶体
polymer crystallite 高分子微晶

crystallization [ˌkrɪstəlaɪˈzeɪʃən]

Definition: form or cause to form crystals. 结晶 jié jīng;晶化 jīng huà

Origin: 1655-1665; crystall- + -ization.

Example:

During this crystallization minute grains of a metal such as iron have been flowed together to

form a concentration.
在此结晶过程中，细小的金属(例如铁)颗粒汇流到一起就形成一个富集体。

Extended Terms:

accumulative crystallization 聚集结晶(晶体)
adductive crystallization 加成化合物的结晶分离操作
annealing crystallization 退火结晶
batch crystallization 分批结晶
bulk crystallization 大量结晶

crystallization differentiation [ˌkrɪstəlaiˈzeiʃən ˌdifərenʃiˈeiʃn]

Definition: 结晶分异作用 jié jīng fēn yì zuò yòng

crystallization：form or cause to form crystals. 结晶 jié jīng；晶化 jīng huà
differentiation：the act of making or becoming different in the process of growth or development. 变异 biàn yì，变化 biàn huà

Extended Term:

crystallization-differentiation deposit 结晶分异矿床

crystallization index (CI) [ˈkrɪstəlaiˈzeiʃən ˈindeks]

Definition: 结晶指数 jié jīng zhǐ shù

crystallization：form or cause to form crystals. 结晶 jié jīng；晶化 jīng huà
index：a figure in a system or scale representing the average value of specified prices, shares, or other items as compared with some reference figure. 指数 zhǐ shù

crystallization interval [ˌkrɪstəlaiˈzeiʃən ˈintəvəl]

Definition: 结晶范围 jié jīng fàn wéi

crystallization：form or cause to form crystals. 结晶 jié jīng；晶化 jīng huà
interval：an intervening time or space. 间隔 jiàn gé

crystallization schistosity [ˌkrɪstəlaiˈzeiʃən ʃisˈtɔsiti]

Definition: 结晶片理 jié jīng piàn lǐ

crystallization：form or cause to form crystals. 结晶；晶化 jié jīng, jīng huà
schistosity：having a laminar structure like that of schist. 片岩结构的 piàn yán jié gòu de

crystalloblastic series [ˈkristələuˌblæstik ˈsiəriːz]

Definition: (Geology) a series of metamorphic minerals ordered according to decreasing formation energy, so crystals of a listed mineral have a tendency to form idioblastic outlines at surfaces of contact with simultaneously developed crystals of all minerals in lower positions. (地质) 变晶系 biàn jīng xì

crystalloblastic texture [ˈkristələuˌblæstik ˈtekstʃə]

Definition: 变晶结构 biàn jīng jié gòu

crystalloblastic: *crystallo-* means crystals; *blastic* menas related to or of recrystallization. 变晶质的 biàn jīng zhì de

texture: the arrangement of the particles or constituent parts of any material, as wood, metal, etc., as it affects the appearance or feel of the surface; structure, composition, grain, etc. 结构 jié gòu

Example:
The sample G36 is of oily luster and translucence, where the tremolite crystals are fine and closely interlocking, being micro-fibrous interlaced crystalloblastic texture.
样品 G36 具典型的油脂光泽、半透明外观特征，其透闪石晶体细小、紧密交织，为显微纤维交织变晶结构。

crystallographic preferred orientation
[ˌkristələuˈgræfik priˈfəːd ˌɔːrienˈteiʃən]

Definition: 结晶学的选择配向性 jié jīng xué de xuǎn zé pèi xiàng xìng

crystallographic: related to or of the branch of science concerned with the structure and properties of crystals. 晶体学的 jīng tǐ xué de

preferred: preferred above all others and treated with partiality. 优先的 yōu xiān de

orientation: the relative physical position or direction of something. 方向 fāng xiàng, 方位 fāng wèi

crystal settling [ˈkristəl ˈsetliŋ]

Definition: 晶体沉淀 jīng tǐ chén diàn

crystal: a piece of a homogeneous solid substance having a natural geometrically regular form with symmetrically arranged plane faces. 晶体 jīng tǐ

settling: a gradual sinking to a lower level. 沉淀物 chén diàn wù

cummingtonite [ˈkʌmɪŋtəˌnaɪt]

Definition: a mineral occurring typically as brownish fibrous crystals in some metamorphic rocks. It is a magnesium-rich iron silicate of the amphibole group. 镁铁闪石 měi tiě shǎn shí

Origin: early 19th century: named after Cummington, a town in Massachusetts, US, + -ite.

Extended Term:
cummingtonite-amphibolite 镁铁闪煌岩

cumulates [ˈkjuːmjuleɪts]

Definition: an igneous rock formed by gravitational settling of particles in a magma 堆积岩 duī jī yán

Origin: mid-16th century. (as a verb in the sense "gather in a heap"): from Latin cumulat- "heaped", from the verb cumulare, from cumulus "a heap". Current senses date from the early 20th century.

Example:
In respect of mineralogy and petrology as well as strontium isotope, the Pl-lherzolite exhibits some features of transition from the cumulates to the mantle peridotite.
在矿物成分、化学成分及锶同位素等方面，含长二辉橄榄岩具有介于地幔橄榄岩和堆晶岩之间的某些过渡特征。

Extended Terms:
cumulate complexes 堆积杂岩
mutual cumulates 联合累积量

cumulate texture [ˈkjuːmjuleɪt ˈtekstʃə]

Definition: 堆积结构 duī jī jié gòu
cumulate: gather together and combine. 堆积 duī jī; 积累 jī lěi
texture: the arrangement of the particles or constituent parts of any material, as wood, metal, etc., as it affects the appearance or feel of the surface; structure, composition, grain, etc. 结构 jié gòu

Example:
There are over 30 mafic and ultramafic intrusions in the area and Cu-Ni sulfide ore bodies were found in No. 1 and No. 7 ultramafic intrusive bodies which exhibit obvious zoning and cumulate

texture.
区内露出 30 多个镁铁和超镁铁质岩体，其中 1 号和 7 号超镁铁岩体中赋存铜镍硫化物矿（床）体。

cumulative curve ['kju:mju,leitiv kə:v]

Definition: any curve expressing the results of combining successive relative density fractions or size fractions. 累积曲线 lěi jī qū xiàn

Example:
Such classes are separated by "natural breaks" in the frequency distribution or the analogous cumulative curve.
这样的级可由频率分布或类似的累积曲线中的"自然断点"划分开来。

Extended Terms:
cumulative curve chart 累积曲线图
probability cumulative curve 概率累积曲线
size cumulative curve 粒度分布曲线

cumulus crystal ['kju: mjuləs 'kristəl]

Definition: 堆晶结构 duī jīng jié gòu

cumulus: cloud forming rounded masses heaped on each other above a flat base at fairly low altitude. 积云 jī yún

crystal: a piece of a homogeneous solid substance having a natural geometrically regular form with symmetrically arranged plane faces. 晶体 jīng tǐ

Example:
The chemical composition of clinopyroxene suggested that lherzolite had experienced partial melting, and the cumulus crystal complex and volcanic lava had the same characteristics as arc volcanic and basalt formed at the oceanic bottom.
单斜辉石的成分反映二辉橄榄岩经历过部分熔融作用，堆晶杂岩和基性熔岩具有火山弧和洋底玄武岩同样的特征。

Extended Term:
cumulus crystal complex 堆晶杂岩

current ripple ['kʌrənt 'ripl]

Definition: 流水波痕 liú shuǐ bō hén

current: a body of water or air moving in a definite direction, especially through a surrounding body of water or air in which there is less movement. 水流 shuǐ liú, 气流 qì liú

ripple: a thing resembling such a wave or series of waves in appearance or movement 波纹 bō wén; 波动 bō dòng

cyclothem [ˈsaikləθəm]

Definition: In geology, cyclothems are alternating stratigraphic sequences of marine and non-marine sediments. (地质)旋回层 xuán huí céng

Example:
Based on the investigation of cyclothem in section the author presents a new way for the study of paleosediments and paleoenvironments.
作者以沉积剖面旋回层为基础，提出了研究古代浅海沉积环境和古沉积的新方法。

Extended Terms:
ideal cyclothem 理想旋回层
mega cyclothem 巨旋回层
mini-cyclothem 微型韵律层理
model cyclothem 模式旋回层
positive cyclothem 正回层

dacite [ˈdeisait]

Definition: a grey volcanic rock containing plagioclase and quartz and other crystalline minerals 英安岩 yīng ān yán

Origin: The word dacite comes from Dacia, a province of the Roman Empire which lay between the Danube River and Carpathian Mountains (now modern Romania) where the rock was first described.

Example:
Volcanicity was violent during the Early Carboniferous in Santanghu region. The volcanic rocks consist of basalts, andesites and dacite liparite, with andesites playing the main part.
三塘湖地区早石炭世时期火山活动强烈，发育以安山岩类为主的玄武岩、英安岩、英安流纹岩的岩石组合。

Extended Terms:
dacite liparite 英安流纹岩
biotite dacite 黑云英安岩

hypersthene dacite 紫苏英安岩
mica-dacite 云母英安岩
pyroxene dacite 辉石英安岩

dating ['deitiŋ]

Definition: the activity of establishing how old an object or substance is, often with the use of sophisticated scientific techniques. 测定年龄 cè dìng nián líng

Extended Terms:
dating by obsidian 黑曜岩水化年代测定法
dating method 纪年法
absolute dating 绝对年代测定
isotopic dating 同位素定年
radiometric dating 放射性测定年代

daughter nuclide ['dɔːtə 'njuːklaid]

Definition: 子体核素 zǐ tǐ hé sù

daughter: a nuclide formed by the radioactive decay of another. 子体 zǐ tǐ
nuclide: a distinct kind of atom or nucleus characterized by a specific number of protons and neutrons. 核素 hé sù

Example:
The daughter nuclides of radon and thoron tend to become attached.
氡和钍射气的子体易被气溶胶吸附。

decarbonation [diːˌkɑːbəˈneiʃən]

Definition: to remove carbon dioxide from. 脱碳作用 tuō tàn zuò yòng

Origin: 1825-1835; de- + carbonate.

Example:
It is well-known that the particle size poses the main impact on the decarbonation rate. However the study in this respect is quite limited.
众所周知石灰石的粒度是影响热解速率的主要因素，但缺乏这方面的基础性的仔细研究。

Extended Terms:
decarbonation reactions 去碳反应
actual degree of decarbonation 实际分解率

apparent degree of decarbonation 表观分解率
Propylene carbonate solvent for decarbonation 丙碳脱碳溶剂
residue decarbonation 渣油脱碳

decay constant [diˈkeiˈkɔnstənt]

Definition: the reciprocal of the decay time. 衰变常数 shuāi biàn cháng shù

Origin: 1930-1935.

Example:
The present paper pionted out that in the measurement of the number of atoms of a radioactive isotope in a sample at time t_0, the error introduced through the inaccuracy of its decay constant (λ) can be reduced to a much smaller value by the careful selection of the time of measurement or measurements of its radioactivity at around one life time ($1/\lambda$) after t_0 and doing the calculation.
本文指出在测量一个样品中某种放射性同位素在 t_0 时刻下的原子数目时，只需精心选择测量其放射性的时刻于 t_0 后约该同位素的一个寿期($1/\lambda$)进行一次或多次测量并计算，则由它的衰变常数(λ)的不准确性而引进的误差可以被降低到较小值。

Extended Terms:
biological decay constant 生物衰变常数
exponential decay time constant 指数衰减时间常数
decay time constant 时间常数
radioactive decay constant 放射性衰变常数

decompression melting [ˌdiːkəmˈpreʃən ˈmeltiŋ]

Definition: melting of hot mantle rocks due to decrease of pressure during migration toward Earth's surface through convection. 减压熔融 jiǎn yā róng róng

Example:
The origin of abyssal peridotites results from a combination of decompression melting of upwelling mantle and olivine crystallization as a result of cooling of acending melts passing through the cold thermal boundary layer atop the mantle.
深海橄榄岩的成因是洋脊下上隆地幔降压熔融作用和熔体上升通过地幔最上部的热边界层时发生冷却，结晶出橄榄石的联合作用的结果。

decussate [diˈkʌseit]

Definition: arranged in opposite pairs, each pair being at right angles to the pair below. 交

互对生的 jiāo hù duì shēng de

Origin: mid-17th century (as a verb): from Latin decussatus, past participle of decussare "divide crosswise", from decussis (describing the figure X, i. e. the Roman numeral for the number 10), from decem "ten".

Example:

Leaves decussate, linear-lanceolate to lanceolate-oblong, 7-30 × 3-5 mm, base obtuse [to cuneate], apex obtuse to retuse.

叶交互对生，线状披针形到披针形长圆形，7-30 乘以 3-5 毫米，基部钝（到楔形），先端钝到微凹。

Extended Terms:

decussate leaves 交互对生的树叶
decussate structure 交错构造
decussate texture 交叉结构
opposite decussate leaves 交互对生叶

decussate structure [diˈkʌseit ˈstrʌktʃə]

Definition: 交错构造 jiāo cuò gòu zào

decussate: arranged in opposite pairs, each pair being at right angles to the pair below. 交互对生的 jiāo hù duì shēng de

structure: the arrangement of and relations between the parts or elements of something complex. 结构 jié gòu；构造 gòu zào

dedolomitization [diː͵dɔləmitaiˈzeiʃən]

Definition: a metamorphic process in which the magnesium in dolomitic rock forms new minerals, as brucite and forsterite, and the calcium forms calcite. 去白云石化作用 qù bái yún shí huà zuò yòng

Origin: de- + dolomitization.

deep shelf facies [diːp ʃelf ˈfeiʃiːz]

Definition: 深水陆架相 shēn shuǐ lù jià xiàng

deep: extending far down from the top or surface. 深的 shēn de
shelf: a submarine bank, or a part of the continental shelf. 大陆架 dà lù jià，陆棚 lù péng
facies: the character of a rock expressed by its formation, composition, and fossil content.

相 xiàng

defects in crystal structure [ˈdiːfekts in ˈkristəl ˈstrʌktʃə]

Definition: 晶体结构缺陷 jīng tǐ jié gòu quē xiàn

defect：a shortcoming, imperfection, or lack. 缺点 quē diǎn
crystal：a piece of a homogeneous solid substance having a natural geometrically regular form with symmetrically arranged plane faces. 晶体 jīng tǐ
structure：the arrangement of and relations between the parts or elements of something complex. 结构 jié gòu；构造 gòu zào

deformation [ˌdiːfɔːˈmeiʃn]

Definition: (Geology, Mechanics) a change in the shape or dimensions of a body, resulting from stress; strain (地质学, 机械学) 变形作用 biàn xíng zuò yòng

Origin: 1400-1450.

Example:
This paper analyzed the difference between the temperature filed and thermal deformation in dynamic characters for spindle components.
文章对机床主轴组件温度场和热变形动态特性之间的差异进行了分析。

Extended Terms:
abutment deformation 桥台变形
deformation resistance 变形阻力
plastic deformation 塑性形变
restrained deformation 约束变形

degree of oxidation (OX°) [diˈgriː əv ˌɔksiˈdeiʃn]

Definition: 氧化度 yǎng huà dù

degree：the amount, level, or extent to which something happens or is present. 程度 chéng dù
Oxidation：(Chemistry) the process or result of oxidizing or being oxidized. (化学) 氧化(作用) yǎng huà (zuò yòng)

degree of freedom [diˈgriː əv ˈfriːdəm]

Definition: 自由度 zì yóu dù

degree: the amount, level, or extent to which something happens or is present. 程度 chéng dù
freedom: the power or right to act, speak, or think as one wants without hindrance or restraint. 自由 zì yóu

> Example:

The principle suggests that a nonlinear vibration system with n degrees of freedom has no less than n nonlinear modes that correspond to n linear modes of the corresponding linear vibration system, no matter the nonlinear vibration system possesses either similar normal modes or dissimilar ones.
非线性模态对应原理指出：无论非线性振动系统具有相似模态还是具有非相似模态，n 个自由度的非线性振动系统至少具有 n 个非线性模态，且这 n 个非线性模态形式上对应于该非线性振动系统对应的线性振动系统的 n 个线性模态。

> Extended Terms:

degree of freedom analysis 自由度分析
economizing degrees of freedom 节约自由度
minimum degree of freedom 最少自由度
spatial degree of freedom 空间自由度
vibrational degree of freedom 振动自由度

dehydration [ˌdiːhaɪˈdreɪʃən]

> Definition: losing a large amount of water. 脱水 tuō shuǐ

> Origin: late 19th century: from de-(expressing removal) + Greek hudros, hudr-"water".

> Example:

The progressive evolution of hydrous minerals in eclogite suggests that a portion of surface water may be carried deep into the mantle in a subduction zone, and that the continuous dehydration of water-bearing eclogite provides a H_2O source for the partial melting of the subducting slab.
榴辉岩中含水矿物的不断行进替代证明在俯冲带中有部分地表水被携带至地幔，而连续的脱水作用也为俯冲板片的部分熔融提供了水源。

> Extended Terms:

dehydration reaction 脱水反应
mechanical dehydration 机械脱水
submersion dehydration 浸渍脱水
sun dehydration 日晒干燥
thermal dehydration 加热脱水
ultra-sonic dehydration 超声脱水

dehydration melting [ˌdiːhaiˈdreiʃən ˈmeltiŋ]

Definition: 脱水熔融作用 tuō shuǐ róng róng zuò yòng

dehydration: losing a large amount of water. 脱水 tuō shuǐ

melting: the process whereby heat changes something from a solid to a liquid. 熔化 róng huà

delamination [diːˌlæmiˈneiʃən]

Definition: the separation of a primordial cell layer into two layers by a process of cell migration 分层 fēn céng

Origin: 1875-1880; de- + lamination.

Example:
The wear mechanisms of un-implanted samples were mainly delamination, while those of the implanted ones were mainly plastic deformation, the change of such wear mechanisms was caused by the change of the structure of the composite by the proton implantation.
未注入样品的磨损主要表现为脱层剥落，而注入样品主要表现为塑性变形，这种磨损机理的变化是由于质子注入引起的复合材料的结构变化。

Extended Terms:
delamination belts 滑脱带
delamination detection 分层损伤诊断
delamination resistance 抗脱层性能
Bond Delamination 粘合层剥层测量
continental delamination 岩石圈脱壳沉降

delta [ˈdeltə]

Definition: a triangular tract of sediment deposited at the mouth of a river, typically where it diverges into several outlets. 三角洲 sān jiǎo zhōu

Origin: mid-16th century: originally specifically as the Delta (of the River Nile), from the shapeof the Greek letter.

Example:
Sediment deposited by flowing water, as in a riverbed, flood plain, or delta.
冲积层河床、洪水淹没的平原或三角洲中的流水淤积所产生的沉积层。

Extended Terms:
delta connected 三角形连接
delta control 三角形舵

levee delta 堤状三角洲
pearl delta 珠江三角洲

delta facies [ˈdeltə ˈfeiʃiiːz]

Definition: 三角洲相 sān jiǎo zhōu xiàng

delta: a triangular tract of sediment deposited at the mouth of a river, typically where it diverges into several outlets. 三角洲 sān jiǎo zhōu

facies: the character of a rock expressed by its formation, composition, and fossil content. 相 xiàng

Origin:

Example:

Good reservoirs mainly pebbled sandstone and sandstone in fluvial facies and delta facies influence by unconformity and fault.
最有利的储层主要为在不整合和断层的影响范围内的河流、三角洲相含砾砂岩、砂岩储层。

Extended Terms:

delta facies deposition 三角洲沉积
delta facies reservoir 三角洲相储层
delta facies sequence 三角洲沉积序列

delta front facies [ˈdeltə frʌnt ˈfeiʃiiːz]

Definition: 三角洲前缘相 sān jiǎo zhōu qián yuán xiàng

delta: a triangular tract of sediment deposited at the mouth of a river, typically where it diverges into several outlets. 三角洲 sān jiǎo zhōu

front: the side or part of an object that presents itself to view or that is normally seen or used first; the most forward part of something. 前部 qián bù, 前面 qián miàn

facies: the character of a rock expressed by its formation, composition, and fossil content. 相 xiàng

Example:

Putaohua reservoir is delta-front facies sedimentary sand with not only high porosity and high permeability reservoirs but also shale-bearing, calcium-bearing and thin interbed reservoirs.
葡西地区葡萄花油层为三角洲前缘相沉积砂体,既有岩性纯、孔渗好的储层,又有含泥、含钙、薄互层的储层。

delta plain facies [ˈdeltə pleɪn ˈfeɪʃiːz]

Definition: 三角洲平原相 sān jiǎo zhōu píng yuán xiàng

delta: a triangular tract of sediment deposited at the mouth of a river, typically where it diverges into several outlets. 三角洲 sān jiǎo zhōu

plain: a large expanse of fairly flat dry land, usually with few trees. 平原 píng yuán

facies: the character of a rock expressed by its formation, composition, and fossil content. 相 xiàng

Example:

The reservoir type is delta plain facies distributary channel sand body, and the lithology is mainly composed of fine arcose with a compact cementation.
储层为三角洲平原分流河道砂体，岩性主要为细粒长石砂岩，胶结致密。

dendritic crystal growth [denˈdrɪtɪk ˈkrɪstəl grəʊθ]

Definition: 枝晶生长 zhī jīng shēng zhǎng

dendritic: consisting of crystalline dendrites. 有枝状结晶的 yǒu zhī zhuàng jié jīng de

crystal: a piece of a homogeneous solid substance having a natural geometrically regular form with symmetrically arranged plane faces. 晶体 jīng tǐ

growth: the process of increasing in physical size. 生长 shēng zhǎng

Example:

Isothermal dendritic crystal growth in Al-4.5%Cu alloy is simulated by using the phase-field model. The influence of the mesh size, anisotropy strength and interface thickness on dendritic growth is investigated.
用相场法模拟了 Al-4.5%Cu 合金的等温枝晶生长过程，分析了网格尺寸、各向异性强度和界面厚度对枝晶生长过程的影响。

density [ˈdensəti]

Definition: the degree of compactness of a substance. 密度 mì dù

Origin: early 17th century: from French densité or Latin densitas, from densus "dense".

Extended Terms:

abnormal density 反常密度
absolute density 绝对密度
acceptor density 受主密(浓)度

acoustic impedance density 声阻抗密度，声阻抗率
actual flux density 有效(磁)通量密度
air-dry density 气干密度

density current [ˈdensəti ˈkʌrənt]

Definition: 异重流 yì zhòng liú

density: the degree of compactness of a substance. 密度 mì dù
current: a body of water or air moving in a definite direction, especially through a surrounding body of water or air in which there is less movement. 水流 shuǐ liú, 气流 qì liú

Example:
Density current has some advantages and disadvantages to treatment efficiency of secondary sedimentation tank.
异重流对二沉池处理效果的影响有利有弊。

Extended Terms:
equilibrium current density 平衡电流密度
exchanging current density 交换电流密度
ion current density 离子流密度
polarized current density 极化电流密度
transition current density 转移电流密度

depleted mantle [diˈpliːtid ˈmæntl]

Definition: 亏损地幔 kuī sǔn dì màn

depleted: no longer sufficient. 亏损 kuī sǔn
mantle: the region of the earth's interior between the crust and the core, believed to consist of hot, dense silicate rocks 地幔 dì màn

Example:
It is implied that magma might be derived from a depleted mantle.
说明岩浆来源于亏损地幔源区。

depolymerization [ˌdiːˌpɔliməraiˈzeiʃən,-riˈz-]

Definition: break (a polymer) down into monomers or other smaller units. 解聚作用 jiě jù zuò yòng

Origin: 1890-1895; de- + polymerize.

> Example:

The degree of depolymerization is characterized by the crystallization and thermal stability of undepolymerized PET residue.
从残留 PET 的结晶度和热稳定性间接体现了解聚反应的程度。

> Extended Terms:

depolymerization reaction 解聚反应
acid depolymerization 酸解聚(作用)
dry depolymerization 干法
radical depolymerization 自由基解聚合
thermal depolymerization 热解聚(作用)(法)

deposition [ˌdepəˈzɪʃən]

> Definition: the geological process by which material is added to a landform or land mass. 沉积作用 chén jī zuò yòng

> Origin: late Middle English：from Latin depositio (n-)，from the verb deponere (see deposit)．

> Example:

All sediment deposition is not injurious.
所有的泥沙淤积并不都是有害的。

> Extended Terms:

deposition rate 沉积率
depositing tank 矿泥沉淀箱
coastal deposition 海岸沉积
estuarine deposition 港湾沉积作用
euxinic deposition 静海沉积

depositional basin [ˌdepəˈzɪʃənəl ˈbeisən]

> Definition: 沉积盆地 chén jī pén dì

depositional：of or related to deposition. 沉积的 chén jī de

basin：a place where the earth's surface is lower than in other areas of the world. 盆地 pén dì；凹地 āo dì

> Example:

The acreage of Nansha area of South China Sea is $82.6 \times 10^4 km^2$, where hydrocarbon basin shad developed with good petroleum geology conditions, the main depositional basins include Zhengmu, Beikang, Nanweixi and Zhongjiangnan.

南沙海域面积为 82.6×10^{4}km^2，含油气沉积盆地发育，石油地质条件优越，其中大型盆地有曾母、北康、南薇西、中建南等。

depositional environment [ˌdepəˈzɪʃənəl ɪnˈvaɪərənmənt]

Definition: In geology, sedimentary depositional environment describes the combination of physical, chemical and biological processes associated with the deposition of a particular type of sediment. （地质）沉积环境 chén jī huán jìng

Example:

This study documents linkages between depositional environments and clinoform attributes.

这一研究记录了沉积环境与斜坡地形属性之间的联系。

depth zone [depθ zəun]

Definition: 深度带 shēn dù dài

depth: the distance from the top or surface to the bottom of something. 深度 shēn dù

zone: an area or a stretch of land having a particular characteristic, purpose, or use, or subject to particular restrictions. 地带 dì dài, 区域 qū yù

Example:

The experimental data clearly demonstrated that there was an increasing linear relationship between mixing zone depth and rainfall intensity.

试验结果表明了混合层深度与降雨强度呈线性增加关系。

Extended Term:

depth zone of earth 地球深带

derivative magma [dɪˈrɪvətɪv ˈmæɡmə]

Definition: 派生岩浆 pài shēng yán jiāng

derivative: of something which is based on another source. 派生的 pài shēng de, 衍生的 yǎn shēng de

magma: hot fluid or semi-fluid material below or within the earth's crust from which lava and other igneous rock is formed by cooling. 岩浆 yán jiāng

Example:

The magma is originated from the mantle and may be the "derivative magma" rich in volatile

and sodium, formed by the differentiation of basalt magma.
熔浆来源于幔源，可能是玄武岩浆分异的、富挥发和富钠质的"派生岩浆"。

descending plate [diˈsendiŋ pleit]

Definition: 下倾板块 xià qīng bǎn kuài

descending: coming down or downward. 下降的 xià jiàng de
plate: each of the several rigid pieces of the earth's lithosphere which together makes up the earth's surface. 板块 bǎn kuài

Example:
On Earth, andesites typically are formed when descending plates mix water into subterranean molten rock.
在地球上，安山岩通常是板块的下沉使水混入地下熔融的岩石而形成。

descriptive petrology [diˈskriptiv piːˈtrɔlədʒi]

Definition: 描述岩石学 miáo shù yán shí xué，岩相学 yán xiāng xué，岩类学 yán lèi xué

descriptive: serving or seeking to describe. 描述的 miáo shù de
petrology: the branch of science concerned with the origin, structure, and composition of rocks. 岩石学 yán shí xué

desert facies [ˈdezət ˈfeiʃiiːz]

Definition: 沙漠相 shā mò xiàng

desert: a dry, barren area of land, especially one covered with sand, that is characteristically desolate, waterless, and without vegetation. 沙漠 shā mò
facies: the character of a rock expressed by its formation, composition, and fossil content. 相 xiàng

Example:
Luohe stage and Luohandong stage are two major sedimentation stages of desert facies.
其中，洛河期和罗汉洞期是沙漠发育的两个主要时期。

desilication [diːˌsiliˈkeiʃən]

Definition: any process that removes silicon from a material 脱硅作用 tuō guī zuò yòng

Example:
Based on these, research directions of desilication in sodium aluminate solution are presented.

在此基础上指出了铝酸钠溶液脱硅的研究方向。

Extended Terms:

desilication reaction 脱硅反应
desilication roasting 焙烧脱硅
desilication product 脱硅产物
desilication speed 脱硅速度
desilication time 脱硅时间
desilication yield 脱硅率
deep desilication 深度脱硅

destructional delta deposit [di'strʌkʃənəl 'deltə di'pɔzit]

Definition: 破坏性三角洲沉积 pò huài xìng sān jiǎo zhōu chén jī

destructional: causing great and irreparable harm or damage. 破坏(性)的 pò huài (xìng) de
delta: a triangular tract of sediment deposited at the mouth of a river, typically where it diverges into several outlets. 三角洲 sān jiǎo zhōu
deposit: a layer or body of accumulated matter. 沉积物 chén jī wù, 沉淀 chén diàn

destructive margin [di'strʌktiv 'mɑːdʒin]

Definition: 消减型边缘 xiāo jiǎn xíng biān yuán

destructive: causing great and irreparable harm or damage. 破坏(性)的 pò huài (xìng) de
margin: the edge or border of something. 边 biān, 边沿 biān yán

detrital [di'traitəl]

Definition: waste or debris of any kind 碎屑的 suì xiè de

Origin: late 18th century (in the sense "detrition"): from French détritus, from Latin detritus, from deterere "wear away".

Example:
Therefore, the terrigenous detrital materials have several places of origin being the major feature of the Paleo zoic disposition of the Ordos Basin.
因此,陆源碎屑沉积物具有多物源供给是鄂尔多斯盆地晚古生代沉积的一个主要特点。

Extended Terms:

detrital buildup 碎屑物堆积
detrital calcite particles 方解石碎屑颗粒

detrital clay matrix 碎屑黏土基质
detrital deposit 碎屑沉积
detrital sediment 碎屑沉积物

detrital rock [diˈtraitəl rɒk]

Definition: 碎屑岩 suì xiè yán

detrital: waste or debris of any kind. 碎屑的 suì xiè de

rock: the solid mineral material forming part of the surface of the earth and other similar planets, exposed on the surface or underlying the soil. 岩石 yán shí；岩 yán

Example:

Triassic ammonites, fossil plants and radiolaria occurred in normal sedimentary micritic limestone, detrital rock and silica rock and represented the sediment age of the Bayankelashan Group.
其中三叠纪的菊石、古植物、放射虫等标准化石分子产于正常沉积的泥晶灰岩、碎屑岩、硅质岩等岩层中，代表巴颜喀拉山群的沉积时代。

Extended Term:

fine grained detrital rock 细粒碎屑岩

detritus [diˈtraitəs]

Definition: a geological term used to describe particles of rock derived from pre-existing rock through processes of weathering and erosion. 岩屑 yán xiè

Origin: 1785-1795.

Example:

Detritus usually consists of gravel, sand and clay.
岩屑通常是由砂砾、沙和黏土组成的。

Extended Terms:

detritus chamber 沉砂池
detritus slide 土砾滑动，土石崩滑

devitrification [diːˌvitrifiˈkeiʃən]

Definition: the formation of small crystals in a glass as a result of slow cooling from the molten state. 脱玻化作用 tuō bō huà zuò yòng

Example:

The formation and intensity of devitrification are related not only to the rock composition, but also to the formation and development of cracks.

脱玻化的发生及其强弱与岩石成分、层位和裂隙发育程度有密切联系。

Extended Terms:

devitrification glaze 失透釉

devitrification nuclei 析晶晶核

devitrification of glass 玻璃闷光

---

diabase ['daiəbeis]

Definition: a dark-gray to black, fine-textured igneous rock composed mainly of feldspar and pyroxene and used for monuments and as crushed stone. 辉绿岩 huī lǜ yán

Origin: mid-19th century (originally denoting diorite): from French, formed irregularly as if from di-"two" + base "base" (thus "rock with two bases", referring to the base minerals of diorite), but associated later perhaps with Greek diabasis "transition".

Example:

Diabase connected depth and shallow and caused deep auriferous metallogenetic fluid through rifted fractured zone uplifting and precipitation.

辉绿岩起着沟通深部与浅部构造，并使深部的含矿流体沿挤压破碎带上升、沉淀而成矿的作用。

Extended Terms:

diabase amphibolite 辉绿闪岩

diabase facies 辉绿岩相

diabase spessartite 辉绿闪斜煌斑岩

essexite diabase 碱辉辉绿岩

glassy diabase 玻质辉绿岩

olivine diabase 橄榄辉绿岩

diabase porphyrite ['daiəbeis 'pɔːfirait]

Definition: 辉绿玢岩 huī lǜ bīn yán

diabase: a dark-gray to black, fine-textured igneous rock composed mainly of feldspar and pyroxene and used for monuments and as crushed stone. 辉绿岩 huī lǜ yán

porphyrite: a rock with a porphyritic structure; as, augite porphyrite. 玢岩 bīn yán

> Example:

Diabase porphyrite was the splitting product of passive margin of Gondwana Continent in Late Jurassic.
辉绿玢岩是冈瓦纳被动陆缘在晚古生代裂解的产物。

diabasic texture [ˌdaiəˈbeisik ˈtekstʃə]

> Definition: 辉绿结构 huī lǜ jié gòu

diabasic: denoting igneous rock in which the inter-stices between the feldspar crystals are filled with discrete crystals or grains of pyroxene. 辉绿岩的 huī lǜ yán de
texture: the arrangement of the particles or constituent parts of any material, as wood, metal, etc., as it affects the appearance or feel of the surface; structure, composition, grain, etc. 结构 jié gòu

diagenesis [ˌdaiəˈdʒenisis]

> Definition: the physical and chemical changes occurring in sediments between the times of deposition and solidification. 成岩作用 chéng yán zuò yòng

> Origin: 1885-1890.

> Example:

The early diagenesis mainly influences the development of primary pore and the late diagenesis mainly influences the development of secondary pore.
早期成岩作用主要影响原生孔隙的发育，晚期成岩作用影响次生孔隙的发育。

> Extended Terms:

chemical diagenesis 化学成岩作用
coal diagenesis 煤成岩酌
peat diagenesis 泥炭岩化作用
retrograde diagenesis 退化成岩作用

diagenetic breccia [ˌdaiədʒəˈnetik ˈbretʃiə]

> Definition: 成岩角砾岩 chéng yán jiǎo lì yán

diagenetic: the physical and chemical changes occurring during the conversion of sediment to sedimentary rock. 成岩作用 chéng yán zuò yòng
breccia: a coarse-grained sedimentary rock made of sharp fragments of rock and stone cemented together by finer material. Breccia is produced by volcanic activity or erosion, including frost

shattering. 角砾岩 jiǎo lì yán

diagenetic concretion [ˌdaiədʒəˈnetik kənˈkriːʃən]

Definition: 成岩结核 chéng yán jié hé

diagenetic: the physical and chemical changes occurring during the conversion of sediment to sedimentary rock. 成岩作用 chéng yán zuò yòng

concretion: a rounded mass of compact concentric layers within a sediment, built up around a nucleus such as a fossil. 结核 jié hé

diagenetic dolomite [ˌdaiədʒəˈnetik ˈdɔləmait]

Definition: 成岩白云岩 chéng yán bái yún yán

diagenetic: the physical and chemical changes occurring during the conversion of sediment to sedimentary rock. 成岩作用 chéng yán zuò yòng

dolomite: a white, reddish, or greenish mineral consisting of calcium magnesium carbonate. 白云岩 bái yún yán

diagnostic metamorphic mineral [ˌdaiəgˈnɔstik ˌmetəˈmɔːfik ˈminərəl]

Definition: 特征变质矿物 tè zhēng biàn zhì kuàng wù

diagnostic: characteristic of a particular species, genus, or phenomenon. 特有的 tè yǒu de

metamorphic: denoting rock that has undergone transformation by heat, pressure, or other natural agencies, e. g. in the folding of strata or the nearby intrusion of igneous rocks. (岩石)变质的 (yán shí) biàn zhì de

mineral: a solid inorganic substance of natural occurrence. 矿 kuàng, 矿物 kuàng wù

diallage peridotite [ˈdaiəlidʒ ˌperiˈdəutait]

Definition: 异剥辉石橄榄岩 yì bō huī shí gǎn lǎn yán

diallage: a green, brown, gray, or bronze-colored clinopyroxene characterized by prominent parting parallel to the front pinacoid a (100). 异剥辉石 yì bō huī shí

peridotite: any of several coarsegrained, dark igneous rocks consisting mainly of olivine and other ferromagnesian minerals. 橄榄岩 gǎn lǎn yán

diamond [ˈdaiəmənd]

Definition: a hard, bright, precious stone which is clear and colourless used in jewellery and for cutting very hard substances. 钻石 zuàn shí, 金刚石 jīn gāng shí

Origin: Middle English: from Old French diamant, from medieval Latin diamas, diamant-, variant of Latin adamans (see adamant).

Example:

The wetting mechanism of metal alloy containing strong carbide-forming elements to diamond surface and the technique of diamond surface metallization are fundamental to the manufacture of single-layer brazing diamond tools.
含强碳化物形成元素合金对金刚石表面浸润的机理与金刚石表面金属化技术是钎焊制造单层金刚石工具的技术基础。

Extended Terms:

diamond coating 金刚石涂层
diamond rough 钻石原石
diamond semiconductor 金刚石半导体
diamond shovel 钻石铲
simulated diamond 人造钻石

diapir [ˈdaiəpiə]

Definition: a domed rock formation in which a core of rock has moved upward to pierce the overlying strata. 底辟 dǐ pì, 挤入构造 jǐ rù gòu zào

Origin: early 20th century: from Greek diapeirainein "pierce through".

Example:

After the river channel shifted, the discharging center of pressure was generated from the seabed erosion, on which the diapir came into being.
河道改变位置后，海底快速冲刷，在最大冲刷中心也是最大压力释放中心刺穿体形成。

Extended Terms:

diapir penetration 底辟刺穿作用
diapir pluton 底辟深成岩体
growth diapir 生长底辟
immature diapir 不完全底辟构造
piercement diapir 刺穿挤入

diapiric intrusion [ˌdaiəˈpirik inˈtruːʒən]

Definition: 底辟侵入 dǐ pì qīn rù

diapiric: a domed rock formation in which a core of rock has moved upward to pierce the overlying strata. 底辟 dǐ pì, 挤入构造 jǐ rù gòu zào

intrusion: the action or process of forcing a body of igneous rock between or through existing formations, without reaching the surface. 侵入 qīn rù

diapiric rise [ˌdaiəˈpirik rais]

Definition: 底辟上升 dǐ pì shàng shēng

diapiric: a domed rock formation in which a core of rock has moved upward to pierce the overlying strata. 底辟 dǐ pì, 挤入构造 jǐ rù gòu zào

rise: move from a lower position to a higher one; come or go up. 升起 shēng qǐ, 上升 shàng shēng

diaspore [ˈdaiəspɔː]

Definition: a native aluminium oxide hydroxide, α-AlO(OH), crystallizing in the orthorhombic system and isomorphous with goethite. 硬水铝石 yìng shuǐ lǚ shí

Origin: 1795-1805; from Greek diasporá.

Example:
Bauxite ore in the Northwest Henan Province is generally sedimentary diaspore type in Benxi Formation of Carboniferous Series.
豫西北地区铝土矿均为硬水铝石型沉积矿床，赋存于中石炭统本溪组。

Extended Terms:
calcined diaspore 熟水铝石
diaspore refractory 水铝石耐火材料
diaspore weight 种子重量
mangan diaspore 锰水铝石

diastem [ˈdaiəstem]

Definition: a minor hiatus in an orderly succession of sedimentary rocks. 沉积暂停期 chén jī zàn tíng qī

Origin: 1850-1855; from Greek diástēma.

Extended Term:
isochronous diastem 同时小间断

diatom [ˈdaiətəm]

Definition: a single-celled alga which has a cell wall of silica. Many kinds are planktonic, and extensive fossil deposits have been found. 硅藻 guī zǎo

Origin: mid-19th century: from modern Latin Diatoma (genus name), from Greek diatomos "cut in two", from diatemnein "to cut through".

Example:
In Tengchong, the diatom was occurred in several Cenozoic lake basin formed by the intensive structural movement after Eogene.
古近纪之后,强烈的构造运动在腾冲形成数个新生代湖盆,成为硅藻生存场所。

Extended Terms:
diatom insulator 硅藻土绝缘物
diatom ooze 硅藻软泥
diatom structure 硅藻结构
diatom test 硅藻壳
marine benthic diatom 海洋底栖硅藻
marine planktonic diatom 海洋浮游硅藻

diatomite [daiˈætəmait]

Definition: a soft very fine-grained whitish rock consisting of the siliceous remains of diatoms deposited in the ocean or in ponds or lakes. It is used as an absorbent, filtering medium, insulator, filler, etc. 硅藻土 guī zǎo tǔ

Origin: late 19th century: from diatom + -ite1.

Example:
With the diatomite as carrier, the vanadium catalyst for sulphuric acid production features high strength, activity and lifetime.
用该硅藻土生产的硫酸钒催化剂具有强度高、活性好、寿命长的特点。

Extended Terms:
diatomite brick 硅藻土砖
diatomite filter 硅藻土滤池
asbestos diatomite 石棉硅藻土
kieslguhr diatomite 硅藻土

diatreme [ˈdaiəˌtriːm]

Definition: a long vertical pipe or plug formed when gas-filled magma forced its way up through overlying strata. 火山爆发口 huǒ shān bào fā kǒu

Origin: early 20th century: from dia-"through" +Greek trēma"perforation".

Example:

The diatreme breccia pipes belong to volcanic crater phase and volcanic conduit phase, and consist of breccias and binding material.
爆破角砾岩筒属火山口相和火山管道相，脉岩属次火山相。

Extended Term:

phreatic diatreme 蒸气(火山)道

differentiation index(DI) [ˌdifərenʃiˈeiʃn ˈindeks]

Definition: The sum of the normative constituents (Q + Ab + Or + Ne + Kp + Lc in an igneous rock, where Q = quartz, Ab = albite, Or = orthoclase, Ne = nepheline, Kp = kaliophilite, and Lc = leucite.) The index, defined in 1960 by two American petrologists, Thornton and Tuttle, seeks to quantify the degree of differentiation a rock has undergone. The greater the degree of differentiation, the more enriched the rock is in felsic minerals and hence the higher the differentiation index. 分异指数 fēn yì zhǐ shù

Example:

The characteristics of the chemical composition of the tin-bearing porphyriesare are rich in silicon and alkali contents, but poor in calcium, magnesium and iron with the differentiation index (DI)>90 and the solidification index (SI)<3, belonging to tension granite.
华南某些含锡斑岩体的化学成分以富硅、贫钙、镁、铁、富碱为特征，其分异指数(DI)一般在90%以上，固结指数低于3，属张裂性花岗岩。

Extended Terms:

differentiation index of cytoplasm 细胞质分化指数
environment trial differentiation index 环境试验鉴别指数
genetic differentiation index 遗传分化指数

diffusion [diˈfjuːʒən]

Definition: the spreading of something more widely. 扩散 kuò sàn

Origin: late Middle English (in the sense "pouring out, effusion"): from Latin diffusio(n-), from diffundere"pour out".

Example:

The importance of diffusion processes can be gauged by comparing the distances which a particle travels under the influence of gravity and diffusion.
扩散过程的重要性可以通过比较粒子在扩散和重力影响下经过的距离来衡量。

> Extended Terms:

active diffusion 有效浸筛，有效扩散
alloy diffusion 合金扩散
ambipolar diffusion 双极扩散
anomalous diffusion 反常扩散
atmospheric diffusion 大气扩散
base contact diffusion 基极接触扩散
bilateral diffusion 双向扩散，双扩散

diffusion coefficient [diˈfjuː ʒən ˌkəuiˈfiʃənt]

> Definition: 扩散系数，漫射系数 kuò sàn xì shù, màn shè xì shù

diffusion: the spreading of something more widely. 扩散 kuò sàn
coefficient: a multiplier or factor that measures some property. 率 lǜ; 系数 xì shù

> Example:

The experimental results show that surface mortar layer can significantly reduce the chloride ion diffusion coefficients of concrete.
结果表明，致密的表面砂浆层能显著地降低混凝土本体的氯离子扩散系数。

> Extended Terms:

apparent diffusion coefficients 外显扩散系数
effective diffusion coefficients 有效扩散系数
error diffusion coefficients 误差扩散系数
partial diffusion coefficients 粒子扩散系数
solvent diffusion coefficients 溶剂扩散系数
tracer diffusion coefficients 示踪扩散系数

diffusive metasomatism [diˈfjuː siv ˌmetəˈsəumətizəm]

> Definition: 扩散交代作用 kuò sàn jiāo dài zuò yòng

diffusive: the spreading of something more widely. 扩散 kuò sàn, 散播 sàn bō
metasomatism: change in the composition of a rock as a result of the introduction or removal of chemical constituents. 交代作用 jiāo dài zuò yòng

dike [daik]

> Definition: an intrusion of igneous rock cutting across existing strata. 岩脉 yán mài, 岩墙 yán qiáng

> Example:

Zircon U-Pb dates suggest that the Anjiayingzi granite was formed at 132~138 Ma, and the rhyolite dike that cut gold lodes was formed at 125~127 Ma. Therefore, gold mineralization in the area is constrained at 126~132 Ma.
锆石 U-Pb 年龄表明，安家营子花岗岩成型时代为 132Ma~138 Ma，而穿切矿体的流纹斑岩岩脉的成岩时代为 125~127 Ma，进而金矿成矿时代可以限定在 126Ma~132Ma。

> Extended Terms:

basalt dike 玄武岩脉
border dike 边缘堤带(岩脉缘)
breccia dike 角砾岩脉
composite dike 复合岩墙
cone dike 锥形岩脉
relict dike 残余岩脉

dike rock [daik rɔk]

> Definition: a tabular body of igneous rock that cuts across adjacent rocks or cuts massive rocks. 脉岩 mài yán

> Example:

Rock dike inserted into quartz sandstone and was then weathered and denuded to form a precipice.
石英砂岩受岩脉穿插，岩脉被风化剥蚀后形成绝壁。

> Extended Terms:

lamprophyric dike rock 煌斑脉岩
undifferentiated dike rock 未分异脉岩

diopside-anorthite system [daiˈɔpsaid æˈnɔːθait ˈsistəm]

> Definition: 透辉石-钙长石体系 tòu huī shí gài cháng shí tǐ xì

diopside：a mineral occurring as white to pale green crystals in metamorphic and basic igneous rocks. It consists of a calcium and magnesium silicate of the pyroxene group, often containing iron and chromium. 透辉石 tòu huī shí

anorthite：a calcium-rich mineral of the feldspar group, typically white, occurring in many basic igneous rocks. 钙长石 gài cháng shí

system：a set of things working together as parts of a mechanism or an interconnecting network. 系统 xì tǒng

diopside [daiˈɔpsaid]

Definition: a mineral occurring as white to pale green crystals in metamorphic and basic igneous rocks. It consists of a calcium and magnesium silicate of the pyroxene group, often also containing iron and chromium. 透辉石 tòu huī shí

Origin: early 19th century: from French, formed irregularly from di-"through" + Greek opsis."aspect", later interpreted as derived from Greek diopsis "a view through".

Example:
Laozhuang Tremolite-diopside Deposit is a superior large deposit firstly found and evaluated in Henan Province.
南召老庄透闪石-透辉石矿是河南省首次发现与评估的一个特大型矿床。

Extended Terms:
diopside kersantite 透辉云斜煌岩
diopside granitite 透辉黑云花岗岩
diopside porcelain 透辉石质瓷
chrome diopside 铬透辉石
diopside solid solution 透辉石固溶体
Dushan diopside 独山玉

diopside zone [daiˈɔpsaid zəun]

Definition: 透辉石带 tòu huī shí dài

diopside: a mineral occurring as white to pale green crystals in metamorphic and basic igneous rocks. It consists of a calcium and magnesium silicate of the pyroxene group, often also containing iron and chromium. 透辉石 tòu huī shí

zone: an area or stretch of land having a particular characteristic, purpose, or use, or subject to particular restrictions. 地带 dì dài, 区域 qū yù

diorite [ˈdaiərait]

Definition: any of various dark, granite-textured, crystalline rocks rich in plagioclase and having little quartz. 闪长岩 shǎn cháng yán

Origin: early 19th century: coined in French, formed irregularly from Greek diorizein "distinguish" + -ite.

Example:
The dark dikes associated with mineralization in the Jinchanggouliang mine area, are mostly fine

diorite-quartz diorites and diorite porphyrite-dacite porphyries.
金厂沟梁金矿区中与成矿伴生的暗色脉岩主要是细晶闪长岩石英闪长岩、闪长玢岩英安斑岩类。

Extended Terms:

diorite slate 闪绿石板瓦
diorite soil 闪绿岩土壤
gabbro diorite 辉长闪长岩
hornblende diorite 角闪闪长岩
pyroxene diorite 辉石闪长岩
quartz diorite 石英闪长岩

dioritic porphyrite [ˌdaiəˈritik ˈpɔːfirait]

Definition: 闪长玢岩 shǎn chǎng bīn yán

diorite: any of various dark, granite-textured, crystalline rocks rich in plagioclase and having little quartz. 闪长岩 shǎn chǎng yán

porphyrite: a rock with a porphyritic structure; as, augite porphyrite. 玢岩 bīn yán

discontinuity [ˌdiskɔnˈtinjuːəti]

Definition: a distinct break in physical continuity or sequence in time, or sharp difference of characteristics between parts of something. 不连续性 bù lián xù xìng

Origin: late 16th century: from medieval Latin discontinuitas, from discontinuus.

Example:

An example which utilized different plane joint elements to simulate the different sorts of discontinuity interface in rock mass was given, and the analysis result indicates that if only the constitutive model of a material is confirmed, and some special element types are adopted, with the basic principle and approach of the finite element to simulate rock materials, a more satisfactory result can be obtained.

给出了运用不同的节理单元对岩体中不同的不连续面进行模拟分析的实例,分析的结果表明只要确定了材料的模型,采用比较特殊的单元类型,仍然可以按照有限元的基本思想和步骤来对岩石类材料进行数值模拟,并可以取得较为满意的结果。

Extended Terms:

absorption discontinuity 吸收不连续性,吸收曲线连续性中断
contact discontinuity 接触不连续,接触间断
derivative discontinuity 导数的不连续性
finite discontinuity 有限不连续性

Gutenberg discontinuity 古登堡间断面
removable discontinuity 可移不连续性

discontinuous reaction [ˌdiskən'tinjuəs ri'ækʃən]

Definition: 不连续反应 bù lián xù fǎn yìng

discontinuous: having intervals or gaps. 不连续的 bù lián xù de；有间断的 yǒu jiàn duàn de
reaction: a chemical process in which two or more substances act mutually on each other and are changed into different substances, or one substance changes into two or more other substances. 反应 fǎn yìng

Example:

The results show that the discontinuous reaction proceeded not only by a conventional cellular reaction controlled by the migrating of the grain boundary, but also a autocatalytic cellular reaction occurring in the grain. The different microstructure lamellar α Al+η Zn and island like η Zn distributed in α Al matrix would be formed by both mechanisms.
结果表明：(Zn Al40)合金非连续沉淀既可发生受晶界迁移控制的常规胞状反应，形成片层相间的 α Al+η Zn 平衡组织，又可在晶内自发形核生长，发生自催化胞状反应，形成 α Al 基体上分布岛状的 η Zn 平衡组织。

Extended Term:

discontinuous reaction series 不连续反应系列

discordance intrusive body [di'skɔːdəns in'truːsiv 'bɔdi]

Definition: 不整合侵入体 bù zhěng hé qīn rù tǐ

discordance: an arrangement of rock strata in which the older underlying ones dip at a different angle from the younger overlying ones; unconformity. (岩层的)不整合(yán céng de) bù zhěng hé
intrusive: of or relating to igneous rock that is forced while molten into cracks or between other layers of rock. 侵入的 qīn rù de
body: the whole physical structure and substance of a human being, animal, or plant. 体 tǐ

discordant intrusion [di'skɔːdənt in'truːʒən]

Definition: 不整合侵入 bù zhěng hé qīn rù, 不谐和侵入 bù xié hé qīn rù

discordant: disagreeing or incongruous. 不一致的 bù yí zhì de；不协调的 bù xié tiáo de
intrusion: the action or process of forcing a body of igneous rock between or through existing

formations, without reaching the surface. 侵入 qīn rù

discordia line [diˈskɔːdiə lain]

Definition: 不谐和曲线 bù xié hé qū xiàn，不一致曲线 bù yí zhì qū xiàn

discordia: disagreeing or incongruous. 不一致的 bù yí zhì de；不协调的 bù xié tiáo de
line: a long, narrow mark or band. 线 xiàn

disequilibrium growth [ˌdisikwiˈlibriəm grəuθ]

Definition: 不平衡生长 bù píng héng shēng zhǎng

disequilibrium: a loss or lack of equilibrium or stability, especially in relation to supply, demand, and prices. 不平衡 bù píng héng；不稳定 bù wěn dìng；失调 shī tiáo
growth: the process of increasing in physical size. 生长 shēng zhǎng

disequilibrium melting [ˌdisikwiˈlibriəm ˈmeltiŋ]

Definition: 不平衡熔融 bù píng héng róng róng

disequilibrium: a loss or lack of equilibrium or stability, especially in relation to supply, demand, and prices. 不平衡 bù píng héng；不稳定 bù wěn dìng；失调 shī tiáo
melting: the process whereby heat changes something from a solid to a liquid. 熔化 róng huà

dish structure [diʃ ˈstrʌktʃə]

Definition: 碟状构造 dié zhuàng gòu zào

dish: a shallow, typically flat-bottomed container for cooking or serving food. 碟 dié，盘 pán
structure: the arrangement of and relations between the parts or elements of something complex. 结构 jié gòu；构造 gòu zào

Example:
The border ribs can be formed by the edges of a fusion which immediately solidifies into a dish-shaped structure.
边缘肋条可立即由固化成碟型结构的熔融边缘形成。

dispersal pattern [daiˈspəːsəl ˈpætən]

Definition: 分散类型 fēn sàn lèi xíng，散布形式 sàn bù xíng shì

dispersal: the action or process of distributing or spreading things or people over a wide area. 分散 fēn sàn，分布 fēn bù

pattern: a regular and intelligible form or sequence discernible in certain actions or situations. 方式 fāng shì，模式 mó shì

dissolution [ˌdisəˈljuːʃən]

Definition: the action or process of dissolving or being dissolved. 溶解作用 róng jiě zuò yòng

Origin: late Middle English: from Latin dissolutio(n-), from the verb dissolvere.

Example:
The Ordovician carbonate sediments (rocks) have experienced, includes compaction, dolomitization, cementation, replacement, dissolution and dedolomitization.
奥陶系碳酸盐沉积物(岩)经历的成岩作用有压实、白云石化、胶结、交代、溶蚀和去白云石化作用。

Extended Terms:
dissolution porosity 溶蚀孔隙度
dissolution velocity 溶解速度
abnormal dissolution 异常溶出(解)
batch dissolution 分批(间歇)溶解
fumeless dissolution 无烟溶解

distally steepened ramp facies [disˈtəli ˈstiːpənd ræmp ˈfeiʃiːz]

Definition: 末端变陡缓坡相 mò duān biàn dǒu huǎn pō xiàng

distally: situated away from the centre of the body or from the point of attachment. 末梢的 mò shāo de，末端的 mò duān de

steepen: become or cause to become steeper. (使)变陡峭(shǐ) biàn dǒu qiào

ramp: a slope or inclined plane for joining two different levels, as at the entrance or between floors of a building. 斜面 xié miàn，斜坡 xié pō

facies: the character of a rock expressed by its formation, composition, and fossil content. 相 xiàng

distribution coefficient [ˌdistriˈbjuːʃən ˌkəuiˈfiʃənt]

Definition: 分配系数 fēn pèi xì shù

distribution: the action of sharing something out among a number of recipients. 分发 fēn fā；分

配 fēn pèi

coefficient: a multiplier or factor that measures some property. 率 lǜ; 系数 xì shù

Example:

The results of the experiment are that the absorption amount is small and the distribution coefficients are respectively 0.10, 0.15, 0.19 cm³/g according to the different ratio of sand and water (1.0, 0.5, 0.2).
实验结果表明：砂层对(阿特拉津的)吸附量小，不同固液比（1.0、0.5、0.2）时的分配系数分别为 0.10, 0.15, 0.19cm³/g。

Extended Terms:

apparent distribution coefficient 表观分配系数
effective distribution coefficient 有效分布系数
equal-energy distribution coefficient 等能分布系数
equilibrium distribution coefficient 平衡分布系数
interface distribution coefficient 界面分配系数

divariant assemblage ['divaiərənt ə'semblidʒ]

Definition: 双变组合 shuāng biàn zǔ hé

divariant: Di- is a prefix meaning twice; two; double; a *variant* is something a little different from others of the same type. 双变 shuāng biàn

assemblage: a form of art involving the assembly and arrangement of unrelated objects, parts, and materials in a kind of sculptured collage. 集合 jí hé

divariant equilibrium ['divaiərənt ˌiːkwi'libriəm]

Definition: 双变平衡 shuāng biàn píng héng

divariant: Di- is a prefix meaning twice; two; double; a *variant* is something a little different from others of the same type. 双变 shuāng biàn

equilibrium: a state in which opposing forces or influences are balanced. 平衡 píng héng; 均衡 jūn héng

divariant reaction ['divaiərənt ri'ækʃən]

Definition: 双变反应 shuāng biàn fǎn yìng

divariant: Di- is a prefix meaning twice; two; double; a *variant* is something a little different from others of the same type. 双变 shuāng biàn

reaction: a chemical process in which two or more substances act mutually on each other and are changed into different substances, or one substance changes into two or more other substances 反应 fǎn yìng

divergence [daiˈvəːdʒəns]

Definition: the scalar product of the operator and a given vector, which gives a measure of the quantity of flux emanating from any point of the vector field or the rate of loss of mass, heat, etc., from it. 发散 fā sàn, 离散 lí sàn

Origin: 1650-1660.

Example:
Slope efficiency of 0.56 W/A and strip red output power as high as 100 mW, vertical and parallel far field divergence angle of 31° and 9° respectively are obtained for 20μm ridge laser device without coating.
制得的未镀膜 20μm 脊型条形红色激光器的输出功率达到 100 mW, 斜率 0.56W/A, 垂直和平行远场发散角分别为 31°和 9°。

Extended Terms:
divergence angle 扩张角
divergence loss 发散损耗

divergent margin [daiˈvəːdʒənt ˈmaːdʒin]

Definition: 离散大陆边缘 lí sàn dà lù biān yuán

divergent: tending to be different or develop in different directions. 不同的 bù tóng de; 有分歧的 yǒu fēn qí de
margin: the edge or border of something. 边 biān, 边沿 biān yán

divergent plate boundary [daiˈvəːdʒənt pleit ˈbaundəri]

Definition: 离散板块边界 lí sàn bǎn kuài biān jiè

divergent: tending to be different or develop in different directions. 不同的 bù tóng de; 有分歧的 yǒu fēn qí de
plate: each of the several rigid pieces of the earth's lithosphere which together make up the earth's surface. 板块 bǎn kuài
boundary: a line which marks the limits of an area; a dividing line. 分界线 fēn jiè xiàn

Example:
The rifts therefore represent divergent plate boundaries where seafloor spreading is taking place.

这样，大洋中脊就成为板块分离边界，海底扩张就发生在这里。

dolerite [ˈdɒlərait]

Definition: a dark, medium-grained igneous rock, typically with ophitic texture, containing plagioclase, pyroxene, and olivine. It typically occurs in dykes and sills. 粗玄岩 cū xuán yán

Origin: mid-19th century: from French dolérite, from Greek doleros "deceptive".

Examples:

① The minerals are disseminated in the schist and hornfelsed schist near the dolerite contact.
这些矿物也浸染在靠近粗玄岩接触带的片岩和角岩片岩中。

② Young basic intrusives, which took place in late Mesozoic times, formed numerous minor dolerite dykes.
这些早期的侵入性活动，发生在中生代晚期，并形成了一些粗玄岩的沟壑。

Extended Terms:

alkali dolerite 碱性粒玄岩
quartz dolerite 石英粗玄岩
tholeiitic dolerite 拉斑玄武岩质粒玄岩

dolomite [ˈdɒləmait]

Definition: a white, reddish, or greenish mineral consisting of calcium magnesium carbonate. Source: sedimentary rocks. Use: building stone, cement, fertilizers. 白云岩 bái yún yán

Origin: late 18th century. From French, after Déodat de Dolomieu (1750-1801), French geologist.

Example:

As a result, through increasing sinter basicity and adopting low-moisture sintering, the adverse effect of dolomite can be eliminated.
结果认为通过提高烧结矿碱度，实行低水烧结，可以消除白云石粉对烧结生产的影响。

Extended Terms:

dolomite plaster 白云石灰浆
calcite dolomite 灰质白云岩
ferroan dolomite 含铁白云石
friable dolomite 易碎白云岩

dolostone [ˌdəʊləˈstəʊn]

Definition: rock consisting of dolomite. 白云岩 bái yún yán

Origin: mid-20th century: from dolomite + stone.

Example:
Crystalline dolostone is formed by seepage-reflux dolomitization and burial dolomitization. 晶粒白云岩有两种成因：一为回流渗透白云石化；二为深埋藏白云石化。

Extended Terms:
dolostone dolst 白云灰岩
calcitic dolostone 方解石白云岩
cyanobacteria dolostone 兰细菌白云岩
dolomite dolostone 白云岩
epigenetic dolostone 后成白云岩

double-chain silicate [ˈdʌbl-tʃeɪn ˈsɪlɪkɪt]

Definition: 双链硅酸盐 shuāng liàn guī suān yán

double: consisting of two equal, identical, or similar parts or things. 双的 shuāng de, 成双的 chéng shuāng de

chain: a sequence or series of connected elements. 一连串 yì lián chuàn, 一系列 yí xì liè

silicate: a salt in which the anion contains both silicon and oxygen, especially one of the anion SiO_4^{2-}. 硅酸盐 guī suān yán

double-diffusive convection [ˈdʌbl-dɪˈfjuːsɪv kənˈvekʃən]

Definition: 双扩散对流 shuāng kuò sàn duì liú

double: consisting of two equal, identical, or similar parts or things. 双的 shuāng de, 成双的 chéng shuāng de

diffusive: the spreading of something more widely. 扩散 kuò sàn, 散播 sàn bō

convection: the movement caused within a fluid by the tendency of hotter and therefore less dense material to rise, and colder, denser material to sink under the influence of gravity, which consequently results in transfer of heat. 对流 duì liú

Example:
The results showed that the density variation induced by double-diffusive convection is significant to the formation and evolution of A-segregates.

结果表明,糊状区中由于热溶质双扩散对流造成的密度变化是形成 A 偏析的根本原因。

double-diffusive system [ˈdʌbl diˌfjuːsiv ˈsistəm]

Definition: 双扩散系统 shuāng kuò sàn xì tǒng

double: consisting of two equal, identical, or similar parts or things. 双的 shuāng de, 成双的 chéng shuāng de

diffusive: the spreading of something more widely. 扩散 kuò sàn, 散播 sàn bō

system: a set of things working together as parts of a mechanism or an interconnecting network 系统 xì tǒng

ductile [ˈdʌktail]

Definition: able to be deformed without losing toughness; pliable, not brittle. 延性的 yán xìng de

Origin: Middle English (in the sense malleable): from Latin ductilis, from duct-"led", from the verb ducere

Example:

This linear, narrow, highly strained belt is a ductile shear zone. The rocks within this zone have been suffered intense deformation which is accomplished entirely by ductile flow.
这个线状的狭窄高应变带是一条韧性剪切带,带内的岩石曾遭受过强烈的变形,而这种变形则是由韧性流动所完成的。

Extended Terms:

ductile fracture 延性断裂
ductile iron 延性铁

ductile deformation [ˈdʌktail ˌdiːfɔːˈmeiʃən]

Definition: 韧性变形 rèn xìng biàn xíng, 塑性变形 sù xìng biàn xíng

ductile: able to be deformed without losing toughness; pliable, not brittle. 延性的 yán xìng de
deformation: the action or process of changing in shape or distorting, especially through the application of pressure. 变形 biàn xíng

Example:

Early ductile deformation is characterized by dislocation glide, twinning displacement and recrystallization, forming schist and mylonite.

早期韧性变形作用是以位错滑移、双晶作用和重结晶作用等变形机制为主，形成各种类型的片糜岩。

Extended Terms:

ductile-brittle deformation 延性-脆性形变
ductile deformation episode 塑性形变幕
ductile shear deformation 韧性剪切变形

ductility [dʌkˈtiləti]

Definition: a mechanical property used to describe the extent to which materials can be deformed plastically without fracture. 延性 yán xìng，韧性 rèn xìng

Origin: 1300-1350.

Example:

These testing beams behaved good ductility behavior after they reached ultimate strength.
试验梁达到极限承载力以后，表现出良好的延性。

Extended Terms:

cold ductility 可冷锻性
hot ductility 热延性，热塑性
impact ductility 冲击韧性
longitudinal ductility 纵向延性
notch ductility 缺口试样，断口收缩率，缺口延性

dunite [ˈdʌnait]

Definition: a green to brownish coarse-grained igneous rock consisting largely of olivine. 纯橄榄岩 chún gǎn lǎn yán

Origin: mid-19th century：from the name of Dun Mountain, New Zealand, + -ite.

Example:

The dunite, on the other hand, is more than 95 percent olivine—the mineral left behind as melt rises through the mantle.
另一方面，纯橄榄岩则有95%以上都是橄榄石，是熔体上行通过地幔后所遗留下来的矿物质。

Extended Term:

ilmenite dunite 钛铁橄榄岩

dyke [daik]

Definition: an intrusion of igneous rock cutting across existing strata. 岩墙 yán qiáng, 岩脉 yán mài

Origin: Middle English (denoting a trench or ditch): from Old Norse dík, related to ditch. It has been influenced by Middle Low German dīk "dam" and Middle Dutch dijc "ditch, dam".

Example:
Summing up the experiences in engineering-geological investigation of the Yellow River embankment, the paper studies and demonstrates systematically the investigation methods for solving three major problems of historic dyke breach.
通过总结黄河堤防工程地质勘察的经验，本文就黄河大堤历史口门工程地质勘察方法问题进行了系统的研究，针对历史口门工程地质勘察所需解决的三大问题，提出了有效的勘察方法。

Extended Terms:
dyke rock 脉岩
basalt dyke 玄武岩岩墙
injection dyke 贯入岩墙

dynamic metamorphism [daiˈnæmik ˌmetəˈmɔːfizəm]

Definition: 动力变质作用 dòng lì biàn zhì zuò yòng

dynamic: characterized by constant change, activity, or progress. 动态的 dòng tài de; 不断变化的 bú duàn biàn huà de

metamorphism: alteration of the composition or structure of a rock by heat, pressure, or other natural agency. 变质作用 biàn zhì zuò yòng

Example:
After the studying of mineralogy in detail, it is clear that the mineralizing process was closely related to the tectonic dynamic metamorphism.
矿物学的研究表明，整个成矿过程与构造动力变质密切相关。

Extended Term:
regional dynamic metamorphism 区域动力变质作用

earth [əːθ]

Definition: the surface of the world as distinct from the sky or the sea. 地球 dì qiú, 陆地 lù dì

earthquake

Origin: Old English eorthe, of Germanic origin; related to Dutch aarde and German Erde.

Example:

The cultivated land in the world is only 9% of the land on the earth at present, which islocated mostly in the scope of 30°~60° north and south latitude.
目前世界上已耕作的土地，只占地球全部土地的9%，大多在南北纬30°~60°范围内。

Extended Terms:

earth-based 地面的
earth-bath 泥浴
earth-deposits 土堆
earthdin 地震
earthfall 塌方

earthquake [ˈəːθkweik]

Definition: a sudden violent shaking of the ground, typically causing great destruction, as a result of movements within the earth's crust or volcanic action. 地震 dì zhèn

Origin: 1300-1350; Middle English erthequake.

Example:

The basic idea of earthquake early warning and the state of the art of earthquake early warning system and earthquake emergency control system are summarized.
对地震预警的基本思想以及地震预警和应急控制系统的国内外建设现状进行综述。

Extended Terms:

dislocation earthquake 断层地震
folding earthquake 褶皱地震
local earthquake 局部地震
palintectic/plutonic earthquake 深源地震
shallow-focus earthquake 浅源地震

East African Rift [iːst ˈæfrikən rift]

Definition: The East African Rift is an active continental rift zone in eastern Africa that appears to be a developing divergent tectonic plate boundary. 东非裂谷带 dōng fēi liè gǔ dài

Example:

While ocean ridges, such as the East Pacific Ridge, the Atlantic Ridge and the East African Rift, are under tension.
而东太平洋洋脊、大西洋洋脊及东非裂谷处应力状态均表现为拉张。

Extended Terms:

East African Rift Valley 东非裂谷
East African Rift System 东非裂谷系
East African Rift Zone 东非裂谷带

East Pacific Rise [iːst pəˈsifik rais]

Definition: The East Pacific Rise is a mid-oceanic ridge, a divergent tectonic plate boundary located along the floor of the Pacific Ocean. It separates the Pacific Plate to the west from (north to south) the North American Plate, the Rivera Plate, the Cocos Plate, the Nazca Plate, and the Antarctic Plate. 东太平洋海隆 dōng tài píng yáng hǎi lóng

Example:

The East Pacific Rise, west of South America, has a high separation rate but no axial rift.
南美西部东太平洋海隆具有高的分离速率，但是没有轴部裂谷。

Extended Term:

East Pacific Rise geothermal belt 东太平洋中脊地热带

---

eclogite [ˈeklədʒait]

Definition: a rock consisting of a granular aggregate of green pyroxene and red garnet, often containing kyanite, silvery mica, quartz, and pyrite. 榴辉岩 liú huī yán

Origin: 1815-1825, from Greek eklogē a selection (see eclogue) + -ite.

Example:

Dr. Xu Zhiqin, a Chinese geologist, first discovered coesite in the eclogite in the west of Tianzhu Mountain in 1987.
1987 年，我国地质学家许志琴首先在天柱山西边榴辉岩中发现了柯石英。

Extended Terms:

eclogite sphere 榴辉岩圈
amphibole-eclogite 闪榴辉岩
basalt eclogite transformation 玄武岩榴辉岩转化
metamorphic eclogite 变质榴辉岩
omphacite-eclogite 绿辉榴辉岩

---

eclogite facies [ˈeklədʒait ˈfeiʃiːz]

Definition: 榴辉岩相 liú huī yán xiàng

eclogite: a rock consisting of a granular aggregate of green pyroxene and red garnet, often containing kyanite, silvery mica, quartz, and pyrite. 榴辉岩 liú huī yán

facies: the character of a rock expressed by its formation, composition, and fossil content. 相 xiàng

effective viscosity [iˈfektiv viˈskɔsəti]

Definition: 有效黏度 yǒu xiào nián dù

effective: successful in producing a desired or intended result. 有效的 yǒu xiào de

viscosity: a quantity expressing the magnitude of such friction, as measured by the force per unit area resisting a flow in which parallel layers unit distance apart have unit speed relative to one another. 黏度 nián dù；黏滞度 nián zhì dù

Example:

The effective viscosity decreases with the increase of shear rate and temperature, while increases with pressure and foam quality.

其有效黏度随剪切速率、温度的增高而减小，随压力、泡沫质量的增大而增大。

Extended Terms:

effective viscosity coefficient 等效黏性系数
effective viscosity tensor 有效黏滞率张量
average effective viscosity 平均有效黏度
mean effective viscosity 平均有效黏度

effusive eruption [iˈfjuːsiv iˈrʌpʃən]

Definition: Effusive eruptions are characterised by the outpouring of lava onto the ground. 溢流喷发 yì liú pēn fā

effusive: poured out when molten and later solidified. 喷发的 pēn fā de
eruption: an act or instance or erupting. 爆发 bào fā；喷发 pēn fā

ejecta [iˈdʒektə]

Definition: material that is forced or thrown out, especially as a result of volcanic eruption, meteoritic impact, or stellar explosion. 喷出物 pēn chū wù

Origin: late 19th century: from Latin, things "thrown out", neuter plural of ejectus "thrown out", from eicere.

Example:

Highly reflective material with a steeper slope (seen as bright orange in this enhanced-color

view) is exposed in the ejecta near the crater's rim.
陡峭斜坡形的高反光物质(在这张增强色视图中的亮橙色)暴露在陨石坑边缘的喷出物中。

Extended Terms:
accessory ejecta 早成同源喷出物
accidental ejecta 异源喷出物
capillary ejecta 毛管喷出物
fragmentary ejecta 喷屑
juvenile ejecta 初生喷出物

electron microprobe [iˈlektrɔn ˈmaikrəuprəub]

Definition: instrument that analyzes the chemistry of very small spots by bombarding the sample with electrons and measuring the X-rays produced. 电子探针 diàn zǐ tàn zhēn

Example:
The magnetite compositions of the Changjiang River sediments were analyzed by electron microprobe.
运用电子探针分析了长江干流和主要支流河漫滩沉积物中磁铁矿的元素组成。

Extended Terms:
electron microprobe analyser 电子显微探针分析仪
electron microprobe analysis 微区电子探针分析
electron microprobe analyzer 电子探针分析器
electron microprobe X-ray analyzer 电子显微探针 X 射线分析仪
electron beam microprobe 电子束显微探头
scanning electron microprobe 扫描电子探针

eluvial facies [iˈljuːviəl ˈfeiʃiːz]

Definition: 残积相 cán jī xiàng

eluvial: of or related to a product of the erosion of rocks that has remained in its place of origin. 残积层的 cán jī céng de
facies: the character of a rock expressed by its formation, composition, and fossil content 相 xiàng

embayment [imˈbeimənt]

Definition: a recess in a coastline forming a bay. 海湾 hǎi wān

Origin: 1805-1815; embay + -ment.

enderbite

Example:
The primary coastline of the port was bedrock embayment and composed of Mesozoic granite.
三亚港原始岸线为基岩港湾海岸，由中生代花岗岩构成。

Extended Terms:
fault embayment 断层湾
landlocked embayment 闭塞潟湖

enderbite [ˈendəˌbait]

Definition: In geology, enderbite is primarily an igneous rock of the charnockite series, consisting essentially of quartz, antiperthite (or perthite), orthopyroxene (usually hypersthene) and magnetite, and is equivalent to an orthopyroxene bearing tonalite. It is named for its occurrence in Enderby Land, Antarctica. （地质）紫苏斜长花岗岩 zǐ sū xié cháng huā gǎng yán

endogenetic sedimentary rock [ˌendəʊdʒiˈnetik ˌsediˈmentəri rɔk]

Definition: 内源沉积岩 nèi yuán chén jī yán

endogenetic: of rocks formed or occurring beneath the surface of the earth. 内成的 nèi chéng de
sedimentary: (of rock) that has formed from sediment deposited by water or air. （岩石）沉积形成的 chén jī xíng chéng de
rock: the solid mineral material forming part of the surface of the earth and other similar planets, exposed on the surface or underlying the soil. 岩石 yán shí; 岩 yán

energy index [ˈenədʒi ˈindeks]

Definition: 能量指数 néng liàng zhǐ shù

energy: the property of matter and radiation which is manifest as a capacity to perform work. 能量 néng liàng
index: a figure in a system or scale representing the average value of specified prices, shares, or other items as compared with some reference figure. 指数 zhǐ shù

Example:
These microfacies types indicate varying energy index and water depth.
这些微相类型可指示不同的能量指数及水深。

Extended Terms:
energy elasticity index 能源弹性系数

energy use index 能源使用指数
tensile energy absorption index 抗张能量吸收指数

enrichment [inˈritʃmənt]

Definition: make (someone) wealthy or wealthier. 富集作用 fù jí zuò yòng

Origin: late Middle English (in the sense "make wealthy"): from Old French enrichir, from en-"in" + riche"rich".

Example:
During Tertiary period, the secondary enrichment happened to the gold in weathering crust of planation surface.
在第三纪，平面上的风化壳中的金经历了重要的次生富集作用。

Extended Terms:
cathode layer enrichment 阴极层富集
downward enrichment 次生富集
excessive enrichment 过度浓缩；过高富集
filtration enrichment 过滤浓缩法
isotopic enrichment 同位素浓缩
mineral enrichment 矿物富集作用

enterolithic structure [ˌenterəuˈliθik ˈstrʌktʃə]

Definition: 盘肠构造 pán cháng gòu zào

enterolithic: In biology and medicine, the prefix *entero-* refers to the intestine; *lithic* is of, relating to, or composed of stone. 内变形的 nèi biàn xíng de
structure: the arrangement of and relations between the parts or elements of something complex. 结构 jié gòu；构造 gòu zào

eogenetic stage [ˌiəudʒiˈnetik steidʒ]

Definition: 始成岩阶段 shǐ chéng yán jiē duàn

eogenetic: formed earlier than the surrounding or underlying rock formation. 早期成岩的 zǎo qī chéng yán de
stage: a point, period, or step in a process or development. 阶段 jiē duàn

epidote [ˈepidəut]

Definition: a lustrous yellow-green crystalline mineral, common in metamorphic rocks. It consists of a basic, hydrated silicate of calcium, aluminium, and iron. 绿帘石 lǜ lián shí

Origin: early 19th century: from French épidote, from Greek epididonai "give additionally".

Example:
In addition, only pentlandite, coffinite and the epidote in the alteration zone of volcanogenic hydrothermal gold deposit are typomorphic minerals.
同时指出只有镍黄铁矿、水硅铀矿和火山热液型金矿床蚀变带中的绿帘石为标型矿物。

Extended Terms:
epidote-amphibolite facies 绿帘石角闪岩相
epidote-mica schist 绿帘石云母片岩
epidote-tremolite schist 绿帘石-透闪石片岩
albite-epidote-amphibolite facies 钠长绿帘闪岩相
manganese epidote 红帘石

epidote-amphibolite facies [ˈepidəut-æmˈfibəlait ˈfeiʃiːz]

Definition: one of the major divisions of the mineral facies classification of metamorphic rocks, encompassing rocks that formed under moderate temperature (500°F-750°F, or 250℃-400℃) and pressure conditions. This facies grades into the greenschist facies under less intense metamorphic conditions and into the amphibolite facies with greater temperature and pressure. Typical minerals include biotite, almandine garnet, plagioclase, epidote, and amphibole. Chlorite, muscovite, staurolite, and chloritoid may also occur. 绿帘角闪岩相 lǜ lián jiǎo shǎn yán xiāng

epigenesis [ˌepiˈdʒenisis]

Definition: ore deposition subsequent to the original formation of the enclosing country rock. 后生作用 hòu shēng zuò yòng

Origin: mid-17th century: from epi-"in addition" + genesis.

Example:
Diagenesis and epigenesis are major factors to control reservoir development.
成岩作用及成岩后生作用是控制本区储集层发育的主要因素。

> Extended Term:

epigenesist theory 后生说

epigenetic dolomite [ˌepidʒi'netik 'dɔləmait]

> Definition: 后生白云岩 hòu shēng bái yún yán

epigenetic: formed later than the surrounding or underlying rock formation. 后生的 hòu shēng de; 后成的 hòu chéng de

dolomite: a white, reddish, or greenish mineral consisting of calcium magnesium carbonate. 白云岩 bái yún yán

epitaxial overgrowth [ˌepi'tæksiəl 'əuvəgrəuθ]

> Definition: 异轴增生 yì zhóu zēng shēng

epitaxial: the natural or artificial growth of crystals on a crystalline substrate that determines their orientation. 取向附生 qǔ xiàng fù shēng, 外延附生 wài yán fù shēng

overgrowth: excessive growth. 生长过度 shēng zhǎng guò dù

equigranular texture [iːkwi'grænjulə 'tekstʃə]

> Definition: 等粒结构 děng lì jié gòu

equigranular: pertaining to the texture of rocks whose essential minerals are all of the same order of size. 等粒状的 děng lì zhuàng de

texture: the arrangement of the particles or constituent parts of any material, as wood, metal, etc., as it affects the appearance or feel of the surface; structure, composition, grain, etc. 结构 jié gòu

equilibrium [ˌiːkwi'libriəm]

> Definition: a state in which opposing forces or influences are balanced. 平衡 píng héng; 均衡 jūn héng

> Origin: early 17th century (in the sense "well-balanced state of mind"): from Latin aequilibrium, from aequi-"equal" + libra"balance".

> Example:

In the range AB, the equilibrium is neither stable nor unstable, that is why it is called neutral equilibrium.

171

在 AB 范围内，这个平衡既不是稳定平衡也不是不稳定平衡，这就是为什么叫做随遇平衡的原因。

Extended Terms:

acetic-alkali equilibrium 酸碱平衡
amplitude equilibrium 振幅平衡
complete equilibrium 完全平衡；不可逆平衡
comprehensive equilibrium 综合平衡
conductive equilibrium 传导平衡
constant equilibrium 恒常平衡

equilibrium crystallization [ˌiːkwiˈlibriəm ˌkristəlaiˈzeiʃən]

Definition: 平衡结晶 píng héng jié jīng

equilibrium：a state in which opposing forces or influences are balanced 平衡 píng héng；均衡 jūn héng

crystallization：form or cause to form crystals. 结晶 jié jīng；晶化 jīng huà

Example:

The authors have given the new quantitative models of different magmatic processes, including batch melting, fractional melting, equilibrium crystallization, fraction alcrystallization and magma mixing, and identification methods of these processes.
作者给出了不同岩浆作用过程的几种新的定量模型（批武熔融模型、分馏熔融模型、平衡结晶模型、分馏结晶模型及岩浆混合模型）和鉴别方法。

equilibrium melting [ˌiːkwiˈlibriəm ˈmeltiŋ]

Definition: 平衡熔融 píng héng róng róng

equilibrium：a state in which opposing forces or influences are balanced. 平衡 píng héng；均衡 jūn héng

melting：the process whereby heat changes something from a solid to a liquid. 熔化 róng huà

Example:

A single glass-transition temperature was observed in the DSC scanning trace of the blend with a weight ratio of 10/90. Besides, the equilibrium melting point of PLLA decreased with the increasing PEG.
由 DSC 扫描可以观察到在重量百分比为 10/90 时有一单玻璃转移温度，而且，随着 PEG 的增加，PLLA 的熔点会下降。

Extended Terms:

equilibrium melting point 平衡熔点

equilibrium melting point theory 平衡熔点理论

erosion rate [iˈrəuʒən reit]

Definition: 侵蚀速率 qīn shí sù lǜ

erosion: the process of eroding or being eroded by wind, water, or other natural agents. 侵蚀 qīn shí

rate: the speed with which something moves, happens, or changes. 速率 sù lǜ

Example:

The results also show that the lower velocity of particle, the lower erosion rate, even if the concentration of particles is very high.
颗粒冲刷速度较低时,即使颗粒浓度很高,侵蚀率也并不是很大。

Extended Terms:

annual erosion rate 年土壤流失率
rural erosion rate 农田水土流失率
rate of coastal erosion 海岸侵蚀速率
delivery rate of erosion 冲刷输沙率

eruption [iˈrʌpʃən]

Definition: an act or instance or erupting. 喷发 pēn fā

Origin: late Middle English: from Old French, or from Latin eruptio(n-), from the verb erumpere.

Example:

The island was convulsed by the eruption.
该岛因火山爆发而激烈震动。

Extended Terms:

area eruption 区域喷发
fissure eruption 裂缝喷发
paroxysmal eruption 阵发性喷发
plateau eruptions 高原喷发
Plinian eruption 普林尼式火山喷发
solar eruption 太阳爆发

eruption rate [iˈrʌpʃən reit]

Definition: 喷发速率 pēn fā sù lǜ

eruption: an act or instance or erupting. 喷发 pēn fā
rate: the speed with which something moves, happens, or changes. 速率 sù lǜ

eruptive rock [iˈrʌptiv rɔk]

Definition: rock formed from a volcanic eruption. Igneous rock reaches the earth's surface in a molten condition. 喷出岩 pēn chū yán

Example:
In this region, the tectonic faults with NS direction develop, neutral-acid intrusive rocks and alkaline basalt eruptive rocks form two nearly NS direction magmatic rock belts.
在此地区内近南北向隐伏断裂发育，中酸性侵入岩、碱性玄武喷发岩构成近南北向岩浆构造带。

essential mineral [iˈsenʃəl ˈminərəl]

Definition: a primary mineral whose presence in an igneous rock is essential to defining the root name of that rock. For example, plagioclase and the pyroxene augite are essential to defining the root name "gabbro" and hence are essential minerals. 主要矿物 zhǔ yào kuàng wù

Example:
The enrichment of Fe in the ash is the reason of fouling and slagging during combustion of CWS, but Na and Fe are the essential mineral elements for the ash deposition of CS.
Fe 的富集是造成水煤浆沾污结渣的根本原因，Na 和 Fe 是引起黑液浆沾污结渣的主要矿物元素。

Extended Term:
essential mineral element 必需矿物质元素

essexite [ˈesikˌsait]

Definition: also called nepheline monzogabbro, a dark gray or black holocrystalline plutonic igneous rock. 碱性辉长岩 jiǎn xìng huī cháng yán

Extended Terms:
essexite aplite 碱辉细晶岩
essexite basalt 碱辉玄武岩
essexite gabbro 碱辉辉长岩
essexite porphyrite 碱辉玢岩

Ethiopian rift [ˌiːθiˈəupiən rift]

Definition: 埃塞俄比亚裂谷 āi sài é bǐ yà liè gǔ

Ethiopian: of or relating to Ethiopia or its people.（与）埃塞俄比亚（有关）的（yǔ）āi sài é bǐ yà（yǒu guān）de

rift: a major fault separating blocks of the earth's surface; a rift valley. 裂谷 liè gǔ

Example:

As the sides of the Ethiopian rift move apart, the gap between them is being plugged with molten rock, which then cools to form new land.
随着埃塞俄比亚裂谷的两岸不断远离，其间的裂缝不断地被岩浆填满。这些岩浆冷却后就形成了新的陆地。

eucrite [ˈjuːkrait]

Definition: a highly basic form of gabbro containing anorthite or bytownite with augite. 钙长辉长岩 gài cháng huī cháng yán

Origin: mid-19th century: from Greek eukritos "easily discerned", from eu- "well" + kritos "separated".

Example:

The two shergottites—GRV 99027 and GRV 99018 show they belong to new type meteorites with significance for researches, and are primary ascertained that GRV 99027 may belong to martian Iherzolite, and GRV 99028 may belong to eucrite.
两块无球粒陨石 GRV 99027 和 GRV 99018 具有较高的研究价值，初步确定 GRV 99027 为二辉橄榄质火星陨石，GRV 99028 为钙长辉长无球粒陨石。

Extended Term:

biotite eucrite 黑云倍长辉长岩

euhedral [juːˈhiːdrəl]

Definition: bounded by faces corresponding to its regular crystal form, not constrained by adjacent minerals. 自形的 zì xíng de

Origin: 1905-1910: eu- + -hedral.

Example:

Deep buried dolostones in study areas mainly consist of fine or medium dolomites, which crystals are euhedral or subhedral.
研究区的深埋藏的白云岩主要由细晶或中晶白云石组成，白云石呈自形或半自形晶。

Extended Terms:
euhedral crystal 自形晶
euhedral-granular 自形粒状
euhedral-granular texture 自形晶粒状结构

eutaxitic [juːˈtæksaitik]

Definition: In igneous petrology, eutaxitic texture describes the layered or banded texture in some extrusive rock bodies. It is often caused by the compaction and flattening of glass shards and pumice fragments around undeformed crystals. 条纹斑状 tiáo wén bān zhuàng

Extended Terms:
eutaxitic structure 条纹斑状构造
eutaxitic tuff 造凝灰岩

eutectic [juːˈtektik]

Definition: relating to or denoting a mixture of substances (in fixed proportions) that melts and solidifies at a single temperature that is lower than the melting points of the separate constituents or of any other mixture of them. 低共熔的 dī gòng róng de; 易熔的 yì róng de

Origin: late 19th century: from Greek eutēktos "easily melting".

Example:
The dimension stability of eutectic and hyper eutectic Al Si alloy pistons in aging at 200℃ over a long period is investigated.
研究了两种共晶和两种过共晶铝硅合金活塞在200℃的长期时效过程中的尺寸稳定性。

Extended Terms:
eutectic mixture 低共熔(混合)物
eutectic temperature 低共熔温度
ternary eutectic 三元低共熔混合物
ternary eutectic chloride 三元共晶氯化物

eutectic point [juːˈtektik pɔint]

Definition: the temperature at which a particular eutectic mixture freezes or melts. 共晶点 gòng jīng diǎn
eutectic: relating to or denoting a mixture of substances (in fixed proportions) that melts and solidifies at a single temperature that is lower than the melting points of the separate constituents

or of any other mixture of them. 低共熔的 dī gòng róng de；易熔的 yì róng de
point：a particular moment in time or stage in a process. 时刻 shí kè，阶段 jiē duàn

> Examples:

①It would be difficult for a manufacturer to evaluate partial or abnormal cycles without knowing the eutectic point and the cycle parameters needed to facilitate primary drying.
不知道产品的共晶点和初步干燥所需的循环参数，生产者很难评估部分循环或非正常循环。
②Typically, the product is frozen at a temperature well below the eutectic point.
特别是，产品的冷冻温度远低于它的共晶点。

> Extended Terms:

eutectic formation point 共熔成形温度
eutectic melting point 共溶点
deep eutectic point 深共晶点
lowest eutectic point 最低共熔点
ternary eutectic point 三元共熔点
three component eutectic point 三元共晶体

eutectoid texture [juːˈtektɔid ˈtekstʃə]

> Definition: 共析结构 gòng xī jié gòu

eutectoid：relating to or denoting an alloy which has a minimum transformation temperature between a solid solution and a simple mixture of metals. 共析(体)的 gòng xī (tǐ) de
texture：the arrangement of the particles or constituent parts of any material, as wood, metal, etc., as it affects the appearance or feel of the surface; structure, composition, grain, etc. 结构 jié gòu

evaporite [iˈvæpərait]

> Definition: a natural salt or mineral deposit left after the evaporation of a body of water. 蒸发盐 zhēng fā yán，蒸发岩 zhēng fā yán

> Origin: 1920s：alteration of evaporate.

> Example:

The widespread progradationally deposited cycles are often capped by oolitic shoals, tidal flat or Sabkha evaporite strata.
广布的加积沉积旋回经常被鲕滩、潮坪或萨勃哈蒸发岩层所覆盖。

> Extended Terms:

evaporite basin 蒸发盐盆地
evaporite deposit 蒸发沉积矿床

evaporite deposition 蒸发盐沉积
evaporite facies 蒸发岩相
anhydrite evaporite 硬石膏蒸发盐
continental evaporite 陆相蒸发岩
deep-water evaporate 深水蒸发岩

evaporite solution breccia [iˈvæpərait səˈljuːʃən ˈbretʃiə]

Definition: 盐溶角砾岩 yán róng jiǎo lì yán

evaporate: a natural salt or mineral deposit left after the evaporation of a body of water. 蒸发盐 zhēng fā yán, 蒸发岩 zhēng fā yán

solution: the process or state of being dissolved in a solvent. 溶解状态 róng jiě zhuàng tài

breccia: a coarse-grained sedimentary rock made of sharp fragments of rock and stone cemented together by finer material. Breccia is produced by volcanic activity or erosion, including frost shattering. 角砾岩 jiǎo lì yán

exchange reaction [ikˈstʃeindʒ riˈækʃən]

Definition: 交换反应 jiāo huàn fǎn yìng

exchange: an act of giving one thing and receiving another (especially of the same type or value) in return. 交换 jiāo huàn; 互换 hù huàn

reaction: a chemical process in which two or more substances act mutually on each other and are changed into different substances, or one substance changes into two or more other substances. 反应 fǎn yìng

Example:
The application of microwave technology in halogen-exchange reaction has many merits, such as short reaction time and high yield.
微波加热在卤素交换氟化反应中的应用结果显示其具有反应时间短和产率高的优点。

Extended Terms:
exchange melt reaction 变熔反应
carbonyl-exchange polymerization reaction 羰基交换聚合反应

exothermic process [ˌeksəuˈθəːmik ˈprəuses]

Definition: 发热过程 fā rè guò chéng

exothermic: accompanied by the release of heat. 放热的 fàng rè de

process: a series of actions or steps taken in order to achieve a particular end. 步骤 bù zhòu，程序 chéng xù；方法 fāng fǎ

Example:

The third stage was a sharp exothermic process, which resulted from the decomposition of the polymer.

第三阶段表现为一尖锐的放热峰，其为聚合物的二次分解所致。

Extended Term:

exothermic ergic process 放热过程

---

exothermic reaction [ˌeksəʊˈθɜːmɪk rɪˈækʃən]

Definition: a chemical reaction accompanied by the evolution of heat. 放热反应 fàng rè fǎn yìng

exothermic: accompanied by the release of heat. 放热的 fàng rè de

reaction: a chemical process in which two or more substances act mutually on each other and are changed into different substances, or one substance changes into two or more other substances. 反应 fǎn yìng

Example:

The result reveals that, the curcumin binding to the unsaturated lipid membrane is an exothermic reaction, and without any external force, it reacts spontaneously.

实验结果显示姜黄素吸附不饱和链脂膜为放热反应，且反应为自然发生，不需任何外力。

Extended Terms:

exothermic chemical reaction 放热化学反应
exothermic nuclear reaction 放热核反应

---

expandable clay [ɪkˈspændəbl kleɪ]

Definition: 膨胀黏土 péng zhàng nián tǔ

expandable: able to expand or be expanded. 可膨胀的 kě péng zhàng de

clay: a stiff, sticky fine-grained earth, typically yellow, red, or bluish-grey in colour and often forming an impermeable layer in the soil. It can be moulded when wet, and is dried and baked to make bricks, pottery, and ceramics. 黏土 nián tǔ，泥土 ní tǔ，陶土 táo tǔ

Example:

Bentonite is an expandable three-layer clay mineral.

皂土是一种可膨胀的三层黏土矿物。

Extended Term:

water-expandable clay 水膨胀黏土

explosive eruption [ikˈspləusiv iˈrʌpʃən]

Definition: a volcanic term to describe a violent, explosive type of eruption. 爆发式喷发 bào fā shì pēn fā

Example:

A volcanic rock formed by the welding together of tuff material from an explosive volcanic eruption.
火山岩在火山喷发后，由凝灰物质聚集形成。

exsolution [ˌeksəˈluːʃən]

Definition: the process in which a solution of molten rocks separate into its constituents upon cooling. 出溶 chū róng

Origin: mid-20th century：(originally as exsolution) from ex- + solution.

Example:

Both clinopyroxene and orthopyroxene show exsolution texture. The exsolution rods generally are parallel each other no matter they are straight or curved.
单斜辉石和斜方辉石均呈现出溶结构。不管是平直还是弯曲，都是平行的。

Extended Terms:

exsolution lamellae 出溶纹层
exsolution structure 固溶体分解结构

exsolution lamellae [ˌeksəˈluːʃən ləˈmeliː]

Definition: 出溶层 chū róng céng

exsolution：the process in which a solution of molten rocks separate into its constituents upon cooling. 出溶 chū róng

lamellae：a thin layer, membrane, scale, or platelike tissue or part, esp. in bone tissue. 薄板 báo bǎn，薄片 báo piàn

Example:

All metamorphic rocks exhibit an equilibrium texture, but exsolution lamellae of orthopyroxene (pigeonite) occur in all clinopyroxenes in mafic granulites.

这些岩石一般呈现出平衡的矿物共生结构,但在镁铁质麻粒岩的单斜辉石中普遍存在斜方辉石(变辉石)出溶片晶。

exsolution texture [ˌeksəˈluːʃən ˈtekstʃə]

Definition: 出溶结构 chū róng jié gòu

exsolution: the process in which a solution of molten rocks separate into its constituents upon cooling. 出溶 chū róng

texture: the arrangement of the particles or constituent parts of any material, as wood, metal, etc., as it affects the appearance or feel of the surface; structure, composition, grain, etc. 结构 jié gòu

Example:

The exsolution texture of sphalerite-chalcopyrite solid solution and the underlying principle and technical method for measuring its decomposition speed are described.
介绍了闪锌矿-黄铜矿固溶体出溶结构的特点,叙述了测定固溶体分解速度的原理、依据和方法。

extensional tectonics [ikˈstenʃənəl tekˈtɔniks]

Definition: Extensional tectonics is concerned with the structures formed, and the tectonic processes associated with, the stretching of the crust or lithosphere. 伸展构造 shēn zhǎn gòu zào

Example:

The upsurge of extensional tectonics is based on the establishment of theory of recent tectonic geology.
伸展构造理论的兴起是建立在近代构造地质学理论基础之上。

extrusive rock [ekˈstruːsiv rɔk]

Definition: Extrusive refers to the mode of igneous volcanic rock formation in which hot magma from inside the Earth flows out (extrudes) onto the surface as lava or explodes violently into the atmosphere to fall back as pyroclastics or tuff. 喷出岩 pēn chū yán

Example:

Both intrusive and extrusive igneous rocks can be further classified according to the differences of chemical and mineral compositions.

根据岩石中的化学成分和矿物成分差异，可以对侵入岩和喷出岩作进一步的详细划分。

Extended Term:

extrusive igneous rocks 喷出岩

fabric [ˈfæbrik]

Definition: In geology, a rock's fabric describes the spatial and geometric configuration of all the elements that make it up. （地质）结构（dì zhì）jié gòu

Origin: late 15th century: from French fabrique, from Latin fabrica "something skilfully produced", from faber "worker in metal, stone, etc." The word originally denoted a building, later a machine or appliance, the general sense being "something made", hence mid-18th century, originally denoting any manufactured material.

Example:

The Na-rich rocks, occurred with the albite breccia bodies or as their angular coarse fragments, are of the albite formed by hydrothermal permeation and metasomatism, in which theamount of hairlike ratile and its arrangement directions clearly show the relict fabric S_0、S_1 and S_2 of protolith.

与钠长角砾岩体相伴或作为其角砾的富钠岩石，系热液渗透交代成因的钠长岩，其中毛发状金红石的多寡和不同方向的排列，较清晰地显示了原岩的残留组构 S_0、S_1 和 S_2。

Extended Term:

fabric tensor 构造张量

fabric analysis [ˈfæbrik əˈnæləsis]

Definition: analysis of the elements that make up the fabric of a rock to determine the response of that rock to stress. In a rock the three-dimensional pattern comprising the distribution, shape, size, and size distribution of crystals or grains constitutes the fabric. 岩组分析 yán zǔ fēn xī

Example:

Rock fabric analysis is very important in the research on the anisotropy of many physical properties in rocks.

岩石组构分析对研究岩石物理性质的各向异性具有十分重要的意义。

Extended Term:

fabric analysis of rock 岩石组构解析

facies ['feiʃiːz]

Definition: the character of a rock expressed by its formation, composition, and fossil content. 相 xiàng

Origin: early 17th century (denoting the face): from Latin, form, appearance, face.

Example:

The accumulation of oil is governed by lateral changes from permeable to impermeable facies. 石油的储集是受从渗透到不渗透的岩相横向变化控制的。

Extended Terms:

facies change 相变
facies sequence 相序
alimentation facies 补给相
continental facies 陆相
eluvial facies 残积相
sedimentary facies 沉积相

facies change ['feiʃiːz tʃeindʒ]

Definition: It is caused by, or reflects, a change in the depositional environment. 相变 xiàng biàn

Example:

Degenerative facies change: under the action of sea floor water (poor in Mn, Ni, Cu but rich in Fe), Mn, Ni and Cu in the primary $\delta\text{-}MnO_2$ were partly carried out by water, followed by precipitation of Fe, thereby forming feruginous $\delta\text{-}MnO_2$ or amorphous ferro-manganese hydrates.

退化相变：具层纹构造的原生 $\delta\text{-}MnO_2$ 在底层水（可能贫 Mn、Ni、Cu 而富 Fe）的作用下，其中一部分 Mn、Ni、Cu 等被水带走而沉淀下 Fe，形成新的富 Fe，$\delta\text{-}MnO_2$ 或非晶质水合铁锰氧化物。

Extended Terms:

facies change belt 相变带
facies change surface 相变面
sedimentary facies change 沉积环境相变

facies of igneous rock ['feiʃiːz ɔv 'ignəs rɔk]

Definition: 火成岩相 huǒ chéng yán xiàng

facies: the character of a rock expressed by its formation, composition, and fossil content. 相 xiàng

igneous: formed by solidification from a molten state. 火成的 huǒ chéng de

rock: the solid mineral material forming part of the surface of the earth and other similar planets, exposed on the surface or underlying the soil. 岩石 yán shí; 岩 yán

facies sequence [ˈfeiʃiːz ˈsiːkwəns]

Definition: 相序 xiàng xù

facies: the character of a rock expressed by its formation, composition, and fossil content. 相 xiàng

sequence: the order that events, actions, etc., happen in or should happen in 顺序 shùn xù; 次序 cì xù

Example:

With the high resolution sequence stratigraphy and the correlation analysis of cores, logs and seismic data, different orders of sequence boundaries and facies sequence are identified and the isochronous stratigraphic framework and the depositional evolutional model of Dongying Formation are built and used to predict the distribution of reservoirs.
为了查明南堡凹陷油气分布规律及勘探前景，利用高分辨率层序地层学方法，结合岩心、测井和地震资料分析，识别南堡凹陷东营组各级层序界面和相序，建立等时地层层序格架和演化模式，综合预测层序地层格架下的沉积相和油气分布。

Extended Terms:

fluvial facies sequence 河流相层序
sedimentary facies sequence 沉积相序

facies series [ˈfeiʃiːz ˈsiəriːz]

Definition: 相系 xiāng xì

facies: the character of a rock expressed by its formation, composition, and fossil content. 相 xiàng

series: a number of things, events, or people of a similar kind or related nature coming one after another. 系列 xì liè

Examples:

①Good reservoirs developed in delta front subfacies, with the kind of diagenetic facies series of compaction-lining chlorite-dissolution-fracture facies.
优质储层发育于三角洲前缘亚相，成岩相类型主要为压实-绿泥石衬边-溶蚀-破裂成岩相。
②The amphiboles of retrogressive metamorphic stages in the granulite and eclogite were mainly formed under the condition of amphibolite facies of medium-pressure facies series.

麻粒岩和榴辉岩中退变质阶段的角闪石主要形成在中压相系角闪岩相的温压条件下。

Extended Terms:

Abukuma type facies series 阿武隈型相系
Barrovian-type facies series 巴罗型相系
jadeite-glaucophane type facies series 硬玉-蓝闪石型相系
low-pressure facies series 低压相系
medium-pressure facies series 中压相系
metamorphic facies series 变质相系

false bedding [fɔːls ˈbediŋ]

Definition: (Geology) another term for cross-bedding. (地质)假层理 jiǎ céng lǐ

Extended Term:

false bedding plane 假层理面

fan delta facies [fæn ˈdeltə ˈfeiʃiːz]

Definition: 扇三角洲相 shàn sān jiǎo zhōu xiàng

fan: a thing that you hold in your hand and wave to create a current of cool air. 扇子
delta: an area of land, shaped like a triangle, where a river has split into several smaller rivers before entering the sea. 三角洲 sān jiǎo zhōu
facies: the character of a rock expressed by its formation, composition, and fossil content. 相 xiàng

Example:

The triassic fan delta facies and the alluvial fan facies are favorable reservoir blocks in thisregion.
三叠系扇三角洲相，冲积扇相是该区有利储集发育的沉积相带。

fast-spreading ridge [ˈfɑːst sprediŋ ridʒ]

Definition: 快速扩张海岭 kuài sù kuò zhāng hǎi lǐng

fast: moving or able to move quickly. 快的 kuài de；迅速的 xùn sù de；敏捷的 mǐn jié de
spread: to extend over a large or increasing area. 扩张 kuò zhāng
ridge: a long narrow hilltop, mountain range, or watershed. 脊 jǐ，岭 lǐng

fault gouge [fɔːlt gaudʒ]

Definition: accumulated crushed rock debris which is the product of the movement of the rocks on each side of a fault. 断层泥 duàn céng ní

Extended Term:
fault gouge clay 断层黏土

fayalite [faiˈaːlait]

Definition: black or brown mineral which is an iron-rich form of olivine and occurs in many igneous rocks. 铁橄榄石 tiě gǎn lǎn shí

Origin: mid-19th century: from Fayal (the name of an island in the Azores) + -ite.

Example:
Nolanda sands contain 70%~80% fayalite mineral component and is a rare mineral resource in natural world.
诺砂中含有70%~80%的铁橄榄石矿物成分，是自然界中罕见的矿产资源。

Extended Terms:
fayalite peridotite 铁橄榄岩
forsterite-fayalite series 镁橄榄石-铁橄榄石系列

feldspar [ˈfeldspɑː]

Definition: an abundant rock-forming mineral typically occurring as colourless or pale-coloured crystals and consisting of aluminosilicates of potassium, sodium, and calcium. 长石 cháng shí

Origin: mid-18th century: alteration of German Feldspat, Feldspath, from Feld "field" + Spat, Spath "spar" (see spar). The form felspar is by mistaken association with German Fels "rock".

Example:
Rock-forming mineral: olivine, pyroxene, hornblende, orthoclase, feldspar, plagioclase, mica, quartz, biotite, calcite and other common metal and nonmetal minerals.
造岩矿物的概念：橄榄石、辉长石、角闪石、正长石、长石、斜长石、云母、石英、黑云母、方解石以及常见金属和非金属矿物。

Extended Terms:
aventurine feldspar 日长石

barium feldspar 钡长石
calcium feldspar 钙长石
glassy feldspar 透长石
half-weathering feldspar 半风化长石
lime feldspar 石灰长石，钙长石
orthoclase feldspar 正长石

feldspathization [ˌfeldspɑːθiˈzeiʃən]

Definition: (Geology) formation of feldspar in a rock usually as a result of metamorphism leading toward granitization. (地质) 长石化作用 cháng shí huà zuò yòng

Example:
Piedmont potassic alteration zone is the product of potash feldspathization, part of the zone is ore body.
矿区山前钾化带是钾长石化的产物，局部地段构成矿体。

Extended Term:
potash feldspathization 钾长石化

feldspathoid [ˈfeldspæθɔid]

Definition: any of a group of minerals chemically similar to feldspar but containing less silica, such as nepheline and leucite. 似长石 sì cháng shí

Example:
The author points out that only the igneous rocks fitting into one of the following conditions can be defined as the alkaline rocks: (1) containing feldspathoid or alkaline melaminerals.
笔者认为符合下列条件之一的火成岩才能定为碱性岩：(1) 含似长石或碱性暗色矿物。

felsic [ˈfelsik]

Definition: of, relating to, or denoting a group of light-coloured minerals including feldspar, feldspathoids, quartz, and muscovite. 长英质的 cháng yīng zhì de

Origin: early 20th century: from feldspar + a contraction of silica.

Example:
All the mafic and felsic volcanic-hosted massive sulfide deposits are therefore defined as the Kunlun type volcanic-hosted massive copper sulfide deposit.
这些基性火山岩型和酸性火山岩型矿床被统称为昆仑式火山岩型块状硫化物铜矿床。

Extended Terms:

felsic magmatism 长英岩浆酌
felsic rock 长英质岩

felsic igneous rock [ˈfelsik ˈigniəs rɔk]

Definition: 长英质火成岩 cháng yīng zhì huǒ chéng yán

felsic：of, relating to, or denoting a group of light-colored minerals including feldspar, feldspathoids, quartz, and muscovite. 长英质的 cháng yīng zhì de

igneous：(of rock) having solidified from lava or magma. 火成的 huǒ chéng de

rock：the solid mineral material forming part of the surface of the earth and other similar planets, exposed on the surface or underlying the soil or oceans. 岩石 yán shí, 岩 yán

felsic index (FL) [ˈfelsik ˈindeks]

Definition: 长英指数 zhǎng yīng zhǐ shù

felsic：of, relating to, or denoting a group of light-coloured minerals including feldspar, feldspathoids, quartz, and muscovite. 长英质的 cháng yīng zhì de

index：a figure in a system or scale representing the average value of specified prices, shares, or other items as compared with some reference figure. 指数 zhǐ shù

felsite [ˈfelsait]

Definition: an aphanitic volcanic rock consisting chiefly of feldspar and quartz in minute crystals. 霏细岩 fēi xì yán

Origin: 1785-1795：fels(par) + -itel.

Example:

The extrusion of silica-rich lava, the Rooiberg felsite, occurred on a grand scale.
富含硅氧的罗依山细岩岩浆大面积喷发。

Extended Terms:

felsite felstone 霏细岩
felsite-pitchstone 霏细松脂岩
minette-felsite 云煌霏细岩
monzonic felsite 二长霏细岩
quartz felsite 石英霏细岩

felsitic texture [fel'sitik 'tekstʃə]

Definition: 霏细岩结构 fēi xì yán jié gòu

felsitic: a fine-grained, light-colored igneous rock, composed chiefly of feldspar and quartz. 霏细岩的 fēi xì yán de

texture: the arrangement of the particles or constituent parts of any material, as wood, metal, etc., as it affects the appearance or feel of the surface; structure, composition, grain, etc. 结构 jié gòu

fenestrae [fi'nestriː]

Definition: a dense, fine-grained, igneous rock consisting typically of feldspar and quartz, both of which may appear as phenocrysts. 沉积岩的缩孔 chén jī yán de suō kǒng

Origin: 1785-1795; fels(par) + -itel.

Extended Term:

crista fenestrae cochleae 蜗窗嵴

fenestral fabric [fi'nestrəl 'fæbrik]

Definition: 格状组构 gé zhuàng zǔ gòu

fenestral: of or relating to or having a fenestra. 格状的 gé zhuàng de

fabric: In geology, a rock's fabric describes the spatial and geometric configuration of all the elements that make it up. (地质)结构(dì zhì) jié gòu

fenite ['fenait]

Definition: a metasomatic alteration associated particularly with carbonatite intrusions and created, very rarely, by advanced carbon dioxide alteration (carbonation) of felsic and mafic rocks. 霓长岩 ní cháng yán

Example:

The fenite is characterized by a zonation pattern in the sequence of shattered zone, aegirine and aegirine-augite zone, and arfvedsonite zone from outside to the center.
霓长岩具有明显的分带特征，由内向外可分为：钠铁闪石黑云母带、霓石霓辉石带、碎裂带。

ferricrete [ˈferikrit]

Definition: a mineral conglomerate consisting of surficial sand and gravel cemented into a hard mass by iron oxide derived from the oxidation of percolating solutions of iron salts. 铁结砾岩 tiě jié lì yán

Origin: The word is derived from the combination of ferruginous and concrete.

ferrogabbro [ˌferəuˈgæbrə]

Definition: a gabbro in which the pyroxene or olivine minerals, or both, are exceptionally high in iron. 铁辉长岩 tiě huī cháng yán

ferrosilite [ˌferəuˈsilait]

Definition: the magnesium endmember of the pyroxene silicate mineral series enstatite ($MgSiO_3$)-ferrosilite ($FeSiO_3$). 铁辉石 tiě huī shí

ferruginous quartzite [feˈruːdʒinəs ˈkwɔːtsait]

Definition: 含铁石英岩 hán tiě shí yīng yán

ferruginous: containing iron oxides or rust. 含铁的 hán tiě de
quartzite: an extremely compact, hard, granular rock consisting essentially of quartz. It often occurs as silicified sandstone, as in sarsen stones. 石英岩 shí yīng yán

ferruginous rock [feˈruːdʒinəs rɔk]

Definition: 铁质岩 tiě zhì yán

ferruginous: containing iron oxides or rust. 含铁的 hán tiě de
rock: the solid mineral material forming part of the surface of the earth and other similar planets, exposed on the surface or underlying the soil or oceans. 岩石 yán shí; 岩 yán

ferruginous shale [feˈruːdʒinəs ʃeil]

Definition: 铁质页岩 tiě zhì yè yán

ferruginous: containing iron oxides or rust. 含铁的 hán tiě de
shale: a type of soft stone that splits easily into thin flat layers. 页岩 yè yán

Example:

The main facies of Suining Formation are ferruginous-shale film, anhydrite, dolomite and calcite.

遂宁组主要的成岩相组合为铁泥膜-硬石膏-白云石-方解石成岩相。

fiamme [fiəm]

Definition: Fiamme are lens-shapes, usually millimetres to centimetres in size, seen on surfaces of some volcaniclastic rocks. 火焰石 huǒ yàn shí

fibroblastic texture [ˈfaibrəublæstik ˈtekstʃə]

Definition: 纤状变晶结构 xiān zhuàng biàn jīng jié gòu

fibroblastic: of, or relating to a fibroblast. 纤维变晶状 xiān wéi biàn jīng zhuàng

texture: the arrangement of the particles or constituent parts of any material, as wood, metal, etc., as it affects the appearance or feel of the surface; structure, composition, grain, etc. 结构 jié gòu

filter pressing [ˈfiltə ˈpresiŋ]

Definition: 滤压(作用) lǜ yā (zuò yòng)

filter: to pass liquid, light, etc., through a special device, especially to remove something that is not wanted. 过滤 guò lǜ

pressing: an act or instance of applying force or weight to something. 按 àn, 压 yā

Example:

By filter pressing, the problem of dehydration was overcome.
采用压滤工艺，解决了脱水难的问题。

Extended Term:

pressing filter machine 压滤机

fire clay [ˈfaiə klei]

Definition: a specific kind of clay used in the manufacture of ceramics, especially fire brick. The fire attribution is given for its refractory characteristics. There are two types of fire clay: flint clay and plastic fire clay. 耐火黏土 nài huǒ nián tǔ

Example:

A refractory brick, usually made of fire clay, used for lining furnaces, fireboxes, chimneys, or fireplaces.

耐火砖，一种通常用耐火黏土做成的耐高温的砖，用以建火炉、火箱、烟囱或壁炉。

Extended Terms:

fire-clay brick 火砖
fire-clay lining 火泥炉衬

fire fountaining [faiə ˈfauntiniŋ]

Definition: 熔岩喷泉 róng yán pēn quán

fire：the flames, light and heat with smoke, that are produced when something burns 火 huǒ
fountain：a strong flow of liquid or of another substance that is forced into the air. 喷泉 pēn quán

Example:

These eruptions include：Hawaiian fire-fountaining, Strombolian explosions, and Plinian falls and ignimbrite.
这些喷发包括：夏威夷火喷泉，斯特隆布利式爆炸，普林尼式瀑布和熔结凝灰岩。

first boiling [fəːst ˈbɔiliŋ]

Definition: 一次沸腾 yī cì fèi téng

first：happening or coming before all other similar things or people；the first. 第一 dì yī
boiling：the application of heat to change something from a liquid to a gas. 沸腾 fèi téng

fissility [fiˈsiliti]

Definition: able to undergo nuclear fission. 易裂性 yì liè xìng, 可裂变性 kě liè biàn xìng

Origin: mid-17th century (in the sense "easily split")：from Latin fissilis, from fiss-"split, cracked", from the verb findere.

Example:

The results show that the quenching medium of water-solubility has effectively solved the difficult problems：such as non-hardening of oil quenching and fissility of water quenching.
生产实践证明，水溶性淬火介质有效地解决了油淬不硬、水淬易裂的难题，避免了淬裂和软点的形成。

Extended Terms:

bedding fissility 层面裂开性
characteristics of fissility 分裂性特征
scale of fissility 劈度表

fission tract [ˈfiʃən trækt]

Definition: 裂变道 liè biàn dào

fission: the act or process of splitting the nucleus of an atom, when a large amount of energy is released. 裂变 liè biàn, 分裂 fēn liè

tract: an area of land, especially a large one. 大片土地 dà piàn tǔ dì; 地带 dì dài

fissure eruption [ˈfiʃə iˈrʌpʃəns]

Definition: the ejection of lava from a crack rather than a vent. 裂隙喷发 liè xì pēn fā

Example:

The island is huge fissure eruption of adjacent formation volcanic island.
该岛是从洋底巨大裂隙喷发形成的火山岛。

flame structure [fleim ˈstrʌktʃə]

Definition: 火焰状构造 huǒ yàn zhuàng gòu zào, 又称泥舌 (also called mud tongue, mud lobe) ní shé

flame: a hot bright stream of burning gas that comes from something that is on fire. 火焰 huǒ yàn; 火舌 huǒ shé

structure: the arrangement of and relations between the parts or elements of something complex. 结构 jié gòu; 构造 gòu zào

Example:

Pearls with a distinct flame structure have a "watered silk" appearance towards the periphery of the sphere.
带有清晰的火焰纹状结构的珠子具有朝着球体外围的波纹绸面。

flaser bedding [ˈflæsə ˈbediŋ]

Definition: a sedimentary bedding pattern created when a sediment is exposed to intermittent flows, leading to alternating sand and mud layers. 脉状层理 mài zhuàng céng lǐ

Example:

The mud-cracking, bird-eye structure, wavy bedding or flaser bedding, nodular structure and bioboringss are characteristics of the tidal flat surroundings instead of the characteristics of the strait.
泥裂、鸟眼构造、波状或脉状层理、瘤状构造、生物钻孔等，都是潮坪环境的特征，而不是海峡特征。

flaser gneiss [ˈflæsə naɪs]

Definition: 压扁片麻岩 yā biǎn piàn má yán

flaser: flatten, battering, press flat. 压扁 yā biǎn

gneiss: a metamorphic rock with a banded or foliated structure, typically coarse-grained and consisting mainly of feldspar, quartz, and mica. 片麻岩 piàn má yán

flint [flɪnt]

Definition: a very hard grayish black fine-grained form of quartz that produces a spark when strucking with steel. It occurs as nodules and bands in chalk. Flint was used in prehistoric times to make tools. 燧石 suì shí

Origin: Old English; related to Middle Dutch vlint and Old High German flins.

Example:

The acousto-optical quality factor for our germanate glass is better than that of heavy flint glasses.

我们的锗酸盐玻璃的声光品质因素优于重燧石玻璃。

Extended Term:

flint spring 火石弹簧

flinty crush rock [ˈflɪnti krʌʃ rɒk]

Definition: 燧石状碎裂岩 suì shí zhuàng suì liè yán

flinty: of, containing, or reminiscent of flint. 含燧石的 hán suì shí de

crush: to break something into small pieces or into powder by pressing hard. 压碎 yā suì; 捣碎 dǎo suì

rock: the solid mineral material forming part of the surface of the earth and other similar planets, exposed on the surface or underlying the soil. 岩石 yán shí; 岩 yán

flocculation [ˌflɒkjuˈleɪʃən]

Definition: a condition in which clays, polymers or other small charged particles become attached and are formed a fragile structure. 絮凝作用 xù níng zuò yòng, 凝聚作用 níng jù zuò yòng

Example:

The flocculation of micro-organism shares a sustainable effect for biological treatment on effluent.

微生物的絮凝性在生物法处理废水过程中是有永续的效应。

Extended Terms:

flocculation agent 絮凝剂
flocculation aid 助凝剂
flocculation morphology 絮凝形态学
flocculation performance 混凝性能
flocculation sedimentation 絮凝沉降
flocculation time 絮凝时间
flocculation value 凝结值;絮凝度
magnetic flocculation 磁团聚
perikinetic flocculation 异向絮凝
selective flocculation 选择性絮凝

floccule [ˈflɔkjuːl]

Definition: a small loosely aggregated mass of flocculent material suspended in or precipitated from a liquid. 絮状物 xù zhuàng wù, 絮凝粒 xù níng lì

Origin: mid-19th century, modern Latin flocculus.

Example:

The PSAF floccule is produced with sulfate aluminum, sulfate iron and water glass as the main material.

以硫酸铝、硫酸铁、水玻璃为主要原料,制备了聚合硅酸铝铁(PSAF)絮凝剂。

Extended Terms:

floccule size 絮凝粒径
floccule structure 絮体结构
bacteria floccule 细菌絮凝体
biological floccule 生物絮状物
electrical floccule 电絮凝物

flood basalt [flʌd ˈbæsɔːlt] = plateau basalt

Definition: 溢流玄武岩 yì liú xuán wǔ yán

flood: a large amount of water covering an area that is usually dry. 洪水 hóng shuǐ;水灾

shuǐ zāi

basalt: a dark, fine-grained volcanic rock that sometimes displays a columnar structure. It is typically composed largely of plagioclase with pyroxene and olivine. 玄武岩 xuán wǔ yán

Example:

In the mafic large igneous province formed by plume magmatism, the granites associated with continental flood basalts are non-orogenic, intraplate and A-type granites rather than S-type. 地幔柱岩浆作用形成的"镁铁质大火成岩省"中，出现的少量花岗岩是非造山、板内或 A-型花岗岩，而不会是 S-型花岗岩。

Extended Term:

Permian flood basalts 二叠纪玄武岩

flood basin deposit [flʌd ˈbeisən diˈpɔzit]

Definition: 河漫盆地沉积 hé màn pén dì chén jī

flood: a large amount of water covering an area that is usually dry. 洪水 hóng shuǐ；水灾 shuǐ zāi

basin: a place where the earth's surface is lower than in other areas of the world. 盆地 pén dì；凹地 āo dì

deposit: a layer or body of accumulated matter. 沉积物 chén jī wù，沉淀 chén diàn

flood plain deposit [flʌd plein diˈpɔzit]

Definition: 河漫滩沉积 hé màn tān chén jī

flood: a large amount of water covering an area that is usually dry. 洪水 hóng shuǐ；水灾 shuǐ zāi

plain: a large expanse of fairly flat dry land, usually with few trees. 平原 píng yuán

deposit: a layer or body of accumulated matter. 沉积物 chén jī wù，沉淀 chén diàn

Example:

Horizontally, the remaining oil in the flood plain deposit is more concentrated than other areas. 从平面上说，河漫滩沉积中的剩余油相对其他区域要更富集些。

flow line [fləu lain]

Definition: the connecting line or arrow between symbols on a flow chart or block diagram. Mark on a molded plastic or metal article made by the meeting of two input-flow fronts during molding. Also known as weld line; weld mark. 流线 liú xiàn

Example:

On the basis of experiment and change of metal flow line, metallographic structure and micro-hardness in deformation zone, this paper researched process, characteristic and mechanism of fine-blanking.
通过实验和变形区流线变化、金相组织和显微硬度的变化，对精冲成形的过程、特点、精冲成形机理进行了研究。

Extended Terms:

flow-line equipment 流水线设备
flow-line map 流线图

flow plane [fləu plein]

Definition: 流面 liú miàn

flow: the steady and continuous movement of something in one direction. 流 liú; 流动 liú dòng
plane: a flat, level, or even surface. 平面的 píng miàn de

Example:

For the free-flow plane working gate slot, the hydraulic characteristics are much more complex than the gate slot under pressure flow.
对于闸后为明流的平板工作闸门门槽，其水力特性比压力段内的闸门门槽复杂得多。

Extended Term:

flow plane map 水流平面图

flow structure [fləuˈstrʌktʃə]

Definition: 流场构造 liú chǎng gòu zào

flow: the steady and continuous movement of something in one direction. 流 liú; 流动 liú dòng
structure: the arrangement of and relations between the parts or elements of something complex. 结构 jié gòu; 构造 gòu zào

Example:

Unique, detailed and regular flow structures in the flow field of detonation have been obtained.
获得了爆震燃烧流场中独特、精细、规则的流场结构。

Extended Terms:

flow event structure 流事件结构
flow measurement structure 量水建筑物

flow topological structure 流动拓扑结构
internal flow field structure 内部流场结构
the flow field structure 流场结构

flowage differentiation [ˈfləuidʒ ˌdifərenʃiˈeiʃən]

Definition: 流动分异作用 liú dòng fēn yì zuò yòng

flowage: the act of flooding; filling to overflowing. 流动 liú dòng, 泛滥 fàn làn
differentiation: the act of making or becoming different in the process of growth or development. 变异 biàn yì, 变化 biàn huà

fluid dynamics [ˈflu(ː)id daiˈnæmiks]

Definition: 流体动力学 liú tǐ dòng lì xué

fluid: a liquid; a substance that can flow. 液体 yè tǐ; 流体 liú tǐ
dynamics: the branch of mechanics concerned with the motion of bodies under the action of forces. 动力学 dòng lì xué

Example:
Fluid dynamics is the nucleus of basin fluid geology.
盆地流体地质学的核心是流体动力学。

Extended Term:
electro-fluid-dynamics 电-流体动力学

fluid inclusion [ˈflu(ː)id inˈkluːʒən]

Definition: 液体包裹体 yè tǐ bāo guǒ tǐ

fluid: a liquid; a substance that can flow. 液体 yè tǐ; 流体 liú tǐ
inclusion: a body or particle recognizably distinct from the substance in which it is embedded. 包裹体 bāo guǒ tǐ

Example:
The fluid inclusion is the primary samples of the basin liquids, and it records much information of the basin fluids during the geological history.
流体包裹体是盆地流体的原始样品，它记录了地质史中盆地流体的很多信息。

fluid overpressure [ˈflu(ː)id ˈəuvəˌpreʃə]

Definition: 流体超压 liú tǐ chāo yā

fluid: a liquid; a substance that can flow. 液体 yè tǐ; 流体 liú tǐ

overpressure: the pressure increase or accumulation when the valve is discharging flow. 超压 chāo yā

Example:

The study on the fluid overpressure shows that there are two fluid overpressure sealing systems which correspond to the two sets of caprocks.
通过流体超压研究，发现两个与两套盖层相对应的流体超压封盖系统。

flute cast [fluː t kɑːst]

Definition: Sole marks are sedimentary structures found on the bases of certain strata, that indicates small-scale (usually on the order of centimetres) grooves or irregularities. 槽模 cáo mó, 又称槽铸型 (also called eroding cast) cáo zhù xíng

Extended Terms:

delicate flute cast 细槽铸型
terraced flute cast 阶状槽铸型

fluvioglacial deposit [ˈfluːviəˈglæʃəl diˈpɔzit]

Definition: 冰水沉积作用 bīng shuǐ chén jī zuò yòng

fluvioglacial: relating to or denoting erosion or deposition caused by flowing meltwater from glaciers or ice sheets. 冰水 bīng shuǐ

deposit: a layer or body of accumulated matter. 沉积物 chén jī wù, 沉淀 chén diàn

Example:

According to the field investigation and research results of predecessors, the evolution feature of the clastic materials is discussed based on the analyses of mineral and chemical compositions, and trace elements of the moraine tills (fluvioglacial deposit), riversand, dunesand and loess (soil) from the southern piedmont of Mount Tuomuer.
本文在野外实地考察和前人工作的基础上，通过对托木尔峰南麓地区冰碛物（冰水沉积物）、河流砂、沙丘砂及黄土（土壤）的矿物成分、化学成分和微量元素的分析，探讨了该地区碎屑物的演化特征。

flysch [fliʃ]

Definition: an association of certain types of marine sedimentary rocks, characteristic of deposition in a foredeep 复理石沉积 fù lǐ shí chén jī

> Example:

The deposition of flysch onto the young floors of chiasmic basins is widely accepted as certain.
在裂陷盆地年青底板上复理石的沉积作用被广泛接受。

> Extended Terms:

flysch sandstone 复理石砂岩
Apenninic flysch 亚平宁复理层
continental flysch formation 陆相复理式建造
parautochthonous flysch 准原地复理层
pre-flysch facies 前复理石相
sandy-conglomeratic flysch 砂质砾岩复理石

foam impression [fəum imˈpreʃən]

> Definition: 泡沫痕 pào mò hén

foam: a mass of very small air bubbles on the surface of a liquid. 泡沫 pào mò
impression: a mark that is left when an object is pressed hard into a surface. 压痕 yā hén

foliation [ˌfəuliˈeiʃən]

> Definition: (Geology) layered texture of rock: a characteristic of metamorphosed rocks in which minerals are aligned in one direction so that the rock can readily be split into thin layers. (地质)叶理(dì zhì) yè lǐ

> Origin: 1615-1625; foliate + -ion.

> Example:

While, the sodium amphibolites in the later stage compatible with the present foliation, might be related to the structure deformation.
晚期的钠质角闪石构成了现存叶理,可能与后期构造变形有关。

> Extended Terms:

foliation structure 叶理构造
bedding foliation 层面叶理;顺层面理
linear foliation 线状叶理
segregation foliation 分凝叶理
tectonic foliation 造构剥理;造构叶理

forearc basin [ˈfɔːrɑːk ˈbeisən]

> Definition: a depression in the sea floor located between an accretionary wedge and a

volcanic arc in a subduction zone, and lined with trapped sediment. 弧前盆地 hú qián pén dì

Example:

They separated the forearc basin into several sub-basins, which imbricate in the background of a forearc basin with sedimentary characteristics of the piggyback basin.
它们将弧前盆地分割成多个呈叠瓦状排列的次级盆地，具有背驮型增生特征。

Extended Terms:

remnant-type forearc basin 残留型弧前盆地
Xigaze forearc basin 日喀则弧前盆地

forearc region [ˈfɔːɑːk ˈriːdʒən]

Definition: 弧前区域 hú qián qū yù

forearc: a depression in the sea floor located between a subduction zone and an associated volcanic arc. 弧前 hú qián

region: a large area of land, usually without exact limits or borders. 地区 dì qū, 区域 qū yù

foreland basin [ˈfɔːlənd ˈbeisən]

Definition: a depression that develops adjacent and parallel to a mountain belt. 前陆盆地 qián lù pén dì

Example:

The geodynamic environment in which foreland basin is developed is rather limited both temporally and spatially in the geological history.
在地质历史上，适于前陆盆地发育的地球动力学环境，在时间和空间上均是相当有限的。

Extended Terms:

foreland deep basin 前渊盆地
foreland flexural basin 前陆挠曲盆地
foreland flysch basin 前陆复理石沉积盆地
foreland like basin 类前陆盆地
foreland molasses basin 前陆磨拉石沉积盆地
broken foreland basin 断裂前陆盆地
continental foreland basin 陆相前陆盆地
Dabashan foreland basin 大巴山前陆盆地
Sichuan Analogous Foreland Basin 四川类前陆盆地

foreshore facies [ˈfɔːʃɔː ˈfeiʃiːz]

Definition: 前滨相 qián bīn xiàng

foreshore: the part of a shore between high-and low-water marks, or between the water and cultivated or developed land. 前滨 qián bīn

facies: the character of a rock expressed by its formation, composition, and fossil content. 相 xiàng

forsterite [ˈfɔːstərait]

Definition: Forsterite (Mg_2SiO_4) is the magnesium, rich end-member of the olivine solid solution series. 镁橄榄石 měi gǎn lǎn shí

Origin: early 19th century. After J. R. Forster (1729-1798), German naturalist.

Example:

The forsterite zirconia material have been produced from the zircon and magnesia, adding graphite, to produce a $Mg_2SiOZrO_2C$.

以锆英砂和电熔镁砂为原料，合成 ZrO_2 有韧力的镁橄榄石材料，进而加入鳞片状石墨，得到 $Mg_2SiO_4(-ZrO_2)C$ 复合材料。

Extended Terms:

forsterite brick 镁橄榄石砖
forsterite ceramics 镁橄榄石陶瓷
forsterite ophicalcite 镁橄榄石大理石
forsterite refractory 镁橄榄石耐火材料

fossil [ˈfɔsəl]

Definition: the remains of an animal or plant preserved from an earlier era inside a rock or other geologic deposit, often as an impression or in a petrified state 化石 huà shí

Origin: mid-16th century. Via French fossile, from Latin fossilis "dug up".

Example:

We see plants and animal fossils, but we never see a fossil of one kind of animal changing into another kind.

我们见过植物和动物化石，但我们从未看见一种动物的化石变成另一种动物的化石。

Extended Terms:

derived fossils 转生化石
facies fossil 指相化石

index fossil 标准化石
living fossil 活化石
trace fossil 踪迹化石

foyaite [fɔiˈeit]

Definition: It is an intrusive rock occuring in alkaline intrusions. It is coarse grained and light grey Foyaite consists mainly of microcline, nepheline and aegirine. 流霞正长岩 liú xiá zhèng cháng yán

Extended Terms:
foyaite-aplite 流霞正长细晶岩
foyaite porphyry 流霞正长斑岩
amphibole-foyaite 闪流霞正长岩
essexite foyaite 碱辉流霞正长岩
mica-foyaite 云母流霞正长岩
orthoclase-free foyaite 无正流霞岩

fractional [ˈfrækʃənl]

Definition: relating to or denoting the separation of components of a mixture by making use of their differing physical properties. 分级的 fēn jí de

Extended Terms:
fractional crystallization 分级结晶
fractional distillation 分级蒸馏
fractional precipitation 分级沉淀

fractional crystallization [ˈfrækʃənl ˌkristəlaiˈzeiʃən]

Definition: a process by which a chemical compound is separated into components by crystallization. In fractional crystallization the compound is mixed with a solvent, heated, and then gradually cooled so that, as each of its constituent components crystallizes, it can be removed in its pure form from the solution. 分离结晶 fēn lí jié jīng

fractional melting [ˈfrækʃənl ˈmeltiŋ]

Definition: 分离熔融作用 fēn lí róng róng zuò yòng

fractional: relating to or denoting the separation of components of a mixture by making use of their differing physical properties. 分级的 fēn jí de

melting: the process whereby heat changes something from a solid to a liquid. 熔化 róng huà

Example:

It is possible that the fractional melting of shocked chondrites would increase the heterogeneity of choudritical metal grains with respect to their distribution.

局部熔融对陨石中金属的富集迁移具有一定的影响，可能使金属分布不均匀性增加。

Extended Term:

fractional melting apparatus 分熔化设备

fractionation [ˌfrækʃəˈneiʃən]

Definition: a separation process in which a certain quantity of a mixture (solid, liquid, solute, suspension or isotope) is divided up in a number of smaller quantities (fractions) in which the changes according to a gradient. 分馏 fēn liú, 分级 fēn jí

Example:

This project is also developing biomass fractionation pretreatments, techniques to reuse enzymes, thermotolerant yeast, and a new generation of high solids bioreactors.

这个计划也包含生物量分离方法、生化酶再用技术、耐热酵母、新代高固体反应器的改良。

Extended Terms:

fractionation effect 分馏效应
air fractionation 风动分级
chromatographic fractionation 层析分离；色层分离；色谱分级
extraction fractionation 萃取分级
gas fractionation 气体分馏
isotope fractionation 同位素分馏
primary fractionation 初分馏
rough fractionation 初步纯化
screen fractionation 筛分

fracture [ˈfræktʃə]

Definition: the physical appearance of a freshly broken rock or mineral, especially as regards the shape of the surface formed. 断层 duàn céng, 断口 duàn kǒu

Origin: late Middle English: from French, or from Latin fractura, from frangere "to break".

Example:

For more than a century researchers have known the basics of how limestone caves form: A tiny

fracture opens in the rock perhaps due to some internal stress and water begins percolating through it.
一个多世纪以来，研究者就懂得石灰岩洞形成的基本原理：可能由于内部的压力，石头上出现了微小的裂缝，然后水开始渗透。

Extended Terms:

fatigue fracture 疲劳断裂
fracture mechanics 断裂力学
fracture mode 断裂模式
fracture strength 破裂强度
fracture surface 断裂面
fracture toughness 断裂韧性
fracture zone 断裂带
bright crystalline fracture 亮晶断口
crystalline fracture 结晶形断口

framework silicate ['freimwə:k 'silikeit]

Definition: 架状硅酸盐 jià zhuàng guī suān yán

framework: the parts of a building or an object that support its weight and give it shape. 框架 kuàng jià，结构 jié gòu

silicate: a compound containing silicon, oxygen and one or more metals. 硅酸盐 guī suān yán

free energy [fri: 'enədʒi]

Definition: thermodynamic quantity equivalent to the capacity of a system to do work. 自由能 zì yóu néng，自由能量 zì yóu néng liàng

Example:

Free energy is a magnetism in which two or more magnets pulse against one another in a continuous spin that is inherently balanced.
自由能量是一种磁场。在这一磁场中，两个或更多个磁体在平衡的持续旋转中形成脉冲对。

Extended Terms:

free radical energy 自由基能量
free surface energy 自由表面能

fresh water phreatic environment [freʃ 'wɔ:tə fri:'ætik in'vaiərənmənt]

Definition: 淡水潜流环境 dàn shuǐ qián liú huán jìng

fresh: containing no salt. 淡的 dàn de;无盐的 wú yán de

water: a liquid without colour, smell or taste that falls as rain, is in lakes, rivers and seas, and is used for drinking, washing, etc. 水 shuǐ

phreatic: relating to or denoting underground water in the zone of saturation (beneath the water table). 潜水 qián shuǐ

environment: the surroundings or conditions in which a person, animal, or plant lives or operates. 环境 huán jìng

fresh water vadose environment [freʃ'wɔːtə 'veidəus in'vaiərənmənt]

Definition: 淡水渗流环境 dàn shuǐ shèn liú huán jìng

fresh: containing no salt. 淡的 dàn de;无盐的 wú yán de

water: a liquid without colour, smell or taste that falls as rain, is in lakes, rivers and seas, and is used for drinking, washing, etc. 水 shuǐ

vadose: relating to or denoting underground water above the water table. 地下水位线以上的 dì xià shuǐ wèi xiàn yǐ shàng de

environment: the surroundings or conditions in which a person, animal, or plant lives or operates. 环境 huán jìng

Example:

The cement minerals and fabrics, peculiar in the fresh water vadose environment, are well-devoloped in dune rocks.
沙丘岩中发育着各种淡水渗流环境中特有的胶结物和胶结组构。

Froude number [fruːd 'nʌmbə]

Definition: The Froude number is a ratio of inertia and gravitational forces. 弗劳德数值 fú láo dé shù zhí

Example:

The experimental results show that within the scope of flow investigated, major factors which influence the mixing are the velocity ratio (V_2/V_1) and density Froude number(F_0), critical condition of stratification stability can be described by a simple formula about V_2/V_2 and F_0, and it is not related to the Reynolds number(Re).
实验结果表明,在所研究的流动范围内,流速比(V_2/V_1)和密度弗劳德数值(F_0)是影响两股流体掺混的主要因素,密度分层流稳定的临界条件可用 F_0 和 V_2/V_1 的简单关系式表示,与雷诺数 R_e 没有直接的关系。

Extended Terms:

distinction Froude number 分界弗劳德数值

density Froude number 密度弗劳德数值
low Froude number 低弗氏数

fugacity [fjuːˈgæsiti]

Definition: a chemical term with units of pressure that is intended to better describe a gas real world pressure than the ideal pressure "P" used in the ideal gas law. 逸度 yì dù

Example:
It is concluded that the fluid forming this type of copper-gold deposit is one high oxygen and sulphur fugacity and low-to-moderate temperature acid fluid.
形成这一类矿床的流体是一种高氧逸度、高硫逸度的中低温酸性流体。

Extended Terms:
fugacity coefficient 逸度系数；逸压系数
oxygen fugacity 氧逸度

gabbro [ˈgæbrəu]

Definition: a usually coarse-grained igneous rock composed chiefly of calcic plagioclase and pyroxene. Also called norite. 辉长岩 huī cháng yán

Origin: mid-19th century：from Latin glaber, glabr-"smooth".

Example:
The intrusive body is consisted mainly of dioritic rocks showing the geochemical characters of calc-alkalic andesite and accompany with minor alkali gabbro.
岩体主要由钙碱性闪长质岩组成，具有钙碱性安山岩的地球化学特点，伴有少量碱性辉长岩。

Extended Terms:
gabbro texture 辉长结构
anorthositic gabbro 钙长辉长岩

gabbro texture [ˈgæbrəuˈtekstʃə]

Definition: 辉长结构 huī cháng jié gòu

gabbro：a usually coarse-grained igneous rock composed chiefly of calcic plagioclase and pyroxene. Also called norite. 辉长岩 huī cháng yán

texture：the arrangement of the particles or constituent parts of any material, as wood, metal, etc., as it affects the appearance or feel of the surface；structure, composition, grain, etc. 结构

jié gòu

gabbro porphyrite ['gæbrəu 'pɔːfirait]

Definition: 辉长玢岩 huī cháng bīn yán

gabbro: a usually coarse-grained igneous rock composed chiefly of calcic plagioclase and pyroxene. Also called norite. 辉长岩 huī cháng yán

porphyrite: a rock with a porphyritic structure; augite porphyrite. 玢岩 bīn yán

gangue [gæŋ]

Definition: In mining, gangue is the commercially worthless material that surrounds, or is closely mixed with, a wanted ore. The separation of the ore from the gangue is a necessary and often significant aspect of mining; it can be a complicated process, depending on the nature of the minerals involved. 脉石 mài shí；尾矿 wěi kuàng

Origin: early 19th century: from French, from German Gang "course, lode"; related to gangl.

Example:
Different particle distribution samples of cement-coal gangue were obtained through mixing different degree of fineness cement-coal gangue.
将不同细度的煤矸石和不同细度的水泥混合，得到具有不同颗粒群分布的水泥-煤矸石试样。

Extended Terms:
gangue froth 脉石泡沫
coal gangue 煤矸石
melted gangue 熔融矿渣

garnet ['gɑːnit]

Definition: a precious stone consisting of a deep red vitreous silicate mineral. 石榴石 shí liú shí

Origin: 13th century. Probably via Middle Dutch garnate from Old French grenat "dark red", from pome grenate "pomegranate", because of its color.

Example:
The ore deposit can be named as subvolcanic hydrothermal garnet-diopside-magnetite deposit.
可把此矿床定名为次火山水热石榴石-透辉石-磁铁矿矿床。

Extended Terms:

garnet zone 石榴石带
common garnet 普通石榴石
crushed garnet 碎石榴石
iron garnet 铁榴石

gas [gæs]

Definition: any substance like air that is neither a solid nor a liquid, for example hydrogen and oxygen are both gases. 气体 qì tǐ

Origin: mid-17th century. Dutch, alteration of Greek khaos "empty space".

Example:
Neptune is the farthest solar system gas giant planets.
海王星是太阳系中最远的气态巨行星。

Extended Terms:

gas chromatography 色层分析;色谱分析法
gas field 天然气田
gas flow 气流;气流量
gas phase 气相, 气态
gas reservoir 气藏;气田;煤气存储器;天然气储层
gas well 气井
coal gas 煤气;煤炭气
natural gas 天然气
oil gas 石油气;油煤气
waste gas 废气

gaseous transfer [ˈɡæsiəs ˈtrænsfəː] = volatile transfer

Definition: 气体搬运 qì tǐ bān yùn

gaseous: like or containing gas. 似气体的 sì qì tǐ de;含气体的 hán qì tǐ de
transfer: to move from one place to another; to move something/somebody from one place to another. (使)转移(shǐ) zhuǎn yí, 搬迁 bān qiān

Extended Term:

gaseous transfer differentiation 气体分异作用

gas exsolution [gæs ˌeksəˈluːʃən]

Definition: 气体出溶 qì tǐ chū róng

gas: any substance like air that is neither a solid nor a liquid, for example hydrogen and oxygen are both gases. 气体 qì tǐ

exsolution: the process in which a solution of molten rocks separate into its constituents upon cooling. 出溶 chū róng

gel texture [dʒel 'tekstʃə] = colloform texture

Definition: 胶状结构 jiāo zhuàng jié gòu

gel: a semisolid colloidal suspension of a solid dispersed in a liquid. 胶状 jiāo zhuàng
texture: the arrangement of the particles or constituent parts of any material, as wood, metal, etc., as it affects the appearance or feel of the surface; structure, composition, grain, etc. 结构 jié gòu

geochron ['dʒiəukrɔn]

Definition: The Geochron was the first world clock to display day and night on a world map, showing the familiar "bell curve" of light and darkness. 地质年代 dì zhì nián dài

Example:
With study on biological stratigraphy, geochron stratigraphy, and lithologic stratigraphy, Permian is re-divided in Sichuan Basin.
通过生物地层、年代地层、岩石地层的研究,对四川盆地二叠纪进行重新划分。

geochronometry [ˌdʒiəukrə'nɔmitri]

Definition: a branch of stratigraphy aimed at the quantitative measurement of geologic time. It is considered also a branch of geochronology. 年代学 nián dài xué

Example:
The isotopic dating and fossil geochronometry are the basic methods of ophiolite geochronometry.
同位素测年及化石年代学法是蛇绿岩年代学研究的基本方法。

Extended Term:
fossil geochronometry 化石地质时代测定法

geologic map [ˌdʒiəu'lɔdʒik mæp]

Definition: A geologic map or geological map is a special-purpose map made to show geological features. Rock units or geologic strata are shown by color or symbols to indicate where they are exposed at the surface. 地质图 dì zhì tú

> Example:

Being coincided with topographic and geologic maps, the anomaly map by this method is a help to the explanation and inference for anomaly and a guide to the exploration.
以此方法绘制的异常图与地形地质图套合后，有利于对异常的解释和推断，也能指导探矿。

> Extended Terms:

geologic sketch map 地质素描图
interpretive geologic map 解释性地质图

geomagnetic dynamo [ˌdʒi(ː)əumæɡˈnetik ˈdainəməu]

> Definition: the dynamo theory proposes a mechanism by which a celestial body such as the Earth or a star generates a magnetic field. The theory describes the process through which a rotating, convecting, and electrically conducting fluid can maintain a magnetic field over astronomical time. 地磁发电机 dì cí fā diàn jī

geopetal structure [dʒiːˈɔpətəl ˈstrʌktʃə]

> Definition: 示底构造 shì dǐ gòu zào

geopetal: Way-up structure, way-up criterion, or geopetal indicator is a characteristic relationship observed in a sedimentary or volcanic rock, or sequence of rocks, that makes it possible to determine whether they are the right way up. 示顶底 shì dǐng dǐ

structure: the arrangement of and relations between the parts or elements of something complex. 结构 jié gòu; 构造 gòu zào

Geophysics [ˌdʒiəuˈfiziks]

> Definition: the branch of earth science that deals with the physics and physical processes of the Earth, especially using noninvasive techniques such as acoustic surveys of the structure of rocks. 地球物理学 dì qiú wù lǐ xué

> Example:

Geophysics is the application of the principles of physics to the study of the earth. The aim of geophysics is to deduce the physical properties of the earth and its internal constitution from the physical phenomena associated with it.

地球物理勘探是应用物理学的原理去研究地球。此方法的目的是通过地球中的一些物理现象去推测地球的物理性质及其内部组分。

Extended Term:

stratigraphic geophysics 地层地球物理学

geotherm [ˈdʒiːəʊθəm]

Definition: a line or surface within or on the earth connecting points of equal temperature. 地等温线 dì děng wēn xiàn

Example:

The lower degrees of mantle melting of the HT and LT2 lavas may be the result of a relatively thicker lithosphere and lower geotherm.
HT 和 LT2 熔岩的地幔熔融程度较低，可能与地幔柱边部的岩石圈相对较厚和地热较低有关。

Extended Terms:

geotherm anomaly 地热异常
geotherm gradient 地温梯度
geotherm resources 地热资源
ancient geotherm 古地温
abnormal geotherm 地热异常
deep geotherm 深层地热
mantle geotherm 地幔地温线

geothermal [ˌdʒiːəʊˈθɜːməl]

Definition: of, relating to, or produced by the internal heat of the earth. 地热的 dì rè de, 地温的 dì wēn de

Example:

Rich geothermal resources are hosted in the Meso-Cenozoic basins in Liaoning Province.
辽宁省中新生代盆地地层中赋有丰富的地热资源。

Extended Terms:

geothermal gas 地热气
geothermal gradient 地温梯度
geothermal heat flow 地热流

geothermal gradient [ˌdʒiːəʊˈθɜːməl ˈgreidiənt]

Definition: the rate at which the Earth's temperature increases with depth, indicating heat flowing from the Earth's warm interior to its cooler surface. Away from tectonic plate boundaries, it is 25℃-30℃ per km of depth in most of the world. 地热梯度 dì rè tī dù

Example:
The influence factors on pattern of wellbore pressure & temperature are production rate, gas composition, geothermal gradient and so on.
井筒压力和温度分布主要受产气量、硫化氢含量(气体组成)、地温梯度等因素影响。

geothermobarometry [ˈdʒiːəʊθəməbəˈrɒmitri]

Definition: Geothermobarometry is the science of measuring the previous pressure and temperature history of a metamorphic or intrusive igneous rocks. 地温压力测定 dì wēn yā lì cè dìng

Example:
The concept and method of high precision relative geothermobarometry, which is not dependent on thermochemical data or activity composition relations of minerals involved, was developed.
出现了不依赖于矿物热力学数据和矿物活度模型的高精度地温压力测定概念和方法。

geothermometer [ˌdʒiːəʊθəˈmɒmitə]

Definition: (Also called geologic thermometer) a thermometer designed to measure temperatures in deep-sea deposits or in bore holes deep below the surface of the earth. 地质温度计 dì zhì wēn dù jì

Example:
The geothermometer is one of the most important methods for reconstructing the thermal history of sedimentary basins.
地质温度计是重构沉积盆地热演化历史的重要手段之一。

Extended Terms:
gas geothermometer 气体地热温标
hydrochemical geothermometer 水化学地热温标

gibbsite [ˈgibsait]

Definition: Gibbsite, $Al(OH)_3$, is one of the mineral forms of aluminium hydroxide. It is often designated as γ-Al(OH)$_3$ (but sometimes as α-Al(OH)$_3$.). It is also sometimes called hydrargillite (or hydrargyllite). 三水铝矿（水铅氧石）sān shuǐ lǚ kuàng (shuǐ qiān yǎng shí)

Origin: early 19th century: named after George Gibbs (1776-1833), American mineralogist, + -ite.

Example:
Ore is in the phase of clay minerals in which gibbsite is less and kaolinite commonly coexists with illite.
矿石处在富黏土矿物阶段，少见三水铝石，一般为伊利石和高岭石共存。

glacial deposit [ˈgleisjəl diˈpɔzit]

Definition: Till or glacial till is unsorted glacial sediment. Glacial drift is a general term for the coarsely graded and extremely heterogeneous sediments of glacial origin. 冰川沉积作用 bīng chuān chén jī zuò yòng

Example:
By means of the sedimentary strata sequences and through analyzing the grain grade and the sedimentary environment, the authors preliminarily consider that whether it is a glacial deposit or the products of the glaciation-gravity flow reconstruction, it must be the alluvial fan of the off-shore slope.
作者从地层层序、粒组分析和沉积环境等沉积学角度分析，初步认为不论它是冰川沉积物或是冰川-重力流改造沉积物，均应属水下岸坡冲积扇的沉积物。

Extended Terms:
glacial boulder deposit 冰川沉积
glacial fluvial deposit 冰水堆积
glacial-lake deposit 冰湖沉积
late glacial deposit 后期冰川沉积物

glacial facies [ˈgleisjəl ˈfeiʃiːz]

Definition: 冰川相 bīng chuān xiàng

glacial: caused or made by glaciers; connected with glacier. 冰河的 bīng hé de; 冰川的 bīng chuān de

facies: the character of a rock expressed by its formation, composition, and fossil content. 相 xiàng

glaciation [ˌgleisiˈeiʃən]

Definition: act of freezing; that which is formed by freezing; ice; the process of glaciating, or the state of being glaciated; the production of glacial phenomena. 冰川作用 bīng chuān zuò yòng; 冰蚀 bīng shí

Example:
During the Pleistocene Ice Age, the ancient drainage systems were almost completely rearranged by glaciation.
在更新世冰期,古水系常被冰川作用完全改造。

Extended Term:
Carboniferous glaciation 石炭纪冰期

glauberite rock [ˈglaubərait rɒk]

Definition: 钙芒硝岩 gài máng xiāo yán

glauberite: a sodium calcium sulfate mineral with the formula $Na_2Ca(SO_4)_2$, which forms as an evaporite. 钙芒硝 gài máng xiāo

rock: the solid mineral material forming part of the surface of the earth and other similar planets, exposed on the surface or underlying the soil. 岩石 yán shí; 岩 yán

glaucophane greenschist facies [ˈglɔːkəˌfein ˈgriːnʃist ˈfeiʃiːz]

Definition: 蓝闪绿片岩相 lán shǎn lǜ piàn yán xiàng

glaucophane: a bluish sodium-containing mineral of the amphibole group, found chiefly in schists and other metamorphic rocks. 蓝闪石 lán shǎn shí

greenschist: also known as greenstone, a general field petrologic term applied to metamorphic and/or altered mafic volcanic rock. 绿片岩 lǜ piàn yán

facies: the character of a rock expressed by its formation, composition, and fossil content 相 xiàng

glaucophane lawsonite schist facies
[ˈglɔːkəˌfein ˈlɔːsəunait ʃist ˈfeiʃiːz]

Definition: 蓝闪石硬柱石片岩相 lán shǎn shí yìng zhù shí piàn yán xiàng

glaucophane: a bluish sodium-containing mineral of the amphibole group, found chiefly in schists and other metamorphic rocks. 蓝闪石 lán shǎn shí

lawsonite: a hydrous calcium aluminium sorosilicate mineral with formula $CaA_{12}Si_2O_7(OH)_2 \cdot$

H_2O. 硬柱石 yìng zhù shí

schist: a type of rock formed of layers of different minerals, that breaks naturally into thin flat pieces. 片岩 piàn yán

facies: the character of a rock expressed by its formation, composition, and fossil content. 相 xiàng

glaucophane schist ['glɔːkəˌfein ʃist]

Definition: (Petrology) metamorphic schist that contains glaucophane. (岩石学) 蓝闪石片岩 lán shǎn shí piàn yán

Extended Terms:

glaucophane schist zone 蓝闪石片岩带
glaucophane-chlorite schist 蓝闪绿泥片岩
glaucophane mica schist 蓝闪云片岩

glowing avalanche ['gləuiŋ 'ævəlɑːntʃ]

Definition: 火山发光云 huǒ shān fā guāng yún

glowing: to produce a dull, steady light. 发出微弱而稳定的光 fā chū wēi ruò ér wěn dìng de guāng

avalanche: a slide of large masses of snow and ice falling down a mountain. 雪崩 xuě bēng

gneiss [nais]

Definition: a coarse-grained high-grade metamorphic rock formed at high pressures and temperatures, in which light and dark mineral constituents are segregated into visible bands. 片麻岩 piàn má yán

Origin: mid-18th century. From German.

Example:

Felsic gneiss is the most widely distributed rock type in Archean metamorphic complex areas. 长英质片麻岩是太古宙变质杂岩区最主要的岩石类型。

Extended Terms:

gneiss soil 片麻岩土
composite gneiss 复片麻岩；混合岩
eyed gneiss 眼状片麻岩

granitic gneiss 花岗片麻岩
layered gneiss 层状片麻岩

gneissic structure [ˈnaisik ˈstrʌktʃə]

Definition: 片麻状构造 piàn má zhuàng gòu zào

gneissic: of or related to gneiss. 片麻岩的 piàn má yán de
structure: the arrangement of and relations between the parts or elements of something complex. 结构 jié gòu；构造 gòu zào

Example:

The rock is offwhite, light-flesh-colored, and it has porphyritic structure and gneissic structure.
岩石为灰白色、浅肉红色，呈似斑状结构和片麻状结构。

gold vein [gəuld vein]

Definition: 金脉 jīn mài

gold: a chemical element which is a yellow precious metal used for making coins, jewellery, ornaments, etc. 金子 jīn zi；黄金 huáng jīn
vein: a thin layer of minerals or metal contained in rock. 矿脉 kuàng mài；矿脉床 kuàng mài chuáng

Example:

Space-time and origin of the gold veins in the ore field were closely related with some lamprophyre dykes.
玲珑金矿田发育的金矿脉以黄铁石英脉为主。

graben structure [ˈgrɑːbən ˈstrʌktʃə]

Definition: 地堑结构 dì qiàn jié gòu

graben: an elongated block of the earth's crust lying between two faults and displaced downward relative to the blocks on either side, as in a rift valley. 地沟 dì gōu, 地堑 dì qiàn
structure: the arrangement of and relations between the parts or elements of something complex. 结构 jié gòu；构造 gòu zào

Example:

During the Late Cretaceous to Paleogene it was manifested as an adjusting boundary of the horst and graben structures, controlling the development of the Chuxian Quanjiao red basin.
晚白垩世—早第三纪时表现为垒堑构造的调整边界，控制着滁全（现滁州全椒县）红色盆地的发展。

grade [greid]

Definition: a gradient or slope 坡度 pō dù

Origin: early 16th century. Latin gradus "step, stage".

Example:

So the relationship between longitudinal gradient, grade length limit, vertical curve and safety is analyzed, which can be referred to improve the safety and optimize the profile design.
根据下坡路段事故统计资料,分析纵坡坡度、坡长、竖曲线和下坡衔接平曲线与道路安全的关系,为提高坡路段安全,优化纵面设计提供参考。

Extended Terms:

grade ability 爬坡能力;上坡能力
straight grade 直坡度

graded bedding [ˈgreidid ˈbediŋ]

Definition: In geology, a graded bed is one characterized by a systematic change in grain or clast size from the base of the bed to the top. (地质)粒序层理(dì zhì) lì xù céng lǐ

Example:

Distinct causes of formation for different graded bedding in debris flow deposits have been analyzed by using the ratio of flow plug.
应用流核比,对泥石流堆积层理进行成因差异分析。

Extended Terms:

biogenic graded bedding 生物粒级层
content graded bedding 含量递变层理
reverse graded bedding 反粒序层理
rhythmic graded bedding 韵律性递变层理

gradient [ˈgreidiənt]

Definition: a measure of change in a physical quantity such as temperature or pressure over a particular distance 梯度 tī dù, 倾斜度 qīng xié dù, 坡度 pō dù

Origin: mid-17th century. Partly from Latin gradient-, present participle of gradi "walk", partly, from grade after quotient.

Example:

Measuring the differences emphasizes anomalies from shallow sources having steep gradients at

the expense of low-gradient variation of the type.
测定两点之间临近的差只需以牺牲低梯度变化为代价而突出浅层源引起的具有陡梯的异常。

> Extended Terms:

gradient descent 梯度下降法
average gradient 平均坡度
good gradient 平缓坡度
hydraulic gradient 水力梯度
potential gradient 势梯度
temperature gradient 温度梯度
velocity gradient 速度梯度

grain boundary [grein ˈbaundəri]

> Definition: the interface between two grains in a polycrystalline material. Grain boundaries disrupt the motion of dislocations through a material, so reducing crystallite size is a common way to improve strength, as described by the Hall-Petch relationship. 晶界 jīng jiè

> Example:

Grain boundaries have an important influence on the strength and ductility of metals.
晶界对金属的强度和延展性有极大的影响。

> Extended Term:

pearlite grain boundaries 珠光体晶界

grain flow [grein fləu]

> Definition: 颗粒流 kē lì liú

grain: a small hard piece of particular substances. 颗粒 kē lì; 细粒 xì lì
flow: the steady and continuous movement of something in one direction. 流 liú; 流动 liú dòng

> Example:

In experiment research, control of serial drive mode of laser sensors, multiplex A/D transform in Visual C++6.0 are realized and the simulation experimental equipment of gas-particles flow in upright pipe are established, moreover, the vertical velocity of quartz grain flow was performed in the experiment; the results show the feasibility of this new optical measurement technology.
在 Visual C++6.0 环境下实现了软件控制激光器的串行激励方式和多路 A/D 转换、建立了气体-颗粒流垂直管段模拟实验装置,并对石英砂颗粒流速度进行测量,实验结果表明该新型光测技术具有应用可行性。

Extended Terms:

clastic grain flow 碎屑流
density modified grain flow 密度颗粒流

grain size [grein saiz]

Definition: 粒度 lì dù

grain: a small hard piece of particular substances. 颗粒 kē lì; 细粒 xì lì
size: how large or small a person or thing is. 大小 dà xiǎo

Example:

The result shows that in micro-scale, inhomogeneous flow of material is obviously emerged, which is more severe with the increase of grain size and stroke.
结果表明，微小尺度下材料呈现出明显的非均匀流动，且这一非均匀流动随晶粒尺寸和变形量的增大而更剧烈。

Extended Terms:

grain-size graph 粒径图
grain-size parameter 粒度参数
grain-size sorting 粒度分选

grain supported [grein səˈpɔːtid]

Definition: 颗粒支撑 kē lì zhī chēng

grain: a small hard piece of particular substances. 颗粒 kē lì; 细粒 xì lì
supported: bear all or part of the weight of; hold up. 支撑 zhī chēng

granite [ˈgrænit]

Definition: a very hard, granular, crystalline, igneous rock consisting mainly of quartz, mica, and feldspar and often used as a building stone. 花岗岩 huā gāng yán

Origin: mid-17th century: from Italian granito, literally "grained".

Example:

The Penshan granitic body is a calc-alkaline pluton composed of muscovite alkali granite, two-mica alkali feldspar granite and biotite adamellite.
彭山岩体是由白云母碱长花岗岩、云母碱长花岗岩和黑云母二长花岗岩组成的钙碱性花岗复式岩体。

Extended Terms:

granite tile 花岗岩砖
black granite 暗色花岗岩
blocky granite 块状花岗岩
binary granite 双云母花岗石

granite greenstone belt ['grænit 'gri:nstəun belt]

Definition: 花岗岩绿岩带 huā gāng yán lǜ yán dài

granite: a very hard, granular, crystalline, igneous rock consisting mainly of quartz, mica, and feldspar and often used as a building stone. 花岗岩 huā gāng yán

greenstone: a greenish igneous rock containing feldspar and hornblende. 绿岩 lǜ yán

belt: an area with particular characteristics or where a particular group of people live. 地带 dì dài; 地区 dì qū

Example:

Jiapigou gold deposit concentration area involves 3 geological units of gold occurrence, the high metamorphic district, granite-greenstone belt and Seluohe Group.
夹皮沟金矿集中区涉及高级变质区、花岗岩-绿岩带、色洛河群三个赋金矿地质单元。

granite porphyry ['grænit 'pɔ:firi]

Definition: 花岗斑岩 huā gǎng bān yán

granite: a very hard, granular, crystalline, igneous rock consisting mainly of quartz, mica, and feldspar and often used as a building stone. 花岗岩 huā gāng yán

porphyry: a hard igneous rock containing crystals, usually of feldspar, in a fine-grained, typically reddish groundmass. 斑岩 bān yán

Example:

The Ulan Uzhur prophyry copper deposit, situated on the western margin of the Qaidam Basin is a porphyry copper deposit controlled by granite porphyry dikes.
乌兰乌珠尔斑岩铜矿位于柴达木盆地西缘，为受花岗斑岩脉控制的斑岩型铜矿。

Extended Terms:

biotite granite porphyry 黑云花岗斑岩
muscovite-granite porphyry 白云花岗斑岩
riebeckite aplite-granite porphyry 钠闪细晶花岗斑岩

granitic gneiss [græˈnitik nais]

Definition: 花岗质片麻岩 huā gǎng zhì piàn má yán

granitic: a very hard, granular, crystalline, igneous rock consisting mainly of quartz, mica, and feldspar and often used as a building stone. 花岗岩 huā gǎng yán

gneiss: a metamorphic rock with a banded or foliated structure, typically coarse-grained and consisting mainly of feldspar, quartz, and mica. 片麻岩 piàn má yán

Example:
In the MORB normalized spider diagram, the granitic gneiss is relatively riched in K, Rb, Th, but evident depleted in Ti, Ta and Nb.
花岗质片麻岩在洋脊玄武岩标准化蛛网图上表现出强烈富集大离子亲石元素（K、Rb、Th）的特点，高场强元素 Ti、Ta 和 Nb 表现为明显的负异常（而 Hf 和 Zr 则为明显正异常）。

granitic migmatitic gneiss [græˈnitik ˈmigmætitik nais]

Definition: 花岗质混合片麻岩 huā gǎng zhì hùn hé piàn má yán

granitic: made of granite. 花岗石的 huā gāng shí de

migmatitic: made of migmatite. 混合岩的 hùn hé yán de

gneiss: a metamorphic rock with a banded or foliated structure, typically coarse-grained and consisting mainly of feldspar, quartz, and mica. 片麻岩 piàn má yán

granitic pegmatite [græˈnitik ˈpegmətait]

Definition: a coarse-grained granite, sometimes rich in rare elements such as uranium, tungsten, and tantalum. 花岗伟晶岩 huā gǎng wěi jīng yán

granitic: made of granite. 花岗石的 huā gāng shí de

pegmatite: a coarsely crystalline granite or other igneous rock with crystals several centimeters to several meters in length. 结晶花岗岩 jié jīng huā gāng yán

Examples:
①Chrysoberyl in some granitic pegmatite is a rare mineral that seldom seen in Altaic area, and it is a valuable precious stone.
新疆阿尔泰地区某些花岗伟晶岩含有金绿宝石，它是一种罕见的稀有元素矿物，而且具有极高的经济价值。
②The Nanping granitic pegmatite is one of the most important rare-metal deposits in China.
福建南平花岗伟晶岩是中国重要的稀有金属花岗伟晶岩之一。

granitic texture [græˈnitik ˈtekstʃə]

Definition: 花岗状结构 huā gǎng zhuàng jié gòu

granitic: made of granite. 花岗石的 huā gāng shí de

texture: It is referred to in geology the physical appearance or character of a rock, such as grain size, shape, arrangement, and pattern at both the megascopic or microscopic surface feature level. 质地 zhì dì; 纹理 wén lǐ; 结构 jié gòu

Example:

The rock shows finegrained granitic texture and consists of perthite, oligoclase, quartz and biotite.
岩石呈细粒花岗结构，矿物成分为条纹长石、更长石、石英和黑云母。

Extended Terms:

granitic vein 花岗岩岩脉
granitics 花岗质岩

granitization [ˌgrænitaiˈzeiʃən]

Definition: the metamorphic conversion of a rock into granite. 花岗岩化 huā gāng yán huà, 花岗岩化作用 huā gāng yán huà zuò yòng

Origin: granitize + -ation.

Examples:

① The gold content decreases with increasing granitization and migmatization or amount of felsic minerals.
含金量随着花岗岩化和混合岩化程度增加而降低。

② The critical position of granite energy and the location of granite in space indicate that the primary cause of its formation is the granitization of sedimentary rocks.
花岗岩的能量临界位置和空间部位说明，其主要成因是沉积岩的花岗岩化。

Extended Terms:

static granitization 静态花岗岩作用; 静态花岗岩
granitization granite 混合花岗岩
synkinematic granitization 造山期花岗岩化作用; 同构造期花岗岩化作用
granitization metamorphism 花岗岩化变质

granitoid [ˈgrænitɔid]

Definition: Granite is a common and widely occurring type of intrusive, felsic, igneous rock. Granites usually have a medium to coarse grained texture. Occasionally some individual crystals (phenocrysts) are larger than the groundmass in which case the texture is known as porphyritic. 花岗岩类 huā gāng yán lèi

Origin: mid-17th century. Italian granito "grainy", Latin granum "seed".

Example:

The Paleoproterozoic Wangjiahui granitoid intrusion, west segment of Wutai mountains, consists mainly of quartz monzodiorite, granodiorite, monzogranite and adamellite.
五台山西段古元古代王家会花岗岩侵入体主要由石英二长闪长岩、花岗闪长岩和二长花岗岩组成。

Extended Terms:

granitoid texture 似花岗岩状结构
granitoidite 花岗状变质岩
graniton 辉长岩

granitoid texture ['grænitɔid 'tekstʃə]

Definition: 花岗状结构 huā gǎng zhuàng jié gòu

granitoid: a field term for a coarse grained felsic igneous rock, resembling granite. 似花岗石的 sì huā gāng shí de

texture: It is referred to the physical appearance or character of a rock, such as grain size, shape, arrangement, and pattern at both the megascopic or microscopic surface feature level. 质地 zhì dì;纹理 wén lǐ;结构 jié gòu

Extended Terms:

granitophile element 亲花岗岩元素
granitoid floor 仿花岗石地面

granoblastic [ˌgrænəu'blæstik]

Definition: Granoblastic is an anhedral phaneritic equi-granular metamorphic rock texture. Granoblastic texture is typical of quartzite, marble and other non-foliated metamorphic rocks without porphyroblasts. 花岗变晶状 huā gǎng biàn jīng zhuàng

Example:

The granulite always contains some hypersthene and >10% of plagioclase and possesses the granoblastic texture and granulitic structure.
麻粒岩应具有花岗变晶结构和麻粒构造,并含有紫苏辉石,以及含>10%的斜长石。

> Extended Terms:

granoblastic fabric 花岗变晶状组构
granoblastic texture 花岗变晶状结构

granoblastic texture [ˌgrænəʊˈblæstɪk ˈtekstʃə]

> Definition: 花岗变晶状结构 huā gǎng biàn jīng zhuàng jié gòu

granoblastic: containing equidimensional grains of metamorphic rock formed by recrystallization. 花岗变晶状 huā gǎng biàn jīng zhuàng

texture: It is referred to the physical appearance or character of a rock, such as grain size, shape, arrangement, and pattern at both the megascopic or microscopic surface feature level. 质地 zhì dì; 纹理 wén lǐ; 结构 jié gòu

> Example:

According to the zircon genetic types, microcosmic structures (the prism-granular blastic texture and granoblastic texture), the Longtang alkaline granite should be metamorphic rock by argillo-arenaceous sediments.
根据锆石的成因类型,结合微观所见的岩石结构(反映沉积变质成因的粒状变晶结构和花岗变晶结构等),表明该碱性花岗岩的成因可能是砂泥质沉积物经变质而成。

granodiorite [ˌgrænəʊˈdaɪərait]

> Definition: Granodiorite is an intrusive igneous rock similar to granite, but contains more plagioclase than potassium feldspar. It usually contains abundant biotite mica and hornblende, giving it a darker appearance than true granite. 花岗闪长岩 huā gǎng shǎn cháng yán

> Origin: late 19th century: granite + diorite.

> Example:

Early Yanshanian intrusions are mainly adamellite, biotite K-feldspar granite with small amount of granodiorite.
燕山早期侵入岩以二长花岗岩、黑云钾长花岗岩为主,其次为花岗闪长岩。

> Extended Terms:

granite 花岗岩
granodiorite 花岗闪绿岩

granophyre [ˈgrænəʊfaɪə]

> Definition: Granophyre (from granite and porphyry) is a subvolcanic rock containing quartz

and alkali feldspar in characteristic angular intergrowths such as those in the accompanying image. 花岗斑岩 huā gǎng bān yán

Origin: late 19th century: granite + porphyry.

Example:

Nanliang complex is mainly composed of porphyritic granite, granophyre, rhyolite-porphyry and rhyolitic agmatite-lava.
南梁岩体主要由斑状花岗岩、流纹斑岩、花斑岩及流纹质角砾熔岩组成。

Extended Terms:

riebeckite granophyre 钠闪花斑岩
granodiorite-granophyre 花岗闪长花斑岩

granophyric texture [ˌgrænəʊˈfirik ˈtekstʃə]

Definition: 花斑结构 huā bān jié gòu

granophyric: made of granophyre. 花斑状 huā bān zhuàng
texture: In geology it referred to the physical appearance or character of a rock, such as grain size, shape, arrangement, and pattern at both the megascopic or microscopic surface feature level. 质地 zhì dì; 纹理 wén lǐ; 结构 jié gòu

Example:

Major minerals of lunar granitic clasts are quartz, K-feldspar and calcic plagioclase with granophyric intergrowths.
月球花岗岩碎屑的主要矿物为石英、钾长石和钙质斜长石，具花斑状结构。

Extended Term:

micro-granophyric texture 微花斑结构

granular texture [ˈgrænjulə ˈtekstʃə]

Definition: 粒状结构 lì zhuàng jié gòu

granular: composed or appearing to be composed of granules or grains 粒状的 lì zhuàng de
texture: In geology it referred to the physical appearance or character of a rock, such as grain size, shape, arrangement, and pattern at both the megascopic or microscopic surface feature level. 质地 zhì dì; 纹理 wén lǐ; 结构 jié gòu

Example:

The ore mainly has cyclopean granular texture and massive and laminar structures.
矿石以镶嵌粒状变晶结构、块状和纹层状构造为主。

Extended Terms:

panidiomorphic granular texture 全自形粒状结构
allotriomorphic granular texture 他形粒状结构
hypidiomorphic granular texture 半自形晶粒状结构
euhedral-granular texture 自形晶粒状结构

granulite [ˈgrænjulait]

Definition: a fine-grained metamorphic rock consisting of similarly sized, interlocking minerals. Unlike most metamorphic rocks, granulites do not exhibit foliation or textural or mineralogical layering. 麻粒岩 má lì yán

Origin: 1840-1850; granule + -ite.

Example:
The high pressure granulite assemblages (Grt-Cpx-Pl-Qtz± Amp ± Rt-Ilm) were formed during decompression under a p-T condition of 9.6-13.7 kbar and 730-870℃.
高压麻粒岩组合(Grt-Cpx-Pl-Qtz±Amp±Rt-Ilm)为退变质作用产物，其形成的变质条件为 $p = 9.6 \sim 13.7$ kbar，$T = 730 \sim 870$℃。

Extended Terms:

granulite terrain 麻粒岩地体
granulite series 麻粒岩系
basic granulite 基性麻粒岩
leucogranulite 浅色麻粒岩
acidic granulite 酸性麻粒岩

granulite facies [ˈgrænjulait ˈfeiʃiːz]

Definition: 麻粒岩相 má lì yán xiàng

granulitie: a fine-grained, granular metamorphic rock in which the main component minerals are typically feldspars and quartz. 粒变岩 lì biàn yán

facies: the character of a rock expressed by its formation, composition, and fossil content. 相 xiàng

Example:
In Archaean era, granulite facies metamorphite series are extensively outcropped in Sa-heqiao, Hebei Province.
冀东迁西洒河桥是太古代麻粒岩相变质岩系广泛出露的地区。

graphic granite [ˈgræfik ˈgrænit]

Definition: 文象花岗石 wén xiàng huā gāng shí

graphic: connected with drawings and design, especially in the production of books, magazines, etc. 书画的 shū huà de; 图样的 tú yàng de

granite: a type of hard grey stone, often used in building. 花岗岩 huā gāng yán; 花岗石 huā gāng shí

Example:

In this paper, the principle and method of the experiment which is used to determine the content of free quartz in the graphic granite by phosphoric acid has been studied.
本文研究了用磷酸溶矿分析测试文象花岗石中游离石英的实验原理和方法。

graphic texture [ˈgræfik ˈtekstʃə]

Definition: Graphic texture is commonly created by exsolution and devitrification and immiscibility processes in igneous rocks. It is called "graphic" because the exsolved or devitrified minerals form wriggly lines and shapes which are reminiscent of writing. 文象结构 wén xiàng jié gòu

Example:

Trachyte is a volcanic rock which typically has graphic texture.
粗面岩是典型的具有文象结构的火山岩。

graphite [ˈgræfait]

Definition: a soft dark carbon that conducts electricity, occurs naturally as a mineral, and is also produced industrially. Use: batteries, lubricants, polishes, electric motors, nuclear reactors, carbon fibers, pencil lead. 石墨 shí mò

Origin: late 18th century: coined in German (Graphit), from Greek graphein "write" (because of its use as pencil "lead").

Example:

The oxidation resistance of oriented graphite is more than the C17B high strength graphite, in particular at below 800℃.
定向石墨较 C17B 高强度石墨具有较强的抗氧化性, 特别是在 800℃以下更为明显。

Extended Term:

colloidal graphite 胶体石墨

gravel ['grævəl]

Definition: a loose aggregation of small water-worn or pounded stones. 砾石 lì shí

Origin: from Old French, diminutive of grave (see grave).

Example:
Concrete is composed of cement, sand and gravel mixed with water.
混凝土是由水泥、沙子和碎石掺上水混合而成的。

Extended Terms:
bank-run gravel 岸流砂砾
beach gravel 海滩砾石
bird's eye gravel 细粒石
cannon-shot gravel 炮弹砾石
cement gravel 胶结砾石
clay gravel 黏土砾石

---

gravitational acceleration [ˌgrævi'teiʃənəl əkˌselə'reiʃən]

Definition: in physics, gravitational acceleration is the specific force or acceleration on an object caused by gravity. In a vacuum, all small bodies accelerate in a gravitational field at the same rate relative to the center of mass. This is true regardless of the mass or composition of the body. 重力加速度 zhòng lì jiā sù dù

Example:
Methods for accurate gravitational acceleration measurement are presented, which are viscous force modified method and viscous force offset each other method.
本文介绍了两种精确测量重力加速度的方法:黏滞阻力修正法和黏滞阴力互相抵消法。

Extended Term:
gravitational acceleration 万有引力加速度

---

gravity anomalies ['græviti ə'nɔməlis]

Definition: 重力异常 zhòng lì yì cháng

gravity: the force that attracts objects in space towards each other, and that on the earth pulls them towards the centre of the planet, so that things fall to the ground when they are dropped. 重力 zhòng lì;地球引力 dì qiú yǐn lì

anomaly: something that deviates from what is standard, normal, or expected. 异常 yì cháng

> Example:

Measurements of physical properties indicate that the rutile-bearing eclogite, due to its very high seismic velocity and density, can produce reflections and gravity anomalies.
岩石物性测定表明，由于金红石榴辉岩的纵波速度高、密度大，会产生反射与重力异常。

> Extended Terms:

positive gravity anomalies 正重力异常
upward continuation of gravity anomalies 重力异常向上延拓

gray gneisses [grei nais]

> Definition: 灰色片麻岩 huī sè piàn má yán

gray: a neutral achromatic color midway between white and black. 灰色 huī sè
gneisses: a metamorphic rock with a banded or foliated structure, typically coarse-grained and consisting mainly of feldspar, quartz, and mica. 片麻岩 piàn má yán

> Example:

A large volume of gray gneisses of tonalitic to trondhjemitic composition exist in the Dengfeng granite-greenstone belt and the Taihua high-grade gneisses terrane, the southern margin of the North China Craton.
华北地块南缘太古宙登封花岗-绿岩地体和太华高级片麻岩地体中的英云闪长-奥氏花岗质和花岗闪长质灰色片麻岩多占据背斜核部或穹窿部位，它们是多期侵入的复合岩体。

Great Dyke [greit daik]

> Definition: the Great Dyke is a linear geological feature that trends nearly north-south through the center of Zimbabwe passing just to the west of the capital, Harare. It consists of a band of short, narrow ridges and hills spanning for approximately. 大矿脉 dà kuàng mài

Greenhouse effect [ˈgriːnhaus iˈfekt]

> Definition: the trapping of the sun's warmth in a planet's lower atmosphere due to the greater transparency of the atmosphere to visible radiation from the sun than to infrared radiation emitted from the planet's surface. 温室效应 wēn shì xiào yìng

> Example:

As the potential technique to reduce greenhouse effect, geological sequestration of CO_2 in deep unminable coal seams is competent for storage of CO_2 released during the forthcoming 26 to 46 years in China, especially in those coal energy bases in North China.

在我国，尤其在华北煤炭能源基地，利用深部不可开采煤层地质封存 CO_2 的技术显示了在减少温室效应方面的巨大潜力，可以处置 26~46 年的 CO_2 排放量。

Extended Term:

greenhouse-effect gas 温室气体

greenschist [ˈgriːnʃist]

Definition: Greenschist—also known as greenstone—is a general field petrologic term applied to metamorphic and/or altered mafic volcanic rock. The green is due to abundant green chlorite, actinolite and epidote minerals that dominate the rock. 绿片岩 lǜ piàn yán

Example:

The Western Tianshan high-pressure metamorphic rocks occur within the greenschist facies country rocks as small discrete blocks, lenses, bands, laminae or intercalated slabs.
西天山高压变质岩呈不连续岩块、凸镜体、条带、薄层或夹层岩片，产于绿片岩相围岩中。

Extended Terms:

greenschist metamorphism 绿片岩变质作用
greenschist facies 绿色片岩相
actinolitic greenschist facies 阳起绿色片岩相；阳起绿片岩相

green schist [griːn ʃist]

Definition: (Petrology) a schistose metamorphic rock with abundant chlorite, epidote, or actinolite present, giving it a green color. 绿片岩 lǜ piàn yán

Example:

Maoduqing gold deposit occurred in the green schist and marble stratum transitional belt of paleoprotozoic Erdaowa Group in Inner Mongolia.
内蒙古卯独庆金矿床存在于古元古代的二道凹岩群的绿片岩层与大理岩层过渡带中。

Extended Terms:

green schist facies 绿片岩相
chlorite green schist 绿泥石绿色片岩
biotite chlorite subfacies(green schist) 黑云绿泥亚相；绿片岩

greenschist facies [ˈgriːnʃist ˈfeiʃiːz]

Definition: Greenschist—also known as greenstone—is a general field petrologic term applied to metamorphic and/or altered mafic volcanic rock. The green is due to abundant green chlorite, actinolite and epidote minerals that dominate the rock. 绿片岩相 lù piàn yán xiāng

Example:

Dynamical metamorphism under greenschist facies is the condition of mineralizing enrichment. 绿片岩相以下的动力变质作用是矿化富集的条件。

greenstone [ˈɡriːnstəun]

Definition: any of various altered basic igneous rocks colored green by chlorite, hornblende, or epidote. 绿岩 lù yán

Origin: 1765-1775; green + stone.

Example:

The greenstone belt type gold deposit in the North China Craton (NCC), one of the most important deposit types in China, is the major source of gold productions and reserves.
华北克拉通中绿岩带型金矿床是我国最重要的矿床类型之一，也是我国黄金产量和储量的主要来源。

Extended Terms:

greenstone belts 绿岩带
granite-greenstone terrain 花岗绿岩区
aquamarine beryl greenstone 绿玉

greisen [ˈɡraizən]

Definition: a light-coloured metamorphic rock consisting mainly of quartz, white mica, and topaz formed by the pneumatolysis of granite. 云英岩 yún yīng yán

Origin: from German, from greissen to split.

Example:

The wolframite in greisen is rich in REE, Nb, Ta, Sc and has low δ Eu values and LREE/HREE and Nb/Ta ratios.
云英岩中黑钨矿富含 REE、Nb、Ta、Sc，其 δ Eu 值和 LREE/HREE、Nb/Ta 比值都较低。

Extended Terms:

feldspar-greisen 长石云英岩
granite-greisen 花岗云英岩
tourmaline-greisen 电气石云英岩
topaz greisen 黄玉云英岩
greisen-lithify 云英岩化

greywacke [ˈgreɪˌwækə]

Definition: Greywacke or Graywacke (German grauwacke, signifying a grey, earthy rock) is a variety of sandstone generally characterized by its hardness, dark color, and poorly-sorted angular grains of quartz, feldspar, and small rock fragments or lithic fragments set in a compact, clay-fine matrix. 杂砂岩 zá shā yán

Origin: late 18th century (as grauwacke): from German Grauwacke, from grau "grey" + wacke. The Anglicized form dates from the early 19th century.

Example:
The sediments are detrital and largely of greywacke facies.
沉积岩为碎屑质，大部分属于杂质砂岩相。

Extended Term:
greywacke schist 杂砂片岩

groove cast [gruːv kɑːs]

Definition: 沟脊模 gōu jǐ mú, 又称沟铸型 gōu zhù xíng
groove: a long narrow cut in the surface of something hard 沟 gōu; 槽 cáo; 辙 zhé; 纹 wén
cast: a shaped container used to make an object. 模子 mú zi; 铸模 zhù mú

Extended Term:
ruffled groove cast 皱状沟铸型

grossular [ˈgrɔsjulə]

Definition: Grossular or grossularite is a calcium-aluminium mineral species of the garnet group with the formula $Ca_3Al_2(SiO_4)_3$, Gemological Institute of America, GIA Gem Reference Guide 1995, ISBN 0-87311-019-6 though the calcium may in part be replaced by ferrous iron and the aluminium by ferric iron. 钙铝榴石 gài lǚ liú shí

Origin: early 19th century: form German Grossularit.

Example:
The end member of garnet is dominated by andradite, with minor grossular and spessartine.
石榴子石端员组分以钙铁榴石为主，伴以少量钙铝榴石和锰铝榴石。

Extended Terms:
grossular garnet / grossularite / grossular 钙铝榴石
green grossular 绿色钙铝榴石

groundmass [ˈgraundmæs]

Definition: the matrix of fine-grained crystalline material in which larger crystals are embedded. 基质 jī zhì

Example:
A fine-grained granite porphyry having a groundmass with irregular intergrowths of quartz and feldspar.
花斑岩，一种细晶的花斑岩，带有不规则交生的石英和长石的基质。

Extended Terms:
sapropelic groundmass 腐泥基质
groundmass, matrix 基质

groundwater [ˈgraundˌwɔːtə]

Definition: water held underground in the soil or in pores and crevices in rock 地下水 dì xià shuǐ

Origin: also ground-water, "water in the ground" 1890, from ground (n.) + water (n.1). Attested from mid-15c. in sense "water at the bottom of a stream."

Example:
Numerical simulation of groundwater flow, involved with wellbore-waterbearing system, is an effective approach in groundwater resource evaluation.
地下水数值模拟是地下水资源评价的有效方法，它涉及井孔-含水系统的问题。

Extended Terms:
groundwater table 地下水位
groundwater level 地下水位；地下水面

guyot [ˈgaiət]

Definition: a flat-topped underwater mountain of a type commonly found in the Pacific Ocean and considered to be an extinct volcano. 平顶海山 píng dǐng hǎi shān

Origin: 1940s: named after Arnold H. Guyot (1807-1884), Swiss geographer.

Example:
Guyots were probably formed by under-water lava spouts.
平顶海山大概形成于水下熔岩喷发。

gypsolith anhydrock=gypsum anhydrite rock
[ˈdʒipsəm ænˈhaidrait rɒk]

Definition: 石膏硬石膏岩 shí gāo yìng shí gāo yán

gypsum: a soft white mineral like chalk that is found naturally and is used in making plaster of Paris. 石膏 shí gāo

anhydrite: a white mineral consisting of anhydrous calcium sulfate. It typically occurs in evaporite deposits. 硬石膏 yìng shí gāo

rock: the hard solid material that forms part of the surface of the earth and some other planets. 岩石 yán shí

gypsum [ˈdʒipsəm]

Definition: a soft white or grey mineral consisting of hydrated calcium sulphate. It occurs chiefly in sedimentary deposits and is used to make plaster of Paris and fertilizers, and in the building industry. 石膏 shí gāo

Origin: late Middle English: from Latin, from Greek gupsos.

Example:

For sea coast location, phosphoric acid plants may be permitted to discharge gypsum into the ocean for disposal.
对于沿海地区,可以允许磷酸将石膏排放入海。

Extended Terms:

anhydrous gypsum 无水石膏
gypsum 玄精石
crystal gypsum 结晶石膏
hemihydrate gypsum 熟石膏,半水石膏
reclaimed gypsum 再生石膏

hafnium [ˈhæfniəm]

Definition: the chemical element of atomic number 72, a hard silver-grey metal of the transition series, resembling and often occurring with zirconium(Symbol: Hf). 铪 hā

Origin: early 20th century; from modern Latin, Hafnia.

Example:

First, we used DC sputter system to deposit hafnium metal and then proceeded with furnace under oxidation at low temperature to prepare HfO_2 thin film.

首先利用直流溅镀法沉积铪金属，接着以低温通氧气的炉管氧化金属铪，而得到氧化铪薄膜。

Extended Terms:

hafnium dust 铪粉末
hafnium base alloy 铪基合金
hafnium carbide 碳化铪
hafnium compound 铪化合物
hafnium isotope 铪同位素

half-life [ˈhɑːflaif]

Definition: the time taken for the radioactivity of a specified isotope to fall to half its original value. 半衰期 bàn shuāi qī

Origin: also halflife, half life, 1864, "unsatisfactory way of living," from half + life; the sense in physics, "amount of time it takes half a given amount of radioactivity to decay" is first attested 1907.

Example:

A radioactive, inert, gaseous isotope of radon, with a half-life of 3.92 seconds.
氡的一种放射性的、惰性的和气态的同位素，其半衰期为3.92秒。

Extended Term:

radioactive half-life 放射性半衰期

halide [ˈhælaid]

Definition: a chemical compound of a halogen with another element or group of atoms. 卤化物 lǔ huà wù

Origin: a compound of a halogen and a metal radical, 1844, from Swedish (Berzelius, 1825), from halo- + chemical suffix -ide.

Example:

The alkali halides can be grown into large crystals of excellent transparency.
碱金属卤化物能生长成具有优良透明性的大晶体。

Extended Terms:

alkyl halides 卤代烃
trichloroacetyl halides 三氯乙酰卤

halogens [ˈhælədʒənz]

Definition: The halogens or halogen elements are a series of nonmetal elements from Group 17 IUPAC Style (formerly: VII, VIIA) of the periodic table, comprising fluorine, (F); chlorine, (Cl); bromine, (Br); iodine, (I); and astatine, (At). 卤素 lǔ sù

Origin: mid-19th century; from halo-; because they readily form salts when combined with metals.

Example:
Halogens also have the necessary absorption in the ultraviolet and rapid dissociation. 卤素也具有吸收紫外线和分解快这两种特性。

Extended Terms:
the halogens 卤素含量
halogens resin 卤素树脂

---

hardground [ˈhɑːdɡraund]

Definition: Carbonate hardgrounds are surfaces of synsedimentarily cemented carbonate layers that have been exposed on the seafloor (Wilson & Palmer, 1992). A hardground is essentially, then, a lithified seafloor. 硬灰岩层 yìng huī yán céng

---

hardground structure [ˈhɑːdɡraund ˈstrʌktʃə]

Definition: 硬底构造 yìng dǐ gòu zào

hardground: a type of etching ground that dries hard yet is easily scratched through. It is used to create linear work on an etching plate. 硬底 yìng dǐ
structure: the way in which the parts of something are connected together, arranged or organized; a particular arrangement of parts. 结构 jié gòu; 构造 gòu zào

---

harzburgite [ˈhɑːzˌbəːɡait]

Definition: (Geology) a plutonic rock of the peridotite group consisting largely of orthopyroxene and olivine. 方辉橄榄岩 fāng huī gǎn lǎn yán, 斜方辉橄岩 xié fāng huī gǎn yán

Origin: late 19th century: from Harzburg, the name of a town in Germany, + -ite.

Example:
The Ophiolite massifs are composed of mantle peridotite that is mainly harzburgite and dunite,

and lacked of crust magmatite within the typical Ophiolite suite.
该蛇绿岩带的岩体由地幔橄榄岩组成，主要岩石类型是方辉橄榄岩和纯橄榄岩，缺少典型蛇绿岩剖面中的洋壳单元。

Extended Terms:
orthopyroxene 斜方辉石类
harrisite 方辉铜矿

hauyne [hə'win]

Definition: Hauyne, haüyne or hauynite is a tectosilicate mineral with sulfate and chloride. It is a feldspathoid and a member of the sodalite group. 蓝方石 lán fāng shí

Extended Terms:
hauyne-basanite 蓝方碧玄岩
hauyne-andesite 蓝方安山岩
hauyne-aplite 蓝方细晶岩

hawaiite [hə'waiiːt / haː'waiiːt]

Definition: an olivine basalt with intermediate composition between alkali olivine and mugearite. 夏威夷岩 xià wēi yí yán

heating phase of metamorphism ['hiːtiŋ feiz əv ˌmetə'mɔːfizəm]

Definition: 加热变质阶段 jiā rè biàn zhì jiē duàn

heating: the process of becoming warmer; a rising temperature. 加热 jiā rè
phase: a stage in a process of change or development. 阶段 jiē duàn；时期 shí qī
metamorphism: alteration of the composition or structure of a rock by heat, pressure, or other natural agency. 变质 biàn zhì

heavy mineral ['hevi 'minərəl]

Definition: an accessory detrital mineral of a sedimentary rock, of high specific gravity, such as magnetite, ilmenite, zircon, rutile. 重矿物 zhòng kuàng wù

Example:
Source analysis in a basin was of great significance of sandstone reservoir prediction, while heavy mineral was an important sign of source region.

盆地物源分析对砂岩储层分析预测具有重要的意义，而重矿物是物源区的重要标志。

hedenbergite ['hedənbəgait]

Definition: Hedenbergite, $CaFeSi_2O_6$, is the iron rich end member of the pyroxene group having a monoclinic crystal system. The mineral is extremely rarely found as a pure substance, and usually has to be synthesized in a lab. 钙铁辉石 gài tiě huī shí

Example:
The results show that, main crystallization phase of glass ceramic was hedenbergite, and the minor crystallization phases were augite and hypersthene.
结果表明：矿渣微晶玻璃的主析晶相为钙铁辉石，次晶相为普通辉石和紫苏辉石。

Extended Term:
ferro hedenbergite 低铁镁铁橄石

helium isotopes ['hiːliəm 'aisəutəup]

Definition: helium: a chemical element with symbol He and atomic number 2. 氦 hài
isotopes: variants of a particular chemical element which differ in neutron number. 同位素 tóng wèi sù；氦同位素 hài tóng wèi sù

Example:
Based on the ~3He/~4He data of the main oil-gas bearing basins in continental China, a systematic study has been made for the first time on the relations between the space distribution of the helium isotopes of natural gas and the tectonic environment.
根据中国大陆主要含油气盆地天然气的~3He/~4He 数据，首次系统研究了天然气中氦同位素空间分布与中国大地构造环境的关系。

hematite ['hemətait]

Definition: hematite, also spelled as hæmatite, is the mineral form of iron (Ⅲ) oxide (Fe_2O_3), one of several iron oxides. Hematite crystallizes in the rhombohedral system, and it has the same crystal structure as ilmenite and corundum. 赤铁矿 chì tiě kuàng

Origin: 15th century. Via Latin, from Greek haimatitēs "blood-like (stone)".

Example:
In the soil, water, modern sediments, hematite, needles is the main iron ore minerals.
在土壤、水体、现代沉积物中，赤铁矿、针铁矿是主要矿物。

Extended Terms:

gray hematite 镜铁矿
ground hematite 赤铁矿粉
hematite rock 赤铁岩

hemipelagic sediment [ˌhemipiˈlædʒik ˈsedimənt]

Definition: hemipelagic sediment is deep-sea sediment that is commonly deposited near continental margins. 半远洋沉积 bàn yuǎn yáng chén jī

hercynite [ˈhəːsənait]

Definition: Hercynite is a spinel mineral with the formula $FeAl_2O_4$. 铁铝尖晶石 tiě lǚ jiān jīng shí

Example:
From all possible members of the spinel groups, hercynite and galaxite, were found to be the most efficient.
在所有可能的尖晶石矿族成员中,铁铝尖晶石和锰铝尖晶石被发现是最有效的(添加剂)。

herringbone cross bedding [ˈheriŋbəun krɔːs ˈbediŋ]

Definition: 人字形交错层理 rén zì xíng jiāo cuò céng lǐ

herringbone：a twilled fabric with a herringbone pattern. 人形状 rén xíng zhuàng
cross：to pass across each other. 交叉 jiāo chā；相交 xiāng jiāo
bedding：the stratification or layering of rocks or other geological materials. 层理 céng lǐ

hieroglyph [ˈhaiərəglif]

Definition: The pictographic symbols of ancient writing systems minerals and rocks which make up the crust of the earth. 象形印痕 xiàng xíng yìn hén

Origin: 1575-1585：from Latin hieroglyphicus, from Greek hieroglyphikós, pertaining to sacred writing.

Example:
Each picture, or hieroglyph, represents either an idea or a sound.
每一图画或每一个象形代表着一种想法或者一种声音。

high-alumina basalt [hai-əˈljuː minə ˈbæsɔːlt]

Definition: may be silica-undersaturated or oversaturated. It has greater than 17% alumina (Al_2O_3) and is intermediate in composition between tholeiite and alkaline basalt; the relatively alumina-rich composition is based on rocks without phenocrysts of plagioclase. 高铝玄武岩 gāo lǚ xuán wǔ yán

Example:
The volcanic rock series mainly belong to alkalic basalt series. However, a small amount of subakalic tholeiite series and high-alumina basalt series are formed by differentiation in the middle-late period of the volcanism.
火山岩系多属碱性玄武岩系列，火山活动的中晚期，分离出少量亚碱性拉斑玄武岩系列和高铝玄武岩系列。

Extended Term:
high alumina basalt 高铝氧玄武岩

high-K [hai kei]

Definition: the term high-K dielectric refers to a material with a high dielectric constant (K) (as compared to silicon dioxide) used in semiconductor manufacturing processes which replaces the silicon dioxide gate dielectric. 高 K gāo k

Example:
The high-K mechanisms and preparation methods of the composites are also discussed, with some suggestions given.
介绍了复合材料的高介电机制、材料制备方法，并给出建议。

Extended Terms:
high-K dielectric 高介电材料
high-pile K 长毛绒针织机

high-pressure [haiˈpreʃə]

Definition: a high-pressure area is a region where the atmospheric pressure at the surface of the planet is greater than its surrounding environment. Winds within high-pressure areas flow outward due to the higher density air near their center and friction with land. 高压 gāo yā

Origin: 1824, of engines, from high (*adj.*) + pressure (*n.*). Of weather systems from 1891; of sales pitches from 1933.

Example:
The main purpose of the thesis is to design a subject to realize the automation of CCH high-

pressure capacitance production process.
本课题的主要目的是设计一个 CCH 片式高压瓷介电容的装配线以实现其生产过程的自动化。

Extended Terms:

high-pressure region 高压区域
low pressure 低压力

high-volcanicity rift [hai-ˌvɔlkənˈisiti rift]

Definition: 大裂谷火山活动 dà liè gǔ huǒ shān huó dòng

high: containing a lot of a particular substance. 含某物多的 hán mǒu wù duō de
volcanicity: A volcano is an opening, or rupture, in a planet's surface or crust, which allows hot magma, ash and gases to escape from below the surface. 火山活动 huǒ shān huó dòng
rift: a large crack or opening in the ground, rocks or clouds. 断裂 duàn liè; 裂缝 liè fèng

high alumina basalt [hai əˈljuː mi nə ˈbæsɔːlt]

Definition: 高铝玄武岩 gāo lǚ xuán wǔ yán

high: of more than normal height; lofty; tall, not used of persons. 高的 gāo de
alumina: an oxide of aluminum, AlO, present in bauxite and clay and found as different forms of corundum, including emery, sapphires, and rubies. 氧化铝 yǎng huà lǚ
basalt: A hard, dense, dark volcanic rock composed chiefly of plagioclase, pyroxene, and olivine, and often having a glassy appearance. 玄武岩 xuán wǔ yán

high flow regime [hai fləu reiˈdʒiːm]

Definition: 高流态 gāo liú tài

high: containing a lot of a particular substance. 含某物多的 hán mǒu wù duō de
flow: the steady and continuous movement of something in one direction. 流 liú; 流动 liú dòng
regime: the statistical pattern of a river's constantly varying (daily) flow rates. 流态 liú tài

Example:

The paper studies the property, the mix proportion design and the formation method, reveals the variation characteristics of physical and mechanical property before and after high flow regime concrete being filled in membrane bag.
本文对模袋混凝土的性能、合比设计及成型方法进行了系统研究，揭示了高流态混凝土充灌模袋前后物理力学性能的变化特点。

high grade metamorphic terrain [hai greid ˌmetəˈmɔːfik teˈrein]

Definition: 高级变质区 gāo jí biàn zhì qū

high: containing a lot of a particular substance. 含某物多的 hán mǒu wù duō de
grade: the quality of a particular product or material. 等级 děng jí; 品级 pǐn jí
metamorphic: of rocks formed by the action of heat or pressure. 变质的 biàn zhì de
terrain: used to refer to an area of land when you are mentioning its natural features, for example, if it is rough, flat, etc. 地形 dì xíng; 地势 dì shì; 地带 dì dài

---

high magnesian calcite [hai mægˈniːʃən ˈkælsait]

Definition: 高镁方解石 gāo měi fāng jiě shí

high: containing a lot of a particular substance. 含某物多的 hán mǒu wù duō de
magnesian: (chiefly of rocks and minerals) containing or relatively rich in magnesium. 氧化镁的 yǎng huà měi de
calcite: a white or colorless mineral consisting of calcium carbonate. It is a major constituent of sedimentary rocks such as limestone, marble, and chalk, can occur in crystalline form (as in Iceland spar), and may be deposited in caves to form stalactites and stalagmites. 方解石 fāng jiě shí

Example:

According to analysis, beachrock is mabe up of terringenous clastic and bioclast, the cement are mainly aragonite and high Mg calcite.
分析测定，海滩岩主要由陆源碎屑和生物碎屑组成，胶结物为文石和高镁方解石.

---

high pressure metamorphism [hai ˈpreʃə ˌmetəˈmɔːfizəm]

Definition: high P (strongly differential), low-modle T; forms in convergent boundary zones. 高压变质作用 gāo yā biàn zhì zuò yòng

Example:

The existence of phengites in the metamorphic siliceous rock indicates that it was formed through middle-high pressure metamorphism.
岩石中多硅白云母的存在显示其曾经受了中-高压的变质作用。

---

hinge fault [hindʒ fɔːlt]

Definition: 枢纽断层 shū niǔ duàn céng

hinge: a piece of metal, plastic, etc., on which a door, lid or gate moves freely as it opens or closes. 铰链 jiǎo liàn；合叶 hé yè

fault：a place where there is a break that is longer than usual in the layers of rock in the earth's crust.（地壳岩层的）断层（dì qiào yán céng de） duàn céng

> Origin:

hinge：late 14 century as "movable joint of a gate or door," not found in Old English, cognate with Middle Dutch henghe "hook, handle."

fault：late 13 cencury, faute, "deficiency," from Old French faute, earlier falte, "opening, gap".

> Extended Terms:

hinge-line fault 转枢线断层
block fault 块断层
branching fault 分支断层

histogram [ˈhistəɡræm]

> Definition:

a statistical graph of a frequency distribution in which vertical rectangles of different heights are proportionate to corresponding frequencies. 直方图 zhí fāng tú

> Origin: 1891, from histo-"tissue" + -gram.

> Example:

Though histograms and pie charts are also graphs, the term usually applies to point plots on a coordinate system.
虽然直方图和圆饼图也是图表，该术语通常用于坐标系的点联曲线图。

> Extended Terms:

histogram equalization 直方图均衡
histogram specification 直方图规格化
colour histograms 颜色分布图

holocrystalline [ˌhɔləuˈkristəlain]

> Definition:

a crystallinity is a solid material whose constituents are arranged in highly ordered microscopic structure, forming a crystal lattice that extends in all direction. 全晶质 quán jīng zhì

> Example:

Olivine basalts with holocrystalline porphyritic texture are mainly found in the Shirreff Point.
橄榄玄武岩主要在史莱夫角出露，为全晶质斑状构造。

Extended Term:

holocrystalline rock 全晶质岩

holohyaline [ˌhɔləuˈhaiəlin]

Definition: a hyaline substance is one with a glassy appearance. 全玻质的 quán bō zhì de

Extended Term:

holohyaline texture 全玻质结构

horizontal bedding [ˌhɔriˈzɔntəl ˈbediŋ]

Definition: 水平层理 shuǐ íng céng lǐ

horizontal: flat and level; going across and parallel to the ground rather than going up and down. 水平的 shuǐ píng de；与地面平行的 yǔ dì miàn píng xíng de

bedding: the stratification or layering of rocks or other geological materials. 层理 céng lǐ

Origin:

horizontal: 1550s, "relating to or near the horizon," from French horizontal, from Latin horizontem.

bedding: later Old English beddinge "bedding, bed covering," from bed.

Example:

The bedding structures include horizontal bedding, parallel bedding, ripple bedding and tabular cross-bedding.

层理构造包括水平层理、平行层理、沙纹层理、板状交错层理。

Extended Term:

horizontal lamination 水平纹层

hornblende [ˈhɔːnblend]

Definition: a dark brown, black, or green mineral of the amphibole group consisting of a hydroxyl alumino-silicate of calcium, magnesium, and iron, occurring in many igneous andmetamorphic rocks. 角闪石 jiǎo shǎn shí

Origin: late 18th century, from German.

Extended Terms:

basaltic hornblende 玄角闪石
hornblende porphyry 角闪斑岩

hornblende andesite 角闪安山岩
hornblende schist 角闪片岩
hornblende geobarometry 角闪石地质压力测定法

hornblende hornfels facies [ˈhɔːnblend ˈhɔːnfelz ˈfeiʃiːz]

Definition: 角闪石 jiǎo shǎn shí; 角岩相 jiǎo yán xiàng

hornblende: a dark brown, black, or green mineral of the amphibole group consisting of a hydroxyl alumino-silicate of calcium, magnesium, and iron, occurring in many igneous andmetamorphic rocks. 角闪石 jiǎo shǎn shí

hornfels: a nonfoliated metamorphic rock of uniform grain size, formed by high-temperature metamorphism, typically formed by contact metamorphism around igneous intrusions. 角页岩 jiǎo yè yán

facies: the character of a rock expressed by its formation, composition, and fossil content. 相 xiàng

hornblende schist [ˈhɔːnblend ʃist]

Definition: 角闪片岩 jiǎo shǎn piàn yán

hornblende: a dark brown, black, or green mineral of the amphibole group consisting of a hydroxyl alumino-silicate of calcium, magnesium, and iron, occurring in many igneous and metamorphic rocks. 角闪石 jiǎo shǎn shí

schist: a type of rock formed of layers of different minerals, that breaks naturally into thin flat pieces. 片岩 piàn yán

Origin:

hornblende: common dark mineral, 1770, from German hornblende.
schist: type of layered metamorphic rock, 1795, from French schiste.

hornblendite [ˈhɔːnblendait]

Definition: a plutonic rock consisting mainly of hornblende. 角闪石岩 jiǎo shǎn shí yán

Examples:

① Most of the smaller hornblendite bodies contain about 5-10 percent magnetite, with only local sporadic enrichment.
在小的角闪岩体中，磁铁矿可占5%~10%，仅有局部的散染状富集。

② Palagioclase hornblendite of the Archeozoic Qianxi Croup Jincahngyu Formation is closely associated with the gold mineralization in eastern Hebei Province.

冀东太古界迁西群金厂峪组斜长角闪岩与金矿关系密切。

Extended Terms:

enstatite hornblendite 顽火角闪石岩
peridotite hornblendite 橄榄角闪石岩
biotite hornblendite 黑云角闪石岩

hornfels [ˈhɔːnfelz]

Definition: Also called hornstone, a hard compact fine-grained metamorphic rock formed by the action of heat from a magmatic intrusion on neighbouring sedimentary rocks. 角岩 jiǎo yán

Example: The Hongniu Cu deposit in "Sanjiang Metallogenetic Zone" is occurred in the skarnization alteration zone in the margin of marble and the hornfels body.
"三江成矿带"内的红牛铜矿，赋存于大理岩边部矽卡岩化蚀变带及角岩体中。

Extended Terms:

pelitic hornfels 泥质角页岩
pelitic hornfels 泥质角页岩
mafic hornfels 镁铁质角岩
magnesian hornfels 镁质角页岩

hornfels texture [ˈhɔːnfelz ˈtekstʃə]

Definition: 角岩结构 jiǎo yán jié gòu

hornfels: a nonfoliated metamorphic rock of uniform grain size, formed by high-temperature metamorphism, typically formed by contact metamorphism around igneous intrusions. 角页岩 jiǎo yè yán

texture: in geology refers to the physical appearance or character of a rock, such as grain size, shape, arrangement, and pattern at both the megascopic or microscopic surface feature level. 质地 zhì dì; 纹理 wén lǐ; 结构 jié gòu

hot spot [hɔt spɔt]

Definition: an area of volcanic activity, especially where this is isolated. 火山活动区 huǒ shān huó dòng qū

Origin: also hotspot, 1888 as a skin irritation; 1931 as "nightclub;" 1938 in the firefighting sense; 1941 as "place of international conflict."

hummocky cross bedding [ˈhʌməki krɔːs ˈbediŋ]

Definition: 丘状交错层理 qiū zhuàng jiāo cuò céng lǐ

hummocky: A hummock (of uncertain derivation; cf. hump or hillock) is a boss or rounded knoll of ice rising above the general level of an ice-field, making sledge travelling in the Arctic and Antarctic regions extremely difficult and unpleasant. 圆丘状的 yuán qiū zhuàng de
cross: to pass across each other. 交叉 jiāo chā; 相交 xiāng jiāo
bedding: the stratification or layering of rocks or other geological materials. 层理 céng lǐ

hyaloclastite [ˌhaiələuˈklæstait]

Definition: Hyaloclastite is a hydrated tuff-like breccia rich in black volcanic glass, formed during volcanic eruptions under water, under ice or where subaerial flows reach the sea or other bodies of water. It has the appearance of angular flat fragments sized between a millimeter to few centimeters. 玻质碎屑岩 bō zhì suì xiè yán

Extended Term:
hyalobasalt 玻质玄武岩

hyalopilitic texture [ˌhaiələuˌpaiˈlitik ˈtekstʃə]

Definition: 玻基交织结构 bō jī jiāo zhī jié gòu

hyalopilitic: a textural term used in petrographic classification of volcanic rocks. 玻晶交织的
texture: the arrangement of the particles or constituent parts of any material, as wood, metal, etc., as it affects the appearance or feel of the surface; structure, composition, grain, etc. 结构 jié gòu

hybrid sedimentary [ˈhaibrid ˌslediˈmentəri]

Definition: 混积岩 hùn jī yán

hybrid: of mixed character; composed of mixed parts. 混合的 hùn hé de
sedimentary: connected with or formed from the sand, stones, mud, etc. that settle at the bottom of lakes, etc. 沉积的 chén jī de; 沉积形成的 chén jī xíng chéng de

hydration [haiˈdreiʃən]

Definition: a chemical compound containing water molecules that can usually be expelled by heating, without decomposition of the compound. 水合作用 shuǐ hé zuò yòng

Example:

If water is the solvent, it is called hydration.
如果以水作溶剂，它便称为水合。

Extended Terms:

osmotic hydration 渗透水化
hydration heat 水合热

hydraulic [haiˈdrɔːlik]

Definition: denoting, relating to, or operated by a liquid moving in a confined space under pressure. 液压的 yè yā de

Origin: early 17th century: via Latin from Greek hudraulikos, from hudro-"water" + aulos "pipe".

Example:

It's a study on shape control features of hydraulic bulging roller and key technology.
这是关于液压胀形轧辊的承载特性及其关键技术的研究。

Extended Terms:

hydraulic motor 液压马达
hydraulic drive 液压传动

hydrodynamic dispersion [ˌhaidrəudaiˈnæmik diˈspəːʃən]

Definition: longitudinal and lateral spreading (at a macroscopic level) of a solute being advected through porous media due to mechanical dispersion and molecular diffusion. 水动力弥散 shuǐ dòng lì mí sàn

Example:

The hydrodynamic dispersion coefficient is a very important parameter for setting up a groundwater quality model and studying the solute migration of water-bearing medium.
水动力弥散系数是建立地下水水质模型并进行地下水溶质运移研究的重要参数。

Extended Term:

coefficient of hydrodynamic dispersion 水动力弥散系数

hydrogen-isotope fractionation [ˈhaidrədʒən-ˈaisəutəup ˌfrækʃəˈneiʃən]

Definition: 氢同位素分馏 qīng tóng wèi sù fēn liú

hydrogen: a chemical element. Hydrogen is a colourless gas that is the lightest of all the elements. It combines with oxygen to form water. 氢 qīng; 氢气 qīng qì

isotope: one of two or more forms of a chemical element with different physical properties(= characteristics) but the same chemical ones. 同位素 tóng wèi sù

fractionation: a process that uses heat to separate a substance into its components. 分馏 fēn liú

Example:

Hydrogen-isotope fractionation between ilvaite and water has been studied experimentally at the temperature range 750℃-250℃.
在750℃~250℃温度范围内,对黑柱石-水之间的氢同位素分馏进行了实验研究。

hydrogen isotope [ˈhaidrədʒən ˈaisəutəup]

Definition: 氢同位素 qīng tong wèi sù

hydrogen: a chemical element. Hydrogen is a colourless gas that is the lightest of all the elements. It combines with oxygen to form water. 氢 qīng; 氢气 qīng qì

isotope: one of two or more forms of a chemical element with different physical properties(= characteristics) but the same chemical ones. 同位素 tóng wèi sù

Example:

In certain fusion processes, among nuclei of the hydrogen isotopes, neutrons of high energy are liberated.
在氢同位素原子核中的某些熔合过程中,会释放出高能中子。

Extended Terms:

exchange and displacement of hydrogen isotopes 氢同位素的交换与置换
hydrogen isotope composition 氢同位素成分
hydrogen-isotope fractionation 氢同位素分馏

hydrogen sulfide [ˈhaidrədʒən ˈsʌlfaid]

Definition: hydrogen sulfide (or hydrogen sulphide) is the chemical compound with the formula H_2S. It is a colorless, very poisonous, flammable gas with the characteristic foul odor of rotten eggs. 硫化氢 liú huà qīng

Extended Terms:

hydrogen-sulfide-forming bacteria 生成硫化氢的细菌
hydrogen sulfide production test 硫化氢试验

hydromica clay [ˌhaidrəˈmaikə klei]

Definition: 水云母黏土 shuǐ yún mǔ nián tǔ

hydromica: an alternate name for illite (illite-A clay mineral of the muscovite mica group, with a lattice structure that does not expand on absorption of water). 伊利石 yī lì shí

clay: a type of heavy, sticky earth that becomes hard when it is baked and is used to make things such as pots and bricks. 黏土 nián tǔ；陶土 táo tǔ

Example:

The minerals in shale are mainly hydromica clay, next to it is kaolinite, halloysite, montomorillonite and little allophane.

页岩中的矿物主要为水云母，次为高岭石、多水高岭石、蒙脱石、少量水铝英石。

Extended Terms:

illite hydromica 伊利水云母
hydromica-schist 水云母片岩
hydromica slate 水云母石板瓦

hydrostatic pressure [ˌhaidrəuˈstætik ˈpreʃə]

Definition: the pressure which is exerted on a portion of a column of fluid as a result of the weight of the fluid above it. 静水压力 jìng shuǐ yā lì

Example:

Below the water table at static equilibrium, hydrostatic pressure potential increases with increasing depth.

在水面以下，当静态平衡时，静水压力随深度而增加。

Extended Terms:

hydrostatic column pressure difference 静液柱压差
hydrostatic excess pressure 超静水压力

hydrothermal alteration [ˌhaidrəuˈθɜːməl ˌɔːltəˈreiʃən]

Definition: the phase changes resulting from the interaction of hydrothermal fluids with pre-existing solid phases. Included are the chemical and mineralogical changes in rocks brought about by the addition or removal of materials through the medium of hydrothermal fluids. 热液蚀变 rè yè shí biàn

Example:

The faulted structure, hydrothermal alteration, gossan, geophysical-geochemical exploration

anomalies are comprehensive ore-hunting indicators.
断裂构造、热液蚀变、铁帽、物化探异常为重要的找矿标志。

> Extended Terms:

hydrothermal activity 水热活动
hygroscopic moisture 潮解水

hydrothermal circulation [ˌhaidrəuˈθəːməl ˌsəːkjuˈleiʃən]

> Definition: Hydrothermal circulation in its most general sense is the circulation of hot water; "hydros" in the Greek meaning water and "thermos" meaning heat. Hydrothermal circulation occurs most often in the vicinity of sources of heat within the Earth's crust. 热液循环 rè yè xún huán

> Example:

Process of hydrothermal circulation can be divided into 3 parts scientifically and reasonably, that is, (1) hydrothermal circulation under the seafloor, (2) hydrothermal circulation in the interface of seafloor and ocean, and (3) hydrothermal fluid entering the ocean.
热液循环过程可以有机地分解为三个子过程：海底以下部分、海底与大洋交界处及热液流体在大洋中的流动。

> Extended Term:

hydrothermal convection system 水热对流系统

Hydrothermal vent [ˌhaidrəuˈθəːməl vent]

> Definition: an opening in the seafloor out of which heated mineral-rich water flows. 热液喷口 rè yè pēn kǒu

> Example:

This paper introduced the species and energy sources of micro-organisms which inhabit in the vicinity of deep-sea hydrothermal vents.
介绍了生长在深海热液喷口周围的微生物的种类及其生长繁殖的能量来源。

> Extended Term:

hydrothermal-vent biotas 热液喷口生物群

hydrous fluid [ˈhaidrəs ˈflu(ː)id]

> Definition: 含水流体 hán shuǐ liú tǐ

hydrous: containing water as a constituent. 含水的 hán shuǐ de

fluid: a substance that has no fixed shape and yields easily to external pressure; a gas or a liquid. 流体 liú tǐ

Example:

Most of the highly evolved granites may belong to F-rich system. The F-rich silicate melt-hydrous fluid system shows distinctive geochemical behaviors.
高度演化花岗岩类多为富 F 的硅酸盐熔体溶液体系，具有鲜明的、不同于其他体系的地球化学特征。

hypabyssal intrusion [ˌhipəˈbisəl inˈtruːʒən]

Definition: 浅成侵入岩 qiǎn chéng qīn rù yán

hypabyssal: (of rock) that solidifies, as a minor intrusion, before reaching the Earth's surface. 浅成 qiǎn chéng

intrusion: the action or process of forcing a body of igneous rock between or through existing formations, without reaching the surface. 侵入 qīn rù

Example:

The volcanism is strong with obvious cycles, and subvolcanism-hypabyssal intrusion happens many times.
区内火山作用强烈旋回明显，期后有多次的地下火山浅成侵入岩浆活动。

Extended Terms:

hypabyssal instrusive 浅成侵入体
hypabyssal rock 半深成岩

hypabyssal rock [ˌhipəˈbisəl rɔk]

Definition: solidifying chiefly as a minor intrusion, especially as a dike or sill, before reaching the Earth's surface. 浅成岩 qiǎn chéng yán

Example:

Ore-bearing breccias can be divided into complete-pipe and compound mineralizing types in the east part of north Huaiyang tectonic zone, in which the former type keeps certain distance from the hypabyssal rock, the latter one is accompanied by the hypabyssal rock in space.
北淮阳构造带东段含矿角砾岩主要有全筒矿化型和复合矿化型两种，前者与浅成岩有一定距离，后者在空间上与浅成岩伴生。

hypersolvus [ˌhaipəˈsɔlvəs]

Definition: In hypersolvus granites, crystallization at relatively low water pressures results in the formation of a single feldspar as opposed to subsolvus granites in which two distinct types of feldspar are present. 超熔线的 chāo róng xiàn de

Example:

The rock mainly consists of sodalite, nepheline, Na, Ca amphibole, aegirine, augite, annite and perthite, without any separate phase of plagioclase, indicating that it is a hypersolvus alkali syenite.
岩石由方钠石、霞石、钠、钙角闪石、霓石、普通辉石、铁黑云母和条纹长石组成, 不含独立相斜长石, 是一种超溶线碱性正长岩。

Extended Term:

hypersolvus crystallization 超熔线结晶

hypersolvus crystallization [ˌhaɪpəˈsɒlvəs ˌkrɪstəlaɪˈzeɪʃən]

Definition: 超熔线结晶 chāo róng xiàn jié jīng

Hypersolvus: In hypersolvus granites, crystallization at relatively low water pressures results in the formation of a single feldspar as opposed to subsolvus granites in which two distinct types of feldspar are present. 超熔线 chāo róng xiàn

crystallization: a mental synthesis that becomes fixed or concrete by a process resembling crystal formation. 结晶 jié jīng

hypersthene [ˈhaɪpəsθiːn]

Definition: a green, brown, or black pyroxene mineral containing iron and magnesium. 紫苏辉石 zǐ sū huī shí

Origin: early 19th century. From French hypersthène "extremely strong (mineral)", from Greek sthenos "strength".

Example:

The results show that, main crystallization phase of glass ceramic was hedenbergite, and the minor crystallization phases were augite and hypersthene.
结果表明: 矿渣微晶玻璃的主析晶相为钙铁辉石, 次晶相为普通辉石和紫苏辉石。

Extended Terms:

hypersthene akerite 紫苏辉正长岩
hypersthene andesite 紫苏安山岩
hypersthene granite 紫苏辉石花岗岩

hypsometry [hipˈsɔmitri]

Definition: Hypsometric is a scientific term relating to the measurement of heights. The term originates from the Greek word "hypsos" meaning height and the word, is from the Greek (métron), "a". 测高法 cè gāo fǎ

Origin: the measuring of altitudes, 1560s.

Example:

Currently, it has a noticeable research advance in passive continental margins, plate boundaries and active continental margins, and continental hypsometry.
目前，它在被动大陆边缘、板块边界和活动大陆边缘，以及大陆测高法方面取得了令人瞩目的研究进展。

Extended Terms:

hypsometer 沸点测高器
altimetry hypsometry 测高法
barometric hypsometry 沸点气压测高法；气压测高术

Iceland [ˈaislənd]

Definition: a Nordic island country in the North Atlantic Ocean. 冰岛 bīng dǎo

Example:

The study on the cause of the color shows that Mn, Pb and CO_2-3 are contributed to the yellow in the Iceland spar, while the purple of the Iceland is caused by Mn, Pb, CO_{32}-and Ni.
对于黄色、紫色冰洲石颜色成因，本文初步研究认为，Mn、Pb 和 CO_{2-3} 影响黄色的形成而 Mn, Pb, CO_{32}-和 Ni 则影响紫色的形成。

Extended Term:

Iceland spar 冰洲石

Icelandic-style eruption [aisˈlændik-stail iˈrʌpʃən]

Definition: 冰岛式喷发 bīng dǎo shì pēn fā

Icelandic: Of or relating to Iceland or its language. 冰岛的 bīng dǎo de；冰岛语的 bīng dǎo yǔ de
style: having the type of style mentioned. 式的 shì de；风格的 fēng gé de
eruption: a sudden outpouring of a particular substance from somewhere. 喷发 pēn fā

Icelandite [aisˈlændait]

Definition: Icelandite is a type of volcanic rock, an iron rich, aluminium poor andesite. 冰岛岩 bīng dǎo yán

Example:
The volcanic rocks are composed of tholeiite and Icelandite.
这些火山岩由拉斑玄武岩和冰岛岩组成。

ice sheets [ais ʃiːts]

Definition: An ice sheet is a mass of glacier ice that covers surrounding terrain and is greater than 50,000 km^2 (20,000 $mile^2$), thus also known as continental glacier. 冰盖 bīng gài

Example:
The ice sheets that cover Greenland and Antarctica now are remnants of this and previous ice ages.
现在覆盖着格陵兰和南极的冰盖就是上述冰块和以前几次冰期的遗留物。

Extended Terms:
floting ice 浮冰
ice shelf 冰架

ichnite footprint [ˈiknait ˈfutprint]

Definition: 动物足迹 dòng wù zú jì

ichnite: a fossilized footprint or track of an animal. 化石足迹 huà shí zú jì
footprint: a mark left on a surface by a person's foot or shoe or by an animal's foot. 脚印 jiǎo yìn; 足迹 zú jì

ichnofossil [ˌiknəuˈfɔsəl]

Definition: trace fossils, also called ichnofossils, are geological records of biological activity. Trace fossils may be impressions made on the substrate by an organism: for example, burrows, borings (bioerosion), footprints and feeding marks, and root cavities. 足迹化石 zú jì huà shí

Example:
Ichnofossil study is especially important in interpreting the depositional environment.
遗迹化石研究具有多方面的意义, 而对沉积环境的解释是其中重要的方面。

Extended Terms:

trace fossil 遗迹化石
ichnology 足迹化石学

idioblastic series [ˌidiəˈblæstik ˈsiəriːz] = crystalloblastic series

Definition: 自形变晶系列 zì xíng biàn jīng xì liè

idioblastic: of, or relating to an idioblast. 自形变晶的 zì xíng biàn jīng de
series: several events or things of a similar kind that happen one after the other. 一系列 yí xì liè；连续 lián xù；接连 jiē lián

idiomorphic crystal [ˌidiəˈmɔːfik ˈkristəl]

Definition: 自形晶 zì xíng jīng

idiomorphic: having normal faces characteristic of a particular mineral; said of crystals in rock that have developed without interference. 自形的 zì xíng de
crystal: a mineral, especially a transparent form of quartz, having a crystalline structure, often characterized by external planar faces. 晶体 jīng tǐ

igneous [ˈigniəs]

Definition: Igneous is a term used for solidified magma. It is also a term used for describing the processes related to the formation of igneous rocks. 火成的 huǒ chéng de

Origin: mid-17th century: from Latin igneus (from ignis "fire") + -ous.

Example:
Because they have been extruded or pored out on the Earth's surface, igneous rocks from volcanoes are called extrusive igneous rocks.
因为这种火山岩是熔岩喷出地球表面而形成的，所以称之为喷出岩。

Extended Terms:

igneous activity 火成活动；岩浆活动
igneous rock 火成岩
igneous complex 火成杂岩

igneous accumulation [ˈigniəs əˌkjuːmjuˈleiʃən]

Definition: 火成堆积 huǒ chéng duī jī

igneous: formed by solidification from a molten state. Used of rocks. 火成的 huǒ chéng de
accumulation: an accumulating or being accumulated; collection. 堆积物 duī jī wù

Example:

They cite this phenomenon as an evidence for an origin by igneous accumulation of hyperaluminous clinopyroxene at high pressure.
他们引用这一现象作为高压环境下高氧化铝单斜辉石的火成堆积起源的证据。

igneous bodies [ˈigniəs ˈbɔdis]

Definition: 火成岩体 huǒ chéng yán tǐ

igneous: of rocks formed when magma(= melted or liquid material lying below the Earth's surface) becomes solid, especially after it has poured out of a volcano. 火成的 huǒ chéng de
bodies: the main or central part of something, especially. a building or text. 主体 zhǔ tǐ

Example:

Interpretation of remote sensing information manifests that main ore-controlling factors in the Panxi area are fault belts and igneous bodies.
控矿构造的遥感信息解译表明攀西地区主要的控矿因素为断裂带和火成岩体。

Extended Term:

sheet-like igneous bodies 席状岩体

igneous environments [ˈigniəs inˈvaiərənmənts]

Definition: 火成岩环境 huǒ chéng yán huán jìng

igneous: of rocks formed when magma(= melted or liquid material lying below the Earth's surface) becomes solid, especially after it has poured out of a volcano. 火成的 huǒ chéng de
environment: the natural world, as a whole or in a particular geographical area, especially as affected by human activity. 环境 huán jìng

Example:

Some tomographic studies have been performed to characterize contaminated igneous environments.
一些断层研究描述了受污染的火成岩环境的特征。

igneous rock [ˈigniəs rɒk]

Definition: a rock type that has been created from super-heated magma. The three main types of rock are igneous, sedimentary, and metamorphic. 火成岩 huǒ chéng yán

Example:

The molten rock material under the Earth's crust, from which igneous rock is formed by cooling.
岩浆地球外壳内部的熔融岩石物质，冷却后形成火成岩。

Extended Terms:

intermediate igneous rock 中性火成岩
alkalic igneous rock 碱性火成岩

igneous rock association [ˈigniəs rɔk əˌsəusiˈeiʃən]

Definition: 火成岩岩石组合 huǒ chéng yán yán shí zǔ hé

igneous: of rocks formed when magma (= melted or liquid material lying below the Earth's surface) becomes solid, especially after it has poured out of a volcano. 火成的 huǒ chéng de
rock: the hard solid material that forms part of the surface of the earth and some other planets. 岩石 yán shí
association: a group of organisms (plants and animals) that live together in a certain geographical region and constitute a community with a few dominant species. 组合 zǔ hé

Example:

The $K_{(60)}$ value of igneous rock association cannot be used to estimate the crustal thickness of the Mesozoic orogenic belts of eastern China.
火成岩组合的 $K_{(60)}$ 值不能用于估算中国东部中生代造山带地壳厚度。

igneous rock series [ˈigniəs rɔk ˈsiəriːz]

Definition: 火成岩系列 huǒ chéng yán xì liè

igneous rock: rocks formed by the cooling and solidifying of molten materials. Igneous rocks can form beneath the Earth's surface, or at its surface, as lava. 火成岩 huǒ chéng yán
series: a number of things, events, or people of a similar kind or related nature coming one after another. 系列 xì liè

igneous rock taxonomy by SiO_2 [ˈigniəs rɔk takˈsɔnəmi]

Definition: 以 SiO_2 为标准的火成岩分类 yǐ SiO_2 wéi biāo zhǔn de huǒ chéng yán fēn lèi

igneous rock: (of rock) having solidified from lava or magma. 火成岩 huǒ chéng yán
taxonomy: (chiefly Biology) the branch of science concerned with classification, especially of organisms; systematics. (尤指生物)分类学 fēn lèi xué; 分类法 fēn lèi fǎ

igneous tectonic assemblage [ˈigniəs tekˈtɔnik əˈsemblidʒ]

Definition: 火成岩构造组合 huǒ chéng yán gòu zào zǔ hé

igneous: formed by solidification from a molten state. Used of rocks. 火成的 huǒ chéng de

tectonic: relating to, causing, or resulting from structural deformation of the Earth's crust. 构造的 gòu zào de

assemblage: a form of art involving the assembly and arrangement of unrelated objects, parts, and materials in a kind of sculptured collage. 集合 jí hé

igneous texture [ˈigniəs ˈtekstʃə]

Definition: 火成岩结构 huǒ chéng yán jié gòu

igneous: of rocks formed when magma(= melted or liquid material lying below the Earth's surface) becomes solid, especially after it has poured out of a volcano. 火成的 huǒ chéng de

texture: in geology refers to the physical appearance or character of a rock, such as grain size, shape, arrangement, and pattern at both the megascopic or microscopic surface feature level. 质地 zhì dì; 纹理 wén lǐ; 结构 jié gòu

Example:

The pale microgranular enclaves with igneous texture (PMEI) could be fragments of a yet-unknown granitic material or unmelted igneous material of an inhomogeneous source.
这些具有火成岩结构的淡色微晶包体可能是未知的花岗岩类物质的粉粒或未融化的非均匀的火成岩物质。

Extended Term:

igneous rock structures 火成岩构造

ignimbrite [ˈignimbrait]

Definition: a volcanic rock consisting of droplets of lava and glass that were welded together by intense heat. 熔结凝灰岩 róng jié níng huī yán

Origin: mid-20th century, from Latin ignis "fire" + imbr-"rain".

Example:

So that it should avoide mixing the difference between rhyolites and acidic ignimbrite on research work.
实际工作中应避免混淆流纹岩与酸性熔结凝灰岩。

ijolite [ˈijəlait]

Definition: ijolite (derived from the first syllable of the Finnish words Ii-vaara, Iijoki, common as geographical names in Finland, and the Greek. Xiflos, a stone), is an igneous rock consisting essentially of nepheline and augite. 霓霞岩 ní xiá yán

Example:
Ijolite and melilitolite intrude peridotite and pyroxenite, while nepheline syenite and carbonatite intrude the ultramafic rocks as well as ijolite.
霓霞岩和黄长石岩侵入橄榄岩和辉石岩，而霞石正长岩和碳酸盐岩侵入超镁铁岩。

Extended Term:
ijolite porphyry 霓霞斑岩

illite [ˈilait]

Definition: a clay mineral of the micagroup containing potassium and aluminum. Source：shale, mudstone. 伊利石 yī lì shí

Origin: mid-20th century, after Illinois.

Example:
The presence of ferric oxide in the ores is key factor resulting in low whiteness of illite.
河南舒山伊利石矿中氧化铁的存在是造成伊利石白度低的主要原因。

Extended Terms:
illite clay 白云，母石黏土
sodium illite 钠伊利石
stripped illite 退化伊利石
illite hydromica 伊利水云母

ilmenite [ˈilmənait]

Definition: a mixed oxide mineral containing iron and titanium. Source：igneous and metamorphic rocks. 钛铁矿 tài tiě kuàng

Origin: early 19th century, after the Ilmen Mountains in the South Urals, Russia.

Example:
Tin, gold, ilmenite, zircon, monazite and kaolin, fire-resistant clay, and so have the potential for exploration.
锡、金、钛铁矿、锆英石、独居石及高岭土、耐火黏土等具有勘探潜力。

Extended Terms:

weathered ilmenite 蚀变钛铁矿

ilmenite bronzitite 钛铁古铜灰岩

immature stage [ˌiməˈtjuə steidʒ]

Definition: 未成熟阶段 wèi chéng shú jiē duàn

immature: behaving in a way that is not sensible and is typical of people who are much younger. 不成熟的 bú chéng shú de; 不够老练的 bú gòu lǎo liàn de

stage: a period or state that something/somebody passes through while developing or making progress. 时期 shí qī; 阶段 jiē duàn; 状态 zhuàng tài

Example:

Diagenesis is an immature stage when mostly carbon dioxide, water, and some methane and heavy hetero-compounds are generated.

成岩作用是一个未成熟阶段,这时多半生成二氧化碳、水、一些甲烷和重质杂化合物。

immobile elements [iˈməubail ˈelimənts]

Definition: 不活动要素 bù huó dòng yào sù

immobile: unable to move. 不能移动的 bù néng yí dòng de; 不能活动的 bù néng huó dòng de

element: a necessary or typical part of something. 要素 yào sù

Example:

In contrast, immobile elements such as Zr、Hf、Nb、Ta、Ti 、Y、REE and Ti show no evidence for significant loss, thus can be used to trace the geochemical nature of protoliths.

非活动性元素如 Zr、Hf、Nb、Ta、Ti、Y、REE 和 Ti 在脱水变质过程中没有受到明显损失,因此可用于追寻原始石器时代地球化学的本质。

incompatible element [ˌinkəmˈpætəbl ˈelimənt]

Definition: Incompatible element is a term used in petrology and geochemistry. 不相容元素 bù xiāng róng yuán sù

Example:

The high-Sr rhyolites, enriched in Ti, Ba, Sr, Co, Ni and depleted in highly incompatible element, such as Rb, Zr and Th, are similar to the those in the continental flood basalt province formed by fractional crystallization.

前者富集 Ti、Ba、Sr、Co 和 Ni 而贫 Rb、Zr 和 Th 等强不相容元素，类似于大陆溢流玄武岩省分异作用形成的流纹岩。

incompatible trace element [ˌinkəmˈpætəbl treis ˈelimənt]

Definition: 不相容微量元素 bù xiāng róng wēi liàng yuán sù

incompatible: two things that are incompatible are of different types so that they cannot be used or mixed together. 不匹配 bú pǐ pèi；不兼容 bú jiān róng

trace: a very small amount of something. 微量 wēi liàng；少许 shǎo xǔ

element: a necessary or typical part of something. 要素 yào sù

Example:
The mantle source is affected by the recent metasomatic enrichment with the characteristics of the concentration of incompatible trace elements of LREE, K, P, Th. U and so on.
其地幔源区受到近期交代富集作用的影响，具有富集 LREE、K、P、Th、U 等不相容微量元素的特征。

index mineral [ˈindeks ˈminərəl]

Definition: An index mineral is used in geology to determine the degree of metamorphism which a rock has experienced. 指示矿物 zhǐ shì kuàng wù

Example:
Amphibole is a major coexisting mineral at pre- and post-eclogite stages and can be regarded as an index mineral because its upper stability limit is less than 2.5 GPa.
角闪石是榴辉岩演化早期和晚期阶段的主要共生矿物，它可以被看做一种指示性矿物，其稳定上限低于 2.5 GPa。

Extended Term:
secondary mineral 次生矿物

Inductively coupled plasma [inˈdʌktivli ˈkʌpld ˈplæzmə]

Definition: An inductively coupled plasma (ICP) is a type of plasma source in which the energy is supplied by electrical currents which are produced by electromagnetic induction. That is, by time-varying magnetic fields. 电感耦合等离子体 diàn gǎn ǒu hé děng lí zǐ tǐ

Example:
A method for determination of 18 impurities in gadolinia by inductively coupled plasma atomic emission spectrometer are presented in this paper.

本文介绍了电感耦合等离子体原子发射光谱法测定氧化钆中的 18 种元素。

Extended Terms:
inductively-coupled plasma spectrometer 电感耦合等离子体光谱仪
inductively coupled plasma quantometer 电感耦合等离子体光量计

induration [ˌɪndjʊəˈreɪʃən]

Definition: the process by which a soft geologic sediment becomes hard. 固结作用 gù jié zuò yòng

Origin: mid-16th century: from Latin indurat-"made hard", from the verb indurare (based on durus "hard").

Example:
The principle of high pressure jet grouting is briefly introduced. The mechanism of underground solidification and induration of cement clay grout is emphatically discussed and tested and vertified in engineering.
简要介绍了高压喷射灌浆新技术的原理,重点论述了水泥黏土浆在地下的固结与硬化机理,并以工程实践作了验证。

inert component [ɪˈnɜːt kəmˈpəʊnənt]

Definition: 惰性组分 duò xìng zǔ fēn
inert: without active chemical or other properties. 惰性的 duò xìng de; 不活泼的 bú huó pō de
component: one of several parts of which something is made. 组成部分 zǔ chéng bù fèn; 成分 chéng fèn; 部件 bù jiàn

Example:
Pyrolysis, discarded printed circuit boards in the degradation of organic resin into small molecules of petroleum products, and at the same time inorganic inert component of the separation.
热解后,废弃印刷电路板中的有机树脂降解成小分子的石油产品,同时与无机惰性组分分离。

Extended Terms:
inert solid component 惰性固体(燃料)组分
determining inert component 确定惰性组分

inertinite [ɪˈnɜːtɪˌnaɪt]

Definition: Inertinite is oxidized organic material or fossilized charcoal. It is found as tiny flakes within sedimentary rocks. The presence of inertinite is significant in the geological record, as it signifies that wildfires occurred at the time that the host sediment was deposited. 惰性煤素质 duò xìng méi sù zhì

Example:

The test results showed that when the vitrainite and inertinite in the sample are individually 65%~75% and 35%~25%, the density of the coal water mixture prepared will be higher and the viscosity will be lower.
结果表明,当样品中镜质组和惰质组的比例为65%~75%和35%~25%时,制得的水煤浆的浓度较高、黏度较低。

infiltration metasomatism [ˌinfilˈtreiʃən ˌmetəˈsəumətizəm]

Definition: 渗透交代作用 shèn tòu jiāo dài zuò yòng

infiltration: a process in which individuals (or small groups) penetrate an area (especially the military penetration of enemy positions without detection). 渗透 shèn tòu

metasomatism: change in the composition of a rock as a result of the introduction or removal of chemical constituents. 交代变化 jiāo dài biàn huà

Example:

Detailed analysis of the whole rock compositions shows that the reaction zonation was formed by infiltration metasomatism that caused significant mass loss in the two alteration zones.
对整个岩石成分的详细分析显示出渗透交代作用导致这两个蚀变区域发生显著质损,从而形成此反应分带。

initial isotopic ratio [iˈnifəl ˌaisəuˈtɔpik ˈreiʃiəu]

Definition: 同位素初始比值 tóng wèi sù chū shǐ bǐ zhí

initial: having to do with, indicating, or occurring at the beginning. 最初的 zuì chū de
isotopic: any of two or more forms of an element having the same or very closely related chemical properties and the same atomic number but different atomic weights (or mass numbers). 同位素的 tóng wèi sù de
ratio: a fixed relation in degree, number, etc., between two similar things; proportion. 比值 bǐ zhí

Example:

If the crust material experienced a Rb-lost or/and Sr-gained event, the slope of the isochron will vary greatly depending on the isotopic ratios of the two components. The slope may result in a very old age and abnormally low initial isotopic ratio, sometimes even a negative slope.

若地壳物质在混合之前经受过 Rb-丢失和(或)Sr-获得，则等时线斜率的变化较大，可能得到很大的年龄和异常低的初始同位素比值，甚至斜率为负值。

injection [inˈdʒekʃən]

Definition: the act or process of introducing fluid under pressure, such as fuel into the combustion chamber of an engine. 注入作用 zhù rù zuò yòng

Origin: 1535-1545; from Latin injectiōn-(s. of injectiō).

Example:
The cause was injection of fluids into deep wells for waste disposal and secondary recovery of oil, and the use of reservoirs for water supplies.
人为引发地震的原因主要是因垃圾掩埋和二次采油而将液体注入深井，以及为了水供给水库的使用。

Extended Terms:
injection moulding 喷射铸造法；注射成形法；注塑；注塑法
injection nozzle 射出喷嘴；注射喷口，压注喷口；喷嘴；注射喷嘴
cement injection 水泥灌注工程；灌注水泥；水泥灌浆；水泥浆灌法
injection welding 注塑焊接；注射焊接
injection head 注射头；注入头；喷头

injection migmatite [inˈdʒekʃən ˈmiɡmətait]

Definition: 注入混合岩 zhù rù hùn hé yán

injection: an act or instance of injecting. 注射 zhù shè
migmatite: a rock of both metamorphic and igneous origin, that exhibits characteristics of both rock types. Migmatites probably form through the heating (but not melting) of rocks in the presence of abundant fluids. 混合岩 hùn hé yán

inner core of Earth [ˈinə kɔː əv əːθ]

Definition: 地球内核 dì qiú nèi hé

inner: located or occurring within or closer to a center. 内部的 nèi bù de
core: the central or innermost portion of the Earth, lying below the mantle and probably consisting of iron and nickel. It is divided into a liquid outer core, which begins at a depth of 2,898 km (1,800 mi), and a solid inner core, which begins at a depth of 4,983 km (3,090 mi). 核心 hé xīn

> Example:

There exists relationship between the Earth's inner core and the Earth's rotation. 地球内核与地球自转有着密切的联系。

instability [ˌɪnstəˈbɪləti]

> Definition: the quality of being unstable, erratic, or unpredictable. 不稳定 bù wěn dìng

> Origin: late Middle English: from French instabilité, from Latin instabilitas, from instabilis, from in-"not" + stabilis.

> Example:

Many dynamic disturbances often lead to local instability of roadway surrounding rock and mining workplace and result in rock burst in coal mining.
煤矿生产中存在着大量的动力扰动，常导致采矿巷道和采场围岩局部失稳，诱发冲击矿压发生。

> Extended Term:

nonlinear instability 非线性不稳定性

interference ripple(s) [ˌɪntəˈfɪərəns ˈrɪpl]

> Definition: 干涉波痕 gān shè bō hén，又称菱形波痕(rhomboid ripple mark) ling xíng bō hén，干扰波痕(interference ripple mark) gān rǎo bō hén

interference: the act of interfering with something, or something that interferes. 干扰 gān rǎo；冲突 chōng tū

ripples: to form or have little waves or undulating movements on the surface, as water or grass stirred by a breeze. 波纹 bō wén

> Example:

We achieved successful reduction of interference ripples through the insertion of index-matching layers on the first and last interfaces.
通过在第一和最后一个界面上插入折射率匹配层，我们成功减少了干涉波痕。

intergranular texture [ˌɪntəˈɡrænjələ ˈtekstʃə]

> Definition: 间粒结构 jiān lì jié gòu

intergranular: occurring between grains. 颗粒间的 kē lì jiān de

texture: the arrangement of the particles or constituent parts of any material, as wood, metal, etc., as it affects the appearance or feel of the surface; structure, composition, grain, etc. 结构

jié gòu

Example:

The study of petrologic characteristics of basalt in the area indicates that the basalt has porphyritic texture and matrix has intergranular texture.
对测区内玄武岩样品的岩石学特征研究表明，玄武岩具斑状结构，基质以间粒结构为主。

Extended Term:

intergranular crack 晶粒间裂缝

intermediate [ˌɪntəˈmiːdiət]

Definition: one rock that is in a middle position or state. 中性岩 zhōng xìng yán

Origin: early 15 century, from Medieval Latin intermediatus "lying between".

internal energy [ɪnˈtɜːnəl ˈenədʒɪ]

Definition: In thermodynamics, the internal energy of a thermodynamic system, or a body with well-defined boundaries, denoted by U, or sometimes E, is the total of the kinetic energy due to the motion of particles (translational, rotational, vibrational) and the potential energy. 内部能源 nèi bù néng yuán

Example:

The internal energy of the Snake and the Euler's Elastic have the same geometric meaning, but the former is easy to optimize.
Snake 的内部能量模型和 Euler 弹性模型的几何意义是相似的，但前者可以直接优化。

Extended Terms:

internal cohesive energy 内(凝)聚能
internal potential energy 内势能
specific internal energy 比内能

intersertal texture [ˌɪntəˈsɜːtəl ˈtekstʃə]

Definition: 间隐结构 jiān yǐn jié gòu

intersertal: applied to an igneous texture, especially well-developed in basalts, in which the wedge-shaped spaces between a meshwork of lath-shaped crystals, such as plagioclase, are filled with glass. 填隙的 tián xì de

texture: the arrangement of the particles or constituent parts of any material, as wood, metal, etc., as it affects the appearance or feel of the surface; structure, composition, grain, etc. 结构 jié gòu; 纹理 wén lǐ

Example:

There is also the difference in the different tectonic element deposition intersertal structure and the deposition type, theactic region mainly grows submarine fan-shaped near bank, the ramp region grows delta, the sunken belt mainly becomes the lacustrine facies deposition.
在不同的构造单元沉积充填结构与沉积类型也存在差异性，陡坡带主要发育近岸水下扇，缓坡带发育三角洲，洼陷带主要为湖沼沉积。

Extended Term:

intersertal ophitic 辉绿填充结构的

intertidal zone facies [ˌɪntəˈtaɪdəl zəʊn ˈfeɪʃiːz]

Definition: 潮间带相 cháo jiān dài xiàng

intertidal: of or pertaining to a shore zone bounded by the levels of low and high tide. 潮间带的 cháo jiān dài de

zone: an encircling band, stripe, course, etc., distinct in color, texture, structure, etc. from the surrounding medium. 地带 dì dài; 地区 dì qū

facies: the characteristics of a rock body or part of a rock body that differentiate it from others, as in appearance, composition, etc. 相 xiàng; 表面 biǎo miàn; 外表 wài biǎo

intra-continental plate volcanism

[ˈɪntrəˌkɒntiˈnentəl pleɪt ˈvɒlkənɪzəm]

Definition: 大陆板块内部火山作用 dà lù bǎn kuài nèi bù huǒ shān zuò yòng

intra: within a boundary. 内部 nèi bù
continental: of a continent. 大陆的 dà lù de
plate: a rigid layer of the Earth's crust that is believed to drift slowly. 板块 bǎn kuài
volcanism: volcanic activity or phenomena. 火山作用 huǒ shān zuò yòng

Extended Term:

continental lithospheric plate 大陆岩石圈板块

intra-oceanic plate volcanism [ˈɪntrə-əʊʃiˈænɪk pleɪt ˈvɒlkənɪzəm]

Definition: 大洋板块内部火山作用 dà yáng bǎn kuài nèi bù huǒ shān zuò yòng

intra: within a boundary. 内部的 nèi bù de

oceanic: designating or of the ecological zone (oceanic zone) beyond the neritic zone in the ocean. 海洋的 hǎi yáng de
plate: a rigid layer of the Earth's crust that is believed to drift slowly. 板块 bǎn kuài
volcanism: volcanic activity or phenomena. 火山作用 huǒ shān zuò yòng

Extended Term:

oceanic plate subduction 大洋板块消减作用

intra-plate volcanism [ˈɪntrə-pleɪt ˈvɔlkənɪzəm]

Definition: 板内火山活动 bǎn nèi huǒ shān huó dòng

intra: within a boundary. 内部的 nèi bù de
plate: 金属板 jīn shǔ bǎn
volcanism: volcanic activity or phenomena. 火山作用 huǒ shān zuò yòng

Example:

The rare earth elements show geochemical characteristics of intra-plate volcanism.
稀土元素具有典型的板内火山岩特征。

intraclast [ˌɪntrəˈklæst]

Definition: a sediment formed by the redeposition of material erodes from an original deposit 内碎屑 nèi suì xiè, 又称盆地碎屑 pén dì suì xiè, 同生碎屑(syngenetic clast) tóng shēng suì xiè

Example:

In the sandstone beds of the transgressive system tract of the Dahongyu Formation, an odd type of sand chip that is similar to the carbonate intraclast may reflect the development of microbial mats in the depositional surface of the Precambrian terrigenous clastic deposits.
在大洪峪组的海侵体系域砂岩层中，一种类似于碳酸盐岩中的"内碎屑"的奇特砂质碎片，代表了前寒武纪海侵砂岩沉积面上曾经发育过微生物席。

Extended Terms:

endoceras 内角石
proto intraclast 原始内碎屑

intraclast limestone [ˌɪntrəˈklæst ˈlaɪmstəʊn]

Definition: 内碎屑灰岩 nèi suì xiè huī yán

intraclast: a sediment formed by the redeposition of material erodes from an original deposit. 内碎屑 nèi suì xiè limestone: a sedimentary rock consisting mainly of calcium carbonate, often composed of the organic remains of sea animals, as mollusks, corals, etc., and used as building stone, a source of lime, etc.; when crystallized by heat and pressure it becomes marble. 石灰岩 shí huī yán

Example:

The rock types include limemud limestones, bioclast limestones, oolites, intraclast limestones and dolomites.
岩石类型包括泥晶灰岩、生物碎屑灰岩、鲕粒灰岩、内碎屑灰岩、白云岩等。

intracontinental basin [ˈɪntrəˌkɒntɪˈnentəl ˈbeɪsən]

Definition: 陆内盆地 lù nèi pén dì

intracontinental: within a continent (especially occupying a large part of a continent). 陆内 lù nèi

basin: a wide, depressed area in which the rock layers all incline toward a central area. 盆地 pén dì

Example:

Junggar basin is regarded as a large-scale composite basin superimposed by intracontinental basin of Meso-Cenozoic and foreland basin of Hercynian.
准噶尔盆地是中-新生代陆内坳陷盆地与海西期前陆盆地相叠加的大型复合叠合盆地。

Extended Terms:

intracontinental sea 陆内海
intracontinental subduction 陆内俯冲作用

intracratonic basin [ˈɪntrəˌkreɪˈtɒnɪk ˈbeɪsən]

Definition: 克拉通内盆地 kè lā tōng nèi pén dì

intracratonic: within a craton. 克拉通内的 kè lā tōng nèi de

basin: a wide, depressed area in which the rock layers all incline toward a central area. 盆地 pén dì

Example:

Ordos is a very stable intracratonic basin that is considered as a prohibited place for deep fluids.
鄂尔多斯是一个非常稳定的克拉通盆地，一直被认为是深部流体禁区。

Extended Terms:

intracratonic rift-style basin 克拉通内裂谷型盆地
intracratonic geosyncline 克拉通内地槽
intracratonic trough 克拉通内坳槽

intracrystalline deformation [ˈintrəˈkrist(ə)lain ˌdiːfɔːˈmeiʃən]

Definition: 晶内变形 jīng nèi biàn xíng

intracrystalline: within or across the crystalline phase or grains of a metal. 晶体内的 jīng tǐ nèi de
deformation: a deforming or being deformed. 变形 biàn xíng

Example:

The preliminary tectonic deformation is recrystallized calcite grain's intracrystalline deformation. 最初的构造变形表现为重结晶方解石颗粒的晶内变形。

Extended Terms:

intracrystalline attack 晶体内侵蚀
intracrystalline failure 晶内断裂；穿晶断裂
intracrystalline fracture 晶内断口
intracrystalline penetration 晶内渗透
intracrystalline pore 晶内孔隙

intraformational conglomerate [ˈintrəˌfɔːˈmeiʃənəl kənˈglɔməreit]

Definition: 层间砾岩 céng jiān lì yán，又称层内砾岩 céng nèi lì yán，同生砾岩 tóng shēng lì yán（syngenetic conglomerate）

intraformational: 层内的 céng nèi de
conglomerate: to form or collect into a rounded or compact mass. 砾岩 lì yán；聚合物 jù hé wù

Extended Term:

intraformational folding 层间褶皱

intraplate extensional basin [ˈintrəpleit ikˈstenʃənəl ˈbeisən]

Definition: 板内拉张盆地 bǎn nèi lā zhāng pén dì

intraplate: within a single tectonic plate. 板块内的 bǎn kuài nèi de
extensional: the amount or degree to which something is or can be extended; ranged extended. 外延的 wài yán de
basin: a wide, depressed area in which the rock layers all incline toward a central area. 盆地 pén dì

Example:

According to crustal activity, sediment filling sequences, sedimentary facies belts and their spatial arrangement, the sedimentary basins in this region may be divided into Yangtze cratonic basin, passive marginal basin on the northern margin of the Yangtze craton, Hunan Jiangxi

intraplate extensional basin and Cathaysian pericratonic basin.
根据晚古生代以来的地壳活动情况、沉积物充填序列、沉积相带及空间配置，该区可以划分为四种沉积盆地，即扬子克拉通盆地、扬子克拉通北缘被动大陆边缘盆地、湘赣板内拉张盆地和华夏克拉通边缘盆地。

intrusion [inˈtruːʒən]

Definition: a body of igneous rock that has moved while molten into older solid rocks with subsequent alteration of those rocks. 侵入岩浆 qīn rù yán jiāng

Origin: 14th century. Directly or via French from medieval Latin intrusion—from Latin intrus-, past participle of intrudere (see intrude).

Example:
Only young magma intrusions contain the amount of heat necessary for the heat supply of an economical geothermal field.
只有年轻的岩浆侵入体才能拥有一个经济性地热田所必需的热能。

Extended Term:
soil intrusions 土壤侵入体

intrusive igneous environment [inˈtruːsiv ˈigniəs inˈvaiərənmənt]

Definition: 火成岩侵入环境 huǒ chéng yán qīn rù huán jìng

intrusive: designating or of igneous rock formed from magma that hardened while still within the earth, moon, etc. 侵入的 qīn rù de
igneous: produced by the action of fire; formed by volcanic action or intense heat, as intrusive or extrusive rock solidified from molten magma or lava. 火的 huǒ de；火成的 huǒ chéng de
environment: a surrounding or being surrounded, something that surrounds; surroundings. 环境 huán jìng

Extended Term:
intrusive igneous rocks 火成侵入岩

intrusive contact [inˈtruːsiv ˈkɔntækt]

Definition: 侵入接触 qīn rù jiē chù

intrusive: designating or of igneous rock formed from magma that hardened while still within the earth, moon, etc. 侵入的 qīn rù de
contact: the act or state of touching or meeting. 接触 jiē chù

Example:

Obvious intrusive contact is observed between the Cambrian epimetamorphic rock system intruding into the eastern flank of the NE Heping anticlinorium and surrounding rocks.
位于北东向和平复式背斜东翼的寒武纪浅变质岩系，与围岩呈明显的侵入接触。

intrusive rock [inˈtruːsiv rɒk]

Definition: igneous rock formed from magma forced into older rocks at depths within the Earth's crust, which then slowly solidifies below the Earth's surface, though it may later be exposed by erosion. igneous intrusions form a variety of rock types. See also extrusive rock. 侵入岩 qīn rù yán

Extended Terms:

acidic intrusive rock 酸性侵入岩
intrusive igneous rock 火成岩侵入体
metamorphic intrusive rock 变质侵入体
intrusive magmatic rock 中酸性侵入岩
intrusive rock belt 侵入岩带

invariance [inˈvɛəriəns]

Definition: a relationship that is not changed by a designated mathematical operation such as the transformation of coordinates. 不变性 bù biàn xìng

Example:

This far-reaching assumption about the invariance of physical laws is a foundation stone of the theory of relativity.
关于物理定律不变性的这一影响深远的假设，是相对论的一个基石。

Extended Term:

topological invariance 拓扑不变性

invariant assemblage [inˈvɛəriənt əˈsemblidʒ]

Definition: 不变组合 bù biàn zǔ hé

invariant: not varying; constant; having the nature of an invariant. 不变的 bù biàn de assemblage: a form of art involving the assembly and arrangement of unrelated objects, parts, and materials in a kind of sculptured collage. 集合 jí hé

ionization potential [ˌaiənaiˈzeiʃən pəuˈtenʃəl]

Definition: The term ionization energy (EI) (of an atom or molecule) is most commonly used to refer to the energy required to remove (to infinity) the outermost electron in the atom or molecule when the gas atom or molecule is isolated in free space and is in its ground electronic state. 电离电位 diàn lí diàn wèi

Example:

The interference effects from matrices K, Na, La, Y and Mg have been investigated in ICP-AES. The effects of matrix on line intensity are quantitatively correlated with excitation potential and ionization potential.

选择不同电离电位的基体元素 K、Na、La、Y 和 Mg，比较研究了它们对分析元素谱线强度的影响，其影响程度与谱线激发电位及基体元素电离电位有定量的相关关系。

Extended Terms:

first ionization potential 第一电离电位
IP (Ionization Potential) 电离电位

ion microprobe [ˈaiən ˈmaikrəuprəub]

Definition: 离子微探针 lí zǐ wēi tàn zhēn

ion: an electrically charged atom or group of atoms, the electrical charge of which results when a neutral atom or group of atoms loses or gains one or more electrons during chemical reactions, by the action of certain forms of radiant energy. 离子 lí zǐ

microprobe: an instrument used to determine the chemical composition at a point on a solid surface. 微探针 wēi tàn zhēn

Example:

Both teams used a high-resolution ion microprobe and a mass spectrometer to analyze the crystals.

两个研究小组都用了高清晰度离子探针和一组分光计来分析水晶。

Extended Terms:

ion microprobe analysis 离子微探针分析
scanning ion microprobe 扫描离子微区探针
ion microprobe mass analyzer, LAMMA 离子探针质量分析器

iron-titanium oxide [ˈaiən-taiˈteiniəm ˈɔksaid]

Definition: 铁钛氧化物 tiě tài yǎng huà wù

iron: a white, malleable, ductile, metallic chemical element that can be readily magnetized, rusts rapidly in moist or salty air, and is vital to plant and animal life; it is the most common and important of all metals, and its alloys, as steel, are extensively used; symbol, Fe 26. 铁的 tiě de

titanium: a silvery or dark-gray, lustrous, metallic chemical element found in rutile and otherminerals and used as a cleaning and deoxidizing agent in molten steel, and in the manufacture of aircraft, satellites, chemical equipment, etc.; symbol, Ti 钛 tài

oxide: a binary compound of oxygen with some other element or with a radical. 氧化物 yǎng huà wù

Extended Term:

iron oxides 氧化铁

iron [ˈaiən]

Definition: a heavy, magnetic, malleable, ductile, lustrous, silvery white metallic element that is present in very small quantities in the blood and is the fourth most abundant element in the Earth's crust. Source: hematite, limonite, magnetite. Use: engineering and structural products. Symbol Fe. 铁 tiě

Origin: Old English īren from German.

Example:

The miners abstract iron from ore.
矿工们从矿石中提取出铁。

Extended Term:

corrugated iron / roofing iron 陨铁

iron formation [ˈaiən fɔːˈmeiʃən]

Definition: 含铁建造 hán tiě jiàn zào

iron: a white, malleable, ductile, metallic chemical element that can be readily magnetized, rusts rapidly in moist or salty air, and is vital to plant and animal life; it is the most common and important of all metals, and its alloys, as steel, are extensively used; symbol, Fe 铁的 tiě de

formation: the way in which something is formed or arranged; structure 形成 xíng chéng

Example:

The major metallogenetic matters came from the multilayered dispersed iron formation in the metamorphic series.

成矿物质来源于变质岩系的多层的分散的含铁建造矿源层中。

Extended Terms:

bedded iron formation 层状含铁建造

banded iron formation 条带状含铁建造；带状铁

iron meteorites [ˈaiən ˈmiːtiəraits]

Definition: consist overwhelmingly of nickel-iron alloys. The metal taken from these meteorites is known as meteoric iron and was one of the earliest sources of usable iron available to humans. 铁陨石 tiě yǔn shí

Example:

The decay constant of 187 Re has been corrected by studying iron meteorites and the new decay constant λ（187 Re）= 1 666×10-11a-1.

通过铁陨石定年修正 187 Re 的衰变常数为：λ(187Re) = 1 6 6 6× 10-11a-1。

Extended Term:

iron manufacture 炼铁

ironstone [ˈaiənstəun]

Definition: any sedimentary rock that contains a large amount of iron ore. 铁矿石 tiě kuàng shí

Origin: 1520s, from iron (n.) + stone (n.). As a type of hard, white pottery, 1825.

Example:

The new GM1000×200 High-pressure Grinding Roller was applied to do serial experiment of crushing ironstone on-the-spot at Gu-Shan mining plant.

新型铁矿石 GM1000×200 高压辊磨机在孤山铁矿厂进行了工业化连续粉碎铁矿石的试验。

Extended Terms:

clay ironstone 泥铁岩；褐铁矿泥铁岩

blackband ironstone 黑菱铁矿

brown clay ironstone 褐泥铁石

ironstone china 铁石硬质陶器

irreversible thermodynamic process

[ˌiriˈvəːsəbl ˌθəːməudaiˈnæmik ˈprəuses]

Definition: 不可逆热力学过程 bù kě nì rè lì xué guò chéng

irreversible: not reversible; 不可逆的 bù kě nì de

thermodynamic: the study of energy conversion between heat and mechanical work, and subsequently the macroscopic variables such as temperature, volume and pressure. 热力学的 rè lì xué de

processes: the act of taking something through an established and usually routine set of procedures to convert it from one form to another, as a manufacturing or administrative procedure. 过程 guò chéng

Example:

The laws of thermodynamics involve a concept called entropy for irreversible thermodynamic processes.
热力学定律包括一个称为熵的有关不可逆热力学过程的概念。

Extended Term:

irreversible process of magnetization 不可逆磁化过程

island arc [ˈailənd ɑːk]

Definition: an island arc is a type of archipelago formed as one oceanic tectonic plate and subducts under another and produces magma at depth below the over-riding plate. 岛弧 dǎo hú

Example:

The geotectonic background of forming the rock sequence includes oceanic ridge, island arc, back-arc basin, and small ocean basin, etc.
其形成的大地构造背景多样：包括洋中脊、弧、后盆地、裂作用导致的小洋盆等（原始岩浆来自于上地幔及下地壳的深熔作用）。

Extended Term:

island-arc-inner arc basin zone 岛弧内弧盆地带

island arc basalt [ˈailənd ɑːk ˈbæsɔːlt]

Definition: 岛弧玄武岩 dǎo hú xuán wǔ yán

island: any piece of sub-continental land that is surrounded by water. 岛屿 dǎo yǔ
arc: a bowlike curved line or object. 弧 hú
basalt: a dark, fine-grained, usually extrusive igneous rock that is more basic than andesite, consisting chiefly of plagioclase feldspars and pyroxene; often found in vast sheets, it is the most common extrusive igneous rock. 玄武岩 xuán wǔ yán

Example:

It is comparable with the geochemical pattern of the typical island arc basalt in the world.

这可与世界典型的岛弧玄武岩的地球化学特性进行比较。

Extended Term:

oceanic island basalt 海岛玄武岩

island arc igneous tectonic assemblage
['ailənd ɑːk 'igniəs tek'tɔnik ə'semblidʒ]

Definition: 岛弧火成岩构造组合 dǎo hú huǒ chéng yán gòu zào zǔ hé

island: a land mass not as large as a continent, surrounded by water. 岛 dǎo
arc: any part of a curve, especially of a circle. 弧 hú
igneous: formed by solidification from a molten state. Used of rocks. 火成的 huǒ chéng de
tectonic: relating to, causing, or resulting from structural deformation of the Earth's crust. 构造的 gòu zào de
assemblage: a form of art involving the assembly and arrangement of unrelated objects, parts, and materials in a kind of sculptured collage. 集合 jí hé

island arc volcanic ['ailənd ɑːk vɔl'kænik]

Definition: 岛弧火山 dǎo hú huǒ shān

island: any piece of sub-continental land that is surrounded by water. 岛屿 dǎo yǔ
arc: a bowlike curved line or object. 弧 hú
volcanic: an opening, or rupture, in a planet's surface or crust, which allows hot magma, ash and gases to escape from below the surface. 火山 huǒ shān

Example:

At last they have formed the tectonic framework, after breakup of the volcanic island arcs, subduction, decline and fall of the back-arc basins and arc-microcontinent and arc-arc collision had finished.
经过了长期而复杂的微大陆和火山弧的裂解、弧后盆地的消减衰亡及弧-陆和弧-弧碰撞等构造演化，最终形成了今天所见到的这种构造样式。

Extended Term:

island-arc volcanic rocks 岛弧火山岩

isobaric condition [ˌaisəu'bærik kən'diʃən]

Definition: 等压条件 děng yā tiáo jiàn

isobaric: having equal weights or pressures. 等压的 děng yā de
condition: a proposition upon which another proposition depends, such as "if-then" statements.

条件 tiáo jiàn

Example:

Based on both Biot's consolidation theory under isothermal conditions and thermo-mechanical theory under isobaric conditions, the physical meanings of coupling coefficients and their expression equations are given.

以等温条件下水-力耦合的 Biot 渗透固结理论和等压条件下热-力耦合的热弹性理论两种退化形式为基础，给出耦合系数的物理意义及其函数表达式。

Extended Term:

isobaric spin 同位旋

isobaric diagram [ˌaisəuˈbærik ˈdaiəgræm]

Definition: 等压图 děng yā tú

isobaric: having equal weights or pressures. 等压的 děng yā de

diagram: a sketch, drawing, or plan that explains a thing by outlining its parts and their relationships, workings, etc. 图表 tú biǎo；图解 tú jiě

Extended Term:

Isobaric T-X diagrams 等压 T-X 图

isochemical series [ˌaisəuˈkemik(ə)l ˈsiəriːz]

Definition: 等化学系列 děng huà xué xì liè

isochemical: 等化学的 děng huà xué de

series: a group or number of related or similar persons, things, or events coming one after another; sequence; succession. 系列 xì liè；连续 lián xù

isochron [aiˈsɔkrən]

Definition: a line on a diagram or map connecting points relating to the same time or equal times. 等时线 děng shí xiàn

Origin: late 17th century (as an adjective in the sense "isochronous"): from Greek isokhronos, from isos "equal" +khronos "time".

Example:

Sm-Nd isochron age of 1047±44.5Ma, εNd =-8.99 and r=0.998 are obtained.

以全岩样品测定得到 1047± 44.5 Ma 的 Sm-Nd 等时线年龄，εNd 值-8.99，r=0.998。

Extended Terms:

secondary isochron 二次等时线
Isochron Method 等时法
whole rock isochron 全岩等时线
isochron dating method 等时线测年法

isochron diagram [aiˈsɔkrən ˈdaiəgræm]

Definition: 等时线图 děng shí xiàn tú

isochron: a set of initial conditions for the system that all lead to the same long-term behaviour. 等时线 děng shí xiàn

diagram: a sketch, drawing, or plan that explains a thing by outlining its parts and their relationships, workings, etc. 图表 tú biǎo; 图解 tú jiě

Example:

The arithmetic means of the ($\sim(87)$Sr/$\sim(86)$Sr) and ($\sim(87)$Rb/$\sim(86)$Sr) ratios of every individual intrusion in a granitic belt or complex in an isochron diagram may distribute along a straight line.
一个花岗岩带或复式岩基中各个岩体的$\sim(87)$Sr/$\sim(86)$Sr 和$\sim(87)$Rb/$\sim(86)$Sr 的算术平均值，在等时线图中可获得一条直线。

isograd [ˈaisəgræd]

Definition: In geology, an isograd is a plane of constant metamorphic grade in the field; it separates metamorphic zones of different metamorphic index minerals. 等变质线 děng biàn zhì xiàn

Example:

It dramatically maps the charnockites and is able to delineate the orthopyroxene isograd.
它明确标识出紫苏花岗岩，可以描画出斜方辉石的等变质线。

Extended Terms:

biotite isograd 黑云母等变度
isograd rocks 等级岩

isolated platform facies [ˈaisəleitid ˈplætfɔːm ˈfeiʃiːz]

Definition: 孤立台地相 gū lì tái dì xiāng

isolated: 孤立的 gū lì de

platform: a raised horizontal surface of wood, stone, or metal. 平台 píng tái; 月台 yuè tái

isopach [ˈaisəpæk]

Definition: a line on a map or diagram connecting points beneath which a particular stratum or group of strata has the same thickness. 等厚线 děng hòu xiàn

Origin: early 20th century: from iso-"equal"+Greek pakhus"thick".

Example:
In essence, this isopach map represents the simulated paleotopographic surface of the eroded Mississippi.
实质上,这张等厚图代表着遭到侵蚀的密西西比河的模拟古地形面。

Extended Terms:
effective isopach 等有效厚度图
isopach map 等厚图;等厚线图

isostasy [aiˈsɔstəsi]

Definition: the equilibrium that exists between parts of the Earth's crust, which behaves as if it consists of blocks floating on the underlying mantle, rising if material (such as an ice cap) is removed and sinking if material is deposited 地壳均衡 dì qiào jūn héng

Origin: late 19th century: from iso-"equal"+Greek stasis"station".

Example:
Then, the crust will lift about 2 km by the isostasy to compensate the crust for a loss of mass after thermal expansion.
随之而来的地壳均衡运动会造成 2 千米左右的地壳上升,用以补偿膨胀后地壳的质量亏损。

Extended Terms:
local isostasy 局部均衡
lithosphere isostasy 岩石圈均衡
glacio isostasy 冰河地壳均衡
isostasy gravity anomaly 均衡重力异常

isotherm [ˈaisəuθəːm]

Definition: a curve on a diagram joining points representing states or conditions of equal temperature. 等温线 děng wēn xiàn

Origin: mid-19th century: from French isotherme, from Greek isos "equal" +thermē "heat".

Example:

An isotherm is a line on a map that joins locations having the same mean temperatures.
等温线是在地图上把具有相同平均温度的地方连接起来的线。

Extended Terms:

zero isotherm 零度等温线；冰点等温线
partition isotherm 分配等温线
solubilization isotherm 加溶等温线
saturation isotherm 饱和等温线

isothermal condition [ˌaɪsəʊˈθɜːməl kənˈdɪʃən]

Definition: 等温条件 děng wēn tiáo jiàn

isothermal: of or indicating equality or constancy of temperature. 等温的 děng wēn de
condition: a proposition upon which another proposition depends, such as "if-then" statements.
条件 tiáo jiàn

Example:

The research results show that the bainite transformation is accelerated by strains under isothermal condition, the nose temperature of bainite transformation is 500 ℃, Bs is 600 ℃.
研究表明，在等温条件下，变形促进了贝氏体相变，贝氏体转变的鼻尖温度为500℃，Bs 为600℃。

isothermal diagram [ˌaɪsəʊˈθɜːməl ˈdaɪəɡræm]

Definition: 等温图 děng wēn tú

isothermal: of or indicating equality or constancy of temperature. 等温的 děng wēn de
diagram: a sketch, drawing, or plan that explains a thing by outlining its parts and their relationships, workings, etc. 图表 tú biǎo；图解 tú jiě

Extended Term:

isothermal phase diagram 等温线图

isotopic [ˌaɪsəʊˈtɒpɪk]

Definition: each of two or more forms of the same element that contain equal numbers of protons but different numbers of neutrons in their nuclei, hence differ in relative atomic mass but not in chemical properties; in particular, a radioactive form of an element. 同位素 tóng wèi sù

Origin: early 20th century. From iso- +Greek topos "place"; because isotopes of the same name occupy the same place in the periodic table.

Example:

Any of several isotopic forms of water, especially deuterium oxide, that consists chiefly of molecules containing heavy hydrogen and is used as a moderator in certain nuclear reactors. 几种水的同位素形式之一，尤指氧化氘（氘水），它主要由含有重氢的分子组成，在某些核反应堆中被用作催化剂。

Extended Terms:

isotopic indicator 同位素示踪剂;同位素指示器
isotopic spin 同位旋;同位素自转

isotopic thermometry [ˌaisəuˈtɔpik θəˈmɔmitri]

Definition: 同位素测温法 tóng wèi sù cè wēn fǎ

isotopic: any of two or more forms of an element having the same or very closely related chemical properties and the same atomic number but different atomic weights (or mass numbers): U-235, U-238, and U-239 are three isotopes of uranium. 同位素的 tóng wèi sù de thermometry: measurement of temperature. 温度测定法 wēn dù cè dìng fǎ

Extended Terms:

calcite-dolomite thermometry 方解石-白云石测温法
electric thermometry 电测温法
gas thermometry 气体测温法

I-type granite [ai-taip ˈɡrænit]

Definition: I 型花岗岩 [ai xíng huā gāng yán]

type: a person, thing, or event that represents or symbolizes another, especially another that it is thought; will appear later; symbol; token; sign. 类型 lèi xíng; 品种 pǐn zhǒng
granite: a very hard, coarse grained, gray to pink, intrusive igneous rock, composed mainly of feldspar, quartz, mica, and hornblende. 花岗岩 huā gāng yán

Example:

The I-type granite, widely distributed in the Yongxing Area of Pucheng County, presents an evolution sequence of the composition and structure. 浦城永兴地区分布大面积 I 型花岗岩，具有成分和结构演化序列。

jadeite [ˈdʒeidait]

Definition: a usually greenish pyroxene mineral consisting of sodium aluminum silicate. Source: metamorphic rocks. Use: ornaments, jewelry. 硬玉 yìng yù

Example:

The jadeite's quality with cataclastic texture is also poor because its hardness with this texture will decrease.
具有碎裂结构的翡翠硬度将略有降低，因而也会影响翡翠的质量。

Extended Terms:

Talifu jadeite 大理府翡翠
pure jadeite 纯翡翠

---

jasper [ˈdʒæspə]

Definition: a red, iron-bearing chalcedony. Use: jewelry, ornaments. 碧玉 bì yù

Origin: 13th century. Via Anglo-Norman jaspre from Latin iaspid—from Greek íaspis "jasper".

Example:

A hard black stone, such as jasper or basalt, formerly used to test the quality of gold or silver by comparing the streak left on the stone by one of these metals with that of a standard alloy.
一种如碧玉或玄武岩的坚硬黑色石头，先前用来检验金或银的质量，用这些金属在石头上留下的划痕与标准纯度进行比较。

Extended Terms:

Egyptian jasper 埃及碧石
striped jasper 缟碧石
opal jasper 碧玉蛋白石

---

jasper rock [ˈdʒæspə rɔk]

Definition: 碧玉岩 bì yù yán

jasper: a kind of porcelain developed by Wedgwood, having a dull surface in green, blue, etc., with raised designs, usually in white. 碧玉 bì yù；墨绿色 mò lǜ sè
rock: a large mass of stone forming a peak or cliff. 岩石 yán shí

Example:

The overlying late Proterozoic ophiolite suite is composed of peridotite-serpentinite, gabbro, pyrolite, cordierite, plagiodiorite, diabase, spilite-kerasophyte, siliceous jasper rock.

其上的新元古代蛇绿岩套，是由 10 亿年左右的橄榄-蛇纹岩、辉橄岩、辉长岩、堇青石，斜长闪长岩、辉绿岩、细碧角斑岩、硅质-碧玉岩组成。

Extended Terms:
iron jasper rock 铁碧玉岩
opal jasper 碧玉蛋白石

kaersutite [ˈkɑːəsjutait]

Definition: Kaersutite is a dark brown to black amphibole mineral with formula：$NaCa_2(Mg_4Ti)Si_6Al_2O_{23}(OH)_2$. 钛闪石 tài shǎn shí

Example:
A mass of pyrolite xenolith (such as peridotite and pyroxenide) and high-pressure mineral megacrysts (such as anorthoclase, augite and kaersutite) are found in cone sheets and BS volcanic rocks.
锥状岩席和 BS 火山岩包含有大量的地幔岩石包体（橄榄岩类和辉石岩类）和高压矿物巨晶（歪长石、普通辉石和钛闪石）。

kalsilite [ˈkælsəlait]

Definition: Kalsilite ($KAlSiO_4$) is a vitreous white to grey feldspathoidal mineral that is found in some potassium-rich lavas, such as from Chamengo Crater in Uganda. 六方钾霞石 liù fāng jiǎ xiá shí

Extended Term:
tetra kalsilite 四型钾霞石

kaolinite [ˈkeiəlinait]

Definition: a white or gray aluminosilicate clay mineral. Source：kaolin, altered feldspars— $Al_2Si_2O_5(OH)_4$ 高岭石 gāo lǐng shí

Example:
Burial diagenesis can cause the kaolinite group of minerals to be deformed, transformed or destroyed.
埋藏成岩作用能使高岭石类矿物新生变形、转化或消失。

Extended Terms:
kaolinite clay 高岭土
authigenic kaolinite 自生高岭石

kaolinitic clay [ˈkeiəlinaitic klei]

Definition: a clay mineral of the kaolin group.
高岭石黏土 gāo lǐng shí nián tǔ，又称高岭土(kaolin) gāo lǐng tǔ，俗称瓷土(porcelain clay，china clay)cí tǔ

Extended Term:
kaolinized zone 高岭石化带

Karst breccia [kɑːst ˈbretʃiə]

Definition: 岩溶角砾岩 yán róng jiǎo lì yán，又称溶洞角砾岩 róng dòng jiǎo lì yán，喀斯特角砾岩 kā sī tè jiǎo lì yán

Example:
The characteristics of fault karst breccia are the combination of fault breccia and Karst breccia.
断溶角砾岩兼具断层角砾岩和岩溶角砾岩的特点。

Extended Terms:
gypsum karst breccia 膏溶角砾岩
Karst basin 岩溶盆地；喀斯特盆地
Karst collapse 岩溶陷落

Karst topography [kɑːst təˈpɔɡrəfi]

Definition: Karst topography is a landscape shaped by the dissolution of a layer or layers of soluble bedrock, usually carbonate rock such as limestone or dolomite. 喀斯特地形 kā sī tè dì xíng

Example:
The Yunnan-Guizhou Plateau is one of the areas where the Karst topography has developed most completely and typically in the world.
云贵高原是世界上喀斯特地貌发育最完美、最典型的地区之一。

Extended Term:
Karst features 岩溶地形

---

katungite [ˈkɑːtʌndʒiːt]

Definition: an alkaliultrabasic rock with more patash than soda, the essential miner as being melilite and olivine; the potash may be present in potash-rish nepheline, kaliophipite, leucite zeolite and/or glass. Special varieties are distinguished as leucite-katungite, biotite-katungite, etc. 白橄黄长岩 bái gǎn huáng cháng yán

Example:

The kamafugite is a rare type of ultra-potassic rocks, and a rock series contained generally that rock types of alnoite, olivine-melilitite, mafurite, katungite and wugantite and so on.
钾霞橄黄长岩是一种非常稀少的超钾质火山岩，也是一个岩石系列，包括黄长煌斑岩、橄榄石黄长岩、橄辉钾霞岩、白橄黄长岩和乌干达岩等多种岩石类型。

Extended Term:

kauaiite 碱明矾橄辉闪长岩

Kenya Rift [ˈkenjə rift]

Definition: 肯尼亚裂谷 kěn ní yà liè gǔ

Kenya: a country in eastern central Africa, on the Indian Ocean; formerly a British crown colony & protectorate, it became independent & a member of the Commonwealth (1963). 肯尼亚 kěn ní yà

rift: an opening caused by or as if by splitting; cleft; fissure. 裂缝 liè fèng

Example:

The rock assemblage resembles that of Kenya rift valley, being at the early stage in valley evolution.
其岩石组合与肯尼亚裂谷相似，处于裂谷演化的早期阶段。

kimberlite [ˈkimbəlait]

Definition: a form of igneous rock, found especially in South Africa, composed mainly of peridotite and often containing diamonds. 金伯利岩 jīn bó lì yán

Origin: late 19th century. After Kimberley, town in South Africa.

Example:

The kimberlite is one of mother rock of diamond.
金伯利岩是金刚石的成矿母岩之一。

Extended Terms:

kimberlite pipe 角砾云橄岩管
primary kimberlite 金刚石原生矿
kimberlite formation 角砾云母橄榄岩

kinetics of metamorphic reactions
[kiˈnetiks əv ˌmetəˈmɔːfik riˈækʃənz]

Definition: 1) a variety of metamorphism brought about by heat and pressure developed by and during regional shearing. 2) Deformation without much chemical reconstitution. 变质反应动力学 biàn zhì fǎn yìng dòng lì xué

Extended Terms:
Kinetics of Parallel Reaction 平行反应之动力学
Kinetics of Consecutive Reaction 连贯反应动力学

Krafla volcano [Krafla vɔlˈkeinəu]

Definition: Krafla is a caldera of about 10 km in diameter with a 90 km long fissure zone, in the north of Iceland in the Mývatn region. Its highest peak reaches up to 818 m and it is 2 km in depth. There have been 29 reported eruptions in recorded history. 卡拉夫拉火山 kǎ lā fū lā huǒ shān

Krakatau volcano [ˌkrækəˈtau vɔlˈkeinəu]

Definition: small island & volcano of Indonesia, between Java & Sumatra: 2,667 ft (813m) 喀拉喀托火山 kā lā kā tuō huǒ shān

kyanite [ˈkaiənait]

Definition: a bluish aluminosilicate mineral found as thin-bladed crystals or in masses. Source: metamorphic rocks. Use: gems, refractory. 蓝晶石 lán jīng shí

Origin: late 18th century. From Greek kuan(e)os "dark blue".

Example:
Kyanite will assist the initiation of lifting the veils of illusion that surround oneself and others to get the truth of the patterning, karma or dance underneath.
蓝晶石将协助提升者揭开围绕自身和他人的幻觉面纱，获得在此之下模式、力或舞蹈的真相。

Extended Terms:
kyanite zone 蓝晶石带
calcined kyanite 煅烧蓝晶石

kyanite zone [ˈkaiənait zəun]

Definition: a zone rich with kyanite. 蓝晶石带 lán jīng shí dài

Example:

The higher greenschist facies and the lower amphibolite facies are respectively composed of almandine zone and staurolite-kyanite zone.
高绿片岩相和低角闪岩相分别由铁铝榴石带和十字石-蓝晶石带构成。

labradorite [ˈlæbrədɔːrait]

Definition: a variety of plagioclase feldspar, the color of which shifts between blue and green depending on the angle it is seen from. 拉长石 lā cháng shí

Example:

Albite and labradorite both belong to plagioclase series, but they are two different varieties.
钠长石和拉长石都属斜长石系列，但却是两个不同的宝石品种。

Extended Terms:

labradorite moonstone 拉长月光石
labradorite sunstone 拉长日光石

laccolith [ˈlækəliθ]

Definition: a mass of igneous rock formed from magma that did not find its way to the surface but spread laterally into a lenticular body, forcing overlying strata to bulge upward. 岩盖 yán gài

Origin: 1875-1880; late 19th century: from Greek lakkos "reservoir" + -lith, from Greek lákko(s) pond + -lith.

Example:

Occurrence: dike, sill, laccolith, lopolith, stock and batholith.
产状的定义：岩脉、岩床、岩盖、岩盆、岩株和岩基。

Extended Terms:

laccolithic 岩盖的
laccoliths 凝结的岩盘
composite laccolith 复合岩盖
symmetric laccolith 对称岩丘；对称岩盖
eruptive laccolith 喷发岩盖

lacustrine facies [ləˈkʌstrain ˈfeiʃiːz]

Definition: 湖泊相 hú pō xiāng

Example:

The seismic amplitude percentages lower than valve values indicate the range of the deep lacustrine facies.
低于阀值的地震振幅百分比指示了这些深湖相的平面展布范围。

Extended Terms:

salt lacustrine facies 咸化湖泊
deep lacustrine facies 深湖相
saline lacustrine facies 咸水湖泊

---

lag deposit [læg diˈpɔzit]

Definition: A coarse-grained residue left behind after finer particles have been transported away, due to the inability of the transporting medium to move the coarser particles. 滞留沉积 zhì liú chén jī

Example:

The Coal Measures Strata are divided into 1 super-sequence and 3 sequences on the basis of paleostructure surface, widespread overlap unconformability, coal seam, and channel lag deposit.
根据古构造运动面、大面积超覆界面、煤层和河床滞留沉积将含煤岩系划分为1个超层序和3个层序。

Extended Terms:

channel lag 河底滞留沉积
channel lag deposit 河道滞后沉积

---

lagoon facies [ləˈguːn ˈfeiʃiːz]

Definition: 湖相 hú xiàng

Extended Terms:

lagoon facies soft soil 泻湖相软土
vegetated lagoon facies 沼泽化泻湖相
coastal lagoon facies 海岸泻湖相

lahar [ˈlɑːhɑː]

Definition: 1) a landslide of wet volcanic debris on the side of a volcano; 2) the deposit of mud or land so formed. 火山泥流 huǒ shān ní liú(爪哇语)

Origin: 1925-1930; from Javanese: lahar, lava.

Examples:

① Lahar deposits of 1 000 a B. P. eruption at Changbaishan volcano and their hazards.
长白山火山1000年前火山泥流堆积及其灾害

② The eruptive pattern of the Chifeng phase is: Plinian column(the Chifeng pumice fall deposits) → pyroclastic flow(the Changbai pyroclastic flow deposits) → lahar(the Erdaobaihe lahar deposits) mainly triggerd by pyroclastic flow.
赤峰期喷发模式为：普林尼式喷发柱(赤峰空落浮岩层)—火山碎屑流(长白火山碎屑流层)—火山泥流(二道白河火山泥流层)，主要由火山碎屑流诱发火山泥流。

laminae [ˈlæməniː]

Definition: (Geology) a layer of sediment or sedimentary rock, only a small fraction of an inch (less than a centimeter) in thickness. 细层 xì céng(又称纹层)

Origin: 1650-1660; from Latin "lame".

Example:

Its equidimensional grain size tends to mask the laminae.
它的粒度均一会掩盖纹层层理。

Extended Terms:

laminated quartz 纹层石英
dark laminae 暗色纹层

laminar flow [ˈlæmɪnə fləu]

Definition: the flow of a viscous fluid in which particles of the fluid move in parallel layers, each of which has a constant velocity but is in motion relative to its neighboring layers. 片零流 piàn líng liú

Extended Terms:

laminar fracture 层断裂
laminar burning 层流燃烧
laminar flow covering 层流罩
laminar flow tunnel 层流式隧道

laminar flow state 层流流态
slope laminar flow 坡面薄层水流

lamination [ˌlæmɪˈneɪʃən]

Definition: 1) act or process of laminating; the state of being laminated. 层压 céng yā; 2) laminated structure; arrangement in thin layers. 纹理 wén lǐ; 3) the formation of layers in something. 层裂 céng liè

Example:
The lamination range resulting from a deep-hole blasting is greatest. When the charge of a blast hole is 5-40 kg, the lamination range reaches 7.36-14.73 m.
深孔爆破引起的岩体层裂范围最大，当单孔装药量 5~40kg 时，其值可达 7.36~14.73m。

Extended Terms:
lamination defect 分层缺陷
lamination range 层裂范围
annual lamination 年纹理
cross lamination 交错纹理
lamination coating 硅钢片漆；层状涂膜
solventless lamination 无溶剂层压
skin lamination 表皮分层

lamprophyre [ˈlæmprəˌfaɪə]

Definition: a porphyritic igneous rock consisting of a fine-grained feldspathic groundmass with phenocrysts chiefly of biotite. 煌斑岩 huáng bān yán

Origin: late 19th century: from Greek lampros "bright, shining" + porphureos "purple".

Example:
Lamprophyre is widely distributed in the Qinling area. Most of the lamprophyre alteration-type gold deposits are discovered in the south place of both sides of the Huixian-Chenxian basin.
秦岭地区煌斑岩广泛分布，大部分蚀变煌斑岩型金矿发现于徽成盆地南侧。

Extended Terms:
lamprophyre facies 煌斑岩相
lamprophyre veins 煌斑岩脉
lamprophyres 煌斑岩
lamproite 钾镁煌斑岩类
lamprophyric texture 煌斑结构

Landsat satellite [ˈlændsæt ˈsætəlait]

Definition: 陆地卫星 lù dì wèi xīng

Landsat: a series of artificial satellites that monitor the Earth's resources by photographing the surface at different wavelengths. The resulting images provide information about agriculture, geology, ecological changes, etc. 地球资源卫星 dì qiú zī yuán wèi xīng

satellite: an artificial body placed in orbit round the Earth or another planet in order to collect information or for communication. 人造卫星 rén zào wèi xīng

Examples:

①The Landsat satellite detects lineaments on the Earth's surface very clearly.
地球资源卫星非常清楚地探测了地球表面的地貌。
②The first Earth resources technology satellite, Landsat, was launched in 1972.
第一个地球资源技术卫星——地球资源卫星是1972年发射的。

Extended Terms:

landsat data 陆地卫星数据
landsat image 卫星影像

lanthanide contraction [ˈlænθənaid kənˈtrækʃən]

Definition: 镧族元素收缩 lán zú yuán sù shōu suō

lanthanide: any of the series of fifteen metallic elements from lanthanum to lutetium in the periodic table. 镧族元素 lán zú yuán sù

contraction: the process of becoming smaller. 收缩 shōu suō

Example:

It is evident from these data that the crystal structure determines the frequency position, and clearly shows the effect of the lanthanide contraction in these series of compounds.
这些数据表明振动频率取决于晶体结构并揭示了镧族元素收缩效应对振动频率的规律性影响。

Extended Terms:

lanthanide series 镧系
lanthanide mixtures 镧系混合物

lapilli [ləˈpilai]

Definition: rock fragments ejected from a volcano. 火山砾 huǒ shān lì

Origin: mid-18th century (in the general sense "stones, pebbles"): via Italian from Latin, plural of lapillus, diminutive of lapis "stone".

Example:

In the course of formation of the Labrador megacryst lapilli, the temperature and the cooling rate are the major factors controling their crystallization and growth.
在拉布拉多半岛巨晶火山砾形成过程中,温度和冷却速率乃是控制其结晶生长的主导因素。

Extended Terms:

lapillus 火山砾(单数)
accretionary lapilli 增生火山砾
lapilli cone 火山砾锥
crystal lapilli 晶体火山砾

Laramie complex ['lærəmi 'kɔmpleks]

Definition: 拉勒米杂岩 lā lè mǐ zá yán

Laramie: a university town in southeast Wyoming. 拉勒米(地名) lā lè mǐ
complex: a compounded rock. 杂岩 zá yán

large-ion lithophile elements ['liθəufail 'elimənt]

Definition: element of large ionic radius and with a valency of 1 or 2. 大离子亲石元素 dà lí zǐ qīn shí yuán sù

Example:

The enrichment of large-ion lithophile elements should be attributed to the dewatering or melting of the subducting plate.
大离子亲石元素的富集应归因于消减板块的脱水作用或熔融作用。

larvikite ['lɑ:vikait]

Definition: it is a variety of monzonite, notable for the presence of handsome, thumbnail-sized crystals of feldspar. 歪碱正长岩 wāi jiǎn zhèng zhǎng yán

Extended Terms:

碱性正长斑岩 larvikite porphyry
碱性英辉正长岩 larvikite akerite

large igneous province（LIP） [lɑːdʒ ˈɪgnɪəs ˈprɔvɪns]

Definition: large igneous province（LIP）is extremely large accumulations of igneous rocks—either intrusive, extrusive, or both—which are found in the Earth's crust. 大火成岩区 dà huǒ chéng yán qū

Examples:

①Our data show that spatial petrogeochemical variations exist in the volcanic rocks of the Tianshan large igneous province.
我们的研究揭示，天山大火成岩区的火山岩中存在空间上的岩石地球化学变化。
② The Tianshan rift-related volcanic rocks make up a large igneous province erupted during the Carboniferous. Early Permian period in northwestern China.
中国西北部石炭纪——早二叠世喷发的天山裂谷火山岩系构成了一个大火成岩区。

Extended Term:

Emeishan large igneous province 峨眉山大规模火成区

Larsen index（LI） [ˈlɑːsən ˈɪndɛks]

Definition: 拉森指数 lā sēn zhǐ shù

Larsen：E. S. Larsen 拉森（人名）lā sēn
index：an index is a system by which changes in the value of something and the rate at which its changes can be recorded, measured, or interpreted. 指数 zhǐ shù

latent heat of fusion [ˈleɪtənt hiːt əv ˈfjuːʒən]

Definition: also known as the heat of fusion or specific melting heat, is the amount of thermal energy which must be absorbed or evolved for 1 mole of a substance to change states from a solid to a liquid or vice versa. 熔化潜热 róng huà qián rè

Examples:

①Secondly, the released latent fusion heat was obtained through measurement of the fraction of solid during welding by means of liquid quench test.
其次，通过液淬法测试金属的固相分数随温度变化得到了凝固潜热释放率。
②A calorimeter based on the quasi-steady state theory was developed to measure the latent heat of fusion and specific heat capacity of substance from 250K to 400K at atmosphere pressure.
基于准稳态理论，建立了准稳态量热器，可用于常压下250K~400K物质的溶解热、固相和液相比热容的实验测量。

Extended Term:

volumetric heat of latent fusion 体积替热

latite [ˈleitait]

Definition: A not visibly crystalline rock of volcanic origin composed chiefly of sodic plagioclase and alkali feldspar with subordinate quantities of dark-colored minerals in a finely crystalline to glassy groundmass. 安粗岩 ān cū yán; 二长安山岩 èr cháng ān shān yán

Example:

The study of petrology, mineralogy and geochemistry suggests that augite latite and augite monzonite are cognate, belonging to the upper and root parts of the volcanic dome respectively. 岩石、矿物地球化学的研究表明,辉石粗安岩与辉石二长岩同源,分别为火山穹隆的上部和根部。

Extended Terms:

latite phonolite 安粗响岩
biotite latite 安粗黑云母
latite phonolite 二长安山响岩
biotite latite 黑云二长粗安岩
latite series 安粗岩系

lattice structure [ˈlætis ˈstrʌktʃə]

Definition: In mineralogy and crystallography, crystal structure is a unique arrangement of atoms or molecules in a crystalline liquid or solid. 格子状构造 gé zi zhuàng gòu zào

Example:

Lattice structure is ubiquitous in nature.
格形结构是自然界中普遍存在的一种结构形态。

Extended Terms:

space lattice structure 空间晶格结构
layer lattice structure 层状晶格结构
lattice unit 单位晶格

laumontite facies [ləuˈmɔntait ˈfeiʃiːz]

Definition: 浊沸石相 [zhuó fèi shí xiāng]

laumontite: a mineral, of a white color and vitreous luster, with the chemical formula

$CaAl_2Si_4OI_2 \cdot 4H_2O$. It is a hydrous silicate of alumina and lime. Exposed to the air, it loses water, becomes opaque, and crumbles. 浊沸石 zhuó fèi shí

facies: the general appearance, aspect, or nature of anything. 相 xiàng; 表面 biǎo miàn

lava [ˈlɑːvə]

Definition: hot molten or semi-fluid rock erupted from a volcano or fissure, or solid rock resulting from cooling of this. 熔岩 róng yán

Origin: mid-18th century. From Italian (originally Neapolitan).

Examples:

① Lava spewed forth from the volcano.
熔岩从火山喷吐出来了。

② In many volcanoes the lava bubbles quietly, tossing lava into the air at times, giving off gases, and, in general, being fairly harmless.
许多火山的熔岩静静地在冒泡,有时喷入空中,释放出气体,一般是无害的。

Extended Terms:

lava sheet 熔岩席;熔岩床
basic lava 基性熔岩
basaltic lava 玄武岩火山石
acidic lava 酸性熔岩
amygdaloidal lava 杏仁状熔岩
aphrolithic lava 块熔岩
aqueous lava 泥熔岩
Aso lava 阿苏熔岩

lava cemented pyroclastic rock [ˈlɑːvə siˈmentid ˌpairəuˈklæstik rɒk]

Definition: welded pyroclastic rock. 熔结碎屑岩 róng jié suì xiè yán

lava dome [ˈlɑːvə dəum]

Definition: roughly circular mound-shaped bulge that builds up from the slow eruption of viscous felsic lava from a volcano. 熔岩穹 róng yán qióng

Example:

Increased lava flow during the past day and a new lava dome forming at the peak triggered immediate concerns.

在过去的几天里，岩浆不断地涌出，新的火山堆在一触即发的火山顶形成。

Extended Term:

volcanic lava dome 火山岩圆顶

lava flow [ˈlɑːvə fləu]

Definition: a mass of flowing or solidified lava. Lava is molten rock expelled by a volcano during an eruption. This molten rock is formed in the interior of some planets, including Earth, and some of their satellites. When first erupted from a volcanic vent, lava is a liquid at temperatures from 700℃ to 1,200℃ (1,300°F to 2,200°F). 熔岩流 róng yán liú

Examples:

①With a hiss of steam, lava flows into the Pacific Ocean in Hawaii Volcanoes National Park, Hawaii.

在夏威夷火山国家公园内，伴随着蒸汽的嘶嘶声，熔岩流向太平洋。

②Some of Earth's largest ancient lava flows lie below the Atlantic Ocean seafloor not far from New York City.

一些地球上最大型的古熔岩流位于离纽约市不远的大西洋海底之下。

Extended Terms:

non-explosive lava flows 无爆炸性熔岩流

mud flow 泥流

non-explosive lava flows 非爆炸性的火山岩流

law magnesian calcite [lɔː mægˈniːʃən ˈkælsait]

Definition: 低镁方解石 dī měi fāng jiě shí

layered intrusion [ˈleiəd inˈtruːʒən]

Definition: a layered intrusion is a large sill-like body of igneous rock which exhibits vertical layering or differences in composition and texture. 层状侵入 céng zhuàng qīn rù

Example:

They show layered intrusion features. The major rock-forming minerals include olivine, orthopyroxene, clinopyroxene, plagioclase and amphibole.

它们呈现出层状侵入特性。其主要造岩矿物为橄榄石、斜方辉石、单斜辉石、斜长石和角闪石。

Extended Terms:

layered basic intrusion 层状基性侵入体

layered 层状
layered foundation 成层地基
layered intrusion body 层状侵入体

layering [ˈleiəriŋ]

Definition: the presence or formation of layers in sedimentary or igneous rock. 分层 fēn céng

Examples:

① According to theoretical analysis, an optimal layering algorithm is obtained with defined energy model.
通过理论分析，获得了给定能量模型下的最优分层算法。

② The attitude of deposit, highness of ore heap, bio-leaching technology, geochemical layering are the dominant influencing factors of the liquid arrangement.
矿床产状、矿堆高度、细菌浸矿技术、地球化学分层是布液方式的主要影响因素。

Extended Terms:

lithologic layering 岩性分层法
layer-parallel fluid flow 层平行流体流动

lead isotope [liːd ˈaisətəup]

Definition: 铅同位素 qiān tong wèi sù

lead: a heavy bluish-grey soft ductile metal, the chemical element of atomic number 82. It has been used in roofing, plumbing, ammunition, storage batteries, radiation shields, etc., and its compounds have been used in crystal glass, as an anti-knock agent in petrol, and (formerly) in paints (Symbol: Pb), 铅 qiān

isotopes: 同位素 tong wèi sù

Origin: 1910-1915; iso- + -tope from Greek tópos place.

Example:

In the study of the lead isotopes of a particular orebody or a special ore district, anomalous leads frequently constitute a quite complicated problem whose interpretation is rather difficult.
在铅同位素研究中，异常铅是经常遇到并难于解释的问题。

Extended Terms:

lead isotope ratios in geochronology 地球年代学铅同位素比
lead metal 铅金属

lenticular bedding [lenˈtikjulə ˈbediŋ]

Definition: flaser bedding 透镜状层理 tòu jìng zhuàng céng lǐ

lenticular: shaped like a lentil or biconvex lens 透镜的 tòu jìng de

bedding: mattresses and bedclothes. (建筑)基床(jiàn zhù) jī chuáng

Example:

The sedimentary structure mainly includes unidirectional and bidirectional cross-bedding, interference and curved ripple mark, and wavy, flaser and lenticular bedding.
沉积构造主要包括发育双向和单向交错层理、曲线形和干涉波痕及波状、脉状、透镜状层理。

Lesser Antilles [ˈlesə ænˈtiliːz]

Definition: part of the Antilles, which together with the Bahamas, the Cayman Islands, the Turks and Caicos Islands, and Greater Antilles form the West Indies. The islands are a long partly volcanic island arc, most of which wrap around the eastern end of the Caribbean Sea on the western boundary with the Atlantic Ocean, and some of which lie on the southern fringe of the sea just north of South America.
小安第列斯群岛 xiǎo ān de liè sī qún dǎo(拉丁美洲群岛，南北美大陆间的安第列斯群岛的一部分)

Examples:

①a member of an American Indian peoples of NE South America and the Lesser Antilles.
东北南美洲和较小的安第列斯群岛的美国印第安族人。
②a group of islands in the Lesser Antilles, just north of Venezuela that are administered by the Netherlands.
委内瑞拉北部的属荷兰的位于小安第列斯群岛的一组小岛。

Extended Terms:

Lesser Antilles Islands / Antilles, Lesser 小安第列斯群岛

Antilles, Greater 大安第列斯群岛

Lesser Antilles island arc 小安第列斯岛弧

leucite [ˈljuːsait]

Definition: a gray or white potassium aluminosilicate, typically found in alkali volcanic rocks. 白榴石 bái liú shí

Origin: late 18th century. From Greek leukos "white".

> Examples:

①Research included leucite preparation and ceramic frit preparation matching leucite.
课题研究包括白榴石晶体的制备与白榴石相匹配的烤瓷熔块的制备。
②Two ways were used to prepare PFM powders, one was to add different proportions of leucite to the ceramic frit, and the other was to grain-process the ceramic frit.
采用在烤瓷熔块中加入不同比例的白榴石和对烤瓷熔块进行晶化处理的两种途径制备金属烤瓷粉。

> Extended Terms:

leucite syenite 白榴正长岩
leucite trachyte 白榴粗面岩
leucite basalt 白榴石玄武岩
leucite phonolite 白榴响岩

leucitite [ˈluːsitait]

> Definition: extrusive igneous rock, coloured ash gray to nearly black, that contains leucite and augite as large, single crystals (phenocrysts) in a fine-grained matrix (groundmass) of leucite, augite, sanidine, apatite, titanite, magnetite, and melilite. 白榴岩 bái liú yán; 白榴石岩 bái liú shí yán

> Extended Terms:

leucitohedron 白榴石体
leucitoid 白榴石状
olivine leucitite 橄榄白榴岩
melilite-leucitebasalt 黄长白榴玄武岩
melilite-leucitite 黄长白榴岩

leucocratic [ˌljuːkəˈkrætik]

> Definition: (of igneous rocks) light-coloured because of a low content of ferromagnesian minerals. 以浅色矿物质为主要成分的 yǐ qiǎn sè kuàng wù zhì wéi zhǔ yào chéng fèn de

> Example:

Leucocratic gabrro, located at Rushan in Sulu orogenic belt, was formed in Late Mesozoic (ca. 120 Ma) and contains SiO_2 of 53%~55%, lower MgO of 3.6%~4.9%, enriched LREE, LILE.
位于苏鲁造山带的乳山浅色辉长岩形成于晚中生代(约120Ma)，SiO_2 含量为53%~55%，MgO含量较低，为3.6%~4.9%，富集轻稀土元素(LREE)。

Extended Terms:

leucocrate / leucocratic rock 淡色岩
leucocratic granites 浅色花岗岩
leucocratic igneous rocks 浅色火成岩

leucocratic granites [ˌljuːkəˈkrætik ˈɡrænit]

Definition: a granite containing less than 5%, by volume of dark minerals. 浅色花岗岩 qiǎn sè huā gāng yán

Example:
The granite gneiss complex in eastern Jiaodong peninsula can be divided into three kinds of lithologies, (1) quartz dioritic tonalitic granodioritic gneiss, (2) adamellitic granitic gneiss; and (3) leucocratic granite.
胶东东部地区的元古宙花岗岩——片麻岩杂岩，可分为(1)石英闪长质-英云闪长质-花岗闪长质片麻岩，(2)二长花岗质-钾长花岗质片麻岩和(3)淡色花岗岩三套岩石。

Extended Terms:

leucogranite 淡色花岗岩
leucobasalt 淡色玄武岩
leucolite 淡色岩

leucocratic rock [ˈljuːkəˈkrætik rɒk]

Definition: rock of light-colored as applied to igneous rock containing 0%-50% dark-colored minerals. 浅色岩 qiǎn sè yán 又称淡色岩 dàn sè yán

Example:
Composed mainly of anorthosite and plagiogranite, the leucocratic rocks show distinct Eu positive anomaly.
这种浅色岩主要成分是钙长石和斜长花岗岩，呈现显著的正 Eu 异常。

leucosome [ˈljuːkəsəm]

Definition: coarse-grained, quartz of eldspathic vein, varying in thickness from a few centimetres to a metre or two, and found as a high-grade metamorphic production rocks of pelitic to psammitic composition. 淡色体 dàn sè tǐ

Example:
Limited of leucosome in a migmatite and effects of progressive partial melting on strain

partitioning
混合岩中浅色体的有限迁移及其对变形分解的影响

Extended Terms:
leucosphenite 淡钡钛石
Leucogranite 淡色花岗岩
leucobasalt 淡色玄武岩
leucolite 淡色岩
leuco-sapphire 无色蓝宝石

lever rule ['levə ruːl]

Definition: a tool used to determine weight percentages of each phase of a binary equilibrium phase diagram. It is used to determine the percent weight of liquid and solid phases for a given binary composition and temperature that is between the liquidus and solidus. 杠杆定则 gàng gǎn dìng zé

Origin: Middle English: from Old French levier, leveor, from lever "to lift".

Examples:
① The lever rule is applied in two-phase regions of ternary (as well as binary) systems.
杠杆规律(除了二元体系之外)还适用于三元体系的二相区。
② While using lever rule, the fundamental law of phase diagram and use of standard terminology must be followed.
应用杠杆定律时,必须遵循相图的基本规律并使用标准术语。

Extended Terms:
lever roller 杠杆滚轮
lever safety valve 杠杆安全阀,杠杆式安全阀

lherzolite ['lhɜːzəulait]

Definition: a coarse-grained rock containing minerals high in iron and magnesium that is believed to originate in the Earth's mantle. 二辉橄榄岩 èr huī gǎn lǎn yán

Origin: after the Lherz massif, French Pyrenees.

Example:
The gem-grade olivines exist in lherzolite enclaves and some xenocrysts.
玄武岩中构成宝石的橄榄石产于二辉橄榄岩包体中和少量的捕虏晶中。

Extended Terms:
hornblende lherzolite 角闪二辉橄榄岩

plagioclase lherzolite 斜长石二辉橄榄岩

Li (lithium) [ˈliθiəm]

Definition: the chemical element of atomic number 3, a soft silver-white metal. It is the lightest of the alkali metals (Symbol: Li) [化]元素锂 lǐ

Origin: early 19th century: from lithia + -ium

Examples:

① Da Chaidan Salt lake belongs to a new type of borate salt lake which brine contains sodium, potassium, (lithium), and magnesium chlorides, sulfates and borates.
大柴旦盐湖是典型的新类型硼酸盐盐湖，常年性湖表卤水中含钠、钾、(锂)镁的氯化物，硫酸盐和硼酸盐。

② There is a promising use of polymer electrolyte for lithium ion batteries.
聚合物电解质在锂离子电池方面具有广泛的应用前景。

Extended Terms:

lithium carbonate 碳酸锂盐
lithium metasilicate 硅酸锂
lithium metavanadate 偏钒酸锂
lithium mica 锂云母
lithium iodide 碘化锂

lichen [ˈlaikən]

Definition: a simple slow-growing plant which typically forms a low crust-like, leaf-like, or branching growth on rocks, walls, and tree. 地衣类 dì yī lè; 使长满地衣 shǐ zhǎng mǎn dì yī

Origin: early 17th century: via Latin from Greek leikhēn.

Examples:

① Lichens produce a wide array of both primary and secondary metabolites (lichen substances).
地衣可以产生很多初生和次生代谢物(地衣物质)。

② At the same time we have discovered that, in this region lichens species taking changes with the grows of the altitude.
调查中我们还发现该地区地衣随着海拔高度的升高而有所不同，北坡和南坡也有明显的差异。

Extended Terms:

lichen tundra 地衣冻原

crustose lichens 壳状地衣
foliaceous lichens;foliose lichens 叶状地衣

liesegang ring [ˈliːzgæŋ riŋ]

Definition: a series of usually concentric bands of a precipitate that are separated by clear spaces and that are often formed in gels by periodic or rhythmic precipitation called also Liesegang Phenomenon. 利泽冈环 lì zé gāng huán, 李四光环 lǐ sì guāng huán

Origin: ring, old English hringan, of Germanic origin, perhaps imitative.

Extended Term:

martensite rings 马氏体环带

light-REE [lait-riː]

Definition: 轻稀土 qīng xī tǔ

light: not very great in amount, degree, quality or intensity. 轻的 qīng de
REE: Rare Earth Element. 稀土 xī tǔ

Origin: old English lēoht, līht (noun and adjective), līhtan (verb), of Germanic origin; related to Dutch licht and German licht, from an Indo-European root shared by Greek leukos "white" and Latin lux "light".

Examples:

①Through analysing the petrochemistry, trace elements and light-REE of the intrusive rocks, we provide the model that the emplacement mechanism of plutonic mass is of the heat-light-balloon-expansive origin.
对岩体的岩石化学、微量元素、轻稀土元素进行分析,提出深成岩体就位机制为热轻气球膨胀型的模式。

②The chondrite-normalized distribution of the rare earth element (REE) shows light-REE enrichment for the majority of samples.
磷灰石中稀土元素含量对于球粒状陨石的标绘表明,绝大多数磷灰石样品对于轻稀土元素有富集现象。

light mineral [lait ˈminərəl]

Definition: 轻矿物 qīng kuàng wù

light: not very great in amount, degree, quality or intensity. 轻的 qīng de
mineral: a mineral is a substance such as tin, salt, or sulphur that is formed naturally in rocks

and in the earth. Minerals are also found in small quantities in food and drink. 矿物 kuàng wù

Example:

The content of light mineral in sand sample is higher than the average value in other desert in China by 1.76%, on the contrary, the content of heavy mineral is lower than that in other dune by 1.76%.
沙样的轻矿物含量比我国其他沙漠轻矿物平均值多1.76%，而重矿物少1.76%。

Extended Terms:

light white mineral oil 轻质液状石蜡
light mineral ripple 轻的矿石粒子

lignite ['lignait]

Definition: a soft brownish coal showing traces of plant structure, intermediate between bituminous coal and peat. 褐煤 hè méi

Origin: early 19th century: coined in French from Latin lignum "wood" + -ite

Examples:

① Lignite and peat show acid when mixing with water.
褐煤和泥炭遇水后呈现酸性。
② Lignite, iron ore, diatomite, etc. will be the future direction of the transition.
褐煤、铁矿、硅藻土等将是今后的转型方向。

Extended Terms:

waxy lignite 含蜡褐煤
immature lignite 未成熟褐煤
woody lignite 木煤;木煤体;木质褐煤;木状褐煤
lignite benzine 褐煤汽油

lime mud [laim mʌd]

Definition: Micrite is a sedimentary rock formed of calcareous particles ranging in diameter from 0.06 to 2 mm (0.002 to 0.08 inch) that have been deposited mechanically rather than from solution. 灰泥 huī ní，又称灰泥基质 (lime mud matrix) huī ní jī zhì

Example:

The genesis of mud mound (also called as micrite mound), which is mainly composed of lime mud, only a few bioclast, has still not been resolved.
微晶丘主要由灰泥组成，仅见少量生物碎屑，其成因尚无圆满解释。

Extended Terms:

lime base mud 灰基泥浆

limestone

lime treated mud 石灰处理泥浆
lime mudrock 灰泥岩

limestone ['laimstəun]

Definition: a sedimentary rock consisting mainly of calcium that was deposited by the remains of marine animals. 石灰石 shí huī shí

Origin: 1515-1525；lime+stone.

Examples:
① Caves form when water infiltrates limestone.
水渗透石灰石则形成洞穴。
② India's major mineral resources include coal, iron, manganese, mica, bauxite, titanium, chromite, limestone and thorium.
印度的主要矿产资源有煤、铁、锰、云母、铝土矿、钛、铬铁矿、石灰岩和钍。

Extended Terms:
limestone flux 石灰石溶剂
yellow limestone 乳黄石灰岩
limestone assimilation 石灰石同化
limeclast 灰岩屑

limestone assimilation ['laimstəun əˌsimi'leiʃən]

Definition: 石灰岩同化 shí huī yán tóng huà

Example:
The results of an experimental study of limestone assimilation by hydrated basaltic magmas in the range 1,050-1,150℃, 0.1-500MPa are reported.
这份报告描述了在1,050℃~1,150℃，0.1Mpa~500MPa条件下进行含水玄武岩岩浆的石灰岩同化实验研究的结果。

Extended Term:
dolomitic limestone 白云质灰岩

limonite ['laimənait]

Definition: a widely occurring iron oxide ore; a mixture of goethite and hematite and lepidocrocite. 褐铁矿 hè tiě kuàng

Origin: early 19th century: from German Limonit, probably from Greek leimōn "meadow" (suggested by the earlier German name Wiesenerz, literally "meadow ore").

Examples:

① Their colors are also generated after the formation of jadeite crystals, often found in the red layer, caused by disseminated limonite.
它们的颜色也是硬玉晶体生成后才形成的,常常分布于红色层之上,是由褐铁矿浸染所致。

② Gold are mainly hosted in limonite and quartz.
主要载金矿物是褐铁矿和石英。

Extended Terms:

pitchy limonite 沥青褐铁矿
limonite deposit 褐铁矿矿床
limonite ore 褐铁矿砂

lineation [ˌliniˈeiʃən]

Definition: the line that appears to bound an object. 线理 xiàn lǐ

Origin: late Middle English: from Latin lineatio(n-), from lineare "make straight".

Examples:

① In the north Dabieshan domain, early foliation and N-S-trending compressional lineation are deformed by N-verging folds coeval to the syn-exhumation ductile structures of the central Dabieshan domain.
大别山北部为构造堆叠系,早期的面理和南北向的挤压线理被轴面北倾的褶皱所改造,这种褶皱对应于大别山中部同折返期的韧性变形的隆升构造。

② A sheath fold looks like a scabbard or human's tongue. Its fold axial plane or flattened plane is parallel to the foliation and there is strong stretching lineation on the sheath fold.
鞘褶皱看起来像一把鞘或是像人的舌头,它的褶皱轴面或压扁面是平行岩石中的片理面的,鞘褶皱上有强烈的拉伸线理。

Extended Terms:

substratal lineation 层底线状印痕
primary lineation 原生线理
lineation in metamorphic rock 变质岩线理
cataclastic lineation 碎裂线理

liptinite [ˈliptinait]

Definition: A hydrogen-rich maceral group consisting of spore exines, cuticular matter,

resins, and waxes; includes sporinite, cutinite, alginite, and resinite. Also known as liptinite
类脂组 lèi zhī zǔ, 壳质煤素质 ké zhì méi sù zhì, 膜煤素 mó méi sù

Extended Term:

particulate liptinite 颗粒状类脂体

liquation [laiˈkweiʃən]

Definition: to heat (a metal, etc.) in order to separate a fusible substance from one less fusible. 熔融 róng róng

Example:

The authors concluded that differentiation, liquation and self-metasomatism of granitic magma are main factors for tin-polymetallic mineralization in the area.
作者总结道，花岗质岩浆的分异、熔离、自交代变质作用是该区锡多金属矿床成矿的主因。

Extended Terms:

cooling liquation 冷却凝析
liquation deposit 熔析矿床
liquation refining 熔析精炼
liquation crack 液化裂纹
liquation of filler metal 钎料熔析

liquid composition [ˈlikwid ˌkɔmpəˈziʃən]

Definition: 液晶组成 yè jīng zǔ chéng

liquid: a liquid is a substance which is not solid but which flows and can be poured, for example, water. 液晶 yè jīng

composition: a mixture of ingredients. 组成 zǔ chéng

Example:

And the composition of liquid crystal material is a kind of organic compounds, that is, as the center posed by carbon compounds.
而液晶的组成物质是一种有机化合物，也就是以碳为中心所构成的化合物。

Extended Term:

liquid(molten) steel composition 钢水成分

liquid immiscibility [ˈlikwid ˌimisliˈbiliti]

Definition: the separation of a homogeneous liquid into two or more immiscible liquid fractions. 液体不混溶性 yè tǐ bù hùn róng xìng

Extended Terms:
liquid filling 液体充填
liquid inclusion 液体包裹体，液态包裹体，液体内含物

liquidus [ˈlikwidəs]

Definition: (Chemistry) a curve in a graph of the temperature and composition of a mixture, above which the substance is entirely liquid. 液相线 yè xiàng xiàn

Origin: Latin, literally "liquid".

Examples:
① Finally, the prospect of near-liquidus moulding technology is proposed.
最后，展望一下近液相线法成形工艺的发展前景。
② The liquidus temperature increases with increasing Ag % of the solder.
合金之液相线温度随银含量增加而提高。

Extended Terms:
liquidus temperature 液线温度
liquidus line 液相线
liquidus projection 液相面投影图
liquidus temperature 液相线温度
liquidus mineral 液相线矿物

liquidus temperature [ˈlikwidəs ˈtempəritʃə]

Definition: TL or Tliq, is mostly used for glasses, alloys and rocks. It specifies the maximum temperature at which crystals can co-exist with the melt in thermodynamic equilibrium. 液相线温度 yè xiàng xiàn wēn dù (岩浆在液相线下的温度)

Example:
Bath temperature, liquidus temperature and cryolite ratio are important parameters in Hall-Heroult for aluminum electrolysis.
电解温度、电解质液相线温度和电解质分子比是铝电解过程中重要的工艺参数。

Extended Term:
full liquidus temperature：完全液化温度

listric fault [ˈlistrik ˈfɔːlt]

Definition: 犁状断层 lí zhuàng duàn céng，铲状断层 chǎn zhuàng duàn céng，上凹曲面断

层 shàng āo qǔ miàn duàn céng
listric: in the shape of a plough or a shovel 犁状的 lí zhuàng de
fault: (Geology) a crack in the Earth's crust resulting from the displacement of one side with respect to the other. 断层 duàn céng

> Examples:

① The sand bodies in the down faults blocks of the marginal listric fault often mainly consists of subaqueous fan in transgressive systems tract.
犁状断裂边界控制沉积中心(或生油洼陷)的分布发育和演化，湖侵体系域发育，以湖底扇砂体为主。
② A new finite element method (FEM) in consideration of inhomogeneous stress field is introduced to study the dislocation of a listric normal fault zone's displacement and stress fields.
本文利用新的有限元方法研究了铲状正断层带在非均匀应力场下错动引起的位移场和应力场。

> Extended Term:

reverse listric fault 铲状逆断层

litharenite [liˈθərəˈnīt]

> Definition: a sandstone that contains more than 25% detrital rock fragments, and more rock fragments than feldspar grains. 岩屑砂屑岩 yán xiè shā xiè yán

> Examples:

① The sandstone in upper wall is mainly the feldspathic litharenite, and in the lower wall it is mainly the arkose.
上盘以长石岩屑砂岩为主，下盘以长石砂石为主。
② The Jurassi reservoir in Wu-Xia area is dominated by lithic sandstone and feldspathic litharenite under alluvial fan, fan delta and braided river sedimentary environments.
乌夏地区侏罗系储集层主要是冲积扇、扇三角洲以及辫状河沉积环境下形成的岩屑砂岩及长石岩屑砂岩。

lithic sandstone [ˈliθik ˈsændstəun]

> Definition: Lithic sandstones, or lithic arenites, or litharenites, are sandstones with a significant (>5%) component of lithic fragments, though quartz and feldspar are usually present as well, along with some clayey matrix. 岩屑砂岩 yán xiè shā yán

> Extended Terms:

inequigranular lithic sandstone 不等粒岩屑砂岩
miscellaneous lithic sandstone 混合岩屑砂岩
lithic fragment types 岩屑类型

lithic petrofacies 岩屑岩性相

lithification [ˌlɪθɪfɪˈkeɪʃən]

Definition: (chiefly Geology) transform a sediment or other material into stone. (地质) 岩化作用 (成岩作用) yán huà zuò yòng (chéng yán zuò yòng)

Origin: late 19th century: from Greek lithos "stone" + -fy.

Examples:

① The second is the constructive diagenesis are high quality reservoir and favorable distribution region of large oil-gas fields such as secondary corrosion facies, dolomite lithification facies, weathering leaching facies, chlorite thin film cementation facies and TSR facies;
次生溶蚀相、白云岩化相、风化淋滤相、绿泥石薄膜胶结相和 TSR 相等建设性成岩作用是优质储层及大油气田分布的有利区。

② Cleats are the endogenetic fracture systems which are widely distributed in the coal bed and are of great significance for the coal-bed gas to be produced, being the natural fractures formed by desiccation, coalification, lithification and tectonic force.
割理是广泛存在于煤层中对煤层气的产出具有重要意义的内生裂隙系统，是煤层经过干缩作用、煤化作用、岩化作用和构造压力等各种过程中形成的天然裂隙。

Extended Terms:

lithification lithify 岩化
shock-lithification 冲击石化作用

lithofacies map [ˈlɪθəʊˌfeɪʃiːz mæp]

Definition: a map showing the areal variation of selected lithological attributes of a stratigraphic unit. 岩相地图 yán xiàng dì tú

Example:

It is the first time to apply this quantitative method to compile the quantitative lithofacies area palaeogeography maps in the Tarim area.
这种定量的岩相古地理图在塔里木地区还是首次出现。

lithophile element [ˈlɪθəʊfaɪl ˈelɪmənt]

Definition: pertaining to elements that have become concentrated in the silicate phase of meteorites or the slag crust of the earth. Pertaining to elements that have a greater free energy of oxidation per gram of oxygen than iron. Also known as oxyphile. 亲石元素 qīn shí yuán sù

> Examples:

①Uranium is a lithophile element.
铀是亲石元素。
② The meta-alkaline intermediate volcanic rocks aresimilar to mugearites in geochemical characteristics, having rich alkalinity($K_{-2}O+Na_{-2}O$), high value $K_{-2}O/Na_{-2}O$, and higher concentration of big ion lithophile elements(Rb, Sr, Ba, P).
偏碱性中性火山岩在地球化学特征上与橄榄安粗岩相似,具有富碱($K_{-2}O+Na_{-2}O$)、$K_{-2}O/Na_{-2}O$ 比值高以及富集大离子亲石元素(Rb、Sr、Ba、P)等特征。

> Extended Terms:

geomorphic elements 地貌要素保存
rich lithophile element 富大离子亲石元素

lithosphere [ˈliθəusfiə]

> Definition: the solid part of the earth consisting of the crust and outer mantle. 岩圈 yán quān, 陆界 lù jiè, 地壳 dì qiào

> Origin: 1885-1890; litho- + -sphere.

> Examples:

① The lithosphere is divided into a few dozen plates of various sizes and shapes, in general the plates are in motion with respect to one another.
岩石圈被划分为几十个大小不同形状各异的板块,一般而言这些板块都处于相对运动之中。
② The Earth is made up of three concentric zones: the lithosphere (solid earth), the hydrosphere(water layer), and the atmosphere.
地球由三个同心圈带组成：土壤岩石圈,水圈,大气圈。

> Extended Terms:

lunar lithosphere 月球岩石圈
lithosphere slab 岩石圈板片
lithosphere plate 岩石圈板块
lithosphere stretching 岩石圈伸展
lithosphere rifting 岩石圈张裂

lithospheric [ˌliθəsˈferik]

> Definition: in the rigid outer part of the earth, consisting of the crust and upper mantle. 岩石圈的 yán shí quān de

> Origin: 1885-1890; litho- + -sphere.

> Examples:

① The compositions of the minerals and peridotites, especially trace elements of clinopyroxene (Cpx) can well reflect the characteristics of the subcontinental lithospheric mantle (SCLM).
橄榄岩及其中矿物的组成，特别是单斜辉石的微量元素，可以很好地揭示岩石圈地幔性质。
② This project mainly focuses on the lithospheric strength of the Okinawa Trough (OT).
本课题的研究重点是冲绳海槽地区的岩石圈强度特征。

> Extended Term:

lithospheric splitting 岩石圈分裂

---

lithospheric mantle [ˌlɪθəsˈferɪk ˈmæntl]

> Definition: 岩石圈地幔 yán shí quān dì màn

lithospheric: in the rigid outer part of the earth, consisting of the crust and upper mantle. 岩石圈的 yán shí quān de
mantle: the layer of the earth between the crust and the core. 地幔 dì màn

> Examples:

① This altered lithospheric mantle constituted the main body of Cenozoic lithospheric mantle in the eastern North China Block.
改造后的岩石圈地幔成为华北地块东部新生代岩石圈地幔的主体。
② We apply a kinematic simulation method to perform modeling for the global lithospheric stress field induced by mantle flow.
我们利用动力学模拟方法研究地幔对流对于大尺度岩石圈内部应力场形成的作用。

> Extended Terms:

mantle rock 地幔岩，风化层
outer mantle 外地幔

---

lithostatic pressure [ˌlɪθəsˈferɪk ˈpreʃə]

> Definition: also called overburden pressure, confining pressure or vertical stress, it is the pressure or stress imposed on a layer of soil or rock by the weight of overlying material. 岩压 yán yā, 岩石静压力 yán shí jìng yā lì, 静岩压力 jìng yán yā lì

> Example:

Lithostatic pressure is the stress exerted on a body of rock in the Earth's crust by surrounding rock, which increases with depth below the Earth's surface.
岩体静压是地壳内部岩体受周围岩石施加的应力，随地表下深度的增加而增加。

> Extended Terms:

lithostatic fluid pressure 静岩流体压力
near-lithostatic fluid pressure 近静岩流体压力
lithostatic load 岩石静载荷
Lithostatic pressure gradient 岩压力梯度

littoral zone [ˈlitərəl zəun]

> Definition: the zone betwen extreme high and low tide limits. 滨海地区 bīn hǎi dì qū

> Example:

Ore beds are located in two horizons of middle Carboniferous shallow and littoral facies strata and early Permian continental lacustrine and swamp facies strata.
它在本区有两个层位，一个是中石炭世的浅海和滨海相地层，另一个是早二叠世的陆相湖沼沉积地层。

> Extended Term:

littoral neritic facies 滨浅海相

load structure [ləud ˈstrʌktʃə]

> Definition: proposed as an alternative to load cast to emphasize the difference in category between this and other casts. 负荷构造 fù hè gòu zào，又称重荷模（load cast）zhòng hè mó，重荷铸型 zhòng hè zhù xíng

> Example:

Because of highly improved strength and stiffness, Profiled Bat-Concrete structure can be applied to wide-span structure, heavily load structure and high rise building.
由于型钢混凝土结构强度、刚度的明显提高，所以可以应用于大跨、重荷、高层及超高层建筑中。

> Extended Terms:

load bearing structure 承载构筑物
load striation structure 负载擦痕构造
structure dead load 结构静负载

local equilibrium [ˈləukəl ˌiːkwiˈlibriəm]

> Definition: 局部平衡 jú bù píng héng

local：belonging or relating to a particular area or neighbourhood, typically exclusively so. 当地

的 dāng dì de

equilibrium: a state in which opposing forces or influences are balanced. 平衡 píng héng; 均衡 jūn héng

Examples:

① Discussion on the Local Equilibrium in Intermediate HIC
关于中能重离子碰撞中的局域平衡讨论
② Also, the entropy inequality was given according to the assumption of local equilibrium of non-equilibrium thermodynamics.
此外，还根据非平衡热力学理论的局域平衡假设建立了描述生长变形体热力学过程的熵不等式。

Extended Terms:

local current 局部电流
local magnetic moment 局部磁矩
local thermodynamic equilibrium 局部热力学平衡

local metamorphism [ˈləukəl ˌmetəˈmɔːfizəm]

Definition: metamorphism genetically connected with the intrusion (or extrusion) of magmas. 局部变质作用 jú bù biàn zhì zuò yòng

lopolith [ˈlɔpəliθ]

Definition: a large saucer-shaped intrusion of igneous rock; A large, bowl-shaped body of igneous rock intruded between layers of sedimentary rock. Lopoliths are usually connected to a dike and typically tens of kilometers thick and hundreds of kilometers wide. 岩盆 yán pén

Origin: early 20th century: from Greek lopas "basin" + -lith.

Example:

Zhongposhanbei rock body lies in the middle belt of Beishan rift zone in Xinjiang and covers an area of 180km² in shape of interconnected lopolith.
中坡山北镁铁质岩体位于新疆北山裂谷带的中带，岩体形态为相互联通的岩盆状，出露面积约180km²。

Extended Terms:

lopolithic sill 岩盆式岩床
rock lopolith 岩盆

low-velocity zone [ləu-viˈlɔsəti zəun]

Definition: a layer or zone in the earth in which the velocity of seismic waves is slightly lower than in the layers above and below. The asthenosphere is thought to be such a zone. 低速区 dī sù qū

Examples:
① The asthenosphere corresponds approximately to the low-velocity zone.
软流圈，大体上相当于低速带。
② It is found that the seismic low-velocity zone lies wholly within the plates.
现在发现地震低速带全部分布在板块之中。

Extended Terms:
crustal low-velocity zone 地壳低速带
low velocity layer 低速带
low-volcanicity rifts 低裂谷火山活动

lower mantle [ˈləuə ˈmæntl]

Definition: the deeper part of the mantle. 下地幔 xià dì màn

Example:
Based on the results from both experimental petrology and mineral physics, scientists have worked out silicate perovskite as the major phase in the lower mantle.
依据高温高压实验岩石学和矿物物理测试结果，科学家们提出硅酸盐钙钛矿是下地幔最主要的矿物相。

Extended Term:
upper mantle 上地幔

luminescence petrography [ˈluː miˈnesns piˈtrɔgrəfi]

Definition: 发光岩石学 fā guāng yán shí xué

luminescence: the emission of light by a substance that has not been heated, as in fluorescence and phosphorescence. 发冷光 fā lěng guāng
petrography: the branch of science concerned with the composition and properties of rocks. 岩石学 yán shí xué

Extended Terms:
luminescence property 发光性能
cathode luminescence 阴极发光

electrochemistry luminescence 电化学发光

lump [lʌmp]

Definition: A lump of something is a solid piece of it. 团块 tuán kuài, 又称葡萄石 (grapestone) pú táo shí、巴哈马石 (bahamite) bā hā mǎ shí

Origin: middle English: perhaps from a Germanic base meaning "shapeless piece"; compared with Danish lump "lump", Norwegian and Swedish dialect lump "block, log", and Dutch lomp "rag".

Example:
The best technological conditions was lump stone coal ∶ anthracite ∶ lime ∶ clay = 70 ∶ 10 ∶ 10 ∶ 5. The fluorine fixing ratio was 91.9%.
最佳制备工艺条件为石煤块∶无烟煤粉∶石灰∶黏土=70∶10∶10∶5,固氟率达91.9%。

Extended Term:
lump coal 块煤

lump limestone [lʌmp ˈlaimstəun]

Definition: 团块灰岩 tuán kuài huī yán

lump: A lump of something is a solid piece of it. 团块 tuán kuài

limestone: a whitish-coloured rock which is used for building and for making cement. 石灰岩 shí huī yán

lunar [ˈljuːnə]

Definition: of or relating to or associated with the moon; lunar surface; lunar module. 月的 yuè de, 月球的 yuè qiú de, 农历 nóng lì

Origin: late Middle English: from Latin lunaris, from luna "moon".

Examples:
① The next total lunar eclipse occurs February 21, 2008, and will be visible from the Americas, Europe and Asia.
下一次月蚀将在2008年2月21日发生,届时美洲、欧洲、亚洲地区的人们都会看到。
② The lunar impact kicked up at least 25 gallons (95 liters) of water and that's only what scientists can see, Colaprete said.
科拉普雷特说,月球的撞击激起了至少25加仑(95公升)的水,而这只是科学家可以看到的。

> Extended Terms:

lunar orbiter 月球探测器
lunar flood 月亮周期与地球引力引起的潮汐
lunar corona 月华
lunar eclipses 月食
lunar flood basalts 月球玄武岩

lutetium [ljuːˈtiːʃiəm]

> Definition: a trivalent metallic element of the rare earth group; usually occurs in association with yttrium. 镥 liù

> Origin: early 20th century: from French lutécium, from Latin Lutetia, the ancient name of Paris, the home of its discoverer.

> Examples:

① Lanthanide: Any of the series of 15 consecutive chemical elements in the periodic table from lanthanum to lutetium (atomic numbers 57-71).
镧系：周期表中从镧到镥(原子序数 57~71)15 种按顺序排列的化学元素系列。
② Thulium, ytterbium and lutetium carbonates are hydrated basic carbonates or oxycarbonates.
重稀土铥、镱、镥的碳酸盐为碱式盐或碳酸氧化物。

> Extended Terms:

lutetium oxide 氧化镥
lutetium nitrate 硝酸镥

maar [mɑː]

> Definition: flat-bottomed volcanic crater that was formed by an explosion; often filled with water. n. 低平火山口 dī píng huǒ shān kǒu；小火山口 xiǎo huǒ shān kǒu

> Origin: early 19th century: from German dialect, originally denoting a kind of crater lake in the Eifel district of Germany.

> Examples:

① Pyroclastic base-surge deposits of volcanic cluster in the Longgang region of Jilin Province is one of the few neoteric Maar deposits which are preserved in good condition in China.
吉林龙岗火山群火山碎屑基浪堆积是中国少数保存较好的、近代喷发的低平火山区之一。
② This is a crater lake of a Maar volcano.
这是低平火山形成的火山湖。

Extended Term:

mud maar 泥火山口

Macquarie Island ophiolite [ˌmæˈkwɔːri ˈailənd ˈɔfiəlait]

Definition: 麦夸里岛蛇绿岩 mài kuā lǐ dǎo shé lǜ yán

Macquarie Island: a UNESCO World Heritage Site, lies in the southwest Pacific Ocean, about halfway between New Zealand and Antarctica. 麦夸里岛 mài kuā lǐ dǎo

ophiolite: a section of the Earth's oceanic crust and the underlying upper mantle that has been uplifted and exposed above sea-level and often emplaced onto continental crustal rocks. 蛇绿岩 shé lǜ yán

mafic [ˈmæfik]

Definition: relating to, denoting, or containing a group of dark-coloured, mainly ferromagnesian minerals such as pyroxene and olivine. 镁铁质的 měi tiě zhì de

Origin: early 20th century: blend of magnesium and a contracted form of ferric.

Examples:

① Their geochemical characteristics of trace elements and Nd and Sr isotopes show that these mafic rocks erupted in a back-arc basin.
根据镁铁质岩的层位关系、岩浆来源和微量元素特征，表明镁铁质岩石是从该盆地喷发出来的。

② Biotite is a subordinate constituent of the mafic and ultramafic rocks generally.
黑云母一般是镁铁质及超镁铁质岩石中的次要成分。

Extended Terms:

mafic rock 镁铁质岩石
mafic margin 镁铁质边缘
mafic mineral 镁铁质矿物
mafic igneous rocks 镁铁质火成岩
mafic-ultramafic igneous rocks 超镁铁质岩
mafic dikes 基性岩脉

mafic index [ˈmæfik ˈindeks]

Definition: 镁铁指数 měi tiě zhǐ shù

mafic: (Geology) relating to, denoting, or containing a group of dark-coloured, mainly ferromagnesian minerals such as pyroxene and olivine. （地质）铁镁质的 tiě měi zhì de；镁铁矿石的 měi tiě kuàng shí de

index：a figure in a system or scale representing the average value of specified prices, shares, or other items as compared with some reference figure. 指数 zhǐ shù

Example:

Even though the mafic index of a rock seems a logical variable for a criterion for naming volcanic rocks, it is not without its problems.
尽管在火山岩石命名标准中，镁铁指数是一个合理的变量，但这一指数本身存在一些问题。

mafic mineral [ˈmæfik ˈmɪnərəl]

Definition: a mineral that is composed predominantly of the ferromagnesian rock-forming silicates. In general, any dark mineral. 镁铁质矿物 měi tiě zhì kuàng wù

Example:

The mode of occurrence of the ilmenite suggests that it is the first mafic mineral to precipitate from the magma.
钛铁矿出现的这一模式表明这是从岩浆中沉淀形成的首种镁铁质矿物。

Extended Terms:

mafic margin 镁铁质边缘
mafic component 镁铁成分
mafic rock 镁铁质岩石

mafic rock [ˈmæfik rɔk]

Definition: in geology, any igneous rock dominated by the silicates pyroxene, amphibole, olivine, and mica. These minerals are high in magnesium and ferrous iron, and their presence gives mafic rock its characteristic dark colour. It is usually contrasted with felsic rock. Common mafic rocks include basalt and gabbro. 镁铁质岩 měi tiě zhì yán

Example:

The khondalite suites dropped into lower crust, then metamorphicose took place, the high-pressure granulite intruded upward along butt zone, and depressure took place, it occured as mingled rock belts together with the khondalite suite and mafic rock.
孔兹岩系卷入下地壳，发生变形变质，高压麻粒岩沿对接带上侵发生减压，高压麻粒岩与孔兹岩岩片、镁铁质岩一同，以混杂岩带的形式产出。

Extended Terms:

mafic igneous rock 镁铁质火成岩

mafic-ultramafic rock belt 镁铁-超镁铁质岩带

mafurite [ˈmɑːfəˌrait]

Definition: 马浮石 mǎ fú shí, 橄辉钾霞岩 gǎn huī jiǎ xiá yán

Example:

The kamafugite is a rare type of ultra-potassic rocks, and a rock series contained generally that rock types of alnoite, olivine-melilitite, mafurite, katungite and wugantite and so on.

钾霞橄黄长岩是一种非常稀少的超钾质火山岩,也是一个岩石系列,包括黄长煌斑岩、橄榄石黄长岩、橄辉钾霞岩、白橄黄长岩和乌干达岩等多种岩石类型。

magma [ˈmæɡmə]

Definition: hot fluid or semi-fluid material below or within the earth's crust from which lava and other igneous rock is formed by cooling. 岩浆 yán jiāng

Origin: late Middle English (in the sense "residue of dregs after evaporation or pressing of a semi-liquid substance"): via Latin from Greek magma (from massein "knead").

Examples:

① When rainwater seeps into cracks in the mountain, the magma turns it into steam and forces it out again.

当雨水渗入山的裂缝中时,岩浆会立刻把它变成水蒸气散发掉。

② Sometimes, however, the top layer of magma can cool, and form a hard crust over the rest of the molten rock.

然而有时候,岩浆的顶层可能会冷却下来,然后形成一层坚硬的外壳,覆在其它的熔岩上面。

Extended Terms:

magma glassashes 岩浆玻璃灰

magmatic assimilation 岩浆同化

parental magma 母岩浆

magma series 岩浆系列

magmatic ore deposits 岩浆矿床

emulsive magma 乳浊岩浆

primary magma 原始岩浆

rhyolitic magma 流纹岩岩浆

magma chamber [ˈmæɡmə ˈtʃeimbə]

Definition: a larger reservoir in the crust of the earth that is occupied by a body of magma. 岩浆库 yán jiāng kù

Example:
As the land above the magma chamber collapses, immense gray clouds called pyroclastic flows burst out horizontally all around the caldera.
当岩浆库上方的陆地崩陷，被称为火山碎屑流的巨大灰色云雾，会突然由火山臼的周围向外水平散开。

Extended Term:
magma reservoir 岩浆房

magma series [ˈmæɡmə ˈsiəriːz]

Definition: 岩浆系列 yán jiāng xì liè
magma：molten rock in the earth's crust. 岩浆 yán jiāng
series：a number of things or events that come one after the other. 系列 xì liè

Example:
The basic lava of Neoproterozoic-early Cambrian in the Tianshan and its adjacent regions include two magma series, they are tholeiitic and alkaline basaltic magma series.
天山地区新元古代-早寒武世玄武质熔岩包括拉斑玄武质和碱性玄武质两个主要岩浆系列。

Extended Terms:
magma evolution series 岩浆演化系列
magma inclusion 岩浆包裹体
oceanic magma series 大洋岩浆系

magmatic differentiation [ˈmæɡmətik ˌdifəˌrenʃiˈeiʃn]

Definition: The formation of different igneous rocks from a parent magma. Some petrologists regard differentiation and magmatic differentiation as synomymous terms, while others, especailly German petrologists, distinguish magmatic differentiation from crystallization differentiation. 岩浆分异作用 yán jiāng fēn yì zuò yòng

Extended Terms:
magmatic diapirism 岩浆底辟作用
magma thermodynamic structure 岩浆热动力构造

magmatic processes ['mæɡmətic 'prɔsesiz]

Definition: 岩浆作用 yán jiāng zuò yòng

magmatic: of hot fluid or semi-fluid material below or within the earth's crust from which lava and other igneous rock is formed by cooling. 岩浆的 yán jiāng de

processes: particular courses of action intended to achieve a result. 作用 zuò yòng

Examples:

① Magma dynamics describes quantitatively magmatic processes by means of the fluid dynamics method, on which basis theoretical models are constructed.
岩浆动力学利用流体力学的方法定量描述岩浆过程，建立理论模型。

② Combining melt inclusion data with conventional petrological and geochemical information and experimental petrology can increase our ability to model magmatic processes.
将熔体包裹体数据和常规的岩石学、地球化学和实验岩石学信息综合一体，可以提高我们模拟岩浆作用过程的能力。

magmatism ['mæɡmətizm]

Definition: the motion or activity of magma. 岩浆作用 yán jiāng zuò yòng

Examples:

① The geological structures of the region is simple with only weak magmatism.
本区域地质构造简单，岩浆活动微弱。

② The main factors to control gold deposits are source beds, magmatism and structures.
控制金矿储备的主要因素包括矿源层、岩浆作用和各种构造。

Extended Terms:

preorogenic magmatism 造山前的岩浆作用
syntectonic magmatism 同构造岩浆作用
felsic magmatism 长英岩浆作用
arc magmatism 岛弧岩浆作用

magnesia ferro ratio (M/F, m/f) [mæɡ'niːʃə 'ferə 'reiʃiəu]

Definition: 镁铁比值 měi tiě bǐ zhí

magnesia: a white hygroscopic solid mineral. 氧化镁 yǎng huà měi

ferro: a chemical element with symbol Fe (from Latin: ferrum) and atomic number 26. 铁 tiě

ratio: a relationship between two things when it is expressed in numbers or amounts. 比值 bǐ zhí

magnesian schist [mæɡˈniːʃən ʃist]

Definition: 镁质片岩 měi zhì piàn yán

magnesian: a chemical element with symbol Mg and atomic number 12. 镁 měi
schist: a medium-grade metamorphic rock[1] with medium to large, flat, sheet-like grains in a preferred orientation (nearby grains are roughly parallel). 片岩 piàn yán

magnesite [ˈmæɡnəsait]

Definition: a black magnetic mineral consisting of magnesium carbonate; a source of magnesium. 菱镁矿 líng měi kuàng

Examples:

① Based on the mineralized and mineralogical feature of magnesite in Iran, the paper sets forth the process of "one fine grinding-positive floatation" for lower grade magnesite.
根据伊朗某地菱镁矿的成矿特性和矿石工艺矿物学特性，提出了低品位菱镁矿选矿脱硅"一次磨细-正浮选"的工艺。

② At present, there are more than 100 varieties of mineral resources, in which, such resources as Iron, Boron, Magnesite and Diamond, etc., have a leading reserves in China.
目前，已发现的矿藏有100多种，其中，铁、硼、菱镁石、金刚石等矿藏储量均居中国首位。

③ The production of NK and NMg fertilizers of various compositions from urea and finely crystalline potassium sulfate or urea and caustic magnesite powder was optimized.
用尿素和细晶硫酸钾，或用尿素和轻烧菱镁砂颗粒制造含多种成分的氮钾化肥和氮镁化肥的工艺得以优化。

magnetic field [mæɡˈnetik fiːld]

Definition: a region around a magnetic material or a moving electric charge within which the force of magnetism acts. 磁场 cí chǎng

Examples:

① When the magnetic field is at its strongest, and sunspots at their most plentiful, they cluster close to the equator.
磁场最强的时候，就是黑子数量最多的时候，同时，它们也更趋向群集于太阳赤道的位置。

② Hours to days before the flare release, excess energy become stored in the magnetic field.
在耀斑释放前几小时至几天，多余的能量就在磁场中储存起来。

Extended Terms:

magnetic leakage field 漏磁场
magnetic force 磁化力

magnetic susceptibility fabrics 磁化率面料

magnetite [ˈmæɡnitait]

Definition: an oxide of iron that is strongly attracted by magnets. 磁铁矿 cí tiě kuàng

Origin: mid-19th century: from magnet + -ite.

Examples:

① Ore minerals are mainly magnetite, chalcopyrite and azurite, etc.
矿石矿物主要为磁铁矿、黄铜矿、蓝铜矿等。
② Magnetite and hematite both are the high thermal-stable magnetic carriers.
磁铁矿和赤铁矿同为高温稳定性磁性载体。

Extended Terms:

magnetic valve 电磁阀
magnetite refractory 镁质耐火材料
magnetite sand 磁矿砂

major element [ˈmeidʒə ˈelimənt]

Definition: any chemical found in great quantity in the rocks of the earth's crust. Compare minor element. 主要元素 zhǔ yào yuán sù，常量元素 cháng liàng yuán sù

Example:

Pb as a major element in heavy metals that contaminated groundwater, the remediation methods in use include physical screening, extraction and in-situ remediation.
铅是地下水重金属污染中的主要元素，目前的修复技术主要分为物理屏蔽法、抽出处理法和原位修复法。

majorite [ˈmeidʒərait]

Definition: 镁铁榴石 měi tiě liú shí

Examples:

① Nearing the earth's surface, the pressure in the mantle becomes too weak to maintain the majorite, which then decomposes.
接近地球表面，地幔的压力变得过于薄弱，不足以维持镁铁榴石，然后镁铁榴石分解。
② As Professor Dr. Christian Ballhaus from the Mineralogical Institute at the Bonn University explains, "The higher the pressure, the more oxygen can be stored by majorite".
正如波恩大学矿物学研究所的教授克里斯汀·鲍尔博士解释的那样："压力越高，镁铁榴石

可贮藏的氧气越多。"

major minerals [ˈmeidʒə ˈminərəls]

Definition: 主要矿物质 zhǔ yào kuàng wù zhì

Example:

The major minerals include calcium, phosphorus, potassium, sodium, chloride, magnesium, and sulfur.
主要矿物质包括钙、磷、钾、钠、氯化物、镁和硫黄。

manganese rock [ˈmæŋgəniːz rɔk]

Definition: 锰质岩 měng zhì yán

manganese: a hard brittle grey polyvalent metallic element that resembles iron but is not magnetic; used in making steel; occurs in many minerals. 锰 měng
rock: material consisting of the aggregate of minerals like those making up the Earth's crust. 岩 yán

Example:

The intraclastic structure and gravity-flow sedimentation of the manganese rock in Taojiang manganese ore deposit, Hunan Province, are studied in this paper.
该论文研究了湖南省桃江锰矿矿床的内碎屑结构和锰岩石重力流沉积。

Extended Terms:

manganese nodule 蓄锰结核
manganese dioxide 二氧化锰

mantle [ˈmæntl]

Definition: the region of the earth's interior between the crust and the core, believed to consist of hot, dense silicate rocks (mainly peridotite). 地幔 dì màn

Origin: Old English mentel, from Latin mantellum "cloak"; reinforced in Middle English by Old French mantel.

Examples:

① The mesosphere comprises the rest of the mantle below the asthenosphere.
中间圈包括岩流圈以下地幔的其余部分。
② However, geologists can wave base message, a judge also divided into the mantle below the crust, the earth's core.

不过，地质学家可以根据地震波传递的信息，判断出地壳之下还可分为地幔、地核。

Extended Terms:

mantle of Earth 地球地幔
origins of mantle 地幔起源

---

mantle plume [ˈmæntl pluːm]

Definition: eep-seated upwelling of magma within the earth's mantle. 地幔柱 dì màn zhù，地幔热缕 dì màn rè lǚ，羽状地幔流 yǔ zhuàng dì màn liú

Origin：1350-1400；earlier plome, plume.

Examples:

① Were Large Igneous Provinces caused by Mantle Plumes?
大火成岩区是地幔柱作用引起的吗？
② The relation between the mantle plume and the generation of inorganic petroleum is discussed.
文中讨论了地幔柱与无机成因油气生成的关系。

Extended Terms:

deep mantle plume 深地幔柱
mantle melting 地幔熔融

---

mantle source [ˈmæntl sɔːs]

Definition: 幔源 màn yuán

mantle：the region of the earth's interior between the crust and the core, believed to consist of hot, dense silicate rocks (mainly peridotite). (地质)地幔 dì màn
source：a place, person, or thing from which something comes or can be obtained. 来源 lái yuán

Example:

Geochemical studies show that magma of these two series originated from the same impoverished mantle source, with the source rock being spinel peridotite.
地球化学研究证明，这两个系列的岩浆起源于同一贫化型地幔源区，源岩类型为尖晶石橄榄岩。

Extended Terms:

enriched mantle source 富集地幔源
mantle source liquid 幔源流体

mantle source rock 幔源岩
mantle source gas 幔源气

mantle wedge [ˈmæntl wedʒ]

Definition: 地幔楔体 dì màn xiē tǐ

wedge: a wedge of something such as fruit or cheese is a piece of it that has a thick triangular shape. 楔体 xiē tǐ

Example:

In order to better understand the influence of the mantle wedge relating to subducting, the peridotites from the southern Mariana fore arc is checked in this paper.

为了深入了解俯冲作用对弧下地幔楔体橄榄岩的影响，本文对马里亚纳岛弧南部前缘的橄榄岩进行了研究。

Extended Terms:

mantle tomour 地幔鼓胀
mantle transform fault 地幔转换断层

marble [ˈmɑːbl]

Definition: metamorphosed limestone, consisting chiefly of recrystallized calcite or dolomite, capable of taking a high polish, occurring in a wide range of colors and variegations and used in sculpture and architecture. 大理岩 dà lǐ yán

Origin: Middle English, from Old French marbre, from Latin marmor, from Greek marmaros.

Extended Terms:

serpentine marble 蛇纹石大理石
onyx marble 条纹状大理石
Parian marble 帕罗斯岛大理石
ruin marble 废墟大理石；角砾状大理岩；块结大理石

mare [mɛə]

Definition: large, level basalt plain on the surface of the moon, appearing dark by contrast with highland areas. 海 hǎi；月海 yuè hǎi

Origin: mid-19th century: special use of Latin mare "sea"; these areas were once thought to be seas.

Examples:

① This is lava that flowed out smoothly onto the lunar surface before solidifying, forming dark plains that early skywatchers mistakenly took for seas (mare in Latin).

熔岩流在凝固之前顺利地流到月球表面上，形成了颜色暗淡的平原，早期的天空观察者误以为是月球上的海洋（"海"在拉丁语中是 mare），称其为月海。

② Some of these surfaces are mare deposits that have been covered by blankets of highland debris—layers of ejected rock spread by the impacts that created the moon's basins.

这类表面有些是被多层高地遗屑（月球受到撞击而喷出四散的岩石）覆盖的月海矿床。

marginal basin ['mɑːdʒinəl 'beisn]

Definition: 盆地边缘 pén dì biān yuán

marginal: of, relating to, or situated at the edge or margin of something. 边缘的，边沿的 biān yuán de, biān yán de

basin: a circular or oval valley or natural depression on the earth's surface, especially one containing water. （尤指聚水的）凹地；海盆 āo dì, hǎi pén

Examples:

① It is welcome to freely and deeply discuss origin and evolution of marginal basins in the NW Pacific with me.

欢迎深入自由讨论西北太平洋边缘盆地的起源和演化。

② Finally the feature of ocean crust of marginal sea basins was discussed with Japan Sea and South China Sea as examples.

以日本海和南海为例，讨论了边缘海"洋壳基底"性质，认为所谓的"洋壳"实为变薄的陆壳。

margin ['mɑːdʒin]

Definition: the edge or border of something. 边缘 biān yuán

Origin: late Middle English: from Latin margo, margin-"edge".

Example:

Middle Proterozoic Tectonic Morbile Belt and Mineralization in North Continental Margin of the North China Plateform.

华北板块北缘中元古代大陆边缘构造活动带及矿化作用。

Marianas Trench [ˌmæriˈɑːnəz trentʃ]

Definition: the deepest spot in the oceans, more than thirty-five thousand feet deep, near the

Philippines. 马里亚纳海沟 mǎ lǐ yà nà hǎi gōu

Example:

Humans have visited the very bottom of the ocean—the Marianas Trench in the western Pacific, nearly seven miles below the wave tops—only a few times.
位于西太平洋下约 7 英里处，可称为海洋之底的马里亚纳海沟，人类至今探测的次数屈指可数。

Extended Term:

Marianas Trough 马里亚纳海槽

marine facies [məˈriːn ˈfeiʃiːz]

Definition: 海相 hǎi xiàng

marine: of or relating to the sea 海的 hǎi de
facies: a body of rock with specified characteristics, which can be any observable attribute of rocks such as their overall appearance, composition, or condition of formation, and the changes that may occur in those attributes over a geographic area. 相 xiàng

Example:

In the Late Permian, marine facies only occurred in southern Qilian area and the other areas were all characterized by continental deposition.
在二叠纪末，仅在南祁连地区有海相沉积，其他地区均为陆相沉积。

Extended Terms:

marine facies bed 海相层
non-marine facies 陆相
marine facies strata 海相底层

marine phreatic environment [məˈriːn friˈætik inˈvaiərənmənt]

Definition: 海水潜流环境 hǎi shuǐ qián liú huán jìng

phreatic: of or relating to ground water. 潜流的 qián liú de
environment: the natural world of land, sea, air, plants, and animals. 环境 huán jìng

Example:

In the Cambrian-Ordovician, micritic dolomite, algal-laminated dolomite and grains in dolomite resulted from dolomitization in the marine phreatic environment in the penecontemporary period.
寒武系至奥陶系的微晶白云岩、藻纹层白云岩及颗粒白云岩中的颗粒是准同生期海水潜流环境白云石化的产物。

marl [mɑːl]

Definition: an unconsolidated sedimentary rock or soil consisting of clay and lime, formerly used typically as fertilizer. 泥灰岩 ní huī yán

Origin: 14th century. Via Old French marle.

Examples:

① Marlstone is a limestone based clay that acts similar to Marl.
泥灰岩石是与泥灰作用相似的一种以石灰石为基础的黏土。
② Saint-Julien has homogeneous gravel soil with clay, marl, ferrite, gravel in its subsoil.
圣朱利安的沙砾泥土很均匀，在下层土里有黏土、泥灰土、亚铁盐和沙砾。

Extended Terms:

flame marl 焰色泥灰岩
bog marl 沼泽泥灰岩
argillaceous marl 泥质泥灰岩
calcareous marl 灰质泥灰岩
globigerinid marl 抱球虫泥灰岩

Marquesas Islands [mɑːˈkeizəz ˈailəndz]

Definition: a group of volcanic islands in the S Pacific, in French Polynesia. 马克萨斯群岛 mǎ kè sà sī qún dǎo

Examples:

① The Marquesas are among the largest groups of islands in French Polynesia. Jagged mountains and steep cliffs make up most of the landscape.
马克萨斯群岛在法属波利尼西亚群岛的最大的一个岛群中。犬牙交错的群山和陡峭的悬崖构成了这里大部分的地貌。
② To be located in south-central Pacific, including Society Islands, Tuamotus, the Tuamotu Islands, the Marquesas Keys, Gambier Islands, Knives-Rorty Islands (Bath Islands) and Lapparent Island, and so on.
法属波利尼西亚，位于太平洋中南部，主要包括社会群岛、土布艾群岛、土阿莫土群岛、马克萨斯群岛、甘比尔群岛、刀罗蒂里群岛(巴斯群岛)和拉帕岛等。

Mars [mɑːz]

Definition: a small reddish planet which is the fourth in order from the sun and is periodically visible to the naked eye. 火星 huǒ xīng

Examples:

① The thin atmosphere means Mars gets very cold.

大气层稀薄意味着火星十分寒冷。

② The planets of our solar system are Mercury, Venus, Earth, Mars, Jupiter, Saturn, Uranus and Neptune.

我们太阳系的行星有水星、金星、地球、火星、木星、土星、天王星、海王星。

Extended Term:

Mars globe 火星仪

mass balance [mæs ˈbæləns]

Definition: (also called a material balance) is an application of conservation of mass to the analysis of physical systems. By accounting for material entering and leaving a system, mass flows can be identified which might have been unknown, or difficult to measure without this technique. The exact conservation law used in the analysis of the system depends on the context of the problem but all revolve around mass conservation, i.e. that matter cannot disappear or be created spontaneously. 质量平衡 zhì liàng píng héng; 配重 pèi zhòng; 平衡重量 píng héng zhòng liàng

Examples:

① The main models that have been widely used are speciation calculations, mass balance method, and reaction-path simulations.

本文主要使用的模型有组分分布模型、质量平衡模型和反应路径模型。

② A core sample is obtained by boring into the snow—various measurements are then recorded so the scientists can later determine the glaciers' mass balance.

核心样品插入雪中，记录不同的测量，让科学家能判定冰河的质量平衡。

Extended Terms:

glacier mass balance 冰川物质平衡

mass balance analysis 质量平衡分析

mass balance calculations 质量平衡计算

mass flow [mæs fləu]

Definition: also known as mass transfer and bulk flow, is the movement of material matter. 质量流 zhì liàng liú

Example:

In this paper the progress of pressure rising for cryogenic liquid in the heated pipe was analyzed

and the critical mass flow velocity in the transfer process was calculated.
分析了低温液体输入热管道后升压过程，并对输送过程中的临界质量流的速率进行了计算。

Extended Terms:

mass flow coefficient 质量流量系数
mass flow meter 质量流量计
mass flow rate 质量流率
unstable mass flow 变质量流动
mass flow sensors 质量流量传感器

---

massive structure ['mæsiv 'strʌktʃə]

Definition: 块状构造 kuài zhuàng gòu zào

massive: imposing in size or bulk or solidity. 块状的 kuài zhuàng de
structure: a thing constructed; a complex entity constructed of many parts. 构造 gòu zào

Example:

Massive structure and banded structure appear widely in the ore bearing quartz veins.
含矿石英脉除了普通的块状构造以外，还普遍存在条带状构造。

Extended Term:

massive and rigid structure 大体积刚性结构

---

mass spectrometry [mæs spek'trɔmitri]

Definition: an instrumental method for identifying the chemical constitution of a substance by means of the separation of gaseous ions according to their differing mass and charge called also mass spectroscopy. 质谱测定法 zhì pǔ cè dìng fǎ

Examples:

① Mass spectrometry: Analytic technique by which chemical substances are identified by sorting gaseous ions by mass using electric and magnetic fields.
质谱测定法：一种分析方法，利用电场和磁场通过质量来将气体离子分类，从而鉴定化学物质。

② Mass spectrometry (MS) plays an important role in the study of lipoprotein metabolism.
质谱分析技术在脂代谢研究中发挥着重要的作用。

Extended Term:

mass-spectrometric technique 质谱技术

mass transfer

accelerator mass-spectrometry 加速器质谱法

mass transfer [mæs trænsˈfəː]

Definition: the net movement of mass from one location to another. Mass transfer is used by different scientific disciplines for different processes and mechanisms. The phrase is commonly used in engineering for physical processes that involve molecular and convective transport of atoms and molecules within physical systems. 物质移动 wù zhì yí dòng

Examples:
① The paper introduces the basic principles and general steps of mass transfer modeling of water-rock systems and an example is given.
文章介绍了水-岩体系质量传输模拟的基本原理和步骤，并给出了一个应用实例。
② Develops a numerical model to describe the heat and mass transfer process in a packed-type liquid dehumidifier.
建立了带填充物的液体除湿器的传热传质数值模型。

Extended Terms:
mass energy transfer coefficient 质量能量传递系数
mass transfer rate 质量迁移率

matrix [ˈmeitriks]

Definition: a mass of fine-grained rock in which gems, crystals, or fossils are embedded. n. 残余磷块岩砾石 cán yú lín kuài yán lì shí; 基质 jī zhì, 填隙物质 tián xì wù zhì, 充填物 chōng tián wù

Origin: late Middle English (in the sense womb)：from Latin, breeding female, "later womb", from mater, matr-"mother".

Examples:
① This is a research on imbibitions characteristics of rock matrix in Huoshaoshan Oilfield.
这是一个关于火烧山油田基质岩块渗吸特征的研究。
② Matrix interferences were eliminated by the matching method in the preparation of calibration curves.
在制备校正曲线时采用基体匹配法消除了基体干扰。

Extended Terms:
matrix recognition 模式识别
transposed matrix 转置矩阵
bone matrix 骨基质

casting matrix 浇铸基体
cell matrix 细胞基质
chromosome matrix 染色体基质
copper-rich matrix 富铜基体

mature stage [məˈtjuə steidʒ]

Definition: 成熟阶段 chéng shú jiē duàn
mature: having reached full natural growth or development. 成熟的 chéng shú de
stage: any distinct time period in a sequence of events. 阶段 jiē duàn

Example:

Among them, Ⅱ sequence SS1 reflects the initial developed stage of the passive continental margin basin in South China, Ⅱ sequence SS2~SS5 represent the passive continental margin basin from developing stage to mature stage, respectively, Ⅱ sequence SS6 reflects the transition process that South China evolved from passive continental margin basin to foreland basin, and Ⅱ sequence SS7 reflects the initial foreland basin stage of South China.
Ⅱ级层序 SS1 代表了被动陆缘盆地发展的初始阶段，Ⅱ级层序 SS2~SS5 是华南被动大陆边缘盆地由发展到成熟的阶段，Ⅱ级层序 SS6 是华南由被动大陆边缘向前陆盆地演化的过渡时期，Ⅱ级层序 SS7 是华南前陆盆地发展的初始阶段。

Extended Term:

mature stage of product lifecycle 产品生命周期的成熟阶段

maturity [məˈtjuərəti]

Definition: the state, fact, or period of being mature. 成熟度 chéng shú dù

Origin: late Middle English: from Latin maturitas, from maturus (see mature).

Example:

On the basis of thin-section analysis, the area percent of pore of reservoir sandstone ranges from 9.13% to 12.66%, with moderate maturity of composition and texture.
通过薄片分析确定，储集层砂岩有效面孔率为 9.13%~12.66%，结构成熟度和成分成熟度中等。

Extended Terms:

comparative maturity 比较成熟
compositional maturity 成分成熟度
optimum maturity 适宜成熟度

relative maturity 相对成熟度
delayed maturity 延迟成熟

maturity index [məˈtjuərəti ˈindeks]

Definition: 成熟度指数 chéng shú dù zhǐ shù

maturity: the state, fact, or period of being matur. 成熟度 chéng shú dù

index: a system by which changes in the value of something and the rate at which it changes can be recorded, measured, or interpreted. 指数 zhǐ shù

Example:

The crude oils of Chang 20 belong to typical immature oil which have odd even predominance and lower ratios of sterane and terpane maturity index.
长 20 井的原油既具有正烷烃奇偶优势，又具有较低甾萜烷异构体比值，属于典型的低熟原油。

meandering river deposit [miˈændəriŋ ˈrivə diˈpɔzit]

Definition: 曲流河沉积 qǔ liú hé chén jī

meandering: winding or rambling. 曲折的 qū zhé de

river: a large natural stream of water (larger than a creek). 河流 hé liú

deposit: the natural process of laying down a deposit of something. 沉积 chén jī

Example:

The meandering river deposit backslider from the studied area largely in the deposition period of Lianmuxing Formation in the early Cretaceous. The meandering stream was widely developed.
在早白垩纪沉积时期，湖水向南迅速退出研究区，曲流河道迅速发育。

mean free path [miːn friː pɑːθ]

Definition: (Physics) the average distance travelled by a gas molecule or other particle between collisions with other particle. 平均自由程 píng jūn zì yóu chéng

Examples:

① The average distance a photon travels before it is scattered is called the "scattering mean free path".
一个光子在被散射前所传播的平均距离称为"散射平均自由路径"。

② On average, the photon will be scattered by the time it traverses a distance of just one mean free path.

平均而言，光子被散射的时间就是它恰好传播一个平均自由路径的时间。

Extended Term:

average mean-free path 平均自由程

mechanical analysis [miˈkænikəl əˈnæləsis]

Definition: the measurement of the sizes of particles in a soil or clastic sediment by screening, sieving or setting; usually presented as a size-frequency distrubution giving the percentage by weight or number of grains within specified size limits or grades. 机械分析 jī xiè fēn xī

Example:

In this paper, by means of macro-morphology inspection, chemical composition analysis, microstructure examiation, mechanical property testing and X-rays diffraction analysis, the fractured bolt and random sample bolt were analyzed.
本文采用断口宏观形貌观察、化学成分分析、显微组织检验、力学性能则试及腐蚀产物的 X 射线衍射分析等手段，对断裂与部分抽检的卡瓦螺栓进行了综合试验分析。

Extended Terms:

mechanical thermal analysis 静态热机械分析
mechanical impedance analysis system 机械阻抗分析系统
dynamical mechanical thermal analysis 动力机械热分析
mechanical analysis of soil 土壤粒度分析

mechanical differentiation [miˈkænikəl difərenʃiˈeiʃən]

Definition: The process of concentration of minerals during metamorphism caused by mechanical conditions favouring recrystallization. 机械分异作用 jī xiè fēn yì zuò yòng

mechanical dispersion [miˈkænikl dIˈspɜːʃ(ə)n]

Definition: 机械分散 jī xiè fēn sàn

mechanical: relating to or concerned with machinery or tools. 机械的 jī xiè de
dispersion: the act of dispersing or diffusing something. 分散 fēn sàn

Examples:

① The solute in the mobile solution is transported in the forms of mechanical dispersion and convection.
在可流动溶液中，以机械弥散和对流传质为主。

② In this study, mechanical dispersion of raw materials and the flow field characteristics in the turbo air classifier were investigated and analyzed.
本文对物料分散性和涡流空气分级机内流场特性两个方面内容进行了实验研究与分析。

Extended Term:
mechanical dispersion halos 机械分散晕

Medford dike [ˈmedfəd daik]

Definition: 梅德福堤防 méi dé fú dī fáng

Medford: a town in northeastern Massachusetts, USA. 梅德福 méi dé fú
dike: a barrier constructed to contain the flow of water or to keep out the sea. 堤防 dī fáng

mediterranean [ˌmeditəˈreiniən]

Definition: of or characteristic of the Mediterranean Sea, the countries bordering it, or their inhabitants. 地中地槽 dì zhōng dì cáo, 陆间地槽 lù jiān dì cáo, 地中海型 dì zhōng hǎi xíng

Origin: mid 16th century: from Latin mediterraneus "inland" (from medius "middle" +terra "land") + -an.

Example:
The back-arc basins in the western Mediterranean
西地中海的弧后盆地

Extended Term:
Mediterranean climate 地中海气候

megacryst [ˈmegəkrist]

Definition: Fourth-order sedimentary cycle in a hierarchy, with increasing thickness and complexity, from cyces (first order), mesocycles (second order), macrocycles (third order) to megacycles. 巨晶 jù jīng, 大晶体 dà jīng tǐ

Example:
The course of formation of the labrador megacryst lapilli, the temperature and the cooling rate are the major factors controling their crystallization and growth.
托勒巴契克火山拉长石巨晶火山砾形成过程中,温度和冷却速率是控制其结晶生长的主导因素。

Extended Term:
megacrystalline 大晶的

melanite [ˈmelənait]

Definition: velvet-black variety of andradite (garnet). (矿)黑榴石 hēi liú shí

Origin: early 19th century: from Greek melas, melan-"black" + -ite.

Examples:

① The nepheline syenite magma would be evolved in the form of fractional crystallization constrained by the fractionation of calcium-rich pyroxene, aegirine-augite, leucite, Fe-rich biotite and melanite.
霞石正长质岩浆受富钙辉石——霓辉石、白榴石、含 Fe 较高黑云母、黑榴石的分离结晶制约发生分异。

② The paper elucidates the geological characteristics of the melanite aegirine-augite sye uitepegmatite and melanite aegirine-augite nepheline syeuite pegmatite found first in Ji'nan intermediate-basic complex and discusses their origin.
本文介绍了济南北部中基性杂岩体中首次发现的黑榴霓辉正长伟晶岩和黑榴霓辉霞石正长伟晶岩的岩石学特征,并对其成因作了探讨。

Extended Terms:

melanite garnet 黑榴石榴石
melanite-phonolite 黑榴响岩

melanocratic [ˌmelənəuˈkreitik]

Definition: dark-colored, referring to igneous rock containing at least 50%~60% mafic minerals. Also known as chromocratic; melanic. 暗色的 àn sè de

Extended Terms:

melanocratic dike 暗色岩脉
melanocratic rock 暗色岩
melanocratic igneous rocks 暗色火成岩

melanocratic rock [ˈmelənəuˈkreitik rɒk]

Definition: rocks of dark-colored, referring to igneous rock containing at least 50%~60% mafic minerals. Also known as chromocratic; melanic. 深色岩 shēn sè yán

Extended Term:

melanocratic rock series 暗色岩系

melanosome [ˈmelənəʊˌsəʊm]

Definition: an organelle which contains melanin and in which tyrosinase activity is not demonstrable. 暗色体 àn sè tǐ

Examples:

① The influence of nicotinamide on melanosome movement velocity and its distribution in UVA irradiated human skin melanocytes were studied.
研究了烟酰胺对经 UVA 照射后人皮肤黑素细胞中黑素小体运动速率和黑素小体分布的影响。

② Recent study supported this view that this disease is a disease with functional defect of melanocyte and abnormal formation of melanosome.
近年来的研究结果表明该病是一个黑色素细胞功能缺陷、黑素小体形成异常的疾病。

Extended Terms:

compound melanosome 复合黑素体
giant melanosome 大黑素体

melilite [ˈmelilait]

Definition: a sorosilicate mineral group, consisting chiefly of sodium, calcium, and aluminum silicates, occurring in igneous rocks. 黄长石 huáng cháng shí

Origin: 1790-1800.

Examples:

① The practice showed that the blast furnace slags mainly consisted of melilite, cuspidine, bivalent metallic sulphide and Ce-Ca silica.
结果表明：黄长石、枪晶石、二价金属硫化物、铈钙硅石是构成高炉渣的主要矿物。

② Depending on the chemical composition the structural parameters of the melilite vary in the range between gehlenite and mernite. Melilitic separation and melilite as a primary phase are very rare.
黄长石的结构参数，按其化学成分是在钙铝黄长石和镁黄长石之间变化，黄长石作为原始矿物相在碱性矿渣中比较罕见。

Extended Terms:

melilite basalt 黄长玄武岩
melilite-leucitite 黄长白榴岩
melilite ankaratrite 黄长橄霞玄武岩
melilite olibine nephelinite 黄长橄霞岩

melilitite [ˈmeləlitait]

Definition: an extrusive rock that is generally olivine-free and composed of more than 90% mafic mineral such as melilite and augite, with minor amounts of feldspathoids and sometimes plagioclase. 黄长岩 huáng cháng yán

Example:
The kamafugite is a rare type of ultra-potassic rocks, and a rock series contained generally that rock types of alnoite, olivine-melilitite, mafurite, katungite and wugantite and so on.
钾霞橄黄长岩是一种非常稀少的超钾质火山岩，也是一个岩石系列，包括黄长煌斑岩、橄榄石黄长岩、白橄黄长岩、橄辉钾霞岩和乌干达岩等多种岩石类型。

Extended Terms:
biotite melilitite 黑云黄长岩
nepheline melilitite 霞石黄长岩

melting [ˈmeltɪŋ]

Definition: to become liquefied by warmth or heat, as ice, snow, butter, or metal. 溶化 róng huà, 融化 róng huà

Examples:
① Earth changes shall increase in the quarter century ahead, with earthquakes, floods, tornadoes, and the continued melting of ice caps due to global warming.
地球的变化将以地震、水、卷风和因地球变暖而持续融化冰盖的方式，在前方四分之一个世纪内渐增。
② The melting is likely to shift weather patterns, too. More sea ice means colder winters, because frigid winds blowing over ice pick up little heat from the warmer waters below.
冰川融化可能也会改变天气状况。越来越多的海洋冰川意味着冬天更冷，因为大风在冰面上刮，不能从温暖的海水底下带走热量。

Extended Terms:
metal melting 金属熔炼
melting point 熔点
melting pot 熔锅
melting range 熔化范围
melting experiments 熔化实验
melting-point lowering 熔点降低

mercury [ˈmɜːkjəri]

Definition: a small planet that is the closest to the sun in the solar system, sometimes visible to the naked eye just after sunset. 汞 gǒng, 水银 shuǐ yín, 水星 shuǐ xīng

Origin: 1300-1350.

Examples:

① The planets of our solar system are Mercury, Venus, Earth, Mars, Jupiter, Saturn, Uranus and Neptune.
我们太阳系的行星有水星、金星、地球、火星、木星、土星、天王星和海王星。

② Mercury, like Venus, appears to go through phases like the Moon.
像金星一样的水星似乎经历和月亮一样的相位。

Extended Terms:

mercury electrode 水银电极
mercury contamination 汞污染

mesozone [ˈmezəzəun]

Definition: the intermediate depth zone of metamorphism in metamorphic rock characterized by moderate temperatures (300-500℃), hydrostatic pressure, and shearing stress. 中带 zhōng dài

Example:

According to the ore-forming depth and condition of physical-chemical, these deposits may be classified into three class: deep-zone type, mesozone type and epizone type.
依矿床形成深度和物理化学条件等差异, 将其划分为深部带型、中部带型和浅部带型三亚类。

Extended Term:

pluton of mesozone 中带深成岩体

meta-igneous rock [ˌmetəˈigniəs rɔk]

Definition: pertaining to metamorphic rock formed from igneous rock. 元火成岩 yuán huǒ chéng yán

metabasite [ˌmetəˈbəsait]

Definition: a collective term, first used by Finnish geologists, for metamorphosed mafic rock that has lost all traces of its original texture and mineralogy owing to complete recrystallization. 准基性岩 zhǔn jī xìng yán, 变基性岩 biàn jī xìng yán

> Example:

Important Stone Material in Taiwan: Basalt, Andesite, Taiwan Jade, Argillite, Metabasite, Sandstone, Slate, Schist.
台湾地区重要石材分布图：玄武岩、安山岩、台湾玉、硬页岩、变质基性岩、砂岩、板岩、片岩。

> Extended Term:

metabitumite 准沥青岩

metaluminous [ˌmetəˈluːminəs]

> Definition: applied to igneous rocks in which there are fewer molecules of Al_2O_3 than of ($CaO+Na_2O+K_2O$). 准铝质的 zhǔn lǚ zhì de，偏铝质的 piān lǚ zhì de

> Examples:

① Petrochemically, the complex is metaluminous to peraluminous (0.94~1.26), rich in SiO_2, K_2O+Na_2O, poor in CaO, MgO, and is characterized by high $FeO\sim*/MgO$ ratio (with an average of 12.05), which is similar to those of A-type granite and different from the ordinary S- and I-type granites.
该杂岩富硅碱贫钙镁、$K_2O/Na_2O>1$、准铝过铝质（0.94~1.26）；FeO/MgO 比值大（平均 12.05），不同于 S 型和 I 型花岗岩。

② Those of the latter are metaluminous (ACNK=0.88-0.96) and characterized by high REE {$\Sigma REE=(250.33-295.99)\times10^{-6}$} and HFSE (Y, Zr, Nb etc.). These characteristics are quite similar to those of A type granites.
后者为准铝质（ACNK=0.88-0.96），稀土元素（ΣREE：（250.33-295.99）×10-6)和高场强元素（Y，Zr，Nb 等）的含量较高，与 A 型花岗岩的地球化学特征相似。

> Extended Terms:

metaluminous rocks 变铝质岩
metaluminous type 变铝质型

metamorphic [ˌmetəˈmɔːfik]

> Definition: denoting rock that has undergone transformation by heat, pressure, or other natural agencies, e.g. in the folding of strata or the nearby intrusion of igneous rocks. 变质的 biàn zhì de

> Origin: early 19th century: from meta-(denoting a change of condition)+Greek morphē "form"+-ic.

> Examples:

① There are low-grade metamorphic rocks which underwent greenschist facies metamorphism in

the middle and north margin of the Dabie ultrahigh pressure(UHP) metamorphic belt.
大别造山带超高压变质带内部及其北缘，出露仅经过绿片岩相变质作用的浅变质岩系。

② The Early Precambrian rocks are the important constituent of old metamorphic terrains in Liaoning Province, mainly in metamorphic facies higher than greenschist and locally reaching granulite facies.
早前寒武纪变质岩系是辽宁地区古老变质岩系的重要组成部分，变质程度高于绿片岩相，局部可达麻粒岩相。

Extended Terms:

metamorphic sequence 变质顺序
metamorphide 变质褶皱带
metamorphic rock 变质岩
metamorphic reaction 变质反应

metamorphic complex [metəˈmɔːfik ˈkɔmpleks]

Definition: 变质杂岩 biàn zhì zá yán

metamorphic: denoting rock that has undergone transformation by heat, pressure, or other natural agencies, e.g. in the folding of strata or the nearby intrusion of igneous rocks. 变质的 biàn zhì de

complex: a compound of rocks. 杂岩 zá yán

Example:

Felsic gneiss is the most widely distributed rock type in Archean metamorphic complex areas.
长英质片麻岩是太古宙变质杂岩区最主要的岩石类型。

Extended Term:

metamorphic core complex 变质核杂岩体

metamorphic core complex [metəˈmɔːfik kɔːˈkɔmpleks]

Definition: 变质核杂岩体 biàn zhì hé zá yán tǐ

core: the dense central region of a planet, especially the nickel-iron inner part of the earth. (天体的，尤指地球的)核 hé，中心 zhōng xīn

Examples:

① Aim in order to determine whether the metamorphic complex of Xiaoqinling area is Metamorphic core complex or not.
目的为确定小秦岭地区发育的变质杂岩体是否属于变质核杂岩。

② The oceanic core complex of the oceanic crust is obviously different from the metamorphic core complex of the continent.

与大陆变质核杂岩相比，海洋核杂岩具有明显的独特性。

metamorphic cycle [ˌmetəˈmɔːfik ˈsaikl]

Definition: 变质旋回 biàn zhì xuán huí

cycle: a recurring period of a definite number of years, used as a measure of time. 循环 xún huán

Examples:

① In the late stage of the Proterozoic metamorphic cycle, the regional metamorphism of low amphibolite facies took place because of the crust rising and depressurization, which formed the low P metamorphic minerals of andalusite, needle-like sillimanite, wollastonite and cordierite, etc.
在元古变质旋回晚期，由于地壳隆升降压，相应发生了低压角闪岩相的区域变质，从而形成了红柱石、针状矽线石、硅灰石及堇青石等低压变质矿物。

② The Ailao Mts rock group experienced a dynamic transition from medium P regional metamorphism to high P one in the Proterozoic metamorphic cycle.
哀牢山岩群在元古变质旋回经历了中压区域变质向高压区域变质作用的动态转变。

metamorphic differentiation [ˌmetəˈmɔːfik ˌdifərenʃiˈeiʃən]

Definition: the process or processes by which contrasted mineral assemblages develop from an initially uniform parent-rock during metamorphis. 变质分异作用 biàn zhì fēn yì zuò yòng

metamorphic event [ˌmetəˈmɔːfik iˈvent]

Definition: during the active life of a mobile belt there may be several metamorphic events which can be identified from radiometric age groupings and which are separated by tens or hundreds of millions of years. Frequently metamorphic events can be subdivided on structual grounds into metamorphic episodes. 变质事件 biàn zhì shì jiàn

Examples:

① The earlier structures in basement and allochthonous cover are products of M1 metamorphic event at different crustal levels.
基底和异地盖层中的早期构造代表 M1 变质事件在不同地壳层次上的产物。

② The result demonstrates that the Northern Dabie gneiss also underwent Jurassic metamorphic event.
这一结果表明北大别片麻岩也经历了印支期变质作用。

metamorphic facies [metəˈmɔːfik ˈfeiʃiːz]

Definition: a group of metamorphic rock units characterized by particular mineralogicassociations. 变质相 biàn zhì xiāng

Origin: 1810-1820; meta- + -morphic.

Extended Terms:

metamorphic facies series 变质相系列
metamorphic field gradient 变质梯度场
metamorphic fluids 变质流体
metamorphic geotherm 地热变质
metamorphic facies group 变质相组
metamorphic facies series 变质相系

metamorphic fluids [metəˈmɔːfik ˈfluːidz]

Definition: 变质流体 biàn zhì liú tǐ

metamorphic: denoting rock that has undergone transformation by heat, pressure, or other natural agencies, e.g. in the folding of strata or the nearby intrusion of igneous rocks. 变质的 biàn zhì de

fluids: a continuous amorphous substance that tends to flow. 流体 liú tǐ

Origin: early 19th century: from meta-(denoting a change of condition) + Greek morphē "form" + -ic.

Examples:

① The mass change and the mobility of some elements in question result from the effects of metamorphic fluids on the rocks during the regional metamorphism.
岩石质量迁移与元素活动是区域变质过程中变质流体的结果。

② One is hydrothermal metasomatism related to magmatic-derived fluids and another is hydrothermal replacement of metamorphic fluids that possibly derived from formation brines.
滑石的成因模式主要有两类：一为与岩浆热液有关的热液交代成矿模式，另一是起源于盆地热卤水的变质流体交代成矿模式。

metamorphic grade [metəˈmɔːfik ɡreid]

Definition: the intensity of metamorphism, measured by the degree of difference between the parent rock and the metamorphic rock. 变质级 biàn zhì jí

> Example:

The metamorphic grade is high and the gas production is big for the coal rock of Liangshan and Longtan stages in which there is great coal bed gas mainly with absorption way.
梁山煤系和龙潭煤系煤岩变质级较高，生气量大，赋存大量以吸附态为主的煤层气。

metamorphic mineral [metəˈmɔːfik ˈminərəl]

> Definition: 变质矿物 biàn zhì kuàng wù

mineral: a substance such as tin, salt, or sulphur that is formed naturally in rocks and in the earth. Minerals are also found in small quantities in food and drink. 矿物 kuàng wù

> Examples:

① 40 Ar 39 Ar dating of the metamorphic mineral biotite implies that the early event took place at the beginning of the Carboniferous (355 Ma B. P.).
通过对变质矿物黑云母的 40Ar 39Ar 定年研究，确定早期变质变形事件发生在石炭纪初 (355 Ma B. P.)。

② Through resent years' regional surveying, the crystallinity of muscovite is determined of b0 exceeding 0.9040nm by X-ray diffraction way, which featured the phengite as a high pressure type of metamorphic mineral.
通过近年区调工作，用 X 射线粉晶衍射方法，测定泥质变质岩的白云母结晶度 b0>0.9040nm，显示出高压型变质矿物——多硅白云母特征。

> Extended Terms:

metamorphic mineral paragenesis 变质矿物共生
metamorphic mineral deposit 变质矿床
metamorphic mineral association 变质矿物共生组合

metamorphic petrology [metəˈmɔːfik piˈtrɒlədʒi]

> Definition: the branch of petrology. The main contents are: to study various features of metamorphic rock, including chemical composition, mineral composition, texture and structure, and geological occurrence; to identify the role of different types of metamorphic rocks and metamorphic conditions and crustal evolution in the development of relations. 变质岩石学 biàn zhì yán shí xué

> Examples:

① Research for geofluids is a hot-point and frontier in the geoscience field, especially, in igneous petrology, metamorphic petrology, geotectology and study of mineral deposits.
流体研究是近年来地学界研究的热点和前沿，它不仅在岩浆岩、构造、变质岩等方面的研究中受到重视，而且在矿床研究领域得到了发展。

② *Contributions to Mineralogy and Petrology* is an international journal that accepts high quality research papers in the fields of igneous and metamorphic petrology, geochemistry and mineralogy.
《矿物学与岩石学研究》是一个国际性的学术期刊，发表有关岩浆和变质岩石学、地球化学、及矿物学领域的高质量学术论文。

Extended Terms:
metamorphic grade 变质级，变质程度
metamorphic hydrology 变质水文
metamorphic mineral facies 变质矿物相

metamorphic reaction [metəˈmɔːfik riˈækʃən]

Definition: a chemical reaction that takes place during the geological process of metamorphism wherein one assemblage of minerals is transformed into a second assemblage which is stable under the new temperature/pressure conditions resulting in the final stable state of the observed metamorphic rock. 变质反应 biàn zhì fǎn yìng

Examples:
① These theoretical and experimental results have important significance for an in-depth study on the feature of metamorphic reaction and the genetic mechanism of mineral phases transformation in granulite facies metamorphism of khondalite series.
该项理论与实验研究对于深化变质反应理论、探讨孔兹岩系矿物相转变成因机制及其动力学过程有着重要的意义。
② The Metamorphic Reaction and Water Activity of Basic Granulite in the Datong-Huai'an Region.
怀安—大同一带基性麻粒岩变质反应序列与水活度。

metamorphic reaction zone [metəˈmɔːfik riˈækʃən zəun]

Definition: zones in the earth's crust characterized by certain definition metamorphic grades and definite metamorphic rock types. 变质反应带 biàn zhì fǎn yìng dài
zone: any area or region considered as separate or distinct from others because of its particular use, crops, plant or animal life, status in time of war, geological features. 地带 dì dài

metamorphic rock [metəˈmɔːfik rɔk]

Definition: one of the major groups of rock that makes up the crust of the Earth; consists of pre-existing rock mass in which new minerals or textures are formed at higher temperatures and greater pressures than those present on the Earth's surface. 变质岩 biàn zhì yán

Examples:

① Types of metamorphic rocks: regional metamorphism, contact metamorphism and dynamic metamorphism.
变质岩类型：区域变质、接触变质和动力变质。

② The Houdidong Au deposit occurred in the Permian, Kedao Formation contact zone between metamorphic rock series and Variscan gabbro body.
后底洞金矿床产于二叠纪柯岛组变质岩系与华力西期辉长岩体内外接触带。

Extended Terms:

regional metamorphic rock 区域变质岩；区域变成岩
thermal metamorphic rock 热变质岩
dynamic metamorphic rock 动力变质岩
contact metamorphic rock 接触变质岩
aureole of contact metamorphism 接触变质圈
Ailaoshan epi-metamorphic rock belt 哀牢山浅变质岩带
metamorphic rank 变质度
regional metamorphism 区域变质作用

metamorphic zone [metə'mɔːfik zəun]

Definition: in geology an area where, as a result of metamorphism, the same combination of minerals occurs in the bed rocks. These zones occur because most metamorphic minerals are only stable in certain intervals of temperature and pressure. 变质带 biàn zhì dài

Example:

The contact metamorphic zone may be further divided into coking coal mixing zone, natural coke zone, coking coal zone and normal coal zone.
接触变质带可进一步划分为岩焦混杂带、天然焦带、焦化煤带、正常煤带。

Extended Terms:

contact metamorphic zone 接触变质带；接触变质带，接触圈

metamorphism [metə'mɔːfizəm]

Definition: a change in the structure or constitution of a rock due to natural agencies, as pressure and heat, esp. when the rock becomes harder and more completely crystalline. 变质作用 biàn zhì zuò yòng

Origin: 1835-1845; meta- + -morphism.

metasomatism

Examples:

① The Sulu-Dabie eclogite, which contains cosine, is the product of high-ultra high pressure metamorphism.
含柯石英的苏鲁-大别榴辉岩是高压-超高压变质带的产物。

② The forming temperature of albite based on its chemical composition is higher than 800℃ and higher than 400℃, suggesting that its formation was related to volcanic effusion and later metamorphism.
钠长石成分计算其形成温度超过800℃和400℃，表明其形成与喷流作用有关，并受到后期变质改造。

Extended Terms:

allochemical metamorphism 异化变质作用；他化变质作用，增减变质作用
destructive metamorphism 破坏变质作用；同温变质；破坏性变质作用
geothermal metamorphism 地热变质；地热变质作用
burial metamorphism 埋藏变质；埋藏变质作用；变质作用；埋深变质作用
constructive metamorphism 增温变质；接力变质；接成变质；递增变质
metasomatic corrosion texture 交代蚕蚀结构
metasomatic edulcoration border texture 交代净边结构
metasomatic graphoid texture 交代似文象结构
metasomatic perforation texture 交代穿孔结构
metasomatic perthitic texture 交代条纹结构
metasomatic porphyritic texture 交代斑状结构
metasomatic pseudomorph texture 交代假象结构
metasomatic relict texture 交代残留结构

metasomatism [ˌmetəˈsəumətizəm]

Definition: change in the composition of a rock or mineral by the addition or replacement of chemicals. 交代作用 jiāo dài zuò yòng

Origin: 19th century：from New Latin; see meta-, somato-.

Example:

The late diagenesis includes filling, devitrification, metasomatism, mechanical impaction and pressure solution, cementation and dissolution.
晚期成岩作用类型为充填作用、脱玻化作用、交代作用、机械压实压溶作用、胶结作用和溶解作用。

Extended Terms:

selective metasomatism 分别变代；选择交代
pneumatolytic metasomatism 气化变质；气成交代作用

lime metasomatism 石灰交代
fluorine metasomatism 氟素交代酌
sodic metasomatism 钠质交代酌

metastable [ˌmetəˈsteibl]

Definition: (Physics)(of a state of equilibrium) stable provided it is subjected to no more than small disturbances. 亚稳的 yà wěn de；相对稳定的 xiāng duì wěn dìng de

Origin: 1895-1900; meta- + stable.

Examples:

① The effect of nitrogen, carbon and manganese on mechanical properties and work-hardening ability metastable austenitic manganese steel was studied.
研究了氮、碳、锰等合金元素对介稳奥氏体锰钢力学性能及耐磨性的影响。

② As they absorb energy, their electrons move up to higher energy levels, tending to accumulate at certain metastable levels.
当它们吸收能量的时候，它们的电子跃迁到高能级层，积聚在某一个亚稳态层。

Extended Terms:

metamorphic core complex 变质核杂岩体
metastable atom 亚稳原子
metastable condition 亚稳状态
metastable phases 亚稳相
metastable states 亚稳态

meteorite [ˈmiːtiərait]

Definition: a piece of rock or metal that has fallen to the earth's surface from outer space as a meteor. Over 90 percent of meteorites are of rock while the remainder consist wholly or partly of iron and nickel. 陨星 yǔn xīng；流星 liú xīng

Origin: 1815-1825; meteor + -ite.

Examples:

① According to the professionals that the meteorite to the moon meteorite, the overall throughout zinc magnesium tablets.
据专业人士探讨认为，该陨石为月亮陨石，整体墨绿色，遍布锌镁反光片。

② If so the single crater area will possibly exceed the 5,000 square kilometers one, the currently publicly recognized meteorite crater group in South Africa.
假如此坑被证实，单坑面积将可能超过目前公认的世界上最大的南非陨石坑群区 5000 平

方公里的总面积。

Extended Terms:

meteorite crater 陨石坑
meteorite fall 陨石雨
meteorite impacts 陨石撞击

methane [ˈmiːθein]

Definition: (Chemistry) a colourless, odourless flammable gas which is the main constituent of natural gas. It is the simplest member of the alkane series of hydrocarbons. *n.* [化] 甲烷 jiǎ wán; 沼气 zhǎo qì

Origin: mid 19th century: from methyl + -ane.

Examples:

① Microbes break apart water molecules in photosynthesis, and the hydrogen can pass like a baton from organic matter to methane and eventually reach space.
微生物在光合作用中分解了水分子,氢可以就像接力棒一样,从有机质传递到甲烷,最终到达太空。

② The explosive potential (energy content per unit mass) of hydrogen is about twice that of the methane in natural gas.
氢的爆炸潜能(每单位质量含有的能量),大约是天然气中甲烷的两倍。

Extended Terms:

dibenzoyl methane 二苯甲酰甲烷
methane detector 沼气检定器
methane determination 沼气测定
methane emission 沼气泄出

micas [ˈmaikəs]

Definition: any member of a group of minerals, hydrous silicates of aluminum with other bases, chiefly potassium, magnesium, iron, and lithium, that separate readily into thin, tough, often transparent, and usually elastic laminae; isinglass. 云母 yún mǔ

Origin: 1700-1710.

Examples:

① The results showed that grain amaranth could efficiently take up K from both soil and micas (biotite and phlogopite).
结果表明,籽粒苋能有效地利用土壤和云母(黑云母和金云母)中的钾。

② The content of clay micas in the soils with higher weathering degree, such as lateritic red soil, red soil, yellowish red soil and yellow soil, was generally low.
风化度较高的赤红壤、红壤、黄红壤和黄壤等土类的粘粒云母含量一般较低。

Extended Terms:
elastic micas 弹性云母
fluor-micas 氟云母类

mica schist [ˈmaɪkə ʃɪst]

Definition: (Petrology) a schist which is composed essentially of mica and quartz and whose characteristic foliation is mainly due to the parallel orientation of the mica flakes. 云母片岩 yún mǔ piàn yán

Example:
There is a suit of mica schist among metamorphic basement of Early-Middle Proterozoic era at Xiangshan Uraniun Mineral Field.
在相山铀矿田的早中元古代变质基底中，出露一套云母片岩。

Extended Terms:
lime-mica-schist 灰云片岩
iron-mica-schist 铁云片岩；铁云母片岩
graphite-mica schist 石墨云母片岩
glaucophane mica schist 蓝闪云片岩
garnet-mica schist 石榴云母片岩

Michigan Basin [ˈmɪʃɪg(ə)n ˈbeɪsn]

Definition: a geologic basin centered on the Lower Peninsula of the US state of Michigan. The feature is represented by a nearly circular pattern of geologic sedimentary strata in the area with a nearly uniform structural dip toward the center of the peninsula. 密歇根盆地 mì xiē gēn pén dì

Examples:
① Michigan basin, a stable craton basin is a huge round basin in North American Continent where the sediment is rather thick, however the structure is very simple.
密歇根盆地是北美大陆上近圆形的大盆地，具巨厚层沉积物，构造作用简单，属于稳定的克拉通内部盆地。
② Distribution of gammacerane in some marine evaporitic environments such as Michigan basin shows that gammacerane, a hypersaline biomarker, is not absolutely associated with hypersaline environment.

密歇根盆地等海相蒸发环境伽马蜡烷的分布表明，伽马蜡烷作为咸水环境的生物标志化合物与高盐度环境并没有绝对的对应关系。

micrite [ˈmikrait]

Definition: [mass noun] (Geology) microcrystalline calcite present in some types of limestone. 泥晶灰岩 ní jīng huī yán；微晶灰岩 wēi jīng huī yán

Origin: 1950s：from micr(ocrystalline) + -ite.

Examples:
① Those of marls facies in deep lake mainly include micrite limestones, marls, micrite limestones or marls with little mud and silt.
半深湖泥灰岩相的主要岩石类型为泥晶灰岩、泥灰岩以及含少量陆屑泥和粉砂的泥晶灰岩或泥灰岩。
② Those of grain banks in shallow lake mainly include ooid limestones with spar or micrite, bioclastic limestones, intraclastic limestones and grain lime-stones.
浅湖颗粒滩有亮晶或泥微晶的鲕粒灰岩、生屑灰岩和内碎屑灰岩，其次为含陆屑颗粒灰岩。

Extended Terms:
micrite enlargement 微晶加大作用
micrite envelope 泥晶套
fossiliferous micrite 含化石泥晶灰岩
micritic aragonite 微晶文石
oolite-bearing micrite 含鲕粒微晶灰岩
fossiliferous micrite 含化石泥晶灰岩
coralgal micrite 珊瑚藻微晶灰岩

micritic limestone [miˈkritik ˈlaimstəun]

Definition: 泥晶灰岩 ní jīng huī yán，又称灰泥岩(lime mudstone) huī ní yán；微晶灰岩(micrite) wēi jīng huī yán
micritic：containing micrite. 含泥晶的 hán ní jīng de
limestone：whitish-coloured rock which is used for building and for making cement. 石灰岩 shí huī yán

Example:
The gamma ray value of reef bafflestone is higher than those of sparrenite and sparrudite, and lower than those of micritic limestone, wackstone, echinoderm grainstone and marlstone.
礁灰岩的自然伽马值高于亮晶砂、砾屑灰岩，而低于泥晶灰岩、棘屑灰岩、泥晶生屑灰岩、泥灰岩的放射性。

Extended Term:

micritic rind 泥晶质壳层

micritization [mikraitaiˈzeiʃən]

Definition: making micritized. 泥晶化 ní jīng huà

Example:

It's a study on the Micritization of Carbonate Grains by Bacteris and Algae—the example of the Early Carboniferous in Baoshan, Western Yunnan.
菌藻对碳酸盐颗粒的泥晶化作用研究——以滇西保山地区下石炭统研究为例。

Extended Terms:

micrite envelope 泥晶套
micritepelsparite 泥晶灰岩

microcline [ˈmaikrəuklain]

Definition: [mass noun] a green, pink or brown crystalline mineral consisting of potassium-rich feldspar, characteristic of granite and pegmatites. 微斜长石 wēi xié cháng shí

Origin: mid-19th century: from German Microklin, from Greek mikros "small"+klinein "to lean" (because its angle of cleavage differs only slightly from 90 degrees).

Examples:

① The effect of albite on activating fly ash is better than microcline's.
钠长石对粉煤灰蒸压制品的激发效果优于钾长石。
② Oxalic acid diluted in hot water will remove the iron oxide from such specimens as microcline, quartz, and fluorite.
草酸稀释后，在热水中，将可消除氧化铁从这些标本为钾长石、石英、萤石。

Extended Terms:

maximum microcline 最大微斜长石
soda microcline 钠斜微长石

microcontinent [ˈmaikrəuˈkɒntinənt]

Definition: an isolated fragment of continental crust forming part of a small crust plat. 微小陆块 wēi xiǎo lù kuài

> Examples:

① At that time, the position of Eurasia was almost the same as today, but Gondwana was near the equator and was an "internal separation" microcontinent then.

当时的欧亚大陆虽然大致在现今位置,而冈瓦纳大陆尚在赤道附近,并且是"内部分隔"的微大陆。

② The discovery of theses fossils may provide direct evidence and new information for the regional stratigraphic classification and the examination of the geological and tectonic evolution of the Tianshuihai microcontinent and the Bayan Har Late Palaeozoic-Mesozoic marginal rift basin.

生物化石的发现,为该套地层时代划分提供了直接的证据,并为进一步详细研究西昆仑微陆块及巴颜喀拉晚古生代—中生代边缘裂陷盆地的地质构造演化提供了新的基础资料。

> Extended Terms:

microplates 微型板块
microcontinental block 微型大陆断块

microcrystalline [ˌmaikrəuˈkristəlain]

> Definition: in carbonate petrology, refers to sediments with crystals between 1 and 10 um or between 4 and 61 um. 微晶(的) wēi jīng (de), 微晶质 wēi jīng zhì

> Examples:

① The spacing of lattice d002 for carbon microcrystal and the microcrystalline sise (Lc002 & La002) were calculated to be 0.3604nm and 1.1174nm & 2.3102nm respectively.

其微晶的面间距(d002)和微晶尺寸(Lc002 及 La002)分别为 0.3604nm 和 1.1174nm 及 2.3102nm。

② It is infeasible for the volcanic rock with microcrystalline and eryptocrystalline matrix to count its mineral content by conventional slice method.

对于基质为微晶质、隐晶质的火山岩,采用常规的薄片显微镜统计矿物含量几乎是不可行的。

micrographic texture [maikrəuˈgræfik ˈtekstʃə]

> Definition: a microscopic-scale variety of graphic texture. 显微文象结构 xiǎn wēi wén xiàng jié gòu

> Example:

In matrix, medium-coarse grains, fine grains and micrographic texture are the main texture.

基质结构有中粗粒、细粒和显微文象结构等。

microperthite [ˈmaikrəuˈpəːθait]

Definition: perthite in which the lamellae are visible only under the microscope. 微条纹长石 wēi tiáo wén cháng shí

Example:

Microperthite is known to occur in the uppermost 1,000 feet of the Upper Zone, but the amount is small.
微条纹长石见于上岩带顶部的 1000 尺范围内, 但含量不多。

Mid-ocean ridge [midˈəuʃn ridʒ]

Definition: a series of mountain ranges on the ocean floor, more than 84,000 kilometers (52,000 miles) in length, extending through the North and South Atlantic, the Indian Ocean, and the South Pacific. According to the plate tectonics theory, volcanic rock is added to the sea floor as the mid-ocean ridge spreads apart. 中洋海脊 zhōng yáng hǎi jǐ

Origin: old English hrycg "spine, crest", of Germanic origin; related to Dutch rug and German Rücken "back".

Examples:

① Iceland is peculiarly volcanic because it is formed by the intersection of a hotspot and a mid-ocean ridge.
冰岛尤其是由火山构成的, 因为它是由一个热点和一个海中间的脊的交点形成的。

② As the plates diverge from a mid-ocean ridge they slide on a more yielding layer at the base of the lithosphere.
当板块从中海脊脱离时, 它们滑向在岩石圈基部较易变形的地层上。

Extended Terms:

mid-ocean ridge basalt (MORB) 洋中脊玄武岩
mid-ocean ridge crest 洋中脊峰线
mid-Alantic ridge 中太平洋海脊
mid-Indian ocean ridge 中印度洋海脊

mid-ocean ridge basalt [mid ˈəuʃən ridʒ ˈbæsɔːlt]

Definition: a general term for an underwater mountain system that consists of various mountain ranges (chains), typically having a valley known as a rift running along its spine, formed by plate tectonics. 大洋中脊玄武岩 dà yáng zhōng jǐ xuán wǔ yán

migmatite [ˈmɪgmətaɪt]

Definition: (Geology) a rock composed of two intermingled but distinguishable components, typically a granitic rock within a metamorphic host rock. [地质] 混合岩 hùn hé yán, 复片麻岩 fù piàn má yán

Origin: early 20th century: from Greek migma, migmat-"mixture" + -ite.

Examples:

① It is considered that the deposit is an anatectic migmatite corundum deposit.
研究认为，本矿床应属深熔型混合岩化作用成因，即"混合岩型"刚玉矿床。

② The formation of migmatite includes partial melting, partial segregation and extraction of melt and partial back and retrograde reactions.
混合岩的形成过程包括部分熔融作用、不同程度熔体分凝与汲取和不同程度的逆反应和退变反应。

Extended Terms:

banded migmatite 带状混合岩
arteritio migmatite 注入混合岩
migmatized metamorphic rock 混合岩化变质岩
banded migmatite 带状混合岩；条带状混合岩
arteritio migmatite 注入混合岩
diffuse migmatite 扩散生成混合岩
streaky migmatite 条痕状混合岩

migmatization [ˌmɪgmətaɪˈzeɪʃən]

Definition: a process of close intermixture of a liquid magmatic fraction with solid rock leading to the formation of migmatites. 混合岩化作用 hùn hé yán huà zuò yòng

Example:

The formation of the gold deposit had undergone the process of migmatization, ductile sheering and magmatic hydrothermal superimposition.
金矿床的形成经历了混合岩化作用、韧性剪切作用及岩浆热液叠加的成矿过程。

Extended Terms:

regional migmatization 区域混合岩化作用
hypo migmatization 深混熔作用

migmatite front [ˈmɪgmətaɪt frʌnt]

Definition: the outer zone into which the process of migmatilization has penetrated. 混合岩

前缘混合岩 qián yuán hùn hé yán

Milankovitch cycle [milæn'kɔvitʃ 'saik(ə)]

Definition: periodic variations in the earth's position relative to the sun as the earth orbits, affecting the distribution of the solar radiation reaching the earth and causing climatic changes that have profound impacts on the abundance and distribution of organisms, best seen in the fossil record of the Quaternary Period (the last 1.6 million years). 米兰科维奇周期 mǐ lán kē wéi qí zhōu qī

Examples:

① The polar ice sheet expands and contacts because of variations in the Earth's orbit (Milankovitch cycles).
由于地球轨道的变更(米兰科维奇旋回)，极地的冰盖不断扩大相连。

② Different types of meter-scale cyclic sequences in stratigraphical records are the outcomes of episodic accumulation of strata related to Milankovitch cycles.
与米兰柯维奇旋回存在成因联系的幕式地层堆积作用的结果，是在地层记录中形成各种类型的米级旋回层序。

mineral assemblage ['minərəl ə'semblidʒ]

Definition: a group of minerals found together in a particular rock. 矿物共生 kuàng wù gòng shēng, 矿物集合 kuàng wù jí hé

Examples:

① The change of the mineral assemblage and content of every ore zone makes the change of the existing state and distribution of Au.
由于各矿带矿物组合及含量的变化，使金的赋存状态及分布也发生相应改变。

② To conclude, the clay mineral assemblage of loess can be used to trace the source of mother material and understand the post-depositional climate conditions.
因此，黄土黏土矿物组合特征不仅反映物源区古环境信息，而且指示了黄土堆积期后的生物气候环境。

Extended Terms:

mineral association 矿物组合
mineral paragenesis analysis 矿物共生分析

mineralogical phase rule [ˌminərə'lɔdʒikəl feiz ruːl]

Definition: a modification of the phase rule of Williard Gibbs, which states: $p = c + 2 - f$, where p is the number of coexinsting phases, c is the number of components and f is the degree of freedom. As the minerals in metamorphic rocks crystallize over a range of temperatures and pressures, the system must have two degrees of freedom. Thus, in the mineralogical phase rule. p=c. 矿物相律 kuàng wù xiāng lǜ

Extended Terms:
Goldschmidt's mineralogical phase rule 哥德施米特矿物相律

Mineralogy [minəˈrælədʒi]

Definition: the scientific study of minerals. 矿物学 kuàng wù xué；矿物学书籍 kuàng wù xué shū jí

Origin: 1680-1690; minera(1) + -logy.

Examples:
① At the same time, mineralogy is also one of the most basic subject of geology.
同时，矿物学又是地质学最基础的学科之一。
② The mineral names and their nomenclature play an important role in the development of mineralogy.
矿物名称在矿物学的发展中起到了重要的作用。

Extended Terms:
experimental mineralogy 实验矿物学
optical mineralogy 光性矿物学
paragenetic mineralogy 共生矿物学
physical mineralogy 物理矿物学

minerals [ˈminərəls]

Definition: any of a class of substances occurring in nature, usually comprising inorganic substances, as quartz or feldspar, of definite chemical composition and usually of definite crystal structure, but sometimes also including rocks formed by these substances as well as certain natural products of organic origin, as asphalt or coal. (mineral 的复数) 矿物 kuàng wù；矿产品 kuàng chǎn pǐn；矿物质 kuàng wù zhì

Origin: 1375-1425.

Examples:
① This kind of clay usually contains many minerals.
这种黏土通常含有许多矿物质。
② We want to buy Clinker, Cement, Minerals, Lead Ore, Zinc Ore Minerals and Refractories.

我们要采购熔结块、水泥、领引矿石、锌矿石、矿物和耐火材料。

Extended Terms:

clay minerals 黏土矿物
minus minerals 负矿物

---

minette [mi'net]

Definition: a syenitic lamprophyre composed chiefly of orthoclase and biotite. *n.* 云煌岩 yún huáng yán

Origin: 1875-1880.

Example:

Outwards from the inner, the ore-bearing rock series of ultrabasic-basic rock → minette → intermediate-acidic porphyry, felsic rock is accompanied by the cogenetic wallrock alteration series of pyrite-carbonatization→pyrite-carbonate-sericitization→silicification and pyrite-phyllic alteration, characterizing the ore deposit types.
随着容矿主岩从内向外, 从超基性岩、基性岩—云煌岩—中酸性斑岩、长英质岩渐次变化, 表征矿床类型的主要围岩蚀变变化系列为:黄铁矿-碳酸盐化→黄铁矿-碳酸盐-绢云母化→硅化和黄铁-绢英岩蚀变, 成因上具有密切的内在联系。

Extended Terms:

quartz minette 石英云煌岩
biotite minette 黑云云煌
nordmarkite minette 英碱正长云煌岩
soda minette 钠云煌岩
pyroxene minette 辉石云煌岩

---

minor component ['mainə kəm'pəunənt]

Definition: 副成分 fù chéng fèn;微量组分 wēi liàng zǔ fèn

minor:inferior in number or size or amount. 微量的 wēi liàng de
components:the components of something are the parts that it is made of. 组成成分 zǔ chéng chéng fèn

Examples:

① The relative standard deviation of the measurements could be controlled to be less than 1% for major components and less than 10% for minor components in the solid samples.
对于样品中的主要组分, 测定的相对标准偏差可控制在1%以内, 对微量组分可控制在10%以内。

② Regarding tailings A and C, the effect of minor components on the early hydration of cement clinker was studied.
针对 A、C 两种尾矿，初步研究了微量元素对熟料矿物早期水化性能的影响。

Extended Term:
minor element 微量元素

minor mineral [ˈmainə ˈminərəl]

Definition: a light-colored, relatively rare or unimportant mineral in an igneous rock; examples are apatite, muscovite, corundum, fluorite, and topaz. 次要矿物 cì yào kuàng wù

Example:
A minor unknown mineral with leaf shape is determined as pyrophyllite using scanning electronic microscope.
在扫描电镜中发现一种微量叶片状未知矿物，确定该矿物为叶蜡石。

Extended Term:
mineral elements 矿质元素

miocrystalline [ˈmaiəˈkristəlain]

Definition: pertaining to the texture of igneous rock characterized by crystalline components in an amorphous groundmass. Also known as hemicrystalline; hypohyaline; merocrystalline; miocrystalline; semicrystalline. 半晶质 bàn jīng zhì

miscibility gap [ˌmisəˈbiləti gæp]

Definition: a miscibility gap is observed at temperatures below an upper critical solution temperature (UCST) or above the lower critical solution temperature (LCST). Its location depends on pressure. In the miscibility gap, there are at least two phases coexisting. 混溶隙 hùn róng xì; 溶混性间隔 róng hún xìng jiàn gé; 不相混溶区 bù xiāng hùn róng qū

Origin: late 16th century: from medieval Latin miscibilis, from Latin miscere "to mix".

Extended Terms:
complete miscibility 完全混溶，完全溶合性
limited miscibility 极限溶混性
liquid-liquid miscibility 液-液互混性
partial miscibility 部分(溶)混性

water miscibility 水混溶性

mobile belt ['məʊbail belt]

Definition: a long, relatively narrow crustal region of tectonic acitivity. [地质]活动带 huó dòng dài

Example:
The mobile belts are distributed along the mountains and foreland while the stable blocks are preserved in the basins.
活动带集中在山系及山前带，盆地内保存稳定地块区。

Extended Terms:
mobile components 活动组分
seismic belt 地震带

Mohorovicic discontinuity (Moho) ['dis,kɔnti'nju(ː)iti]

Definition: usually referred to as the Moho, is the boundary between the Earth's crust and the mantle. The Moho separates both oceanic crust and continental crust from underlying mantle. The Moho mostly lies entirely within the lithosphere; only beneath mid-ocean ridges does it define the lithosphere—asthenosphere boundary. The Mohorovičić discontinuity was first identified in 1909 by Andrija Mohorovičić. 莫霍洛维奇不连续面 mò huò luò wéi qí bù lián xù miàn

molasse [məʊ'lɑːs]

Definition: a soft sediment produced by the erosion of mountain ranges after the final phase of mountain building. n. 磨拉石 mó lā shí；磨砾层 mó lì céng

Origin: from French, perhaps alteration of mollasse, from Latin mollis soft.

Examples:
① Lithologically, the Jingzhushan Formation is made up of a suite of purplish red molasse elastic rocks which were laid down in the fan delta and delta environments.
从岩性上来看，竟柱山组的主要组成为一套磨拉石的紫红色碎屑岩，其沉积环境以扇三角洲和三角洲为主。
② The Early Permian sedimentary configuration in Bogda mountain and its adjacent basins are characterized by normal grading from coarse to fine, being a typical extensional molasse.
博格达山及相邻盆地中的下二叠统的沉积建造总体上具有由粗到细的正粒序特征，为典型

的伸展磨拉石建造。

Extended Terms:
molasse face 磨砾岩相
molasse basin 磨砾岩盆地

mole-percent composition [məʊl pəˈsent ˈkɔmpəˈziʃən]

Definition: percent composition is the percent by mass of each element present in a compound. 莫耳分率组成 mò ěr fēn lǜ zǔ chéng

molecular weight [məˈlekjʊlə weit]

Definition: the average weight of a molecule of an element or compound measured in units once based on the weight of one hydrogen atom taken as the standard or on 1/16 the weight of an oxygen atom, but after 1961 based on 1/12 the weight of the carbon-12 atom; the sum of the atomic weights of all the atoms in a molecule. 分子量 fēn zǐ liàng

Examples:
① Molecular weights are expressed in the same units as atomic weights.
分子的重量和原子的重量用相同的单位表示。
② In most cases, the molecular weights rise to a finite value with time or conversion.
在大多数情况下,分子量随时间或转化率升高至一个有限的值。

Extended Terms:
molecular vibration 分子振动
molecular formula 分子式

mole fractions [məʊl ˈfrakʃ(ə)n]

Definition: the ratio of the number of moles of a given component of a mixture to the total number of moles of all the components. 摩尔分数 mó ěr fēn shù;克分子分数 kè fēn zǐ fēn shù

Examples:
① A new solution method to compositional model is investigated, which is implicit to pressure and saturation of oil and gas, and explicit to mole fractions of components.
提出了一种对压力和饱和度隐式、组分摩尔分数显示的部分隐式组分模型求解方法。
② Poly (butylenes terephthalate)/p-hydroxybenzoic acid copolyesters (PBT/PHB) with different mole fractions of PHB were synthesized from PBT and PHB by in-situ acetylation.
采用原位乙酰化法,合成了对羟基苯甲酸(PHB)摩尔分数不同的 PBT/PHB 共聚酯,研究

了其组成与相转变和液晶性之间的关系。

Extended Terms:

volume fraction 体积分数，体积分率，容积率
vulgar fraction 普通分数，简分数

monazite [ˈmɔnəzait]

Definition: a brown crystalline mineral consisting of a phosphate of cerium, lanthanum, other rare earth elements, and thorium. 独居石 dú jū shí

Origin: mid 19th century: from German Monazit, from Greek monazein "live alone" (because of its rare occurrence).

Examples:

① Tin, gold, ilmenite, zircon, monazite and kaolin, fire-resistant clay, and so on have the potential for exploration.
锡、金、钛铁矿、锆英石、独居石及高岭土、耐火黏土等具有找矿潜力。
② A soft yellowish-white trivalent metallic element of the rare earth group can be recovered from bastnasite or monazite by an ion-exchange process.
一种软质的黄白色三价金属元素，稀土属，能通过离子交换过程从氟碳锑矿或独居石中重新获取。

Extended Term:

monazite sand 独居石砂

monchiquite [ˈmɔnʃikait]

Definition: a lamprophyre composed of olivine, pyroxene, and usually mica or amphibole phenocrysts embedded in a glass or analcime groundmass. [地]沸煌岩 fèi huáng yán

Extended Terms:

nosean-monchiquite 黝方方沸碱煌岩

monocrystalline [ˌmɔnəˈkristəlain]

Definition: a crystal, usually grown artificially, in which all parts have the same crystallographic orientation. 单晶体 dān jīng tǐ; 单晶质 dān jīng zhì; 单晶的 dān jīng de

Examples:

① P-nitrophenol showed no sign of thermochromism. The title compound, p-nitro-phenol hydrate was synthesized in solution. It's monocrystalline was cultivate.

对硝基苯酚是没有热色性的物质,本工作在水溶液中合成了水合对硝基苯酚,且培养出单晶。
② Microfabrication processes widely use CVD to deposit materials in various forms, including monocrystalline, polycrystalline, amorphous, and epitaxial.
微制程大多使用 CVD 技术来沉积不同形式的材料,包括单晶、多晶、非晶及磊晶材料。

Extended Terms:
monocrystalline filament 单晶长丝
monocrystalline gallium 单晶镓
monocrystalline silicon 硅单晶
monocrystalline grains 单晶砂目
monocrystal 单晶

monticellite [ˌmɔntiˈselait]

Definition: a mineral, silicate of magnesium and calcium, $CaMgSiO_4$, belonging to the olivine group and often occurring in contact metamorphosed limestones. 钙镁橄榄石 gài měi gǎn lǎn shí

Origin: 1825-1835; named after T. Monticelli(1758-1846), Italian mineralogist; see-itel.

Example:
Other equilibrium phases containing CaO may be monticellite and anorthite.
其他平衡相有钙镁橄榄石和钙长石。

Extended Terms:
gehlenite 钙铝黄长石
vesecite 钙镁橄黄玄岩
monticellite nephelinite 钙镁橄霞石岩

montmorillonite [ˌmɒntməˈrilənʌit]

Definition: [mass noun] an aluminium-rich clay mineral of the smectite group, containing some sodium and magnesium 蒙脱石 méng tuō shí, 微晶高岭石 wēi jīng gāo lǐng shí

Origin: mid-19th century: from Montmorillon, the name of a town in France, + -ite.

Examples:
① Preparation, Structure and Properties of Poly (Ethylene Terephthalate)/Montmorillonite Nanocomposite
聚对苯二甲酸乙二醇酯/蒙脱土纳米复合材料的制备、结构及性能
② Study on Synthesis, Structures and Properties of Polydimethylsiloxane/ Montmorillonite

Nanocomposites
对聚二甲基硅氧烷/蒙脱土纳米复合材料的合成、结构及性能的研究

Extended Terms:

monticellite nephelinite 钙镁橄霞石岩
monticellite nepheline basalt 钙镁橄霞玄武岩
monomer monticellite 钙橄榄石

---

monzonite ['mɔnzənait]

Definition: a visibly crystalline, granular igneous rock composed chiefly of equal amounts of two feldspar minerals, plagioclase and orthoclase, and small amounts of a variety of colored minerals. 二长岩 èr cháng yán

Origin: 19th century: from German, named after Monzoni, Tyrolean mountain where it was found.

Example:

Zinccopperite is a rare native alloy mineral discovered in quartz monzonite porphyry in the Xifanping area, Yanyuan county, Sichuan province.
西范坪锌铜矿是一种罕见的天然锌铜金属互化物,是锌铜互化物系列中的新变种,发现于四川省盐源县西范坪地区石英二长斑岩中。

Extended Terms:

albite monzonite 钠长二长岩
quartz monzonite 石英二长岩
micro monzonite 微二长岩
monzonite yamaskite 二长玄闪钛辉岩
monzonite granite 二长花岗岩

---

monzonitic granite [ˌmɔnzə'nitik 'grænit]

Definition: 二长花岗岩 èr cháng huā gāng yán

monzonitic: an igneous rock composed chiefly of plagioclase and orthoclase, with small amounts of other minerals. 二长岩的 èr cháng yán de

granite: a very hard, coarsegrained, gray to pink, intrusive igneous rock, composed mainly of feldspar, quartz, mica, and hornblende. 花岗岩 huā gāng yán

Example:

Gangjiang porphyry Cu, Mo deposit occurs in the spotted medium-grain amphibole-biotite monzonitic granite and is controlled by fault and joints and fractures near the fault.

岗讲斑岩型铜钼矿产于续迈单元含斑中粒角闪黑云二长花岗岩中，受断层带及附近节理、裂隙的控制，矿区有弱硅化、钾化、黄铁矿化、青磐岩化、泥化等蚀变。

Extended Terms:

monzonitic texture 二长结构

Moon [muːn]

Definition: the earth's natural satellite, orbiting the earth at a mean distance of 238,857 miles (384,393 km) and having a diameter of 2,160 miles (3,476 km).

This body during a particular lunar month, or during a certain period of time, or at a certain point of time, regarded as a distinct object or entity. *n.* 月亮 yuè liàng;月球 yuè qiú;月光 yuè guāng;卫星 wèi xīng

Example:

In 2003, President Bush announced a plan to return humans to the Moon by 2020 and use the moon as a base to go on to Mars.

2003年，美国总统布什宣布了一项计划，目的是使人类在2020年重新登陆月球，并利用月球作为登陆火星的基地。

Extended Terms:

moonship 月球飞船
moonshot 月球探测器

mortar texture [ˈmɔːtə ˈtekstʃə]

Definition: was used to describe the texture produced by dynamic metamorphim of granties and gneisses. These rocks were characterized by large irrigular grains of quartz and feldspar in a crushed, finely granular aggregate of the same minerals and resemble stones set in mortar. 碎斑结构 suì bān jié gòu

mosaic texture [məuˈzeiik ˈtekstʃə]

Definition: 镶嵌结构 xiāng qiàn jié gòu

mosaic: the process of making pictures or designs by inlaying small bits of colored stone, glass, tile, etc. in mortar. 镶嵌 xiāng qiàn

texture: the arrangement of the particles or constituent parts of any material, as wood, metal, etc., as it affects the appearance or feel of the surface; structure, composition, grain, etc. 结构 jié gòu

> **Example:**

The results indicate that the isotropic optical texture of cokes from Yu-cun coal can be modified by CTP addition to fine mosaic texture.
结果表明，禹村焦的各向同性光学结构通过焦油沥青能改变为细粒镶嵌结构。

> **Extended Terms:**

equant mosaic texture 粒状镶嵌结构
mosaic structure framework 镶嵌格局
mosaic granoblastic texture 镶嵌粒状变晶结构

mount [maunt]

> **Definition:** a land mass that projects well above its surroundings; higher than a hill. 山峰 shān fēng

> **Origin:** "hill, mountain," mid-13th century, from Anglo-French mount, Old French mont "ountain".

> **Extended Terms:**

Mount Ararat volcano 亚拉腊火山
Mount Desert Island 沙漠山岛
Mount Etna volcano 埃特纳火山
Mount Everest 埃佛勒斯峰
Mount Hood volcano 胡德火山
Mount Johnson 约翰逊山
Mount Kenya 肯尼亚山
Mount Kilimanjaro volcano 乞力马扎罗山火山
Mount Mazama 马扎马火山
Mount Pelee Martinique 佩利山
Mount Pinatubo volcano 皮纳图博火山
Mount Rainier volcano 雷尼尔火山
Mount Royal 皇家山
Mount St. Helens 圣海伦火山
Mount Vesuvius volcano 维苏威火山

M type granite [em taip ˈgrænit]

> **Definition:** M 型花岗岩类 M xíng huā gāng yán lèi

type: a person, thing, or event that represents or symbolizes another, esp. another that it is thought will appear later; symbol; token; sign. 类型 lèi xíng

granite: a very hard, coarsegrained, gray to pink, intrusive igneous rock, composed mainly of feldspar, quartz, mica, and hornblende. 花岗岩 huā gāng yán

mud [mʌd]

Definition: soft, sticky matter resulting from the mixing of earth and water. 泥浆 ní jiāng

Origin: late Middle English: probably from Middle Low German mudde.

Examples:

① The potassium salt mud can be used to protect oil zones and reduce the drilling damage to reservoirs.
应用钾酸盐泥浆能保护油层，减小钻井过程中对储层的破坏。
② Potassium-salt mud is a kind of high-salinity mud and its special composition makes a certain effect on electric and SP logs.
钾酸盐泥浆是一种高矿化度泥浆，其特殊的泥浆成分对常规电测曲线、自然电位曲线有一定影响。

Extended Terms:

mud mounds 泥丘
mudflows 土石流
mud pump 泥浆泵
mudrock 泥石
mudstone 泥岩；泥状灰岩

mud ball [mʌd bɔː]

Definition: an accumulation of mud solids that sometimes builds up on the drilling bit during circulation prior to logging. It can present a problem to logging if the ball should become dislodged in the borehole or scrape off on the face of porous and permeable reservoir rock. 泥球 ní qiú

Example:

Combined with other correlated technologies, the adhesive of PF-FNJ was employed to prevent cuttings aggregation forming mud ball which encountered frequently in early development.
防黏结剂 PF-FNJ 及其相关技术应用解决了前期开发中岩屑聚集成泥球的现象。

Extended Term:

armored mud ball 嵌石泥球；贴沙砾泥球

mud crack [mʌd kræk]

Definition: a crack in a deposit of mud or silt resulting from the contraction that accompanies drying. 泥裂 ní liè, 干裂(drying crack, desiccation crack) gān liè

Example:
The large boulders, the boulder piling up boulder, mud crack, lateral accumulation mound and surge deposits demonstrate the debris flows are viscous.
巨大的泥石流漂砾、石背石现象、龟裂现象、侧积堤和龙头堆积证实了这次泥石流为黏性泥石流。

Extended Terms:
crumpled mud crack cast 揉皱泥裂铸型
mud crack cast 泥裂铸型
mud crack polygons 多边形泥裂

mud flow [mʌd fləu]

Definition: mass flow of material impregnated with water either of pyfoclastic partices following volcanic eruptions or of other sediment following heavy rainfall, especaily in arid environments. Used to refer to both the process and the sediment. Strictly the term should refer to flows involving sedimentantly of mud grade as distinct from sand grades and larger. 泥流沉积 ní liú chén jī
mud: wet, soft, sticky earth. 泥 ní

mudstone [ˈmʌdstəun]

Definition: Mudstone (also called mudrock) is a fine grained sedimentary rock whose original constituents were clays or muds. Grain size is up to 0.0625 mm (0.0025 in) with individual grains too small to be distinguished without a microscope. 泥岩 ní yán

Example:
The variety of these mudstone and shales is very complex in lacustrinehigh-frequency cycle.
这些泥岩和页岩在湖相高频旋回中的变化相当复杂。

Extended Terms:
mudstone conglomerate 泥岩;泥砾岩
pebbly mudstone 含中砾泥岩;含砾泥岩

mud supported [mʌd səˈpɔːtid]

Definition: In carbonate petrology, it refers to those sediments in which the grains are not abundant enough to be in contact and support one another but are surrounded by mud. 灰泥支撑 huī ní zhī chēng

Extended Term:
mud-supported biomicrite 灰泥支撑的生物微晶灰岩

mud volcanic structure [mʌd vɔlˈkænik ˈstrʌktʃə]

Definition: conical stucture formed by fine-grained material extruded by volcanic gases, hydrocarbons. 泥火山构造 ní huǒ shān gòu zào

mugearite [muˈgiərait]

Definition: a type of orthoclase-bearing basalt, comprising olivine, apatite, and opaque oxides. The main feldspar in mugearite is oligoclase. [地]橄榄粗安岩 gǎn lǎn cū ān yán

Example:
The rocks include shoshonite, latite, trachyte, mugearite, dacite and rhyolite.
岩石类型主要有钾玄岩、安粗岩、粗面岩、橄榄粗安岩、英安岩和流纹岩。

Extended Terms:
macedonite 橄榄粗面岩
kentallenite 橄榄二长岩

mull [mʌl]

Definition: a soft, thin muslin. 细腐殖质 xì fǔ zhí zhì

Origin: 1790-1800; earlier mulmul, from Hindi malmal.

Example:
This is Gollarnia neckerella (C. Mull.).
这是粗枝藓。

Extended Term:
insect mull 虫生腐殖质土

muscovite [ˈmʌskəvait]

Definition: silver-grey form of mica occurring in many igneous and metamorphic rocks. 白云母 bái yún mǔ

Origin: mid-19th century: from obsolete Muscovy glass (in the same sense) + -ite.

Examples:

① Muscovite is usually colorless but may be light gray, brown, pale green, or rose red.
白云母通常是无色的，但也可以是浅灰色、棕色、淡绿色或玫瑰色。
② The gold are mainly fine grained and invisible under microscope. The muscovite content of the ore is high. The ores are extremely sliming so the flotation recovery rate is low.
金主要以镜下不可见的微细粒金为主，矿石中绢云母含量很高，矿石泥化严重，浮选回收率偏低。

Extended Terms:

muscovite mica 白云母
muscovite schist 白云母片岩
manganobarium muscovite 锰钡白云母
lithian muscovite 锂白云母

Mush zone [mʌʃ zəun]

Definition: a region where the ground wave and sky wave from a transmitter are received at approximately equal signal strength. 木斯区 mù sī qū

mylonite [ˈmailənait, ˈmil-]

Definition: a fine-grained laminated metamorphic rock in which preexisting minerals have been partially pulverized and drawn out into bands. Mylonite forms along geologic faults where shearing and grinding of rocks takes place. 糜棱岩 mí léng yán

Origin: late 19th century: from Greek mulōn "mill" + -ite.

Examples:

① According to the research result, the tectonite consists of mylonite and cataclasite in the fracture zone.
结果表明：断裂带的构造岩以糜棱岩和碎裂岩为主。
② Mylonite is a kind of dynamometamorphic rock. Its particular formation mechanism makes itself a carrier of rich geology information.
糜棱岩是一种动力变质岩，其独特的形成机制使其成为一种具有丰富地质信息的载体。

Extended Terms:

mylonite gneiss 糜棱片麻岩
mylonite zone 糜棱岩带
hyalo-mylonite 玻璃摩棱岩
phyllite-mylonite 千枚搓变岩
mylonite-gneiss 磨变片麻岩；磨岭片麻岩

myrmekite ['mə:mi‚kait]

Definition: describes a vermicular, or wormy, intergrowth of quartz in plagioclase. The intergrowths are microscopic in scale, typically with maximum dimensions less than 1 millimeter. 蠕状石 rú zhuàng shí

Example:

There exists much controversy concerning the genesis of myrmekite, and most geologists hold that myrmekite is of metasomatic origin.
关于蠕状石的成因，许多地质学家提出各种不同的成因假说，多数人认为是交代生成的。

Extended Term:

myrmekitic texture 蠕状结构

mélange ['meilɑ:nʒ]

Definition: mixture; a medley. 混合体 hùn hé tǐ

Origin: from French mélange, from mêler "to mix".

Examples:

① Based on this, the author puts forward an ideal of plate tectonics of oceanic crust→trench→mélange accumulation → arc → front basin → arc (island) from Zaoyang in the south to Tongbaishan in the north.
作者在此基础上提出了枣阳以北至桐柏山由南向北的一条从洋壳—海沟相沉积—混杂岩堆积—孤前盆地沉积—岛孤的理想板块构造模式。
② Tectonically, the Gangrnarco fault is believed to be a tectonic mélange zone, where the structural deformation appears in the upper, middle and lower levels of structures.
在冈玛日一带发现呈岩片状产出的蓝片岩。构造方面，查明冈玛错断裂为一构造混杂岩带，区内存在上、中、下构造层次的变形。

Extended Term:

melange zones 混合区

natron ['neitrən]

Definition: a mineral salt found in dried lake beds, consisting of hydrated sodium carbonate. [矿]天然碳酸钠;tiān rán tàn suān nà;泡碱 pào jiǎn

Origin: late 17th century: from French, from Spanish natrón, via Arabic from Greek nitron (see nitre).

Example:
Saline deposits of salt lake consist of more than 10 salt minerals, but the glauber salt, glaserite, pgrotechnite, hanksite, natron, trona and halite are mainly salt minerals.
盐湖盐类沉积由十余种盐类矿物构成,其中主要盐类矿物为芒硝、钾芒硝、天然碱、泡碱、碳钾钠矾、石盐等。

Extended Terms:
natron lake 泡碱湖
natron saltpeter 钠硝石
natron catapliite 多钠锆石

natural gas ['nætʃrəl gæs]

Definition: flammable gas, consisting largely of methane and other hydrocarbons, occurring naturally underground (often in association with petroleum) and used as fuel. 天然气 tiān rán qì

Origin: gas: mid-17th century: invented by J. B. van Helmont (1577-1644), Belgian chemist, to denote an occult principle which he believed to exist in all matter; suggested by Greek khaos "chaos", with Dutch g representing Greek kh.

Examples:
① Petroleum and natural gas industries—Subsurface safety valve systems—Design, installation, operation and redress.
石油天然气工业——水下安全阀系统——设计,安装,操作和矫正。
② In the ground floor there are oil, natural gas, coal, geothermal, limestone, dolomite, hard iron bauxite and other mineral resources, rich reserves, the exploitation of high value.
地下蕴藏着石油、天然气、煤炭、地热、石灰岩、白云石、硬铁矾土等矿产资源,储量丰富,具有较高的开采价值。

Extended Terms:
natural gas mine 气矿井
natural gas pipeline 天然气管道
natural gas resources 油气资源

nelsonite [ˈnelsənait]

Definition: a group of hypabyssal rocks composed mainly of ilmenite and apatite. 钛铁磷灰岩 tài tiě lín huī yán

Extended Terms:
biotite nelsonite 黑云钛铁磷灰岩
rutile nelsonite 金红钛铁磷灰岩

neodymium [ˌniːəʊˈdimiəm]

Definition: the chemical element of atomic number 60, a silvery-white metal of the lanthanide series. Neodymium is a component of misch metal and some other alloys, and its compounds are used in colouring glass and ceramics(Symbol：Nd). [化]钕 nǚ

Origin: late 19th century：from neo-"new" +a shortened form of didymium.

Example:
Neodymium is a bright, silvery rare-earth metal element, found in monazite and bastnaesite and used for coloring glass and for doping some glass lasers.
钕是一种亮银色的稀土金属元素，见于单独在石和氟碳锑矿中存在，用于给玻璃染色和给一些玻光器上色。

Extended Terms:
neodymium glass 钕玻璃
neodymium boride 硼化钕
neodymium isotopes 钕同位素

neomorphism [ˌniːəʊˈmɔːfizəm]

Definition: is a recrystallization process in sedimentary rocks that changes the size (either larger or smaller) and form (it can include the overgrowth of polymorphous minerals) of crystals in the rock. 新生变形 xīn shēng biàn xíng

Origin: 1965：Some aspects of recrystallization in ancient limestones, Society of Economic Paleontologists and Mineralogists, pp.14-48.

Example:
The pore, cave, and fracture of the limestone are destroyed mainly by neomorphism, campaction, cementation and filling.

灰岩的孔、洞、缝主要为新生变形、压实胶结、充填作用所破坏。

Extended Terms:

coalescive neomorphism 聚结新生变形
aggrading neomorphism 加积新生形变作用
porphyroid neomorphism 残斑新生变形作用

nepheline [ˈnefəlin]

Definition: colourless, greenish, or brownish mineral consisting of an aluminosilicate of sodium (often with potassium) and occurring as crystals and grains in igneous rocks. [矿]霞石 xiá shí

Origin: early 19th century: from French néphéline, from Greek nephelē "cloud" (because its fragments are made cloudy on immersion in nitric acid) + -ine.

Examples:
① The exploitation and utilization of non-metallic ores in Yunnan Province, such as zeolite, diatomaceous, nepheline syenite, wollastonite, bentonite, kaolin and bauxite, etc. were introduced.
介绍了云南省沸石、硅藻土、霞石正长岩、硅灰石、膨润土、高岭土、铝矾土等非金属矿的开发利用情况。
② Using nepheline syenite as a substitute for feldspar, the batch mixtures were prepared on a composition of float glass and are melted into glasses.
以霞石正长岩取代长石，按照浮法玻璃的成分制备配合料并熔制玻璃。

Extended Terms:

nepheline syenite 霞石正长岩；霞长石；高纯霞石粉；霞石长岩
nepheline basalt 霞石玄武岩，橄榄霞石

nepheline syenite [ˈnefəlin ˈsaiənait]

Definition: a holocrystalline plutonic rock that consists largely of nepheline and alkali feldspar. The rocks are mostly pale colored, grey or pink, and in general appearance they are not unlike granites, but dark green varieties are also known. Phonolite is the fine-grained extrusive equivalent. 霞石正长岩 xiá shí zhèng cháng yán

Extended Terms:

orthoclase nepheline syenite 正霞正长岩
cancrinite nepheline syenite 钙钠霞正长岩

rhombic feldspar nepheline syenite 菱长霞石正长岩
microcline nepheline syenite 微斜霞石正长岩
melanite-nepheline syenite 黑榴霞石正长岩

nephelinite ['nefəlinait]

Definition: a fine-grained, dark rock of volcanic origin, essentially a basalt containing nepheline but no feldspar and little or no olivine. 霞石岩 xiá shí yán

Origin: 1860-1865; nepheline + -ite.

Examples:
① These rocks consist of olivine basalt, alkali olivine basalt, basanite and nephelinite, together with alkali diabase, tholeiitic diabase and analcite syenite which have been formed as a result of differentiation and contamination.
主要岩石类型有橄榄玄武岩、碱性橄榄玄武岩、碧玄岩、霞石岩以及由分异作用或混染作用形成的碱性辉绿岩、拉斑辉绿岩和方沸正长岩。
② According to the main rock-forming minerals in them, the Cenozoic volcanic rocks are mainly alkali olivine basalt, basanite and nephelinite.
郯庐断裂带中段的新生代火山岩以碱性系列的碱性橄榄玄武岩、碧玄岩和霞石岩为主，其中又以钠质亚系的岩石为主。

Extended Terms:
nephelinite porphyry 霞斑岩
sodalite nephelinite 方钠霞石岩
biotite nephelinite 黑云霞石岩
olivine nephelinite 橄榄霞石岩
corundum nephelinite 刚玉霞石岩

Newtonian liquid [njuː ˈtəunjən ˈlikwid]

Definition: a simple fluid in which the state of stress at any point is proportional to the time rate of strain at that point; the proportionality factor is the viscosity coefficient. 牛顿液体 niú dùn yè tǐ

Examples:
① Based on the collsion frequencies model and their factors of particles in gas-solid flow, a collsion frequencies model of droplets in Newtonian liquid-liquid flow is put forward.
通过分析气固两相流中颗粒的碰撞率模型及其影响因素，提出牛顿型液-液两相流中的液滴碰撞率模型，并分析了在剪切流场中液滴发生椭球形变形后的碰撞概率。
② The coal-water slurry is a solid-liquid suspension with high viscosity and opacity, which

features property of non-Newtonian fluid, boasting very complex rheological property.
水煤浆是一种高黏性、不透明的液固分散悬浮液,表现出非牛顿特性,其流变特性十分复杂。

Extended Terms:
generalized Newtonian liquid 广义牛顿液体
non-Newtonian liquid 非牛顿液体

Newton's second law [ˈnjuːtənz sekənd lɒ]

Definition: The law that the acceleration of a particle is directly proportional to the resultant external force acting on the particle and is inversely proportional to the mass of the particle. Also known as second law of motion. 牛顿第二定律 niú dùn dì èr dìng lǜ

Example:
Newton's second law is the law of the instantaneous effect.
牛顿第二定律是力的瞬时作用规律。

nickel [ˈnɪkl]

Definition: a silvery-white metal, the chemical element of atomic number 28 (Symbol: Ni).

Origin: mid-18th century: shortening of German Kupfernickel, the copper-coloured ore from which nickel was first obtained, from Kupfer "copper" + Nickel "demon" (with reference to the ore's failure to yield copper). n. 镍 niè

Examples:
① study on Improvement Properties of Lead, Nickel, Aluminum, Zinc and Copper Electrodes
铅、镍、铝、锌和铜电极改性的研究
② study on a New Method for Preparing Special Nickel Powder
特种镍粉制备新方法研究

Extended Terms:
arsenical nickel 红砷镍矿
bismuth nickel 铋镍矿
chromated nickel 铬镍
ferro nickel 铁镍合金,镍铁
malleable nickel 展性镍
powdered nickel 镍粉

nomenclature [nəʊˈmenklətʃə]

Definition: the devising or choosing of names for things, especially in a science or other discipline. *n.* 命名法 mìng míng fǎ;术语 shù yǔ

Origin: early 17th century: from French, from Latin nomenclatura, from nomen "name" + clatura "calling, summoning" (from calare "to call").

Examples:
① GRIN taxonomic data provide the structure and nomenclature for the accessions of the National Plant Germplasm System (NPGS).
GRIN 分类数据库提供了使用国家植物胚质系统所需的结构和术语解释。
② He introduced the "binary nomenclature", i. e. he assigned two names to a plant or animal, the first one for the genus (like a surname) the following second one for the species (like a first name).
他提出了"二进位命名法",即为植物或动物分派了两个名称,第一个叫属(就像人的姓),第二个名称叫种(就像人的名)。

Extended Terms:
Geneva nomenclature 日内瓦命名法
enzyme nomenclature 酶命名法

non-Newtonian liquid [ˈnɔnnjuˈtəunjən ˈlikwid]

Definition: a fluid whose flow behavior departs from that of a Newtonian fluid, so that the rate of shear is not proportional to the corresponding stress. Also known as non-Newtonian system. 非牛顿液体 fēi niú dùn yè tǐ,非牛顿界面流动 fēi niú dùn jiè miàn liú dòng

Example:
Both polyacrylamide (PAM) and body liquid are pseudoplastic fluid, in this paper, PAM aqueous solution has been used to study the diffusion behavior of the nutritious molecules in non-Newtonian fluid.
聚丙烯酰胺与生物体液均为假塑性流体,并且对聚丙烯酰胺的基本性质已经有明确的认识,本文采用聚丙烯酰胺水溶液作为非牛顿流体,研究营养类分子在非牛顿流体中的扩散。

Extended Terms:
non-Newtonian suspension 非牛顿悬浮体
non-Newtonian motion 非牛顿运动
non-Newtonian system 非牛顿流体
non-Newtonian flow 非牛顿流动

nonconservative [ˈnɔnkənˈsəːvətiv]

Definition: (of a vector or vector function) having curl equal to zero; irrotational; lamellar.

adj. 非保守的 fēi bǎo shǒu de;非守恒的 fēi shǒu héng de,不可逆流的 bù kě nì liú de

Origin: 1350-1400.

Examples:

① Two classes of numerical methods based on FD-WENO schemes were recommended for solving nonconservative compressible ideal fluid dynamics equations.
以求解双曲守恒律组的 FD-WENO 格式为基础提出了两类用于求解非守恒可压缩理想流体力学方程组的数值方法。

② In the case of steady flow, it is shown that whether a system is conservative or nonconservative is mainly dependent on the relative magnitudes of the spring constants.
定常流情况下的结果表明,一个系统是保守的还是非保守的主要取决于支承弹簧刚度的比值。

Extended Terms:

nonconcentrated 非集中的
nonconservative force 非守恒力
noncontinuous 不连续的

nordmarkite ['nɔːdmɑːkait]

Definition: a quartz-bearing alkalic syenite that has microperthite as its main component with smaller amounts of oligocase, quartz, and biotite and is characterized by granitic or trachytoid texture. 英碱正长岩 yīng jiǎn zhèng cháng yán

Examples:

① The Muchang A-type granite is a composite intrusive body consisting of the riebeckite nordmarkite, riebeckite granite and aegirine granite.
木厂 A 型花岗岩是由钠闪英碱正长岩、钠闪花岗岩、霓石花岗岩组成的复式岩体。

② The rare earth element geochemical studies demonstrate that the REE distribution patterns of fluorite, calcite and bastnaesite are similar to those of nordmarkite, implying that the deposit is closely related to magma of nordmarkite. In addition, REE of fluid from quartz exhibits distribution pattern of flat curve with negligible Eu and Ce anomalies, showing its deep source origin.
稀土元素地球化学特征研究显示,矿区萤石、方解石、氟碳铈矿稀土元素分布模式与区内英碱正长岩相似,表明牦牛坪稀土矿床成矿与英碱正长岩岩浆活动有关以及石英包裹体流体 REE 曲线平直,Eu 与 Ce 无明显异常,曲线规律性较强,可以认为牦牛坪轻稀土矿床成矿流体稀土元素是深源的。

Extended Terms:

nordmarkite porphyry 英碱正长斑岩
nordmarkite aplite 英碱正长细晶岩

norite [ˈnɔːrait]

Definition: a coarse-grained plutonic rock similar to gabbro but containing hypersthene. *n.* 苏长岩 sū cháng yán

Origin: late 19th century: from Norway + -ite.

Example:

Four different lithofacies are recognized as peridotite+pyroxenite, norite+gabbronorite, gabbro+diabase, and diorite including ultramafic rocks to intermediate rocks from core to margin of the intrusive body.

杂岩体可分为四个岩相带(橄榄岩+辉石岩相，苏长岩+辉长苏长岩相，辉长岩+辉绿岩相，闪长岩相)，包括中性、基性、超基性的多种类型岩石组合而成。

Extended Terms:

cordierite norite 董青苏长岩
enstatite norite 顽火苏长岩
bronzite norite 古铜苏长岩
magnetite norite 磁铁苏长岩
norite bronzitite 苏长古铜岩

normative mineral [ˈnɔːmətiv ˈminərəl]

Definition: a serious of ideal mineral compounds used to express the chemical composition of an igneous rock in mineralogical terms. Normative mineralsall have simpe compositions so do not include any alumious ferromagnesian minerals. 标准矿物 biāo zhǔn kuàng wù

North American plate [nɔːθ əmerikən pleit] 北美板块 běi měi bǎn kuài

North Atlantic [nɔːθ ætˈlæntik] 北大西洋 běi dà xī yáng

North Island [nɔːθ ˈailənd] 北岛(新西兰) běi dǎo (xīn xī lán)

Northern Hemisphere Reference Line

[ˈnɔːð(ə)n ˈhemisfiə ˈref(ə)r(ə)ns lain] 北半球基准线 běi bàn qiú jī zhǔn xiàn

Northern Volcanic Zone [ˈnɔːð(ə)n vɔlˈkænik zəun]

北部火山带 běi bù huǒ shān dài

novaculite [nəuˈvækjulait]

Definition: (Geology) a hard, dense, fine-grained siliceous rock resembling chert, with a high content of microcrystalline quartz. [地质] 均密石英岩 jūn mì shí yīng yán

Origin: late 18th century: from Latin novacula "razor" + -ite.

Example:
The characteristics and genesis of novaculite of the lower permian series in western Jiangxi Province
江西下二叠统茅口组均密石英岩的特征、风化习性及其成因

Extended Term:
novaculite novac 燧石岩

nucleation [ˌnjuːkliˈeiʃən]

Definition: forming a nucleus. *n.* 成核现象 chéng hé xiàn xiàng; 集结 jí jié; 人工降雨作用 rén gōng jiàng yǔ zuò yòng

Origin: 1860-1865; from Lalin nucleātus having a kernel or stone.

Examples:
① By decomposing the precipitation process into some sub processes as mixing, nucleation, growth, ageing and aggregation, the theoretical fundament forming every sub process was analyzed.
将沉淀过程分解为混合、成核、生长、陈化、团聚等子过程，分析了各子过程形成的理论基础。
② As key factors in processing microcellular plastics, the formation of polymergas homogenous phase system, nucleation and bubble expansion controlling have been discussed in this paper.
在微孔塑料加工过程中，聚合物气体均相体系的形成、成核与泡孔长大控制是关键因素。

Extended Terms:
heterogeneous nucleation 非均匀形核
nucleation center 成核中心
nucleation and crystal growth 蒸发结晶

nuclei [ˈnjuːkliai]

Definition: plural form of nucleus. *n.* 核心 hé xīn, 核子 hé zǐ; 原子核 yuán zǐ hé (nucleus

的复数形式)

Origin: 1695-1705; from Latin: kernel, syncopated of nuculus.

Examples:

① In nuclear fusion, a pair of light nuclei unite (or fuse) together, to form a nucleus of a heavier atom.
在核聚变的情况下,一对轻核合并(或聚合)在一起,形成一个较重的原子核。

② Other vertebrates such as fish, reptiles and birds have red cells that contain nuclei that are inactive.
其他的脊椎动物,比如说鱼类、爬行动物和鸟类等含有灭活核的有核红细胞。

Extended Terms:

condensation nuclei 凝结核
distributed nuclei 扩散核

numerical analysis [nju(ː)'merikəl ə'næləsis]

Definition: the study of algorithms that use numerical approximation (as opposed to general symbolic manipulations) for the problems of mathematical analysis. [数]数值分析 shù zhí fēn xī

Examples:

① The course, numerical analysis, plays an important role in computer science and technology education, and it is one of the important teaching reform contents of computer education.
数值分析课程在计算机科学与技术学科中占有非常重要的地位,它已成为计算机专业教学改革的重要内容之一。

② It is found that, from the structural dynamic feature information, the structures like F320 model derrick can not be simplified as simple beam in the process of numerical analysis.
从分析获得的结构动态特性信息中发现,对于这一类F320石油井架模型结构,不能用简单的线性梁模型来简化处理。

Extended Terms:

numerical value analysis 数值分析
numerical method analysis 数值法分析

Nusselt number ['njuːsəl 'nʌmbə]

Definition: a dimensionless number used in the study of mass transfer, equal to the mass-transfer coefficient times the thickness of a layer through which mass transfer is taking place divided by the moleculor diffusivity. Symbolized Num; NNum. Also known as Sherwood number(NSh). 努塞尔特数 nǔ sài ěr tè shù

Examples:

① Effects of thermal boundary condition and aspect ratio on Nusselt number are discussed and analyzed from the view of field coordination.
推导出单面加热时的场协同方程，以此分析了不同截面比下加热边界条件对努塞尔数的影响规律。

②New expressions for calculation of temperature jump efficient and Nusselt number are derived under the condition of uniform heat flux boundary.
根据定热流边界条件，推导了圆形微通道的努塞尔数和温度跳跃的新表达式。

---

nuée ardente [njuː ˈei ɑːˈdɔŋt]

Definition:
an incandescent cloud of gas, ash, and lava fragments ejected from a volcano, typically as part of a pyroclastic flow. 炽热火山云 chì rè huǒ shān yún

Origin:
French, literally "burning cloud".

---

obduction [ɔbˈdʌktʃən]

Definition:
(Geology) the sideways and upwards movement of the edge of a crustal plate over the margin of an adjacent plate. (地质) 逆冲作用 nì chōng zuò yòng, 仰冲 yǎng chōng

Origin:
1970s: from Latin obduct-"covered over", from the verb obducere, from ob-"against, towards"+ducere "to lead".

Examples:
① The main shock occurred on the dextral obduction-type fault zone and the hypocenter depth was about 15km.
地震的主震发生在具有右旋性质的仰冲型断层带上，震源深度为15公里左右。

② The region is typical overriding obduction region, and the research results have representative and catholicness.
该区域作为推覆逆冲的典型地区，其研究成果具有代表性和普遍性。

Extended Terms:
obduction slab 逆冲板块
obduction zone 仰冲带

---

oblique bedding [əˈbliːk ˈbediŋ]

Definition:
a general term for sedimentary layers deposited at an angle to true bedding suefaces. The layer maybe less than 1cm thick or greaterwhen the bedding may be particularized

as cross-lamination and cross-bedding respectively. 斜层理 xié céng lǐ

Example:
Sedimentary structures are mainly massive bedding, large trough cross bedding and oblique bedding with few parallel bedding, wave bedding and horizotal bedding.
沉积构造以块状层理、大型槽状交错层理和斜层理为主,见平行层理、波状层理和极少量的水平层理。

Extended Term:
compound oblique-bedding 复合斜层理

obsidian [ɔbˈsidiən]

Definition: a hard, dark, glass-like volcanic rock formed by the rapid solidification of lava without crystallization. *n.* [矿]黑曜石 hēi yào shí

Origin: mid-17th century: from Latin obsidianus, error for obsianus, from Obsius, the name (in Pliny) of the discoverer of a similar stone.

Example:
Obsidian, a form of volcanic rock valued for its strength and durability, was an important resource for ancient Mesoamerican cultures including the Zapotec and Maya.
黑曜石,是火山石的一种形态,因其坚固耐久而别具价值,在古代中美洲文明,包括萨巴特克和玛雅文明中,都是一种重要的资源。

Extended Terms:
obsidian dating 黑曜石年代测定法
blue obsidian 蓝绿曜岩
obsidian glaive 黑耀石长矛
enchanted obsidian 强化黑曜石
obsidian golem 黑曜石魔像

ocean [ˈəuʃən]

Definition: a very large expanse of sea, in particular, each of the main areas into which the sea is divided geographically. 大洋 dà yáng

Origin: Middle English: from Old French occean, via Latin from Greek ōkeanos "great stream encircling the earth's disc". "The ocean" originally denoted the whole body of water regarded as encompassing the earth's single land mass.

Example:
As the Earth moves under the oceans, tidal waves roll inland first on one side of the ocean,

then as sloshing occurs, over the other side.
随着地球向海洋下面移动,首先,在海洋一侧潮汐会席卷内陆地区,然后在海洋的另一侧海水出现了搅动和泼溅。

Extended Terms:

ocean circulation 海洋环境
ocean currents 洋流
ocean-ridge (MORB) 洋脊,海岭
ocean floor 洋底,大洋底,海底
ocean island basalts 洋岛玄武岩
ocean island tholeiite 洋岛拉斑玄武岩

oceanic [ˌəʊʃiˈænik]

Definition: of or relating to the ocean. adj. 海洋的 hǎi yáng de;海洋产出的 hǎi yáng chǎn chū de;在海洋中生活的 zài hǎi yáng zhōng shēng huó de;广阔无垠的 guǎng kuò wú yín de

Origin: mid-17th century: from ocean + -ic.

Examples:

① Oceanic airspace outside the published OTS is available, subject to application of the appropriate separation criteria and NOTAM restrictions.
OTS 以外的海洋空域是可用的,受相应间隔实施标准和 NOTAM 限制的影响。
② In Triassic period, Qiangbei-Changdu area is a paraplatform area placed between Jinshajiang river oceanic basic and Bangong lake-Nujiang river oceanic basic.
三叠纪,西藏羌北-昌都地区为夹持于金沙江洋盆和班公湖-怒江洋盆之间的一个准地台区。

Extended Terms:

oceanic crust 海洋地壳,大洋地壳
oceanic regions 海洋地区
oceanic magma series 大洋岩浆系
oceanic volcanism 大洋火山
oceanic-island 洋岛
ocean-island basalts (OIBs) 洋岛玄武岩
ocean-ridge basalts 海脊玄武岩

ocean marginal basin [ˈəʊʃən ˈmɑːdʒɪnəl ˈbeɪsən]

Definition: 大洋边缘盆地 dà yáng biān yuán pén dì

ocean: a very large expanse of sea, in particular, each of the main areas into which the sea is divided geographically. 大洋 dà yáng

marginal: at or constituting a border or edge. 边缘的 biān yuán de
basin: types of geological depressions. 盆地 pén dì

Example:

Jurassic and Cretaceous, the marginal ocean basin was in the stage of ocean crust evolution and had a setting similar to mid-ocean ridge.
侏罗-白垩纪为洋盆洋壳演化期，处于类似洋中脊的构造环境。

ocelli [əuˈselai]

Definition: an eyelike spot, as on a peacock feather. 眼斑状 yǎn bān zhuàng

Origin: 1810-1820; from Latin: little eye, dim. of oculus eye.

Example:

The day and night cycle makes the screening pigment granules in cells move horizontally and vertically in relation to the longitudinal axis of the ocelli.
昼夜交替还导致细胞内屏蔽色素颗粒周期性的水平和垂直移动。

oil [ɔil]

Definition: a slippery or viscous liquid or liquefiable substance not miscible with water; oil paint containing pigment that is used by an artist; any of a group of liquid edible fats that are obtained from plants. *n.* 油 yóu; 石油 shí yóu

Origin: 1125-1175.

Examples:

① Some oil fields have a mat of heavy tar at the oil water contact. This mat is produced by the degradation of oil as bottom waters move beneath the oil-water contact. Tar mats cause considerable production problems by inhibiting water from displacing oil.
由于油水界面下面底水的流动造成了原油的降减作用，一些油田的油水界面处常形成一种重油垫，由于重油垫阻止了水对油的替换作用，因而造成了许多严重的生产问题。

② Oil-polluted water and oil mixtures from offshore oil rigs, drilling platforms and oil extraction platforms may not be directly discharged into the sea. When they are discharged after recovery treatment, the oil content of the discharges may not exceed the standards set by the State.
海洋石油钻井船、钻井平台和采油平台的含油污水和油性混合物，不得直接排放；经回收处理后排放的，其含油量不得超过国家规定的标准。

Extended Terms:

lubricating oil 润滑油

mineral oil 矿物油

absorption oil 吸收油

acid oil 酸性油

acid refined oil 酸漂油；酸精制油

acid-refined linseed oil 酸漂亚麻仁油

aeroplane oil 航空润滑油

Olduvai Gorge 奥杜瓦伊峡谷 ào dù wǎ yī xiá gǔ

oligoclase [ˈɔligəukleis]

Definition: a feldspar mineral common in siliceous igneous rocks, consisting of a sodium-rich plagioclase (with more calcium than albite). *n.* [矿] 奥长石 ào cháng shí, 钠钙长石 nà gài cháng shí

Origin: mid-19th century: from oligo-"relatively little"+Greek klasis "breaking" (because thought to have a less perfect cleavage than albite).

Examples:

① The plagioclase contains albite, oligoclase, andesine, laboratories, bytownite and anorthite.
钠长石、更长石、中长石、拉长石、培长石和钙长石属于斜长石大类。

② Iridescent plagioclase in Peristerite area is albite oligoclase.
在晕长石连生区产生晕彩的斜长石为钠奥长石。

Extended Terms:

oligoclase basalt 富钠长石玄武岩，奥长石玄武岩

oligoclase andesite 奥安山岩

oligomictic [ˈɔligəumiktik]

Definition: pertaining to a lake that circulates only at rare, irregular intervals during abnormal cold spells. *adj.* 单岩碎屑的 dān yán suì xiè de；少循环的 shǎo xún huán de

Example:

A oligomictic lake has relatively stable stratification with only rare periods of circulation.
少循环湖有相当稳定的层化且仅循环短暂时期。

Extended Terms:

oligomictic rock 单岩碎屑岩

oligomictic lake 少循环湖

olivine [ˌɔliviːn]

Definition: an olive-green, grey-green, or brown mineral occurring widely in basalt, peridotite, and other basic igneous rocks. It is a silicate containing varying proportions of magnesium, iron, and other elements. *n.* 橄榄石 gǎn lǎn shí; 黄绿 huáng lǜ

Origin: late 18th century: from Latin oliva (see olive) + -ine.

Examples:
① The greater the extent of mantle melting, the more depleted the peridotites, and the higher the excess olivine contents.
地幔熔融程度越高,深海橄榄岩越亏损,则含有越多的过量橄榄石。
② With regard to the oxide minerals, chromite is almost entirely contained in the olivine-rich ultramafic rocks in the lower part of the intrusion.
关于氧化矿物,铬铁矿几乎全部含于侵入体下部的富橄榄石的超镁铁质岩石中。

Extended Terms:
olivine diabase 橄榄辉绿岩
olivine gabbro 橄榄辉长岩
olivine group 橄榄石类

olivine basalt [ˈɔliviːn ˈbæsɔːlt]

Definition: basalt proper. 橄榄玄武岩 gǎn lǎn xuán wǔ yán

Example:
Over the past eight years, the TES instrument has discovered that Martian rocks and sands are composed almost entirely of the volcanic minerals feldspar, pyroxene and olivine—the components of basalt.
TES 仪器在过去八年间发现,火星的岩石与沙几乎全是由构成玄武岩的火山矿物——长石、辉石和橄榄石组成。

Extended Terms:
alkali olivine basalt 碱性橄榄石玄武岩
alkali olivine basalt magma 碱性橄榄玄武岩岩浆

omphacite [ˈɔmfəsait]

Definition: a pale-green variety of pyroxene similar to olivine, found in eclogite. *n.* 绿辉石 lǜ huī shí

Origin: 1820-1830; from Greek Gk omphakītēs green stone, equiv. to omphak-

Examples:

① Eclogites, consisting of garnet-rich matrix and omphacite veins, occur in the Donghai area, northern Jiangsu.
在苏北东海地区出露一种由富石榴子石基质与绿辉石脉组成的榴辉岩。
② The primary amphiboles equilibrated with omphacite and garnet in eclogite, belonging to pargasite and pargasitic hornblende;
与榴辉岩中矿物平衡存在的原生角闪石,为韭闪石或韭闪石质普通角闪石。

Extended Term:

omphacite-eclogite

Ontong java plateau [ˈɔntɔŋ ˈdʒɑːvə ˈplætəv]

Definition: 翁通爪哇海台 wēng tōng zhǎo wā hǎi tái

oncolite [ˈɔŋkəlait]

Definition: Oncolites are sedimentary structures formed out of oncoids, which are layered spherical growth structures formed by cyanobacterial growth. 藻灰结核 zǎo huī jié hé, 又称核形石(oncolite) hé xíng shí

Example:

In South China, the Late Sinian microbiolite (stromatolite, stratifera and oncolite etc.) are widely developed. Among them, the stromatolite is importent.
我国南方晚震旦世广泛发育有由叠层石、层纹石和核形石等组成的微生物岩,其中以叠层石最为普遍。

Extended Terms:

oncolite onc 核形石
oncolith 球状叠层石类

ooid [əuid]

Definition: are small (< 2 mm in diameter), spheroidal, "coated" (layered) sedimentary grains, usually composed of calcium carbonate, but sometimes made up of iron-or phosphate-based minerals. n. 鲕粒 ér lì; 鲕状石 ér zhuàng shí

Examples:

① The Ningxiang-type iron deposit in western Hubei is widely-spreading and of large-size and is characterized by hematite ooid.

鄂西宁乡式铁矿分布广、规模大，普遍发育赤铁矿鲕粒。
② Those of grain banks in shallow lake mainly include ooid limestones with spar or micrite, bioclastic limestones, intraclastic limestones and grain lime-stones.
浅湖颗粒滩有亮晶或泥微晶的鲕粒灰岩、生屑灰岩和内碎屑灰岩，其次为含陆屑颗粒灰岩。

Extended Term:

calcareous ooid 钙质鲕粒

oolitic dolomite [ˌəuəuˈlitik ˈdɔləmait]

Definition: 鲕状白云岩 ér zhuàng bái yún yán, 鲕粒白云岩 ér lì bái yún yán
oolitic: in the shape of ooid. 鲕状的 ér zhuàng de
dolomite: a sedimentary carbonate rock that is composed mostly of an anhydrous carbonate mineral composed of calcium magnesium carbonate, ideally $CaMg(CO_3)_2$. 白云岩 bái yún yán

Example:

High-quality reservoir rocks of oolitic dolomite evolved in 6 stages, and led to the key factors, facies, hydrocarbon charge, corrosin in deep burial, and multistage faulting.
优质鲕粒白云岩储集层经历6个阶段的演化历史，导致沉积相带、烃类充注、深埋藏条件下强烈溶蚀以及多期断层活动。

Extended Terms:

oolitic limestone 鲕状石灰岩
oolitic pelmicrite 鲕状团粒微晶灰岩
oolitic porosity 鲕状孔隙性
oolitic structure 鲕状构造
pseudo-oolitic chert 假鲕状燧石

oolitic limestone [ˌəuəuˈlitik ˈlaimstəun]

Definition: a limestone consisting of many compressed oolites, was formed in the United Kingdom during the Jurassic period, and forms the Cotswold Hills. 鲕状灰岩 ér zhuàng huī yán, 鲕粒灰岩 ér lì huī yán

Example:

The oolitic limestone layer was thought to belong to oolitic beach-facies sediments with good storage performance.
这属于鲕粒滩相沉积，储集性能好。

Extended Term:

deep gray oolitic limestone 深灰色鲕状石灰岩

opal [ˈəupəl]

Definition: a gemstone consisting of a quartz-like form of hydrated silica, typically semi-transparent and showing many small points of shifting colour against a pale or dark ground. *n.* 猫眼石 māo yǎn shí, 蛋白石 dàn bái shí; 乳白玻璃 rǔ bái bō li

Origin: late 16th century: from French opale or Latin opalus, probably based on Sanskrit upala "precious stone" (having been first brought from India).

Examples:
① For many years researchers have been trying to develop a synthetic material with the same light-scattering properties as an opal, by etching patterns into various materials. That approach has failed.
多年来研究者们一直致力于开发一种与猫眼石一样有散射特性的合成材料，他们在各种各样的材料中蚀刻图案，但是都无功而返。
② This process can be observed in the vibrant colours of the precious stone opal, a natural photonic crystal.
在珍贵的蛋白石中可以观察到这一颜色变化的过程，蛋白石是一种天然的光子水晶。

Extended Terms:
opal cement 蛋白石水泥
opaque mineral 不透明矿物
black opal 黑蛋白石

open platform facies [ˈəupən ˈplætfɔːm ˈfeiʃiːz]

Definition: 开阔台地相 kāi kuò tái dì xiàng

open: not closed; accessible; unimpeded. 开阔的 kāi kuò de
platform: the part of a continental craton that is covered by sedimentary rocks. 台 tái
facies: a body of rock with specified characteristics, which can be any observable attribute of rocks such as their overall appearance, composition, or condition of formation, and the changes that may occur in those attributes over a geographic area. 相 xiàng

Example:
The shoal and patch reef sub-facies in territorial seawater shallowing open platform facies are developed in Middle Ordovician Yijianfang Formation.
中奥陶统一间房组沉积时期为区域性的海水变浅的开阔台地浅滩夹点礁沉积。

open system [ˈəupən ˈsistəm]

Definition: 开放系统 kāi fàng xì tǒng

open: not closed. 开放的 kāi fàng de

system: a set of devices powered by electricity, for example a computer or an alarm. 系统 xì tǒng

Examples:

① The complex science emphasizes studying the nonlinear open system and initiates the system to get to the dynamic balance through self-organization.
复杂科学崇尚对非线性开放系统的研究，倡导系统通过自组织达到动态稳定，承认并接受随机性。

② Geographic Environment is a huge and open system, which has quite trenchant arrangement and integration.
地理环境是一个开放的巨大系统，系统本身又具有较为分明的层次性和综合性。

Extended Term:

open ended system 可扩充系统

ophiolite [ˈɔfiəlait, ˈəu-]

Definition: a widespread rock formation containing a mixture of sedimentary, igneous, and metamorphic rocks, thought to be the result of sea-floor rifting or crustal plate collisions. 蛇绿岩 shé lǜ yán

Example:

The ophiolite migmatitic complex is in Hongshishan deep huge rift of the north margin of Tarim slab.
红石山蛇绿混杂岩产生于塔里木板块北缘红石山深大断裂带中。

Extended Terms:

ophiolite complex 蛇绿岩套
ophiolite serpentine 蛇纹石
ophiolite formation 蛇绿岩建造
ophiolite emplacement 蛇绿岩侵位
ophiolite suite 蛇绿岩组

ophitic [ɔˈfitik]

Definition: relating to or denoting a poikilitic rock texture in which crystals of feldspar are interposed between plates of augite. adj. 辉绿岩的 huī lǜ yán de；辉绿岩结构的 huī lǜ yán jié gòu de

Origin: late 19th century: via Latin from Greek ophitēs "serpentine stone" (from ophis "snake") + -ic.

Extended Terms:
ophitic textures 辉绿结构
intersertal ophitic 辉绿填充结构的

ophitic texture [əˈfitik ˈtekstʃə]

Definition: is the association of lath-shaped euhedral crystals of plagioclase, grouped radially or in an irregular mesh, with surrounding or interstitial large anhedral crystals of pyroxene; it is characteristic of the common rock type known as diabase. 含长结构 hán cháng jié gòu; 辉绿岩结构 huī lǜ yán jié gòu

Extended Terms:
taxitic ophitic texture 斑杂辉绿结构
hyalo-ophitic texture 玻璃辉绿岩结构

orbicular textures [ɔːˈbikjulə]

Definition: a structure found in some coarse-grained ignerous rocks which consists of concentric mineral shells of different composition which may or may not have a central xenolithic nucleus. Also described as spheroidal or nodular structures. 圆形火成包体的结构 yuán xíng huǒ chéng bāo tǐ de jié gòu

order of crystallization [ˈɔːdə ɔv ˌkristəlaiˈzeiʃən]

Definition: a phase loosely employed for the order in which the minerals of an igneous rock ceased to crystallize, which may be determined by such textural features as the idimorphism of one mineral to another and the indentation or enclosure of one mineral by another. Such features rarely provide evidence of the order in which minerals began to crystallize. 结晶顺序 jié jīng shùn xù

ore body [ɔː ˈbɔdi]

Definition: 矿体 kuàng tǐ

Examples:
① The uranium ore body occurs in the transitional zone.
铀矿体主产于氧化-还原过渡带中。

② Vertical slicing by filled stopes method is effective to mine fracturing ore body with the roof of moderate inclination.
垂直分条充填采矿方法对于中等倾斜顶板破碎矿体的开采较为有效。

Extended Terms:
lenticular ore body 透镜状矿体
ring like ore body 环状矿体

ore deposits [ɔː diˈpɒzit]

Definition: accumulation of ore. 矿床 kuàng chuáng

Examples:
① Hydrothermal sedimentary silicalite studies are important for ore deposits.
热水沉积硅质岩对于矿床研究具有重要意义。
② Genetic types of ore deposits in this area mainly include skarn type, porphyry type, magmatic hydrothermal type related to intermediated-acidic intrusions, then the volcanic hydrothermal type.
它们的成因类型主要是与中酸性岩浆侵入作用有关的矽卡岩型、斑岩岩型和热液型(包括破碎带蚀变岩型),其次为火山热液型。

Extended Terms:
quarternary exogenetic ore deposits 第四纪外生矿床
Geology of Ore Deposits 矿石沉积地质学

orendite [ˈɔːrəndait]

Definition: a porphyritic extrusive rock containing phlogopite phenocrysts in a nepheline-free reddish-gray groundmass of leucite, sanidine, phlogopite, amphibole, and diopside. 金云白榴斑岩 jīn yún bái liú bān yán

organic lattice [ɔːˈgænik ˈlætis] =growth lattice

Definition: the framework of a reef excluding associated detrital deposits. 生物骨架 shēng wù gǔ jià, 生物格架 shēng wù gé jià

organic matter [ɔːˈgænik ˈmætə]

Definition: the large pool of carbon-based compounds found within natural and engineered, terrestrial and aquatic environments. 有机质 yǒu jī zhì；有机物 yǒu jī wù；有机物质 yǒu jī wù zhì

Examples:

① The possible sources of dissolved organic matter in the ocean include: terrestrial input, phytoplankton release, diffusion from sediments and rainfall etc.
海洋中其可能的来源包括陆源输入、沉积物释放、降雨等。

② Geologists think the loess is organic matter blown down in centuries past from Mongolia and from the west by the great winds that rise in Central Asia.
地质学家认为，这种黄土是有机物质，是许多世纪以来被中亚细亚的大风从蒙古和西方吹过来的。

Extended Terms:

organic particulate matter 有机粒状物
organic sediments 有机沉积物
organic coloring matter 有机色素

organic reef [ɔːˈgænik riːf]

Definition: 生物礁堆积 shēng wù jiāo duī jī

organic: of or relating to an organism, a living entity. 生物的 shēng wù de
reef: a bar of rock, sand, coral or similar material, lying beneath the surface of water. 礁 jiāo

Example:

It is in the area with good preservation and thermal action is procreating that fault trap and organic reef gas pool will be found.
在那些保存条件好的地带，可能发现断层圈闭及潜伏生物礁气藏。

Extended Term:

organic framework reef 生物骨架礁

orogenic [ɔːrəˈdʒenik]

Definition: of a process in which a section of the earth's crust is folded and deformed by lateral compression to form a mountain range. *adj.* 造山的 zào shān de；造山运动的 zào shān yùn dòng de

Examples:

① Precambrian mineral deposits are mainly distributed in paleocontinents especially in North China platform and Yangtze platform, with a few distributed in orogenic belts.
中国前寒武纪矿床主要分布在陆块区，尤以华北陆块、扬子陆块较多，还有一些分布在造

山带。

② The influence on oil and gas, combination types, tectonic origin mechanism, orogenic zone thrust-nappe tectonics and benching thrust-nappe tectonics are summarized in this paper.
本文概述了逆冲推覆构造对油气的影响，组合类型，构造成因机制，造山带逆冲推覆构造和台阶式逆冲推覆构造。

Extended Terms:
orogenic rocks 造山岩
orogenic phase 造山相，造山期，造山幕
orogenic zone 造山带

orthoclase ['ɔ:θəkleis]

Definition: a common rock-forming mineral occurring typically as white or pink crystals. It is a potassium-rich alkali feldspar and is used in ceramics and glass-making. *n.* ［矿］正长石 zhèng cháng shí

Origin: mid-19th century：from ortho-"straight" +Greek klasis "breaking"（because of the characteristic two cleavages at right angles）.

Examples:
① Latites contain plagioclase feldspar（andesine or oligoclase）as large, single crystals（phenocrysts）in a fine-grained matrix of orthoclase feldspar and augite.
安粗岩中包含斜长石（中长石或奥长石），是在正长石和普通辉石组成的细粒基质中的大单晶体（斑晶）。
② Rock-forming mineral, olivine, pyroxene, hornblende, orthoclase, plagioclase, mica, quartz, biotite, calcite and other common metal and nonmetal minerals.
岩石矿物、橄榄石、辉石、角闪石、正长石、斜长石、云母、石英、黑云母、方解石和常见金属和非金属矿物。

Extended Terms:
soda orthoclase 钠正长石
orthoclase feldspar 正长石
iso-orthoclase 正钾长石
germanium-orthoclase 锗正长石

orthoconglomerate [ɔ:θəkən'glɔmərit]

Definition: Conglomerates are sedimentary rocks consisting of rounded fragments and are thus differentiated from breccias, which consist of angular "clasts". 正砾岩 zhèng lì yán

Extended Terms:

orthoclastization 正长石化

orthocumulate 正堆积岩

orthocumulate [ɔːθəˈkjuːmjuleit]

Definition: a cumulate composed chiefly of one or more cumulus minerals plus the crystallization products of the intercumulus liquid. 正堆积岩 zhèng duī jī yán

Example:

All members of cumulus sequence characteristically possess adcumulate and/or orthocumulate textures.
所有堆积岩都具有累积岩和/或正堆积岩纹理特点。

Extended Term:

orthocumulate textures 正堆积结构

orthopyroxene [ɔːθəpaiˈrɔksiːn]

Definition: a mineral of the pyroxene group crystallizing in the orthorhombic system. *n.* [矿]斜方辉石类 xié fāng huī shí lèi

Examples:

① The major rock-forming minerals include olivine, orthopyroxene, clinopyroxene, plagioclase and amphibole.
其主要造岩矿物为橄榄石、斜方辉石、单斜辉石、斜长石和角闪石。

② All metamorphic rocks exhibit an equilibrium texture, but exsolution lamellae of orthopyroxene(pigeonite) occur in all clinopyroxenes in mafic granulites.
这些岩石一般都展示了平衡的矿物共生结构，但在镁铁质麻粒岩的单斜辉石中普遍发育斜方辉石(变辉石)出溶片晶。

Extended Term:

orthopyroxene-andesite 斜方辉安山岩

oscillatory zoning [ˈɔsileɪˌtəri ˈzəʊnɪŋ]

Definition: 斜长石韵律环带 xié cháng shí yùn lǜ huán dài

oscillatory: having periodic vibrations. 韵律的 yùn lǜ de

zoning: the dividing of an area that has particular features or characteristics. 环带 huán dài

Examples:

① The small-scale oscillatory zoning is formed through self-organization with no relevance to the environmental factor.

斜长石中小尺度韵律环带起因于自组织过程，而与环境因素无关。

② This study is based on the physical chemistry elementary principle to build the mathematics modal of the genesis of plagioclase oscillatory zoning.

根据物理化学基本原理，建立了斜长石韵律环带形成的数学模型。

Oslo Graben [ˈɔsləʊ ˈɡrɑːbən]

Definition: Oslo Graben or Oslo Rift is a graben formed during a geologic rifting event in Permian time, the last phase of the Variscan orogeny. The main graben forming period began in the late Carboniferous, which culminated with rift formation and volcanism, with associated rhomb porphyry lava flows. This activity was followed by uplifting, and ended with intrusions about 65 millon years after the onset of the formation. It is located in the area around the Norwegian capital Oslo. 奥斯陆地堑 ào sī lù dì qiàn

osmium [ˈɔzmɪəm]

Definition: the chemical element of atomic number 76, a hard, dense, silvery-white metal of the transition series (Symbol: Os). 锇（符号：Os）é

Origin: early 19th century: modern Latin, from Greek osmē "smell" (from the pungent smell of its tetroxide).

Examples:

① The osmium may use in the industry to make the catalyst.

锇在工业中可以用作催化剂。

② But, the powdery osmium, gradually will be oxidized under the normal temperature, and production perosmic anhydride.

可是，粉末状的锇在常温下就会逐渐被氧化，并且生成四氧化锇。

Extended Terms:

osmium filament 锇丝
osmium compounds 锇化合物
osmium dichloride 二氯化锇
osmium chloride 氯化锇
osmium trichloride 三氯化锇

Ostwald ripening [ˈəustwɔːld ˈraipəŋ]

Definition: an observed phenomenon in solid solutions or liquid sols that describes the change of an inhomogeneous structure over time, i. e., small crystals or sol particles dissolve, and redeposit onto larger crystals or sol particles. 奥斯特瓦尔德成熟 ào sī tè wǎ ěr dé chéng shú

overgrowth [ˈəuvəgrəuθ]

Definition: excessive growth. *n.* 增生 zēng shēng; 繁茂 fán mào; 生长过度 shēng zhǎng guò dù; 过度发育 guò dù fā yù; 加大 jiā dà

Origin: 1595-1605; over- + growth.

Examples:

① Overgrowth quartz, chlorite and illite obviously reduce the permeability and porosity of the reservoir.
次生加大石英、绿泥石、伊利石对储层的孔渗有明显的破坏作用。
② The main diagenesis of SQ_8 are pressure solution, kaolinite corrosion and quartz overgrowth, el al.
SQ_8 的成岩作用主要体现在压溶、高岭石溶蚀、石英次生加大等。

Extended Terms:
overgrowth fault 超生断层
overstepping 过度分化

oversaturated rock [ˈəuvəˈsætʃəˌreitid rɔk]

Definition: oversaturated rocks are rocks containing quartz. 过饱和岩 guò bǎo hé yán

Example:

According to comprehensive research of lithogengesis and lithogeochemisty, all these volcanic rocks have higher alkalinity, alkalis with sub-alkalis series, normal type with akali type have coexisted, and SiO_2 have oversaturated rock with under saturate.
通过岩石学、岩石地球化学分析,该火山岩整体碱度较高,碱性和亚碱性系列共存,正常类型和钾质类型共存,SiO_2 过饱和。

oxbow lake deposit [ˈɔksbəu leik diˈpɔzit]

Definition: 牛轭湖沉积 niú è hú chén jī

oxbow lake: a U-shaped body of water that forms when a wide meander from the main stem of a river is cut off, creating a free-standing body of water. 牛轭湖 niú è hú

deposit: material added to a landform. 沉积 chén jī

oxidation ratio (OX) [ɔksi'deiʃən 'reiʃiəu]

Definition: 氧化率 yǎng huà lǜ

oxidation: (Chemistry) the process or result of oxidizing or being oxidized. (化)氧化(作用) yǎng huà zuò yòng

ratio: the quantitative relation between two amounts showing the number of times one value contains or is contained within the other. 比 bǐ; 比率 bǐ lǜ; 比例 bǐ lì

Example:

In this paper, studies are conducted on flotation and cyanidation of arsenic-bearing and oxidized refractory gold ore whose oxidation ratio and content of arsenic are 99.9% and 0.65% respectively.
本文对原矿含砷 0.65%、矿石氧化率为 99.9% 的难选氧化金矿石进行了浮选和氰化研究。

oxide ['ɔksaid]

Definition: (Chemistry) a binary compound of oxygen with another element or group. n. [化]氧化物 yǎng huà wù

Origin: late 18th century: from French, from oxygène "oxygen" + -ide (as in acide "acid").

Examples:

① The presence of ferric oxide in the ores is the key factor resulting in low whiteness of illite. 氧化铁是造成伊利石白度低的主要原因。

② The catalyst system comprised hydrocarbon and hydrocarbon oxide compounds of aluminum, magnesium, calcium, zinc and other metals, the catalyst promoter involved a number of compounds.
环氧乙烷聚合用催化剂体系包括铝、镁、钙、锌及其他金属的烃化物和烃氧化物, 助催化剂涉及多种化合物。

Extended Terms:

oxidation-reduction reactions 氧化还原反应
iron oxide 氧化铁
alkylene oxide 烯化氧

oxygen [ˈɔksidʒən]

Definition: colourless, odourless reactive gas, the chemical element of atomic number 8 and the life-supporting component of the air. Oxygen forms about 20 percent of the earth's atmosphere, and is the most abundant element in the earth's crust, mainly in the form of oxides, silicates, and carbonates(Symbol: O). *n.* 氧气 yǎng qì, 氧 yǎng

Origin: late 18th century: from French (principe) oxygène "acidifying constituent" (because at first it was held to be the essential component in the formation of acids).

Examples:

① Hydrogen and oxygen atoms can vary in how much they weigh.
氢和氧原子在重量上多少有点变化。
② We introduce a method of examining oxygen pipeline divulgence using oxygen concentration meter.
介绍一种用氧浓度仪检测的方法来检测氧气管道泄漏。

Extended Terms:

dissolved oxygen 溶解氧, 液态氧
oxygen content 氧含量
oxygen fugacity 氧逸度
oxygen isotope fractionation 氧同位素分馏系数
oxygen isotope ratios 氧同位素比率
oxygen isotopes 氧同位素

P-T-t path [piː tiː tiː paːθ]

Definition: PTt 轨迹 guǐ jì

path: the line that it moves along in a particular direction. 轨迹 guǐ jì

Example:

The second metamorphic event took place at about 2000Ma. Its peak P-T condition is 800±20℃ and 0.53-0.68GPa. The P-T-t path in the second stage of metamorphism is a kind of evolution pattern which typically belong to erogenic belt.
第二期变质作用发生于2.0Ga，变质作用峰期温度为800±20℃，压力为0.53~0.68GPa，其P-T-t轨迹属于典型的造山带型演化模式。

Extended Term:

path curve 轨线；轨迹曲线；路径线

Pacific Ocean [pəˈsifik ˈəuʃən]

Definition: the largest of the world's oceans, lying between America to the east and Asia and Australasia to the west. *n.* 太平洋 tài píng yáng

Examples:

① Tsunami are most common in the Pacific Ocean.
太平洋是海啸发生最频繁的区域。

② The corals were discovered at the bottom of the Pacific Ocean, near the Hawaiian Islands.
这些珊瑚是在靠近夏威夷群岛的太平洋底被发现的。

Extended Terms:

Pacific Ocean region 太平洋区
Packstone 泥粒灰岩，粒灰岩

pahoehoe [pəˈhəuihəui]

Definition: [mass noun] (Geology) basaltic lava forming smooth undulating or ropy masses. *n.* [地]绳状熔岩 shéng zhuàng róng yán

Origin: mid-19th century: from Hawaiian.

Example:

This image shows ropy pahoehoe, a type of basaltic lava that forms when solidifying lava meets an obstruction.
这幅照片展示了当玄武质熔岩流动时遇到障碍形成的绳状熔岩。

Extended Terms:

corded pahoehoe 绳状熔岩
festooned pahoehoe 花彩绳状熔岩
pahoehoe lava 绳状熔岩

paired metamorphic belts [pɛəd ˈmetəˈmɔːfik belts]

Definition: 双变质带 shuāng biàn zhì dài

paired: consisting of two identical or similar related things, parts, or elements. 成对的 chéng duì de

metamorphic: denoting rock that has undergone transformation by heat, pressure, or other natural agencies, e.g. in the folding of strata or the nearby intrusion of igneous rocks. 变质的 biàn zhì de

belt: a strip or encircling band of something having a specified nature or composition that is different from its surroundings. 条形(或环形)地带 tiáo xíng (huó huáng xíng) dì dài

Example:

As a result of irregular metamorphism, Sulu fold belt has been divided into two structural zones: eastern (blueschist) and western (gneiss-migmatite) belts that allow us to consider it as a paired metamorphic belt in a system of such belts distinguished by A.
由于变质作用的不规律性，苏鲁褶皱带可分为两个构造带：东部构造带（蓝片岩）和西部构造带（片麻岩混合岩）。根据 A 所识别的变质带系统，可将其作为一个双变质带。

palaeomagnetism [ˌpæliəˈmægnətizəm]

Definition: the branch of geophysics concerned with the magnetism in rocks that was induced by the earth's magnetic field at the time of their formation. *n.* [物] 古地磁学 gǔ dì cí xué

Examples:

① Palaeomagnetism, strain analysis, modeling and sedimentology are main methods used in the research of arcuate structure;
古地磁、应变分析、构造模拟和沉积环境研究是其主要研究方法。
② Segmented relation of major plates in Tengchong and nearby is obtained by tectonic relative relation, time-space and palaeomagnetism. The evolution progress of tectonic is also studied.
借助古地磁研究成果而得出的腾冲及邻区主要板块拼合关系，研究该区地壳构造演化过程。

Extended Terms:

palaeomagnetic time scale 古地磁年代表
palaeomagnetism 古地磁；古磁性
palaeomantle 古地幔

palagonite [pəˈlægənait]

Definition: a brown to yellow altered basaltic glass found as interstitial material or amygdules in pillow lavas. *n.* 橙玄玻璃 chéng xuán bō lí

Examples:

① The color of palagonite ranges from shades of yellow to shades of brown.
橙玄玻璃颜色变化范围从暗黄色系到深棕色系。
② Palagonite is the first stable product of volcanic glass alteration.
橙玄玻璃是火山岩玻璃质化的首个稳定生成物。

Extended Term:

palagonite mud 玄玻质软泥

paleosol [ˈpæliəusɔl]

Definition: a stratum or soil horizon that was formed as a soil in a past geological period. *n.* 古土壤 gǔ tǔ rǎng

Examples:

① The grain size of loess and paleosol sequence is regarded as one of the most important proxies to reconstruct paleoenvironment.
黄土、古土壤剖面中物质粒度是恢复古环境的重要指标之一。

② Sequence boundary could be recognized with stratigraphic contact relationship, the patterns of superimposed channel sand bodies, the alluvial incision and paleosol.
地层接触关系、河道砂体的叠置方式、冲积下切作用以及古土壤面可以用来识别层序边界。

Extended Terms:

paleoshore line variation 古海岸线变化
Palaeostrophia 古凸贝

Palisades sill [ˌpæliˈseidz sil]

Definition: 帕利塞德岩床 pà lì sài dé yán chuáng

Palisades: (also Palisades of the Hudson) cliffs that line the western side of the Hudson River, in New Jersey and in New York, beginning across from New York City in New Jersey and extending north to Newburgh in New York. 帕利塞德 pà lì sài dé
sill: a tabular sheet of igneous rock intruded between and parallel with the existing strata. [地质]岩床 yán chuáng

Example:

Magma chamber recharge is proposed to have occurred at the 27m and 45m levels, when a slightly more-primitive HTQ magma was injected into the Palisades sill chamber.
当稍稍更远古的 HTQ 岩浆注入帕利塞德岩床岩浆库时，岩浆在 27 米和 45 米地层的岩浆库重新流入。

palimpsest mineral [ˈpælimpsest ˈminərəl]

Definition: 变余矿物 biàn yú kuàng wù

palimpsest: Geological features thought to be related to features or effects below the surface. 变余 biàn yú
mineral: an inorganic substance occurring naturally in the earth and having a consistent anddistinctive set of physical properties (e. g., a usually crystalline structure, hardness, color,

etc.) and a composition that can be expressed by a chemical formula; sometimes applied to substances in the earth of organic origin, such as coal. 矿物 kuàng wù

palimpsest structure ['pælimpsest 'strʌktʃə]

Definition: 变余构造 biàn yú gòu zào

palimpsest: geological features thought to be related to features or effects below the surface. 变余 biàn yú

structure: manner of building, constructing, or organizing. 构造 gòu zào

Example:

Their occurrence, composition, enclaves, palimpsest structure and intrusive contact relationships prove that they are ancient granitic plutons which now occur in the form of gneisss. 本文以产状、成分、所含包体、残留的变余结构及侵入关系为依据，论证了它们是以片麻岩形式存在的古老花岗岩体。

palimpsest texture ['palim(p)sest 'tekstʃə]

Definition: 变余结构 biàn yú jié gòu

palimpsest: something reused or altered but still bearing visible traces of its earlier form. 仍然看得出原状痕迹的东西 réng rán kàn de chū yuán zhuàng hén jì de dōng xī

texture: the feel, appearance, or consistency of a surface or a substance. （物质、表面的）结构 jié gòu；构造 gòu zào；纹理 wén lǐ；肌理 jī lǐ

palingenesis [ˌpælin'dʒenisis]

Definition: the repetition by a single organism of various stages in the evolution of its species during embryonic development. 再生作用 zài shēng zuò yòng

Origin: early 19th century; from Greek palin "again, back" + genesis "birth".

Example:

It causes the center palingenesis because the modernization lacks overall systematicness. 现代化缺乏全面系统性，使其在推进过程中出现重心轮回。

Extended Terms:

second birth palingenesis, rebirth regeneration regenesis 再生
a new lease of life freshman palingenesis rebirth renascence 新生

pangaea [pænˈdʒiə]

Definition: a vast continental area or supercontinent comprising all the continental crust of the earth, which is postulated to have existed in late Palaeozoic and Mesozoic times before breaking up into Gondwana and Laurasia. *n.* 泛大陆 fàn dà lù；盘古大陆 pán gǔ dà lù

Origin: early 20th century; from pan- "all" + Greek gaia "earth".

Examples:
① Pangaea was the last supercontinent to form before the present.
盘古大陆是史前形成的最后一块超大陆。
② When continental drift began, Pangaea broke up into Laurasia and Gondwanaland.
当大陆板块漂移开始后，泛大陆分裂成劳亚古大陆和冈瓦纳大陆。

Extended Terms:
profusulinella 原始虫筵；原始纺锤虫
panthalassa 原始大洋

panidiomorphic [pænˌidiəuˈmɔːfik]

Definition: (of igneous rocks) having well-developed crystals. *adj.* [地质] 全同形的 quán tóng xíng de；全自形的 quán zì xíng de

Extended Terms:
panidiomorphic texture 全自形结构
panidiomorphic granular texture 全自形粒状结构

panidiomorphic granular texture [pænˌidiəuˈmɔːfik ˈgrænjulə ˈtekstʃə]

Definition: 全自形粒状结构 quán zì xíng lì zhuàng jié gòu

panidiomorphic: having a completely idiomorphic structure; said of certain rocks. 全自形的 quán zì xíng de

granular: composed or appearing to be composed of granules or grains. 颗粒的 kē lì de

texture: the arrangement of the particles or constituent parts of any material, as wood, metal, etc., as it affects the appearance or feel of the surface; structure, composition, grain, etc. 结构 jié gòu

pantellerite [pænˈtelərait]

Definition: a peralkaline rhyolite. It has a higher iron and lower aluminum composition than comendite. *n.* 碱流岩 jiǎn liú yán

Origin: It is named after Pantelleria, a volcanic island in the Strait of Sicily and the type location for this rock.

Example:

It is found that the volcanic rocks in Mesozoic group are andesite and pantellerite, and the volcanic rocks in Paloogene system are basalts, which shows that olivine and pyroxene had crystallized and differentiated during the evolution of magma.

本区中生界岩石类型主要为安山岩和碱流岩，而火山岩主要是玄武岩，这表明在岩浆的演化过程中曾发生过橄榄石和辉石的结晶分异作用。

Extended Term:

hyalo-pantellerite 玻质碱流岩

paraconglomerate [ˌpærəkən'glɔmərit]

Definition: Conglomerates are sedimentary rocks consisting of rounded fragments and are thus differentiated from breccias, which consist of angular "clasts." 副砾岩 fù lì yán

Extended Terms:

paraconglomerate deposit 副砾岩沉积

paraconformity 泥砾岩

parallel bedding ['pærəlel 'bediŋ] = concordant bedding

Definition: 平行层理 píng xíng céng lǐ

parallel: of lines, planes, surfaces, or objects) side by side and having the same distance continuously between them. (线，平面，物体)平行的 píng xíng de

bedding: a base or bottom layer. 底层 dǐ céng；基层 jī céng

Example:

In parallel or concordant bedding the strata overlay each other like pages of a book.

在平行的或者整合的层面中，地层好像书页一般互相重叠。

partial derivatives ['pɑːʃəl di'rivətivs]

Definition: the derivative with respect to a single variable of a function of two or more variables, regarding other variables as constants. 偏导数 piān dǎo shù；偏微分 piān wēi fēn；偏导函数 piān dǎo hán shù

Examples:

① Mathematically, the MPPs are given by the partial derivatives of the production function.
在数学上，这些边际产量是生产函数的偏导数。
② A vector having coordinate components that are the partial derivatives of a function with respect to its variables.
一个矢量，其坐标分量是一个函数对其变量的偏微分。

Extended Term:

second partial derivatives 第二阶偏微分

partial melting [ˈpɑːʃəl ˈmeltiŋ]

Definition: occurs when only a portion of a solid is melted. For mixed substances, such as a rock containing several different minerals or a mineral that displays solid solution, this melt can be different from the bulk composition of the solid. 部分熔融 bù fen róng róng

Examples:

① The Luobusha podiform chromites locate at metamorphic peridotites that show partial melting and dynamic remelting.
罗布莎豆荚状铬铁矿床产于高度部分熔融和动力重熔的变质橄榄岩中。
② All these demonstrated that Mariana Trough basalts are products of partial melting from a mixed mantle(the contamination of MORB mantle by auckland mantle).
上述特征反映了马里亚纳海槽玄武岩是 MORB 型与岛弧型地幔源不同程度混合后部分融熔的产物。

Extended Terms:

partial melting region 半熔化区
melting with partial oxidation 部分氧化熔炼

partial molar volume [ˈpɑːʃəl ˈməulə ˈvɔljuːm]

Definition: that portion of the volume of a solution or mixture related to the molar content of one of the components within the solution or mixture. 偏摩尔体积 piān mó ěr tǐ jī

partition coefficient [pɑːtiʃən ˌkəuiˈfiʃənt]

Definition: in the equilibrium distribution of a solute between two liquid phases, the constant ratio of the solute's concentration in the upper phase to its concentration in the lower phase. 分配系数 fēn pèi xì shù(等于 distribution coefficient)

Examples:

① Conclusion: The partition coefficient for volatile anesthetic in lipid emulsion accords with volume fraction partition coefficient concept.
结论：挥发性麻醉药在脂肪乳中的分配系数符合容积比分配系数的概念。

② Shake-flask method was used to determine water solubility (Sw) and n-octanol/water partition coefficient (Kow) for 15 substituted aromatic ketone and aldehyde.
本文测定了15种芳香族醛酮化合物的溶解度和正辛醇/水分配系数。

Extended Terms:

isotopic partition coefficient 同位素分配系数
partition coefficient of pesticides 农药分配系数

pascal ['pæskəl]

Definition: the SI unit of pressure, equal to one newton per square metre (approximately 0.000145 pounds per square inch, or 9.9×10^{-6} atmospheres). *n.* 帕斯卡 pà sī kǎ (压力的单位)

Origin: 1950s: named after B. Pascal (see Pascal).

Extended Term:

Pascal triangle 帕斯卡三角

passive continental margin basin
['pæsiv ˌkɒnti'nentəl 'mɑːdʒin 'beisən]

Definition: 被动大陆边缘盆地 bèi dòng dà lù biān yuán pén dì

passive: accepting or allowing what happens or what others do, without active response or resistance. 被动的 bèi dòng de; 消极的 xiāo jí de

continental: of or of the nature of a continent. 大陆的 dà lù de

margin: the edge or border of something. 边缘 biān yuán

basin: a circular or oval valley or natural depression on the earth's surface, especially one containing water. 盆地 pén dì

Example:

The Pearl River Mouth Basin (PRMB), located at South China Sea, is a Cenozoic passive continental margin rift basin with a great potential for oil and gas exploration.
南海珠江口盆地是一个可以开发石油和天然气的新生代被动大陆边缘盆地。

passive margin ['pæsiv 'mɑːdʒin]

Definition: 被动陆缘 bèi dòng lù yuán

passive: accepting or allowing what happens or what others do, without active response or resistance. 被动的 bèi dòng de

margin: the edge or border of something. 边缘 biān yuán

> Origin:

passive: late 14th century, in grammatical sense (opposed to active), Old French passif "suffering, undergoing hardship" (14th century).

margin: middle 14th century, "edge of a sea or lake;" late 14th century.

> Examples:

① There were a basin of passive margin on the northern border of Yangtze Plate and a intracontinental rift basin in Qinglin-Daba and Jiangnan-Dabie regions during Early Palaeozoic. 早古生代扬子板块北缘秦巴地区为被动边缘型盆地，江南大别区为陆内裂谷。

② With the passive margin, development of Nanpanjiang basin in early Triassic, the carbonate slope evolved from the ramp into the prograding sedimentary slope. 早三叠世，盆地处于被动边缘，碳酸盐斜坡经历了由缓坡到进积型沉积斜坡的发展。

pearlite [ˈpəːlait]

> Definition: pearlstone, a variety of obsidian consisting of masses of small pearly globules; used as a filler, insulator, and soil conditioner. 珍珠岩 zhēn zhū yán

> Origin: 19th century: from French, from perle pearl.

> Example:

Non-metallic minerals, especially granite, marble, asbestos, perlite reserves are of up to the best.
非金属矿尤以花岗石、大理石、石棉、珍珠岩储量为最多最好。

> Extended Terms:

crude pearlite 珍珠岩原矿；珍珠岩
perlite powder 珍珠岩粉
perlite structure 珠光体组织
perlite concrete 珍珠岩混凝土
perlite mortar 珍珠岩灰浆

peat [piːt]

> Definition: a brown soil-like material characteristic of boggy, acid ground, consisting of partly decomposed vegetable matter and widely cut and dried for use in gardening and as fuel. n. 泥煤 ní méi；泥炭块 ní tàn kuài；泥炭色 ní tàn sè

Origin: Middle English: from Anglo-Latin peta, perhaps of Celtic origin.

Examples:

① Natural chemical reactions also take place when iron rusts, or when peat is formed.
铁生锈和泥煤的形成也是天然的化学反应。

② Results have shown that the TONSTEINS are all synsedimentary acid volcanic ash, which were turned to relatively pure kaolinitic mudstone through hydrolysis and diagenesis in peat moor environment.
结果表明，它们都是同沉积的酸性火山灰，在泥炭沼泽环境下，经过水解和后期成岩作用，转变成质地较纯的高岭石黏土岩。

Extended Terms:

terrestrial peat 陆相泥炭
limnic peat 湖水泥炭

pebble [ˈpebl]

Definition: a small stone made smooth and round by the action of water or sand. 卵石 luǎn shí

Origin: late Old English, recorded as the first element of papel-stān "pebble-stone", "pebble-stream", of unknown origin. The word is recorded in place names from the early 12th century.

Example:

Through laboratory tests on sandy pebble soil by using the dynamic triaxial test, researches of saturated sandy pebble soil have been done.
通过对砂卵石土室内动三轴试验，对饱和砂卵石土的动力特性进行了较为深入的研究。

Extended Terms:

faceted pebble 棱石
flint pebble 燧石子
heat-exchange pebbles 石子载热体
land pebble 卵石土
nuclear fuel pebble 核燃料芯块
refractory pebbles 耐火石

Peclet number [pekˈlei ˈnʌmbə]

Definition: dimensionless group used to determine the chemical reaction similitude for the scale-up from pilot-plant data to commercial-sized units; incorporates heat capacity, density, fluid velocity, and other pertinent physical parameters. [化]沛克莱数 pèi kè lái shù

Examples:

① A larger kinematic viscosity of the fuel also augments oxidant's Peclet number to improve the performance.
增加阳极燃料动黏滞系数时,阴极氧化物的沛克莱数也将增加,因此性能也较佳。

② Altering aspect ratio of the cross section area to bring down area and bring up Peclet number shall be improving the cell performance greatly.
改变截面积之长宽比时,若能因此减少截面积并且增加沛克莱数,在双重影响下,电池性能也将大幅地提升。

Extended Term:

turbulent Peclet number 紊流沛克莱数

pectinate texture [ˈpektineit ˈtekstʃə]

Definition: 栉壳结构 zhì ké jié gòu, 晶簇状结构(category texture) jīng cù zhuàng jié gòu
pectinate: of an appendage of an insect consisting of or bearing a row of bristles or chitinous teeth. 梳状的 shū zhuàng de; 栉齿状的 zhì chǐ zhuàng de
texture: the feel, appearance, or consistency of a surface or a substance(物质、表面的)结构 jié gòu; 构造 gòu zào; 纹理 wén lǐ; 肌理 jī lǐ

pedalfer [piˈdælfə]

Definition: a soil rich in alumina and iron, with few or no carbonatesn. [地]淋余土 lín yú tǔ;[地]铁铝土 tiě lǚ tǔ

Example:

The pedalfer boundary shifts westward during El Nino episodes and moves eastward during La Nina events.
淋余土边界在厄尔尼诺现象期间向西漂移,在拉尼娜现象期间向东漂移。

Extended Term:

pedestal rock (gour, mushroom rock) 菌状石;风蚀柱

pedocal [ˈpedəkæl]

Definition: a soil rich in carbonates, esp. those of lime. n. 钙层土 gài céng tǔ

Extended Term:

tropical pedocal 热带聚钙土

pedogenic [pidəuˈdʒenik]

Definition: of the process of soil formation. *adj.* 成土的 chéng tǔ de

Origin: 1870-1875; pedo- + -genesis.

Examples:

① The in-situ pedogenic enhancement of ferrimagnetic content is normally believed to be the main reason for the increase of susceptibility in soil units.
成壤过程中形成的亚铁磁性矿物被认为是古土壤磁化率增加的主要原因。

② The results show that grain-size fractal dimension is related to loess and paleosol closely, which can be regarded as good indictor of deposition and pedogenic environment.
计算结果表明粒度的分形维数与黄土和古土壤层位具有良好的对应关系，对沉积和成壤环境具有很好的指示意义。

Extended Terms:

pedogenesis 成土作用
pedogenic environment 成壤环境
pedogenic process 土壤发生过程
pedogenic rock 成土母岩

pegmatite [ˈpeɡmətait]

Definition: any of a class of exceptionally coarse-grained intrusive igneous rocks consisting chiefly of quartz and feldspar; often occurring as dykes among igneous rocks of finer grain. 伟晶岩 wěi jīng yán

Origin: 19th century; from Greek pegma something joined together.

Example:

Alto Ligonha is one of the world famous regions for its pegmatite where rare metals and rare earth, crystals and gemstone resources are abundant.
阿尔特里哥纳是世界著名的伟晶岩区之一，蕴藏着丰富的稀有金属、稀土、水晶和宝石等矿产资源。

Extended Terms:

migmatic pegmatite 混合伟晶岩
calcite pegmatite 方解伟晶岩
ekerite pegmatite 钠闪花岗伟晶岩；钠闪花冈伟晶岩
canadite pegmatite 钠霞正长伟晶岩
kersantite pegmatite 云斜伟晶岩

pegmatitic texture [ˈpegmətaitik ˈtekstʃə]

Definition: 伟晶结构 wěi jīng jié gòu

pegmatitic: a coarse-grained granite, sometimes rich in rare elements such as uranium, tungsten, and tantalum. 伟晶岩的 wěi jīng yán de

texture: the arrangement of the particles or constituent parts of any material, as wood, metal, etc., as it affects the appearance or feel of the surface; structure, composition, grain, etc. 结构 jié gòu

pelagic sediments [peˈlædʒikˈsedimənts]

Definition: Pelagic sediments, also known as marine sediments, are those that accumulate in the abyssal plain of the deep ocean, far away from terrestrial sources that provide terrigenous sediments. 远洋沉积 yuǎn yáng chén jī

Example:

The radiolarian cherts in the Weibei region of Shaanxi Province are a formation of pelagic sediments formed within the North China Platform.
渭北奥陶系的放射虫燧石岩是在稳定的华北地台上形成的远洋沉积.

pele's hair [ˈpɛlez heə]

Definition: a spun volcanic glass formed naturally by blowing out during quiet fountaining of fluid lava. Also known as capillary ejecta; filiform lapilli; lauoho o pele. 火山丝 huǒ shān sī

pelite [ˈpiːlait]

Definition: a sediment or sedimentary rock composed of very fine clay or mud particles. *n.* [地质] 泥质岩 ní zhì yán

Origin: late 19th century: from Greek pēlos "clay", mud + -ite.

Example:

Distinctive from the low-medium pressure metamorphism of pelite, progressive dehydrations proceed even in the decompression process post metamorphic peak of UHP eclogites.
与泥质岩中低压变质演化明显不同的是超高压榴辉岩在峰期以后的减压过程中仍然发生递进脱水作用。

Extended Terms:

pelitic schists 泥质片岩

pelite mudstone 泥岩
peltie pelite 泥质岩

pelitic texture [piˈlitik ˈtekstʃə]

Definition: 泥质结构 ní zhì jié gòu, 又称黏土结构(clay texture) nián tǔ jié gòu
pelitic: any clayey rock, as mudstone or shale. 泥质岩的 ní zhì yán de
texture: the feel, appearance, or consistency of a surface or a substance. (物质、表面的)结构 jié gòu; 构造 gòu zào; 纹理 wén lǐ; 肌理 jī lǐ

Extended Terms:
pelitic siltstone 泥质粉砂岩
pelitic structure 泥岩状结构
pelitic hornfels 泥质角页岩
pelitic metamorphic rocks 泥质变质岩类

pellet [ˈpelit]

Definition: a ball, usually made of stone, formerly used as a cannonball or as a missile firedfrom a catapult. 球粒 qiú lì; 团粒 tuán lì; 球状粒 qiú zhuàng lì

Origin: 14th century. From French pelote "small ball", from Latin pila "ball".

Extended Terms:
pelletizing 球团
pelmicrite 球粒

pellet limestone [ˈpelit ˈlaimstəun]

Definition: 球粒灰岩 qiú lì huī yán, 又称团粒灰岩 tuán lì huī yán
pellet: a small, rounded, compressed mass of a substance. 颗粒状物 kē lì zhuàng wù; 小团 xiǎo tuán; 丸 wán
limestone: a hard sedimentary rock, composed mainly of calcium carbonate or dolomite, used as building material and in the making of cement. 石灰岩 shí huī yán

peloid [ˈpelɔid]

Definition: mud used therapeutically. n. 疗效泥 liáo xiào ní

Origin: from Greek pēl(ós) mud, clay + -oid.

penecontemporaneous [ˌpiːnikənˈtempəreinjəs]

Definition: (of a process) occurring immediately after deposition of a particular stratum. 准同生作用 zhǔn tóng shēng zuò yòng

Origin: early 20th century: from Latin paene "almost" +contemporaneous.

Example:

Penecontemporaneous or seepage-reflux dolomitization is the original mechanism for gypsum-bearing muddy dolomicrite.
准同生白云石化或卤水回流渗透白云石化是泥晶-微晶白云岩(含膏云岩)的形成机制。

Extended Terms:

penecontemporaneous breccias 准同生角砾岩
penecontemporaneous deformatin structure 准同期形变构造
penecontemporaneous deformation 准同时的变形
penecontemporaneous erosion 准同期侵蚀
penecontemporaneous replacement 准同期替换作用

penecontemporaneous deformation structure
[ˌpiːnikənˈtempəreinjəs ˌdiːfɔːˈmeiʃən ˈstrʌktʃə]

Definition: 准同生变形构造 zhǔn tóng shēng biàn xíng gòu zào

penecontemporaneous: (of a process) occurring immediately after deposition of a particular stratum. [地质](进程)准同期的 zhǔn tóng qī de; 近同期的 jìn tóng qī de; 准同时的 zhǔn tóng shí de

deformation: the action or process of changing in shape or distorting, especially through the application of pressure. 受压变形 shòu yā biàn xíng

structure: the arrangement of and relations between the parts or elements of something complex. 结构 jié gòu; 构造 gòu zào

pentlandite [ˈpentlənˌdait]

Definition: a bronze-yellow mineral which consists of a sulphide of iron and nickel and is the principal ore of nickel. n. 镍黄铁矿 niè huáng tiě kuàng;[化]硫镍铁矿 liú niè tiě kuàng

Origin: mid-19th century: from the name of Joseph B. Pentland (1797-1873), Irish traveller, + -ite.

Examples:

① The sulfide assemblage consists largely of pyrrhotite, pentlandite, chalcopyrite, and pyrite.

硫化物集体大部由磁黄铁矿、镍黄铁矿、黄铜矿及黄铁矿组成。
② The disseminated blebs are generally composed of one pyrrhotite individual accompanied by varying amounts of pentlandite and chalcopyrite.
分散的泡体常由磁黄铁矿个体与多少不等的镍黄铁矿及黄铜矿共生。

Extended Term:
cobalt adamite pentlandite 钴水砷锌镍黄铁矿

peralkaline [pəˈrælkəlain]

Definition: (of an igneous rock) containing a higher proportion (taken together) of sodium and potassium than of aluminium. *n.* 过碱性 guò jiǎn xìng; 过碱性岩 guò jiǎn xìng yán

Extended Terms:
peralkaline igneous rock 过碱火成岩
peralkaline magmas 过碱岩浆
peralkaline rocks 过碱性岩

peraluminous [ˌpɜːrəˈluːminəs]

Definition: of igneous rock, having a molecular proportion of aluminum oxide greater than that of sodium oxide and potassium oxide combined. 过铝质岩 guò lǚ zhì yán

Example:
This paper describes in detail the features of mineralogy and mineral chemistry about tourmaline from peraluminous granite in south Tibet(PGST).
本研究讨论了藏南过铝花岗岩中电气石的地质产状、矿物学和矿物化学特征。

Extended Term:
peraluminous rock 过铝质岩

periclase [ˈperikleis]

Definition: a colourless mineral consisting of magnesium oxide, occurring chiefly in marble and limestone. *n.* [矿]方镁石 fāng měi shí

Origin: mid-19th century: from modern Latin periclasia, erroneously from Greek peri "utterly" +klasis "breaking" (because it cleaves perfectly).

Examples:
① Normal magnesia brick is made of Alumina-Magnesia spinel, periclase. It has higher temperature strength, good thermal shock stability, strong resistance to basic slag.

普通镁铝以镁铝尖晶石和方镁石为主要晶相，具有较高的高温强度和较好的热稳定性，抗碱性炉渣侵蚀。
② As to the steel slag, the effect of reducing alkali aggregate reaction is better when alkali reactivity is low and the content of free calcium oxide and periclase is low.
对于钢渣，当碱活性低，游离氧化钙和方镁石的含量低时，抑制碱集料反应的效果愈好。

Extended Terms:
periclase brick 方镁石砖
periclase porcelain 方镁石瓷

peridotite ['perɪdɒtaɪt]

Definition: a dense, coarse-grained plutonic rock containing a large amount of olivine, believed to be the main constituent of the earth's mantle. n. [矿] 橄榄岩 gǎn lǎn yán

Origin: 1895-1900; < F; see peridot, -itel.

Examples:
① Below the lithosphere is the geosphere comprising partly fused peridotite.
岩石圈之下是由部分熔融的橄榄岩所组成的软流圈。
② Peridotite can also be found at the surface in other parts of the world, including some Pacific islands, along the coasts of Greece and Croatia, and in smaller deposits in America.
橄榄岩也可以在地表的其他地方找到，包括一些太平洋岛屿、希腊和克罗地亚的沿岸和美国的一些较小的矿场上。

Extended Terms:
magnetite peridotite 磁铁橄榄岩
ore peridotite 矿石橄榄岩
peridotite shell 橄榄岩壳层
biotite peridotite 黑云橄榄岩
anorthite peridotite 钙长橄榄岩

peripheral basin [pəˈrɪfərəl ˈbeɪsən]

Definition: 周缘盆地 zhōu yuán pén dì

peripheral: of, relating to, or situated on the edge or periphery of something. 外围的 wài wéi de; 边缘的 biān yuán de
Basin: a circular or oval valley or natural depression on the earth's surface, especially one containing water. 盆地 pén dì

Example:

At the earliest Paleogene, Tibet-Tethys was closed down, causing the forming of the southern Tibet peripheral basin system.

从古近纪初开始,西藏特提斯关闭,形成周缘盆地体系(由褶冲带、前渊带、前隆带和隆后盆地等单元构成)。

Extended Terms:

peripheral foreland basin 周缘前陆盆地
peripheral sedimentary basin 边缘沉积盆地

---

peritectic [perəˈtektik]

Definition: of or noting the phase intermediate between a solid and the liquid that results from the melting of the solid. 包晶 bāo jīng

Origin: 1920-1925; peri- + Gk těktikós able to dissolve, akin to těkein to melt.

Examples:

① The Al-based amorphous alloy forms in the peritectic reaction and peritectoid reaction, and its growth velocity is slow.

分析了铝基准晶合金的组织与相变特征,铝基准晶的生成通常为包晶、包析反应,其长大速度较慢。

② The results show that the strict coupled growth between primary and peritectic phases only takes place at the temperature higher than peritectic phase transformation temperature in steady state.

研究结果表明:在定向凝固条件下,包晶系两相严格的共生生长只有在进入稳态且温度高于包晶相变温度时才可能进行。

Extended Terms:

peritectic lines 包晶线
peritectic points 包晶点
peritectic reaction 包晶反应
peritectic structure 包晶组织
peritectic temperature 包晶温度

---

permeability [pɜːmiəˈbiliti]

Definition: the state or quality of a material or membrane that causes it to allow liquids or gases to pass through it. *n.* 渗透性 shèn tòu xìng;[物]透磁率 tòu cí lǜ;导磁系数 dǎo cí xì shù;弥漫 mí màn

> Examples:

① During the solutions, effects of temperature on viscosity, oil saturation and oil-water relative permeability are taken consideration.
在求解过程中,考虑了温度对粘度、含油饱和度和油水相对渗透率的影响。
② For solving porous media multiphase flow equations, capillary pressure, relative permeability, fluid saturation relationships are the most important bases.
毛细压力、相对渗透率与饱和度的关系是求解多孔介质多相流运动方程的基础。

> Extended Terms:

absolute permeability 绝对透气性
capillary permeability 毛(细)管渗透性
gas permeability 透气性
water permeability 透水性

perovskite-spinel transition [pəˈrɔvzkait ˈspinel trænˈsiʒən]

> Definition: 钙钛矿尖晶石过渡 gài tài kuàng jiān jīng shí guò dù

perovskite: yellow, brown, or greyish-black mineral form of calcium titanate with some rare-earth elements, which is used in certain high-temperature ceramic superconductors. 钙钛矿 gài tài kuàng

spinel: a variously colored, crystalline mineral, MgAlO, used as a gem; magnesium aluminum oxide. 尖晶石 jiān jīng shí

transition: the process or a period of changing from one state or condition to another. 过渡 guò dù; 转变 zhuǎn biàn; 变革 biàn gé; 过渡期 guò dù qī; 转型期 zhuǎn xíng qī

perovskite [pəˈrɔvzkait]

> Definition: yellow, brown, or greyish-black mineral form of calcium titanate with some rare-earth elements, which is used in certain high-temperature ceramic superconductors. 钙钛矿 gài tài kuàng

> Origin: 19th century: named after Count Lev Alekseevich Perovski (1792-1856), Russian statesman.

> Examples:

① Layered perovskite-type organic-inorganic hybrids are sort of the assembly of the organic and inorganic units on the molecular scale.
层状类钙钛矿结构的有机-无机化合物由有机、无机组元在分子水平上组合而成。
② Many kinds of well-crystallized nanometer or submicron Ti-based perovskite ceramic

powders were already synthesized successfully by sol-hydrothermal method at lower temperatures.

采用溶胶-水热法,在较低的温度下已成功地合成了多种结晶良好的纳米或亚微米钛基钙钛矿型陶瓷粉体。

Extended Terms:

double perovskite 双钙钛矿

perovskite structure 钙钛矿型结构;钙钛矿结构;钙钛结构矿;钙钛矿

niobium perovskite 含铌钙钛

organic-inorganic perovskite materials 有机-无机类钙钛矿材料

cerium perovskite 含铈钙钛矿

perthite ['pəːθait]

Definition: a variety of feldspar containing irregular bands of albite inmicrocline. 条纹长石 tiáo wén cháng shí

Origin: 1832; named after Perth, Ontario, Canada.

Examples:

① These rocks are composed of ortho-and clinopyroxenes, fayalite, perthite, antiperthite, and quartz, and are characterized by positive ε Nd values (+4.8~+5.9).

紫苏花岗岩由斜方辉石、单斜辉石、铁橄榄石、条纹长石、石英组成,以正 ε Nd 为特征(εNd =+4.8~+5.8)。

② Under the pressure of 510MPa, the perthite in granite begin to disappear at 600℃, and under the 700℃, most of them vanish and turn into orthoclase.

在 510MPa 压力下,花岗岩的条纹长石于 600℃开始消失,至 700℃时它们大量消失并变为正长石.

Extended Terms:

perthite pyroxenite 条纹辉岩;条纹辉石岩

braid perthite 辫状条纹长石

double perthite 复纹长石

perthite syenite 条纹正长岩

microcline-perthite 微斜纹长石

perthitic texture [pəːˈθitik ˈtekstʃə]

Definition: 条纹结构 tiáo wén jié gòu

perthitic: of a texture produced by perthite, exhibiting sodium feldspar as small strings, blebs,

films, or irregular veinlets in a host of potassium feldspar. 条纹长石 tiáo wén cháng shí
texture: the arrangement of the particles or constituent parts of any material, as wood, metal, etc., as it affects the appearance or feel of the surface; structure, composition, grain, etc. 结构 jié gòu

petrochemistry [ˌpetrəuˈkemistri]

Definition: the branch of chemistry concerned with the composition and formation of rocks (as distinct from minerals and ore deposits). 岩石化学 yán shí huà xué, 化学岩石学 (chemical petrology)。

Origin: 1935-1940; petro-2 or petro-1 + chemistry.

Extended Terms:
petrochemistry characteristic 岩石化学特征
petrochemistry industry 石化行业
petroleum and petrochemistry 石油石化

petrogenesis [ˌpetrəˈdʒenisis]

Definition: the formation of rocks, especially igneous and metamorphic rocks. n. 岩石发展学 yán shí fā zhǎn xué, 岩石成因论 yán shí chéng yīn lùn

Origin: 900-1905; petro-1 + -genesis.

Examples:
① Unfortunately, there were different cognition about the era and petrogenesis of Raohe granites by now.
然而关于本区花岗岩的形成时代及成因等方面的问题至今仍存在不同的观点。
② Despite of this, the field geology and the character of pyroclastic rocks remain the basis of the petrogenesis of pyroclastic rocks.
尽管这样，野外地质和火山碎屑岩的岩性特征，仍是火山碎屑岩成因研究的基础。

Extended Terms:
petrogenetic 岩石成因的
petrogenetic grid 成岩格子

petrogenetic grid [ˌpetrəudʒiˈnetik grid]

Definition: 成岩格子 chéng yán gé zi

petrogenetic: of the formation of rocks, especially igneous and metamorphic rocks. [地质]岩

石成因的 yán shí chéng yīn de

grid: a framework of crisscrossed or parallel bars; a grating or mesh. 格子 gé zi

Examples:

① Petrogenetic grid of metamorphic rocks of Zhongtiao Group shows that the regional progressive metamorphic belt are mainly induced by temperature progressing.
中条群变质成岩格子表明：区域递增变质带主要是由温度值升高引起的。

② Petrogenetic grid and T-P calculations indicate that the peak metamorphism took place at temperature of 750-880℃ and pressure of 1.1GPa.
成岩格子和温压计算表明它们的峰期变质作用发生在 750-880℃ 和 1.1GPa 的高温高压条件下。

petrographic characteristics [petrəuˈɡræfik kærəktəˈristiks]

Definition: 岩相学特征 yán xiàng xué tè zhēng

petrographic: the branch of science concerned with the composition and properties of rocks. 岩相学的 yán xiàng xué de; 岩类学的 yán lèi xué de

characteristic: feature or quality belonging typically to a person, place, or thing and serving to identify them. 特性 tè xìng; 特征 tè zhēng; 特色 tè sè

Origin: 1645-1655; < New Latin petrographia. petro-, -graphy.

Example:

On the basis of study on the petrographic characteristics of coal 2-2 in Shenmu Coalfield, three types of peat swamp are recognized.
在研究神木煤田 2-2 煤层煤岩特征的基础上，划分了三种泥炭沼泽类型。

petrographic comparison map [petrəuˈɡræfik kəmˈpærisən mæp]

Definition: 岩比图 yán bǐ tú

petrographic: of the branch of science concerned with the composition and properties of rocks. 岩相学的 yán xiàng xué de

comparison: a consideration or estimate of the similarities or dissimilarities between twothings or people. 比较 bǐ jiào; 对照 duì zhào

map: a diagram or collection of data showing the spatial arrangement or distribution of something over an area. 分布图 fēn bù tú

phase diagram [feiz ˈdaiəgræm]

Definition: 相图 xiàng tú

phase: a distinct period or stage in a process of change or forming part of something's development. *n.* 相 xiàng; 阶段 jiē duàn; 位相 wèi xiàng

diagram: a simplified drawing showing the appearance, structure, or workings of something; a schematic representation. 图解 tú jiě; 简图 jiǎn tú; 示意图 shì yì tú

Example:

The phase diagram analysis of sodium nitrite and sodium nitrate production was made and the control condition of product quality in the production was also put forward.
对亚硝酸钠和硝酸钠生产相图进行了分析，提出了提高亚硝酸钠和硝酸钠产品质量的生产控制条件。

petrographic province [ˌpetrəuˈɡræfik ˈprɒvins]

Definition: a broad area in which similar igneous rocks are formed during the same period of igneous activity. Also known as comagmatic region; igneous province; magma province. 岩石区 yán shí qū

Extended Terms:

circum pacific calc alkaline petrographic province 环太平洋钙碱性岩石区
petrographic configuration 岩相结构
petrographic factor 岩相因子

petrographic suite [ˌpetrəuˈɡræfik swiːt]

Definition: 岩套 yán tào

petrographic: of the branch of science concerned with the composition and properties of rocks. 岩相学的 yán xiàng xué de; 岩类学的 yán lèi xué de

suite: a set of things belonging together, in particular. 一套 yī tào

Example:

The two groups of lavas are regarded as a single petrographic suite in which several continuous and discontinuous mineral reaction series are recognized.
这两组火山岩被视为单一的岩套，其中可见数个持续的和非持续的矿物反应连锁。

Extended Terms:

alkaline suite 碱性岩套
petrographic constituents 煤岩组分

petrography [peˈtrɒɡrəfi]

Definition: the branch of petrology dealing with the description and classification of rocks, especially. by microscopic examination. 岩相学 yán xiàng xué

Origin: 1645-1655；< NL petrographia. See petro-, -graphy.

Extended Terms:

coal petrography 煤相学；煤岩学；煤岩石学，煤岩相学；煤岩学，煤相学
microscopic petrography 显微岩相学
petrography of foils 化石岩石学
petrography character 岩性特征
sedimentary petrography 沉积岩相学

petroleum [pəˈtrəuliəm]

Definition: an oily, thick, flammable, usually dark-colored liquid that is a form of bitumen or a mixture of various hydrocarbons, occurring naturally in various parts of the world and commonly obtained by drilling: used in a natural or refined state as fuel, or separated by distillation into gasoline, naphtha, benzene, kerosene, paraffin, etc. 石油 shí yóu

Origin: late Middle English: from medieval Latin, from Latin petra "rock" (from Greek) + Latin oleum "oil".

Examples:

① Geology, in general, petroleum geology, in particular, still rely on value judgements based on experience an assessment of validity among the data presented.
普遍来说，地质学和石油地质学特别依赖于基于经验的数值判断和对现有资料的有效性评估。
② Using microorganism to degrade petroleum hydrocarbon is an important technology of the bioremediation technology disposing the soil contaminated by petroleum.
石油污染土壤的微生物修复技术是利用微生物来降解土壤中的石油烃类污染物。

Extended Terms:

petroleum gas 石油气；石油气体；石油挥发气体
petroleum oil 石油；石油油料；矿物油
petroleum pitch 石油沥青；石油类柏油脂
petroleum processing 石油加工；石油炼制；炼油
petroleum tank 石油柜；原油罐；油罐；石油舱

petrology [pəˈtrɔlədʒi]

Definition: the scientific study of rocks, including petrography and petrogenesis. 岩石学 yán shí xué

Origin: 1805-1815; petro- + -logy.

> Examples:

① Organic petrology is an important method for studying coal formed oil.
有机岩石学是研究煤成油的重要方法之一。
② Organic petrology, developed on the base of coal petrology and palynology, has played an important role in the exploration and evaluation of oil and gas.
有机岩石学是在煤岩学和孢粉学基础上发展起来的一门学科,在油气勘探和评价中已成为一种重要的研究手段。

> Extended Terms:

chemical petrology 化学岩石学
coal petrology 煤岩学;煤岩石学,煤岩学
igneous petrology 火成岩石学;火成岩岩石学
sedimentary petrology 沉积岩石学;沉积岩岩石学
statistical petrology 统计岩石学

phaneritic igneous rocks [ˌfænəˈritik ˈigniəs rɔks]

> Definition: 显晶火成岩 xiǎn jīng huǒ chéng yán

phaneritic: of the texture of an igneous rock, being visibly crystalline. Also known as coarse-grained; phanerocrystalline; phenocrystalline. 显晶的 xiǎn jīng de

igneous rock: (of rock) having solidified from lava or magma. 火成岩 huǒ chéng yán

> Example:

The sand and gravel deposits of the northern part of Davis County are composed mainly of metamorphic and phaneritic igneous rock particles.
郡北部的砂卵石沉积物主要由变质的和显晶的火成岩微粒构成。

pH [piːeitʃ]

> Definition: the symbol for the logarithm of the reciprocal of hydrogen ion concentration in gram atoms per liter, used to express the acidity or alkalinity of a solution on a scale of 0 to 14, where less than 7 represents acidity, 7 neutrality, and more than 7 alkalinity. 酸碱度 suān jiǎn dù

> Origin: 1909, from German PH, introduced by S. P. L. Sörensen, from P, for German Potenz "potency, power"+H, symbol for the hydrogen ion that determines acidity or alkalinity.

> Example:

The results showed that the important factors with live cell included: salt solution, pH, temperature, oxygen in the extracellular environment.

结果显示：组织细胞外环境中盐溶液、pH 值、温度、氧气均为影响细胞活性的重要因素。

Extended Terms:

optimum pH 最适酸碱度；最适酸度

pH detector 酸碱探测器

pH electrode 酸度计电极；酸碱电极

pH meter 酸碱计；酸碱度表；氢离子计；酸度计

pH phasemeter 相位仪

phacolith [ˈfækəuliθ]

Definition: a minor, concordant, lens-shaped, and usually granitic intrusion into folded sedimentary strata. 岩鞍 yán ān

Origin: 1835-1845; from Greek phakó(s) lentil plant + -lite.

Example:

The intrusire igneous rocks of coal field in Linyi are mainly neutralbasic rocks of Late Yanshanian later period of which dioritic porphyrites are most developed. The forms of intrusive masses are in sill, dike, phacolith, and so on.
临沂煤田的侵入岩主要为燕山晚期的中基性岩类，以闪长玢岩最发育。产状有岩床、岩脉、岩鞍等。

Extended Term:

granite phacolith 花岗岩岩鞍

phaneritic [fænəˈritik]

Definition: of or relating to an igneous rock in which the crystals are so coarse that individual minerals can be distinguished with the naked eye. Phaneritic rocks are intrusive rocks that cooled slowly enough to allow significant crystal growth. 显晶岩的 xiǎn jīng yán de

Origin: 1860-1865; < Greek phaneró(s) visible.

Example:

Most of the phaneritic gold in protogenous ore, and the inclusion gold and fracture gold in micro-gold ore are closely related to pyrite.
原生矿石中绝大部分明金、显微金中的包体金和裂隙金都与黄铁矿密切相关。

Extended Term:

phaneritic gold 明金

phanerocrystalline texture [ˌfænərəuˈkristəlain ˈtekstʃə]

Definition: 显晶质结构 xiǎn jīng zhì jié gòu

phanerocrystalline: of the texture of an igneous rock, being visibly crystalline. Also known as coarse-grained; phanerocrystalline; phenocrystalline. 显晶质(的) xiǎn jīng zhì de
texture: the arrangement of the particles or constituent parts of any material, as wood, metal, etc., as it affects the appearance or feel of the surface; structure, composition, grain, etc. 结构 jié gòu

Origin: 1860-1865; from Greek phaneró(s) visible + Greek krystállinos.

Extended Term:
phanerocrystalline-adiagnostic texture 微晶质结构

phase [feiz]

Definition: a distinct period or stage in a process of change or forming part of something's development. n. 相 xiàng; 阶段 jiē duàn; 位相 wèi xiàng

Origin: early 19th century (denoting each aspect of the moon): from French phase, based on Greek phasis "appearance", from the base of phainein "to show".

Examples:
① The methods used include chemical oxidation, direct oxidation at high temperatures using high oxygen pressures, and the crystal growth of stoichiometric lead dioxide in the solution phase.
所使用的方法包括化学氧化法，在高温下用高压氧直接氧化以及化学计量准确的二氧化铅在溶液相中的晶体生长。
② A new control method of multiunit inverter is described, which determines the triggering pulse for each unit to obtain the sine wave output through multiple-phase shifting superimposition principle.
研究了多重化逆变器的一种新的控制方法，采用逆推的方法确定各单元的触发脉冲，通过多重移相叠加原理得到正弦波输出。

Extended Terms:

phase diagram 相图
phase difference 相位差
phase equilibria 相平衡
phase lag 相位滞后
phase layering 相层理
phase leading 相位超前

phase rule 相律
phase triangle 相三角形

phase diagram [feiz ˈdaiəgræm]

Definition: 相图 xiàng tú

phase: a distinct period or stage in a process of change or forming part of something's development. *n.* 相 xiàng; 阶段 jiē duàn; 位相 wèi xiàng

diagram: a simplified drawing showing the appearance, structure, or workings of something; a schematic representation. 图解 tú jiě; 简图 jiǎn tú; 示意图 shì yì tú

Example:

The phase diagram analysis of sodium nitrite and sodium nitrate production was made and the control condition of product quality in the production was also put forward.
对亚硝酸钠和硝酸钠生产相图进行了分析，提出了提高亚硝酸钠和硝酸钠产品质量的生产控制条件。

phase equilibria [feiz ˌiːkwiˈlibriə]

Definition: 相平衡 xiàng píng héng

phase: a distinct period or stage in a process of change or forming part of something's development. *n.* 相 xiàng; 阶段 jiē duàn; 位相 wèi xiàng

equilibria: a condition in which all acting influences are canceled by others, resulting in a stable, balanced, or unchanging system. 均衡 jūn héng; 平衡 píng héng

Example:

The method proposed could be used to determine the multicomponent solid-liquid phase equilibria which was difficult to obtain.
本文的方法还可以用来预测较难测定的多元固、液混合物的相平衡关系。

Extended Terms:

phase changes and equilibria 相变和平衡
phase equilibria of materials 材料相平衡
quaternary phase equilibria 四相平衡

phase layering [feiz ˈleiəriŋ]

Definition: 相层 xiàng céng

phase: a distinct period or stage in a process of change or forming part of something's development. *n.* 相 xiàng; 阶段 jiē duàn; 位相 wèi xiàng layering: the presence or formation of layers in sedimentary or igneous rock. [地质](沉积岩或火成岩的)层理 céng lǐ; 分层 fēn céng

Example:
The experimentally observed properties can be interpreted based on the model of phase layering and the model of superexchange magnetic ordering.
实验观察到的属性可基于相层模型和超交换磁序模型来解释。

Extended Term:
phase-change layering 相变层理

phase rule [feiz ru: l]

Definition: 相律 xiàng lǜ

phase: a distinct period or stage in a process of change or forming part of something's development. n. 相 xiàng; 阶段 jiē duàn; 位相 wèi xiàng
rule: the normal or customary state of things. 正常情况 zhèng cháng qíng kuàng; 普遍情况 pǔ biàn qíng kuàng; 惯例 guàn lì; 通例 tōng lì

Example:
The factors which influence phase rule is intensity factor, isn't related to relation of amount of different substance.
影响相律的因素是强度性质，与各物质数量间的关系无关。

Extended Term:
mineralogical phase rule 矿物相律; 矿物学相律

phase triangle [feiz ˈtraiæŋgl]

Definition: 三角相图 sān jiǎo xiàng tú

phase: a distinct period or stage in a process of change or forming part of something's development. n. 相 xiàng; 阶段 jiē duàn; 位相 wèi xiàng
triangle: a plane figure with three straight sides and three angles. 三角形 sān jiǎo xíng

Example:
Furthermore, the symmetry of the triangle phase diagram was broken when the sequence of triblock copolymer was varied from ABC to BAC.
此外，嵌段序列的变化还导致了三元相图对称性的破缺。

phengite [ˈfenʤait]

Definition: is a series name for dioctahedral micas of composition $K(AlMg)_2(OH)_2$

(SiAl)$_4$O$_{10}$, similar to muscovite but with addition of magnesium. It is a non-IMA recognized mineral name representing the series between muscovite and celadonite. *n.* 多硅白云母 duō guī bái yún mǔ

Example:
A fresh eclogite sample is selected for U-Ph isotopic dating of zircon and Ar-Ar dating of phengite, and a garnet-bearing muscovite gneiss sample (country rock of eclogite) for Ar-Ar dating of muscovite.
选择新鲜的保存较好的榴辉岩分别进行 U-Ph 和 Ar-Ar 同位素年代学测定，含石榴子石的白云母片麻岩(此榴辉岩的围岩)被选用于白云母的 Ar-Ar 测定。

Extended Terms:
phengite eclogite 多硅白云母榴辉岩
phengite mudstone 多硅白云母泥岩

phenocryst [ˈfiːnəkrɪst]

Definition: a large or conspicuous crystal in a porphyritic rock, distinct from the groundmass. *n.* [地]斑晶 bān jīng; 显斑晶 xiǎn bān jīng

Origin: late 19th century: from French phénocryste, from Greek phainein "to show" + krustallos "crystal".

Example:
The early stage comprises fine grain adamellite and medium coarse grained porphyritic adamellite, while the late stage, it consists of medium grain adamellite with huge phenocryst.
早期由细粒二长花岗岩、含斑中粗粒二长花岗岩构成；晚期由斑状中粗粒二长花岗岩、巨斑状中粒二长花岗岩构成。

Extended Term:
phenocryst settling 斑晶的

phi scale [fəi skeil]

Definition: the general dimensions of the particles or mineral grains in a rock or sediment based on the premise that the particles are spheres; commonly measured by sieving, by calculating setting velocities, or by determining areas of microscopic images. 斐标度 fěi biāo dù

phlogopite [ˈflɔgəpait]

Definition: a brown micaceous mineral which occurs chiefly in metamorphosed limestone and magnesium-rich igneous rocks. *n.* [矿]金云母 jīn yún mǔ

Origin: middle 19th century: from Greek phlogōpos "fiery" (from the base of phlegein "to burn") +ōps, ōp-"face" + -ite.

Examples:
① The shoshonitic magma is considered to be the partial melting product of phlogopite-bearing lherzolite in the enriched Mesozoic lithospheric mantle.
钾玄质岩浆可能是富集的中生代岩石圈地幔——含金云母的二辉橄榄岩部分熔融的产物。
② The results showed that grain amaranth could efficiently take up K from both soil and micas (biotite and phlogopite).
结果表明，籽粒苋能有效地利用土壤和云母(黑云母和金云母)中的钾。

Extended Terms:
bronze mica phlogopite 金云母
phlogopite phlogopitum 金礞石
sodium phlogopite 钠金云母
synthetic phlogopite 人造金云母

phonolite [ˈfəunəlait]

Definition: a fine-grained volcanic igneous rock composed of alkali feldspars and nepheline. *n.* [地]响岩 xiǎng yán; 响石 xiǎng shí

Origin: early 19th century: from phono-"relating to sound" (because of its resonance when struck) + -ite.

Example:
Discovery of leucite phonolite in the Tangra Yumco area, Tibet and its geological significance.
西藏当惹雍错地区白榴石响岩的发现及地质意义。

Extended Terms:
sanidine phonolite 透长响岩
phonolite basalt 响玄岩
phonolite obsidian 响岩质黑曜岩
phonolite trachyte 响岩质粗面岩

phonolitic [fəunəˈlitik]

Definition: a light-colored volcanic rock composed largely of feldspars. 响岩的 xiǎng yán de

Example:
Controlled by the Middle Proterozoic incipient rifting and extensional faulting, in the Miyun-

Pinggu area of the eastern part of Beijing, potassic alkaline basaltic and phonolitic volcanic rocksof the Dahongyu Formation in the rift area are spatially and temporally associated with rapakivigranites and gabbro-diabase(anorthosite) intrusions.

北京东部密云-平谷一带，受中元古代拉张性断裂控制，裂谷断陷区内的大红峪组钾质碱性玄武质与响岩质火山岩在时间及空间上与侵入相球斑花岗岩-辉绿岩共生。

Extended Terms:

phonolitic tephrite 响岩质碱玄岩；玄岩
phonolitic texture 响岩结构；响岩结构
phonolitic trachyte 近响岩质粗面岩

phonolitic tephrite [fəunə'litik 'tefrait]

Definition: 响岩质碱玄岩 xiǎng yán zhì jiǎn xuán yán

phonolitic: a light-colored volcanic rock composed largely of feldspars. 响岩的 xiǎng yán de
tephrite: a group of basaltic extrusive rocks composed chiefly of calcic plagioclase, augite, and nepheline or leucite, with some sodic sanidine. 碱玄岩 jiǎn xuán yán

Example:

The second tookplace during late Tertiary along the Fangshan-Xiaodanyang fault, producingthe phonolitic tephrite followed by the alkali-olivine basalt, which are characterized by the presence of the inclusions of mantle-peridotite and theblocks of underlying rocks, hence implying quite a deep magmatic source.

中期喷发于晚第三纪，主要沿方山-小丹阳断裂分布，先后为响碧玄岩-碱性橄榄玄武岩，多幔源二辉橄榄岩包体、捕虏晶及下伏岩层碎块，生成很深。

phonolitic texture [fəunə'litik 'tekstʃə]

Definition: 响岩结构 xiǎng yán jié gòu

phonolitic: a light-colored volcanic rock composed largely of feldspars. 响岩状的 xiǎng yán zhuàng de

texture: the arrangement of the particles or constituent parts of any material, as wood, metal, etc., as it affects the appearance or feel of the surface; structure, composition, grain, etc. 结构 jié gòu

phosphate ['fɔsfeit]

Definition: any salt or ester of any phosphoric acid, esp a salt of orthophosphoric acid. 磷酸盐 lín suān yán

phosphatic rock

Origin: 18th century: from French phosphat; see phosphorus, -ate.

Extended Terms:

adenosine phosphate 腺苷磷酸；磷酸腺苷
ammonium phosphate 磷酸铵；面团调节剂；发酵剂
rock phosphate 磷酸盐岩；磷矿石
diammonium Phosphate 磷酸氢二铵；磷酸二铵
zinc phosphate 磷酸锌；磷锌白；磷酸锌（二水）；磷酸锌（四水）

phosphatic rock [fɔsˈfætik rɔk]

Definition: 磷质岩 lín zhì yán, 磷块岩（phosphorite）lín kuài yán

phosphatic: (chiefly of rocks and fertilizer) containing or consisting of phosphates. 磷酸盐的 lín suān yán de; 含磷酸盐的 hán lín suān yán de

rock: the solid mineral material forming part of the surface of the earth and other similar planets, exposed on the surface or underlying the soil. 岩石 yán shí; 岩 yán

Example:

The most important naturally occurring phosphate is phosphate of lime or tri-calcium phosphate found as the chief ingredient of phosphatic rock, bones and phosphatic guanos.
自然产生的磷酸盐中以磷酸三钙最为重要，它是磷灰石、骨骼和磷质海鸟粪的主要成分。

phosphorite [ˈfɔsfərait]

Definition: a sedimentary rock sufficiently rich in phosphate minerals to be used as a source of phosphorus for fertilizers. 磷灰石 lín huī shí; 磷矿 lín kuàng

Origin: 1790-1800; phosphor- + -ite.

Example:

In the end, the human reliability at a phosphorite stope in Guizhou was analysed, which can validate the model rationality.
最后，分析了贵州磷灰石采场的人员可靠性，验证了模型的合理性。

Extended Terms:

apatite phosphorite 磷灰石
chemical phosphorite 化学磷灰岩，化学磷块岩
guano phosphorite 粪磷矿，粪化纤核磷灰石
phosphorite nodule 磷质结核；磷灰岩结核，磷钙岩结核
residual phosphorite 残积纤核磷灰石

phyllite [ˈfilait]

Definition: flaky rock with silky sheen: a fine-grained metamorphic rock with a distinctive shiny surface, containing large quantities of mica and resembling slate or schist. 千枚岩 qiān méi yán

Origin: early 19th century. From Greek phullon "leaf".

Example:

In light of comprehensive utilization of tailings from a phyllite-type gold mine in northeast China, a study was made for recover sericite from the tailings.
针对我国东北某千枚岩型金属矿山选矿尾矿综合治理问题，开展了回收绢云母的实验及应用研究。

Extended Terms:

biotite phyllite 黑云千枚岩
calcareous phyllite 钙质千枚岩
feldspar phyllite 长石千枚岩
pelitic phyllite 泥质千枚岩
phyllite slate 千枚板岩

phyllitic structure [fiˈlitik ˈstrʌktʃə]

Definition: 千枚状构造 qiān méi zhuàng gòu zào

phyllitic: of a green, gray, or red metamorphic rock, similar to slate but often having a wavy surface and a distinctive micaceous luster. 千枚状的 qiān méi zhuàng de
structure: the arrangement of and relations between the parts or elements of something complex. 结构 jié gòu; 构造 gòu zào

Example:

When the main metamorphic stage (D_1) first happened, a phyllitic structure (S_1) paralleled to bedding (S_0) had been formed.
主变质期(D_1)发生在先，伴生与层理(S_0)平行的千枚状构造(S_1)然后形成。

phytoplankton [ˈfaitəuˌplæŋktən]

Definition: the photosynthesizing organisms in plankton, mainly unicellular algae and cyanobacteria. 浮游生物 fú yóu shēng wù

Origin: 1895-1900; phyto- + plankton.

> Example:

Phytoplankton provide the basis for the whole marine food chain.
浮游植物为整个海洋的食物链提供了基础。

> Extended Terms:

autotrophic phytoplankton 自营植物性浮游生物
marine phytoplankton 海洋单细胞海藻
phytoplankton bottle 浮游生物采集瓶
phytoplankton bloom 浮游植物大量繁殖；藻华；水华；浮游植物潮

picrite [ˈpikrait]

> Definition: a coarse-grained ultrabasic igneous rock consisting of olivine and augite with small amounts of plagioclase feldspar. 辉石 huī shí；橄榄岩 gǎn lǎn yán

> Origin: 1805-1815; from Gk pikr (ós) bitter + -ite.

> Example:

The recovering work reveals that the initial rocks of amphibolite wall rocks should be basalt, picrite basalt etc.
围岩斜长角闪岩恢复其原岩大致相当于玄武岩、苦橄玄武岩等。

> Extended Terms:

ankaratrite picrite 橄霞玄武苦橄岩
bronzite picrite 古铜苦橄岩
nepheline picrite 霞石苦橄岩
picrite basalt 苦橄玄武岩；苦榄岩
picrite porphyrite 苦橄玢岩

picrite basalt [ˈpikrait bæsɔːlt]

> Definition: is a variety of high-magnesium olivine basalt that is very rich in the mineral olivine. 苦橄玄武岩 kǔ gǎn xuán wǔ yán

> Example:

Major chemical compositions of the greenschists indicate that they were originated from picrite basalt and basalt.
常量元素研究表明，绿片岩的原岩为苦橄玄武岩和玄武岩。

picrite porphyrite [ˈpikrait ˈpɔːfirait]

> Definition: 苦橄玢岩 kǔ gǎn bīn yán

picrite: a medium-to fine-grained igneous rock composed chiefly of olivine, with smaller amounts of pyroxene, hornblende, and plagioclase felspar. 辉石 huī shí
porphyrite: a rock with a porphyritic structure; as, augite porphyrite. 玢岩 bīn yán

Example:

Characteristics of Picrite—Porphyrite in the DaHongShan Area, Hubei Province
湖北大洪山地区苦橄玢岩岩性特征

pigeonite [ˈpidʒinait]

Definition: a monoclinic variety of pyroxene consisting mainly of a mixture of $(MgFe)SiO_3$ and $CaMg(SiO_3)_2$. 易变辉石 yì biàn huī shí

Origin: 1895-1900; named after Pigeon Point, NE Minnesota.

Extended Term:

pigeonite-augitite 易变辉石岩

pillow basalt [ˈpiləu ˈbæsɔːlt]

Definition: 枕状玄武岩 zhěn zhuàng xuán wǔ yán

Example:

However, the total thickness of the sheeted dikes is about three times of that for the pillow basalts. The sheeted dikes should have contributed to the seafloor magnetic anomalies to some extents.
然而，由于枕状岩墙的厚度大约是喷出岩的三倍，其整体的磁化量对海底磁性异常应该有一定的贡献。

pillow lava [ˈpiləu ˈlɑːvə]

Definition: When basalt lava is erupted into the sea, rounded pillow shapes form. 枕状熔岩 zhěn zhuàng róng yán

Example:

Geological and geochemical characteristics show that they are formed in back-arc environment, matching Carboniferous pillow lava from Jianshui area.
地质、地球化学特征显示其形成于弧后盆地环境，与建水地区石炭纪枕状熔岩相配套。

pillow structure [ˈpiləu ˈstrʌktʃə]

Definition: a primary sedimentary structure that resembles a pillow in size and shape. Also known as mammillary structure. 枕状构造 zhěn zhuàng gòu zào

Example:

Pillow structure consists of a suit of deformed concave upward bodies made of silt and sand whose original layering bent round parallel the basal surface and it truncated at the top surface. 枕状构造是砂层中一组呈"凹"形弯曲的变形沉积体，它的原始层平行于枕状体的底面，顶面则是一个平直的截切面。

Extended Term:

ball-and-pillow structure 球-枕状构造

pilotaxitic texture [ˌpailəutækˈsitik ˈtekstʃə]

Definition: 交织结构 jiāo zhī jié gòu

pilotaxitic: pertaining to the texture of the groundmass of a holocrystalline igneous rock in which lath-shaped microlites (usually of plagioclase) are arranged in a glass-free felty mesh, often aligned along the flow lines. 交织的 jiāo zhī de

texture: the arrangement of the particles or constituent parts of any material, as wood, metal, etc., as it affects the appearance or feel of the surface; structure, composition, grain, etc. 结构 jié gòu

Example:

The lava and subvolcanic rocks are usually of porphyritic texture, and the matrix, of intergranular or pilotaxitic texture.
熔岩主要为玄武安山岩和玄武岩，普遍具有斑状结构，基质为间粒、交织结构。

pisolite [ˈpaisəlait]

Definition: (Geology) a sedimentary rock, especially limestone, made up of small pea-shaped pieces. [地质] 豆粒 dòu lì

Origin: early 18th century. < Greek pisos "pea".

Example:

The author has studied these pisolite and oolite and classified them into sixteen types according to the shape, texture, structure and mineral constituents.
笔者研究了下雷锰矿床的锰质豆鲕粒，按其形态、结构构造和矿物成分可划分为 16 种类型。

Extended Terms:

algal pisolite 藻豆粒
pisolite piso 豆状岩

pisolith ['paisəuliθ]

Definition: a pea-size calcareous concretion, larger than an oolith, aggregates of which constitute a pisolite. 豆石 dòu shí

Origin: 1790-1800; see pisolite, -lith.

pit and mound structure [pit ənd maund 'strʌktʃə]

Definition: 坑岗构造 kēng gǎng gòu zào

pit: a large deep hole from which stones or minerals are dug. 矿坑 kuàng kēng; 矿窑 kuàng yáo; 矿井 kuàng jǐng; 坑道 kēng dào

mound: a raised mass of earth, stones, or other compacted material. 小丘 xiǎo qiū; 小山岗 xiǎo shān gǎng

structure: the arrangement of and relations between the parts or elements of something complex. 结构 jié gòu; 构造 gòu zào

Example:
The pit and mound structure as mineral soil disturbance was limited to less than 1/4 in this area. 坑岗构造作为矿质土体扰动局限在少于四分之一的区域。

pitchstone ['pitʃstəun]

Definition: a dark glassy acid volcanic rock similar in composition to granite, usually intruded as dykes, sills, etc. 松脂岩 sōng zhī yán

Origin: C18: translation of German Pechstein.

Example:
Perlite, obsidian and pitchstone are volcanic glass having the same compositions commonly with rhyolite as well as andesite and basalt.
珍珠岩、黑耀岩和松脂岩等玻璃质熔岩以酸性为主，中基性者也常见。

Extended Terms:
ball pitchstone 球状松脂岩
felsite-pitchstone 霏细松脂岩
pitchstone porphyry 松脂斑岩
trachy-pitchstone 粗面松脂岩

PIXE

Definition: PIXE (=particle induced X-ray emission). 粒子诱发 X 射线发射 zhì zǐ jī fā X shè xiàn fā shè；质子激发 X 射线荧光分析 zhì zǐ jī fā X shè xiàn yíng guāng fēn xī

Example:

PIXE is a non-destructive nuclear technique with high sensitivity, which can quantitatively analyze multi-elements at the same time.
它是一种高灵敏度、非破坏性、多元素定量分析的核技术方法。

Extended Terms:

applications of PIXE 粒子诱发 X 射线发射的应用
PIXE study on provenance determination 中国古瓷断源的粒子诱发 X 射线发射研究

placer ['pleisə]

Definition: a surficial mineral deposit formed by the concentration of small particles of heavy minerals, as gold, rutile, or platinum, in gravel or small sands. 砂矿 shā kuàng

Origin: 1835-1845, Americanism；from AmerSp；Spanish：sandbank from Catalan placel, derivation of plaza open place；see plaza.

Extended Terms:

placer deposit 砂矿床；砂积矿床；砂砾矿床
placer gold 砂金；沙金
placer mining 砂矿开采

plagioclase ['pleidʒiəklei]

Definition: a series of feldspar minerals consisting of a mixture of sodium and calcium aluminium silicates in triclinic crystalline form：includes albite, oligoclase, and labradorite. 斜长石 xié cháng shí

Origin: 1865-1870；plagio- + -clase.

Example:

Plagioclase protrudes toward the center area of ringed mountains, this feature weaken as the size of ringed mountains increase.
斜岩露头朝向环形山系中心区，呈对称性分布。该特征随环形山系规模增大而减弱。

Extended Terms:

plagioclase arkose 斜长石砂岩
plagioclase granite 斜长花岗岩
zoned plagioclase 带状斜长石
plagioclase processing 斜长岩处理技术

plagioclase gneiss ['pleidʒiəkleiz naiz]

Definition: 斜长片麻岩 xié cháng piàn má yán

plagioclase: a form of feldspar consisting of aluminosilicates of sodium and/or calcium, common in igneous rocks and typically white. 斜长岩 xié cháng yán

gneiss: a metamorphic rock with a banded or foliated structure, typically coarse-grained and consisting mainly of feldspar, quartz, and mica. 片麻岩 piàn má yán

Example:

Archean metamorphic rock is mainly composed by amphibolite facies of the plagioclase biotite gneiss, amphibolite plagioclase gneiss, biotite change granulite and plagioclase amphibolite, and other components.
太古宙变质岩石，主要由属于角闪岩相的黑云斜长片麻岩、角闪斜长片麻岩、黑云变粒岩和斜长角闪岩等组成。

planar cross bedding ['pleinə krɔːs 'bediŋ]

Definition: 板状交错层理 bǎn zhuàng jiāo cuò céng lǐ

planar: of, relating to, or in the form of a plane. (数)平面的 píng miàn de

cross: go or extend across or to the other side of (a path, track, stretch of water, or area). 穿过 chuān guò; 交错 jiāo cuò

bedding: a base or bottom layer. 底层 dǐ céng; 基层 jī céng

Example:

Sand-sized grains comprising the planar cross-bedding show an obliquely downslope preferred orientation.
板状交错层理包含的沙屑表现出偏倾斜下坡的导向。

Extended Term:

planar cross-stratification 平面交错层

Planetary Geology ['plænitəri dʒi'ɔlədʒi]

Definition: 行星地质学 xíng xīng dì zhì xué

planetary: of, relating, or belonging to a planet or planets. 行星的 xíng xīng de

geology: the science which deals with the physical structure and substance of the earth, their history, and the processes which act on them. 地质学 dì zhì xué

Example:

The sciences of planetary geology have benefited tremendously from these new findings.

行星地质科学从这些新发现中受益匪浅。

plastic deformation [ˈplæstik diːfɔːˈmeiʃən]

Definition: 塑性变形 sù xìng biàn xíng

plastic: a synthetic material made from a wide range of organic polymers such as polyethylene, PVC, nylon, etc., that can be moulded into shape while soft, and then set into a rigid or slightly elastic form. 塑料 sù liào

deformation: the action or process of changing in shape or distorting, especially through the application of pressure. (尤指受压)变形 biàn xíng

Example:

The computation methods for the plastic deformation (drained) or pore pressure (undrained) caused by pure variation of stress Lode angle are also provided.
同时也给出了纯应力洛德角变化引起的土体塑性变形(排水)与孔压生成(不排水)的计算式。

plastic shard [ˈplæstik ʃɑːd]

Definition: 塑变玻屑 sù biàn bō xiè

plastic: a synthetic material made from a wide range of organic polymers such as polyethylene, PVC, nylon, etc., that can be moulded into shape while soft, and then set into a rigid or slightly elastic form. 塑料 sù liào

shard: a fragment of a brittle substance, as of glass or metal. 碎片 suì piàn

plateau basalt [ˈplætəu ˈbæsɔːlt]

Definition: One or a succession of high-temperature basaltic lava flows from fissure eruptions which accumulate to form a plateau. Also known as flood basalt. 高原玄武岩 gāo yuán xuán wǔ yán

Example:

The analysis of geographic formation of soil nutrient spatial difference of plateau basalt was made in the fifth part.
第五部分是高原玄武岩土壤养分空间差异的地理成因分析。

plate convergence [pleit kənˈvəːdʒəns]

Definition: 板块收敛 bǎn kuài shōu liǎn; 板块汇聚 bǎn kuài huì jù

plate: a large block or tabular section of the lithosphere that reacts to tectonic forces as a unit and moves as such. 板块 bǎn kuài

convergence: the process or state of converging. 汇集 huì jí；相交 xiāng jiāo

Example:

Transpression is a result of plate convergence.
横移压缩是板块聚合作用的一种结果。

plate tectonic boundary [pleit tek'tɔnik 'baundəri]

Definition: 板块构造边界 bǎn kuài gòu zào biān jiè

plate: a large block or tabular section of the lithosphere that reacts to tectonic forces as a unit and moves as such. 板块 bǎn kuài

tectonic: of or relating to the structure of the earth's crust and the large-scale processes which take place within it. [地质]地壳构造上的 dì qiào gòu zào shàng de

boundary: a line which marks the limits of an area; a dividing line. 分界线 fēn jiè xiàn，边界 biān jiè；界限 jiè xiàn

plate tectonics [pleit tek'tɔniks]

Definition: 板块结构学 bǎn kuài jié gòu xué

plate: a large block or tabular section of the lithosphere that reacts to tectonic forces as a unit and moves as such. 板块 bǎn kuài

tectonics: of or relating to the structure of the earth's crust and the large-scale processes which take place within it. [地质]地壳构造上的 dì qiào gòu zào shàng de

Example:

The revolution is based on the work of scientists who study the movement of the continents—a process called plate tectonics.
这是科学家们研究大陆运动所得出的结论——这种过程就是"板块结构学"。

platform facies ['plætfɔːm'feiʃiːz]

Definition: 台地相 tái dì xiàng

platform: a raised level surface on which people or things can stand. 台 tái；平台 píng tái

facies: the character of a rock expressed by its formation, composition, and fossil content. [地质]相 xiàng

Example:

The down-faulted stagnant basin facies and carbonate platform facies have close relationship

with lead and zinc deposits.
断陷滞流盆地相及碳酸盐台地相与铅锌矿产有紧密的关系。

platform tidal flat facies ［ˈplætfɔːm ˈtaidəl flæt ˈfeiʃiːz］

Definition: 台地潮坪相 tái dì cháo píng xiàng

platform：a raised level surface on which people or things can stand. 台 tái；平台 píng tái
tidal：of, relating to, or affected by tides. 潮(汐)的 cháo xī de；有潮的 yǒu cháo de；受潮汐影响的 shòu cháo xī yǐng xiǎng de
flat：(of land) without hills. ［陆地］平坦的 píng tǎn de
facies：the character of a rock expressed by its formation, composition, and fossil content. ［地质］相 xiàng

Example:

Middle cambrian AWaTaGe formation denude close to YingMai 36. Analysis of sedimentary facies indicate the formation located in the restricted platform tidal flat facies.
中寒武统阿瓦塔格组剥蚀出露于英买 36 井区附近，沉积相分析表明主要处于局限台地潮坪相区内。

platinum group elements ［ˈplætinəm gruːp ˈelimənts］

Definition: 铂族元素 bó zú yuán sù

platinum：a precious silvery-white metal, the chemical element of atomic number 78. It was first encountered by the Spanish in South America in the 16th century, and is used in jewellery, electrical contacts, laboratory equipment, and industrial catalysts. (Symbol：Pt)［化学元素］铂 bó，白金(符号：Pt) bái jīn
group：a number of people or things that are located close together or are considered or classed together. 组 zǔ；群 qún；类 lèi
element：a part or aspect of something abstract, especially one that is essential or characteristic. 基本组成部分 jī běn zǔ chéng bù fèn；要素 yào sù

Example:

Three sulphide zones with platinum group elements (PGE) mineralization have been discovered in the upper part of the ultramafic rock formations of the Great Dyke, Zimbabwe.
大岩墙超铁镁质岩层上部主要有三个硫化物带，铂族元素(PGE)主要分布在这些矿带中。

pleochroic halo ［ˌpliːəˈkrəuik ˈheiləu］

Definition: 多色晕 duō sè yùn

pleochroic: (of a crystal) absorbing different wavelengths of light differently depending on the direction of incidence of the rays or their plane of polarization, often resulting in the appearance of different colours according to the direction of view. (晶体)多向色性的 duō xiàng sè xìng de；多色的 duō sè de

halo: a circle of white or coloured light round the sun, moon, or other luminous body caused by refraction through ice crystals in the atmosphere. (日、月或其他发光体的)光晕 guāng yùn；晕圈 yùn quān

Example:
Several pleochroic halos of anomalous radii have been found.
数个异常半径的多色晕被发现。

holocrystalline [ˌhɔləuˈkristəlain]

Definition: pertaining to igneous rocks that are entirely crystallized minerals, without glass. 全晶质 quán jīng zhì

Example:
The holocrystalline porphyritic olivine basalts and pyroclastic rocks are the components of pleistocene to recent volcanic activities.
更新世至现代的火山岩为全晶质斑状的橄榄玄武岩和火山碎屑岩。

Extended Terms:
holocrystalline rock 全晶质岩
holocrystalline interstitial 全晶间片状的
holocrystalline porphyritic 全晶斑状的
holocrystalline texture 全晶质结构

Plinian [ˈpliniən]

Definition: geology (of a volcanic eruption) characterized by repeated explosions. 普林尼式 pǔ lín ní shì

Origin: 20th century: named after Pliny the Younger, who described such eruptions.

Example:
A great volcanic eruption with the characteristics of Plinian eruption as happened in Changbai Mountains 800 years ago.
距今800年前，长白山火山曾发生大规模普林尼式喷发。

Extended Term:
Plinian-type eruption 普林尼式喷发

Plinian eruption [ˈpliniən iˈrʌpʃən]

Definition: 普林尼式火山喷发 pǔ lín ní shì huǒ shā pēn fā

Plinian: relating to or denoting a type of a volcanic eruption in which a narrow stream of gas and ash is violently ejected from a vent to a height of several miles. [地质]普林尼式的 pǔ lín ní shì de

eruption: an act or instance or erupting. 爆发 bào fā; 喷发 pēn fā

Example:

The ejecta samples were collected in pumice deposits formed during two major Plinian eruptions.
喷出物样本采自于在两次大型普林尼式火山喷发中形成的浮石沉积物。

plume [pluːm]

Definition: geology also called: mantle plume a rising column of hot, low viscosity material within the earth's mantle, which is believed to be responsible for linear oceanic island chains and flood basalts. 柱 zhù; 热柱 rè zhù; 羽状物 yǔ zhuàng wù

Origin: 14th century: from Old French, from Latin plūma downy feather.

Example:

The results show that the evident characteristic spectra of abnormal materials were presented while these materials were introduced into the plume.
结果表明当发动机有异常杂质进入羽流时,羽流光谱中有该物质的显著特征光谱。

Extended Terms:

fanning plume 扇形烟羽; 扇形烟缕; 成扇形烟缕
lofting plume 高升空中的烟羽; 高耸的烟羽
plume structure 羽毛状构造
smoke plume 烟羽; 烟缕; 拉烟
thermal plume 热卷流; 热柱; 热烟流; 热烟羽

pluton [ˈplutɑn]

Definition: any mass of igneous rock that has solidified below the surface of the earth. 火成侵入体 huǒ chéng qīn rù tǐ; 深成岩体 shēn chéng yán tǐ

Origin: 1935-1940; from Greek Pluton, back formation from plutonisch plutonic.

Example:

Yongtai stratigraphic sequence is a complex pluton, which is composed of 4 plutonic masses and 48 intrusives.

永泰序列系复式深成岩体，由 4 个深成岩体、48 个侵入体组成。

Extended Terms:

areal pluton 区域深成岩体
cogenetic pluton 同源深成岩体
diapir pluton 底辟深成岩体；冲顶构造深成岩体
migma pluton 混浆深成岩
trichter pluton 漏斗状深成岩

plutonic classification [pluˈtɑnik ˌklæsifiˈkeiʃən]

Definition: 深成岩分类 shēn chéng yán fēn lèi

plutonic: relating to or denoting igneous rock formed by solidification at considerable depth beneath the earth's surface. [地质] 深成的 shēn chéng de；火成的 huǒ chéng de
classification: the action or process of classifying something according to shared qualities or characteristics. 分类 fēn lèi

plutonic metamorphism [pluˈtɑnik ˈmetəˈmɔːfizəm]

Definition: 深成变质作用 shēn chéng biàn zhì zuò yòng

plutonic: relating to or denoting igneous rock formed by solidification at considerable depth beneath the earth's surface. [地质] 深成的 shēn chéng de，火成的 huǒ chéng de
metamorphism: alteration of the composition or structure of a rock by heat, pressure, or other natural agency. [岩石] 变质作用 biàn zhì zuò yòng

Example:

Thermal maturation of Permo-Carboniferous coal seams were mainly controlled by geothermal anomaly of Late Mesozoic, on the basis of this geothermal anomaly, different structural units has different metamorphic rank due to plutonic metamorphism and regional magmatic metamorphism.
沁水盆地石炭系-二叠系煤层热演化程度主要受中生代晚期异常地温场控制，在此基础上，深成变质作用和区域岩浆热变质作用是造成沁水盆地石炭系-二叠系煤层变质程度分带性及差异性的原因。

plutonic rock [plutɑnik rɒk]

Definition: a rock formed at considerable depth by crystallization of magma or by chemical alteration. 深成岩 shēn chéng yán

> Example:

In the investigation region, in the northern part of the West Kunlun Mountains calc-alkaline plutonic rocks are exposed.
在西昆仑山北部的调查区内，都有钙-碱性的深成岩体出露。

pluvial fan facies ['pluːvɪəl fæn 'feɪʃiːz]

> Definition: 洪积扇相 hóng jī shàn xiàng

pluvial: relating to or characterized by rainfall. 雨的 yǔ de；多雨的 duō yǔ de
fan: an alluvial or talus deposit spread out in such a shape at the foot of a slope. 坡脚 pō jiǎo, 扇形沉积 shàn xíng chén jī
facies: the character of a rock expressed by its formation, composition, and fossil content. [地质]相 xiàng

> Example:

Underpart S75 layer was pluvial fan facies sediment, and further step subdivided into five kinds of microfacies styles.
下部 S75 层为洪积扇相沉积，并进一步细分为五种微相类型。

pneumatolytic hydrothermal metamorphism
[ˌnjuːmətəʊˈlɪtɪk haɪdrəʊˈθɜːməl metəˈmɔːfɪzəm]

> Definition: 气液变质作用 qì yè biàn zhì zuò yòng

pneumatolytic: the chemical alteration of rocks and the formation of minerals by the action of hot magmatic gases and vapours. [地质]气化 qì huà，气成作用 qì chéng zuò yòng
hydrothermal: of, relating to, or denoting the action of heated water in the earth's crust. 与地热有关的 yǔ dì rè yǒu guān de
metamorphism: alteration of the composition or structure of a rock by heat, pressure, or other natural agency. [地质][岩石]变质作用 biàn zhì zuò yòng

podsol ['pɒdˌzɒl]

> Definition: a type of soil characteristic of coniferous forest regions having a greyish-white colour in its upper leached layers. 灰壤 huī rǎng；灰化土 huī huà tǔ

> Origin: 20th century: from Russian: ash ground, from pod ground + zola ashes.

> Example:

Where this layer is sufficiently impermeable to maintain the A horizon in a waterlogged

condition, the soil is termed a gley podsol.
淋溶层 B 高度的不透性保持了淋溶层 A 处于浸水状况，这种土壤称为灰壤。

Extended Terms:

gley-podsol 潜育灰壤
ground water podsol soil 潜育灰化土；潜水灰壤
podsol soil 灰白土壤
reddish podsol 红灰壤

poikilitic [ˌpɔikiˈlitik]

Definition: Petrography. (of igneous rocks) having small crystals of one mineral scattered irregularly in larger crystals of another mineral. 嵌晶的 qiàn jīng de

Origin: 1830-1840; from Greek poikíl(os) various + -ite + -ic.

Example:

The occurrence of the disequilibrium mineral assemblage and textures and development of acicular apatite crystals and poikilitic quartz and alkali feldspar in the enclaves show the petrographic characteristics of magma-mixing, suggesting that the Mid-Late Jurassic granite formed by crust-mantle magma-mixing.
明显的不平衡矿物组合和结构、发育针状磷灰石晶体、嵌晶状石英和碱性长石，都显示出岩浆混合的岩相学特征，表明该区中、晚侏罗世花岗岩为壳-幔岩浆混合而成的。

Extended Terms:

poikilitic cementation 嵌晶胶结
poikilitic inclusion 嵌晶状包裹体
poikilitic texture 嵌晶状结构；包含结构
seriate poikilitic texture 不等粒嵌晶结构；不等粒嵌晶组织

poikilitic cementation [ˌpɔikiˈlitik ˌsiːmenˈteiʃən]

Definition: 嵌晶胶结 qiàn jīng jiāo jié

poikilitic：(Geology) relating to or denoting the texture of an igneous rock in which small crystals of one mineral occur within crystals of another. [地质]嵌晶结构的 qiàn jīng jié gòu de；嵌晶状的 qiàn jīng zhuàng de
cementation：(chieflyGeology) the binding together of particles or other things by cement. [主地质]黏结 nián jié；胶结作用 jiāo jié zuò yòng

poikilitic texture [ˌpɔikiˈlitik ˈtekstʃə]

Definition: 嵌晶结构 qiàn jīng jié gòu

poikilitic: (Geology) relating to or denoting the texture of an igneous rock in which small crystals of one mineral occur within crystals of another. [地质]嵌晶结构的 qiàn jīng jié gòu de；嵌晶状的 qiàn jīng zhuàng de

texture: the feel, appearance, or consistency of a surface or a substance. (物质、表面的)结构 jié gòu；构造 gòu zào；纹理 wén lǐ；肌理 jī lǐ

Example:

Ore is characterized by replacement remnant texture, poikilitic texture, desseminated structure, veinlet structure and massive structure.
矿石结构构造为交代残余结构、嵌晶结构和星点状、细脉状、网脉状、团块状构造。

poikiloblastic [ˌpɔiˌkiləˈblæstik]

Definition: (of metamorphic rocks) having small grains of one mineral embedded in metacrysts of another mineral. 变嵌晶状的 biàn qiàn jīng zhuàng de；变嵌晶状 biàn qiàn jīng zhuàng

Origin: 1915-1920；< Gk poikiló (s) various + -blastic.

Example:

Poikilitic in the poikiloblastic texture of metamorphic rock most concentrate in the middle part of host and the grain size of crystal is large.
变质岩中包含变嵌晶结构中的嵌晶多数集中于主晶中部，且晶粒大。

Extended Term:

poikiloblastic texture 变嵌晶结构

poikiloblastic texture [ˈpɔiˌkiləˈblæstik ˈtekstʃə]

Definition: 变晶结构 biàn jīng jié gòu

poikiloblastic: (Geology) relating to or denoting the texture of a metamorphic rock in which small crystals of an original mineral occur within crystals of its metamorphic product. [地质]变嵌晶的 biàn qiàn jīng de

texture: the feel, appearance, or consistency of a surface or a substance. (物质、表面的)结构 jié gòu；构造 gòu zào；纹理 wén lǐ；肌理 jī lǐ

point bar deposit [pɔint bɑː diˈpɔzit]

Definition: the accumulation of fluvial sediment at the slip-off slope on the inside of a meander. 凸岸坝沉积 tū àn bà chén jī

Example:

It points out that the lower part of the member is a subaqueous fan and point bar deposit, while the middle part is a fan-delta deposit.
沙三段下部为水下扇和凸岸坝沉积，中部为扇三角洲沉积。

Poisson's ratio [pwɑːsɔns ˈreiʃiəu]

Definition: In a material under tension or compression, the absolute value of the ratio of transverse strain to the corresponding longitudinal strain. 泊松比 pō sōng bǐ

Example:

Seismic responses of unconsolidated sands indicate that Poisson's ratio is related to the parameters of sand with low effective stress.
非固结砂岩的地震响应研究表明泊松比参数与低效应力的砂岩模型参数有关。

polycrystalline [ˌpɔliˈkristəlain]

Definition: (of a rock or metal) composed of aggregates of individual crystals. 多晶的 duō jīng de

Origin: 1920-1925; poly- + crystalline.

Example:

Lattice parameters are the important crystal structural parameters of polycrystalline materials.
晶胞参数是多晶材料晶体结构的重要参数。

Extended Terms:

polycrystalline diamond 多晶金刚石；多晶质钻石
polycrystalline glass 聚晶玻璃
polycrystalline material 多晶物质；多晶体散射；多晶质材料；多晶形材料
polycrystalline ingot 多晶锭；多晶激光器
polycrystalline irons 多晶形铁粉

polygonal texture [pəˈligənəl ˈtekstʃə]

Definition: 多角结构 duō jiǎo jié gòu

polygonal: (Geometry) a plane figure with at least three straight sides and angles, and typically five or more. (几何)多边形 duō biān xíng; 多角形 duō jiǎo xíng

texture: the feel, appearance, or consistency of a surface or a substance.(物质、表面的)结构 jié gòu; 构造 gòu zào; 纹理 wén lǐ; 肌理 jī lǐ

Example:

From the above account it follows that the problem of the space filling in a polygonal texture reduces to the problem of the filling of pyramids.
由上述得出结论,多角结构的空间填充问题归结为金字塔填充的问题。

polyhalite [ˌpɒliˈhælait]

Definition: a sulfate mineral in evaporite deposits [$K_2Ca_2Mg(SO_4)_4 \cdot 2H_2O$] that often occurs with anhydrite and halite. Its name, from the Greek words meaning "many salts," reflects its composition, hydrated sulfates of potassium, calcium, and magnesium. It makes up 7 percent of the rock in the salt deposits at Stassfurt, German, and is also abundant in the salt deposits of the Saratov region of Russia, where certain beds consist of 85 percent polyhalite. The Texas-New Mexico potash region is another noteworthy locality. 杂卤石 zá lǔ shí

Example:

However, so far the polyhalite has not been exploited and utilized due to its poor dissolubility and deep occurrence.
然而,由于杂卤石的难溶性和埋深大,至今还未被开采和利用。

Extended Terms:

polyhalite deposit 杂卤石矿床
polyhalite ore 杂卤石矿
shallow polyhalite 浅层杂卤石

polymerization [ˌpɒlimərai'zeiʃən]

Definition: the act or process of forming a polymer or copolymer, esp. a chemical reaction in which a polymer is formed. 聚合作用 jù hé zuò yòng

Origin: 1875-1880; polymerize + -ation.

Example:

The results showed that properties of products by this method are all superior to aqueous solution polymerization and general inverse suspension polymerization.
结果表明,该法制备的产品各项性能均优于溶液聚合法和一般反相悬浮聚合法制备的产品。

Extended Terms:

addition polymerization 加聚反应；加成聚合；加聚作用
chain polymerization 链(式)聚合；链锁聚合；连锁聚合；链型聚合
condensation polymerization 缩合聚合；缩聚(反应)；缩合聚合(作用)；缩聚
emulsion polymerization 乳液聚合；乳浊聚合；乳化聚合
polymerization unit 标本聚合仪；聚合设备

polymorph ['pɔliˌmɔːf]

Definition: Crystallography. Any of the crystal forms assumed by a substance that exhibits polymorphism. 同质多相变体 tóng zhì duō xiàng biàn tǐ

Origin: 1905-1910; granule + -o- + -cyte.

Example:

Andalusite is a polymorph with two other minerals, Kyanite and Sillimanite.
红柱石属多晶型物，含蓝晶石和硅线石两种不同矿物。

Extended Terms:

polymorph engine 多形体引擎
polymorph study 多晶型物研究，同质多形物研究

porcellanite [pɔːˈselənəs] = porcelanite

Definition: Porcellanite or porcelanite, is a hard, dense rock somewhat similar in appearance to unglazed porcelain. It is often an impure variety of chert containing clay and calcareous matter. 瓷状岩 cí zhuàng yán，又称白陶土(pot clay) bái táo tǔ

Origin: 1520-1530; from French porcelaine, from Italian porcellana orig. +Middle English, from Latin-ita, from Greek-itēs.

Example:

The mineral resources are dominated by feldspar, porcellanite and quartz sand.
矿产以长石、白陶土和石英砂岩为主。

Extended Term:

porcelain jasper (porcellanite) 瓷碧玉

pore velocity [pɔː vəˈlɑsəti]

Definition: 实际流速 shí jì liú sù

pore: a minute opening in a surface, especially the skin or integument of an organism, through which gases, liquids, or microscopic particles may pass. 管孔 guǎn kǒng;（尤指皮肤表面的）毛孔 máo kǒng

velocity: the speed of something in a given direction. 速度 sù dù; 速率 sù lǜ

Example:

In the following derivation, instead of using pore velocity it is advantageous to use seepage or Darcian velocity.
在下面的推导中，不用孔隙速度，而是渗透或达西速度。

porosity [pɔːˈrɔsiti]

Definition: geology, the ratio of the volume of space to the total volume of a rock. 孔隙度 kǒng xì dù

Origin: 14th century: from Medieval Latin porōsitās, from Late Latin porus pore.

Example:

The relation between the chlorinity and porosity may be approximated by a hyperbola.
含氯量和孔隙度之间的关系图式也许近似于一条双曲线。

Extended Terms:

block porosity 基块孔隙度
capillary porosity 毛细孔隙；毛管孔隙度；毛细管孔隙度；微管孔度
cavernous porosity 穴管状孔隙；孔洞孔隙度
structural porosity 结构多孔性；结构孔隙度
total porosity 总孔隙度；总气孔率；总孔隙率

porosity of carbonate [pɔːˈrɔsiti ɔv ˈkɑːbəneit]

Definition: 碳酸盐岩孔隙 tàn suān yán yán kǒng xì

porosity: (of a rock or other material) having minute interstices through which liquid or air may pass. (岩石或其他材料)有孔 yǒu kǒng; 多孔 duō kǒng

carbonate: salt of the anion CO_3^{2-}, typically formed by reaction of carbon dioxide with bases. 酸盐 suān yán

Example:

Neural Net Technique is used to forecasting the Porosity of Carbonate Reservoir.
应用神经网络技术预测碳酸盐岩储层孔隙度。

porous cementation [ˈpɔːrəs ˌsiːmenˈteiʃən]

Definition: 孔隙胶结 kǒng xì jiāo jié

porous: (of a rock or other material) having minute interstices through which liquid or air may pass.(岩石或其他材料)有孔的 yǒu kǒng de; 多孔的 duō kǒng de

cementation: (chiefly Geology), the binding together of particles or other things by cement. [主地质]黏结 nián jié; 胶接作用 jiāo jiē zuò yòng

porphyrite [ˈpɔːfirait]

Definition: a rock with a porphyritic structure, as, augite porphyrite. 玢岩 bīn yán

Origin: 1350-1400; Middle English porfurie, porfirie, from Medieval Latin porphyreum, alteration of Latin porphyrītēs, from Greek porphyrítēs porphyry, short for porphyrítēs líthos porphyritic (i. e., purplish) stone, equivalent to porphyry (os).

Example:
The diorite porphyrite has been regarded as an ore hunting indicator in.
闪长玢岩可作为该区主要的找矿标志。

Extended Terms:
enstatite porphyrite 顽火玢岩
theralite porphyrite 霞斜玢岩

porphyritic [ˌpɔːfiˈritik]

Definition: of, pertaining to, containing, or resembling porphyry, its texture, or its structure. 斑状的 bān zhuàng de

Origin: 1375-1425.

Example:
Yanshanian acid is a mountain of granite, mostly granite and coarse-grained porphyritic granite.
山地属燕山期酸性花岗岩,多为粗粒花岗岩和斑状花岗岩。

Extended Terms:
holocrystalline porphyritic 全晶斑状的
porphyritic diorite 斑状闪长岩
porphyritic granite 斑状花岗岩;斑状花冈岩
porphyritic microgranite 斑微花岗岩;斑微花冈岩
porphyritic texture 斑状结构;斑状组织

porphyritic texture [ˌpɔːfiˈritik ˈtekstʃə]

Definition: 斑状结构 bān zhuàng jié gòu

porphyritic: (Geology) relating to or denoting a rock texture containing distinct crystals or crystalline particles embedded in a compact groundmass. [地质]斑岩的 bān yán de
texture: the arrangement of the particles or constituent parts of any material, as wood, metal, etc., as it affects the appearance or feel of the surface; structure, composition, grain, etc. 结构 jié gòu

Example:

The petrological textures of lava flows consist of porphyritic texture, glomeroporphyritic texture, spherulitic texture, and the groundmass typically consists of microlites and glass.
熔岩流岩石呈斑状结构、聚斑结构、发育球粒结构，基质为微晶和玻璃。

porphyroblast [ˌpɔːfirəˈblæst]

Definition: a large crystal that is surrounded by a finer-grained matrix in a metamorphic rock. Porphyroblasts form by the recrystallization of existing mineral crystals during metamorphism. They are analogous to phenocrysts in igneous rock. 变斑晶 biàn bān jīng

Example:

Early event of metamorphism and deformation (M1D1) is distinguished by studing porphyroblast inclusion trails of garnets and albites, which caused formation of structural framework approximately with E-W trend.
早期变质变形事件(M1D1)可由钠长石和石榴石变斑晶的包裹体痕迹确定，形成近E-W向的变形构造格局。

porphyroblastic texture [pɔːfaiˈrɒblɑːstik ˈtekstʃə]

Definition: 斑状变晶结构 bān zhuàng biàn jīng jié gòu

porphyroblastic: (Geology) a larger recrystallized grain occurring in a finer groundmass in a metamorphic rock. [地质]斑状变晶 bān zhuàng biàn jīng
texture: the arrangement of the particles or constituent parts of any material, as wood, metal, etc., as it affects the appearance or feel of the surface; structure, composition, grain, etc. 结构 jié gòu

Example:

The harzburgite shows massive structure in hand-specimens, and granular, diablastic and porphyroblastic textures in thin-sections.
手标本呈显晶质块状构造，薄片中呈粒状、筛网状及斑状变晶结构。

porphyroid texture ['pɔːfirɒid 'tekstʃə]

Definition: 似斑状结构 sì bān zhuàng jié gòu

porphyroid: metamorphic rock having porphyritic texture. 残斑岩 cán bān yán
texture: the arrangement of the particles or constituent parts of any material, as wood, metal, etc., as it affects the appearance or feel of the surface; structure, composition, grain, etc. 结构 jié gòu

Example:

The structures often observed in Dushan jade include granular texture, cataclastic texture, porphyroid texture, metasomatic texture, nematoblastic texture, and mylonitization structure occasionally.
独山玉常具有粒状结构、碎裂结构、残斑结构、熔蚀交代结构、纤状变晶质结构，偶见糜棱结构。

porphyry ['pɔːfiri]

Definition: a hard igneous rock containing crystals of feldspar in a fine-grained, typically reddish groundmass. 斑岩 bān yán

Origin: late Middle English: via medieval Latin from Greek porphuritēs, from porphura "purple".

Example:

Wenquan molybdenum deposit is a porphyry type mineral deposit related with Indosinian granite.
温泉钼矿是与印支期花岗岩有关的斑岩型钼矿床。

Extended Terms:

ball porphyry 球状斑岩
glass porphyry 玻璃斑晶
norite porphyry 苏长斑岩
quartz porphyry 石英斑岩
quartzfree porphyry 无石英质斑岩

porphyry copper deposit ['pɔːfiri'kɒpə di'pɒzit]

Definition: 斑岩铜矿床 bān yán tóng kuàng chuáng

porphyry: a hard igneous rock containing crystals of feldspar in a fine-grained, typically reddish groundmass. 斑岩 bān yán copper: a red-brown metal, the chemical element of atomic number 29 (Symbol: Cu). 铜(符号: Cu) tóng

deposit: a layer or body of accumulated matter. 沉积物 chén jī wù; 沉淀 chén diàn

Example:
Dexing Copper Mine in Jiangxi Province is the biggest open-pit non-ferrous metal mine in China, and it is a porphyry copper deposit.
江西德兴铜矿是我国最大的有色露天金属矿，属斑岩型铜矿山。

post tectonic crystallization [pəust tek'tɔnik ˌkristəlai'zeiʃən]

Definition: 后构造期结晶作用 hòu gòu zào qī jié jīng zuò yòng

tectonic: (Geology) of or relating to the structure of the earth's crust and the large-scale processes which take place within it. [地质]地壳构造上的 dì qiào gòu zào shàng de
crystallization: the process of forming or causing to form crystals. 结晶 jié jīng; 晶化 jīng huà

potassic [pəu'tæsik]

Definition: of, pertaining to, or containing potassium. 钾的 jiǎ de; 含钾的 hán jiǎ de

Origin: 19th century: New Latin potassa potash.

Example:
Piedmont potassic alteration zone is the product of potash feldspathization, part of the zone is ore body.
矿区山前钾化带是钾化蚀变的产物，局部地段构成矿体。

Extended Terms:
potassic fertilizer 钾肥；钾肥料；钾肥制造；钾素肥料
potassic glass 钾玻璃
potassic nitrate 硝酸钾
potassic shale 钾质页岩
potassic zone 钾化带

potassic alteration [pəu'tæsik ˌɔːltə'reiʃən]

Definition: 钾化 jiǎ huà

potassic: the chemical element of atomic number 19, a soft silvery-white reactive metal of the alkali-metal group (Symbol: K). (化学元素)钾(符号: K) jiǎ
alteration: the action or process of altering or being altered. 变更 biàn gēng; 改变 gǎi biàn

potassium-argon dating [pəˈtæsiəm ˈɑːɡɒn ˈdeitiŋ]

Definition: dating of archeological, geological, or organic specimens by measuring the amount of argon accumulated in the matrix rock through decay of radioactive potassium. 钾氩测年 jiǎ yà cè nián

potential temperature [pəˈtenʃl ˈtempəritʃə]

Definition: the temperature that would be reached by a compressible fluid if it were adiabatically compressed or expanded to a standard pressure, usually 1 bar. 位势温度 wèi shì wēn dù; 位温 wèi wēn

Example:
Mixed-layer height determined by acoustic echo was higher than that determined by potential temperature profiles.
分析结果表明，声雷达回波所反映出的大气温度高于位温廓线。

preferred crystallographic orientation
[priˈfɜːd ˌkristələˈɡræfik ˌɔːrienˈteiʃən]

Definition: 择优取向 zé yōu qǔ xiàng

preferred: being a favorite; favored, favorite, popular, well-liked. 优先的 yōu xiān de
crystallographic: of the branch of science concerned with the structure and properties of crystals. 晶体的 jīng tǐ de; 晶体学的 jīng tǐ xué de
orientation: the relative physical position or direction of something. 方向 fāng xiàng; 方位 fāng wèi

Example:
The degree of preferred crystallographic orientation parallels to light reflectivity from the surface of the electrodeposit.
锌镀层择优取向程度与镀层表面反射率成平行关系。

Precambrian [priːˈkæmbriən]

Definition: of, denoting, or formed in the earliest geological era, which lasted for about 4 000 000 000 years before the Cambrian period. 前寒武纪 qián hán wǔ jì

Origin: 1860-1865; pre- + Cambrian.

Example:

Precambrian mineral deposits are mainly distributed in paleocontinents especially in North China platform and Yangtze platform, with a few distributed in orogenic belts.
中国前寒武纪矿床主要分布在陆块区，尤以华北陆块、扬子陆块较多，较少分布在造山带。

Extended Terms:

precambrian geology 前寒武纪地质
precambrian research 前寒武纪研究
precambrian shield 前寒武纪地盾
precambrian strata 前寒武纪地层
precambrian time 先寒武纪时期

precipitation [priˌsipi'teiʃən]

Definition: the precipitating of a substance from a solution. 沉淀 chén diàn

Origin: 1425-1475.

Example:

Result from the study indicated that the removal efficiency of phosphate by precipitation was satisfactory.
由本研究之结果可知，以沉淀法去除含磷酸盐的废水的效果非常好。

Extended Terms:

adsorption precipitation 吸附沉淀；吸附沈淀
affinity precipitation 亲和沉淀
carrier precipitation 载体沉淀（作用）
chemical precipitation 化学沉淀；化学沉积
differential precipitation 示差沉淀

preferred orientations [pri'fɜːd ˌɔːrien'teiʃən]

Definition: 优选方位 yōu xuǎn fāng wèi；择优取向 zé yōu qǔ xiàng

preferred: being a favorite: favored, favorite, popular, well-liked. 优先的 yōu xiān de
orientation: the relative physical position or direction of something. 方向 fāng xiàng；方位 fāng wèi

prehnite pumpellyite facies ['preinait pʌm'peliˌait 'feiʃiːz]

Definition: 葡萄石 pú táo shí；绿纤石相 lǜ xiān shí xiàng

prehnite: a light-green to white mineral sorosilicate crystallizing in the orthorhombic system and

generally found in reniform and stalactitic aggregates with crystalline surface. 葡萄石 pú táo shí
pumpellyite: a greenish epidote-like mineral that is probably related to clinozoisite. Also known as lotrite; zonochlorite. 绿纤石 lǜ xiān shí

facies: the character of a rock expressed by its formation, composition, and fossil content. [地质]相 xiàng

Example:

According to this phase equilibrium, the metamorphic pressure and temperature conditions of prehnite pumpellyite facies of metabasite in Saertuohai ophiolite of Xinjiang are 325℃-335℃ and 0.45-0.475GPa.
根据该相平衡计算了新疆萨尔托海蛇绿岩的变质 PT 条件为 T = 325℃ ~ 335℃, P = 0.45 ~ 0.475GPa。

pressure-release melting ['preʃə ri'liːs 'meltiŋ]

Definition: 压力释放熔化 yā lì shì fàng róng huà

pressure-release: the outward-expanding force of pressure which is released within rock masses by unloading, as by erosion of superincumbent rocks or by removal of glacial ice. 压力释放的 yā lì shì fàng de

melting: to be changed from a solid to a liquid state especially by the application of heat. 熔化 róng huà

Example:

These observations cannot be explained by pressure-release melting.
压力释放熔化无法解释这些观测结果。

pressure solution ['preʃə sə'luːʃən]

Definition: Pressure solution is a deformation mechanism that involves the dissolution of minerals at grain to grain contacts into an aqueous pore fluid in areas of relatively high stress and either deposition in regions of relatively low stress. 压溶作用 yā róng zuò yòng

Example:

The late diagenesis includes filling, devitrification, metasomatism, mechanical impaction and pressure solution, cementation and dissolution.
晚期成岩作用类型为充填作用、脱玻化作用、交代作用、机械压实压溶作用、胶结作用和溶解作用。

pre tectonic crystallization [ˌpri tekˈtɒnik ˌkrɪstəlaɪˈzeɪʃən]

Definition: 前构造期结晶作用 qián gòu zào qī jié jīng zuò yòng

tectonic: (Geology) of or relating to the structure of the earth's crust and the large-scaleprocesses which take place within it. (地质)地壳构造上的 dì qiào gòu zào shàng de
crystallization: the process of forming or causing to form crystals. 结晶 jié jīng; 晶化 jīng huà

primary dolomite [ˈpraɪməri ˈdɒləmaɪt]

Definition: 原生白云岩 yuán shēng bái yún yán

primary: the Primary or Palaeozoic era. 古生代 gǔ shēng dài
dolomite: a translucent mineral consisting of a carbonate of calcium and magnesium, usually also containing iron. 白云岩 bái yún yán

Example:
According to the different dolomitization stage and mechanism, all the current main dolomite can be classified into four styles: primary dolomite, autogenetic dolomite, detrital dolomite and detrital dolostone in the limestone, diagenetic dolomite and epigenetic dolomite.
根据成因阶段和机理的不同，当前各种主要的白云岩可分为四个类型：原生白云石、同生白云岩、成岩白云岩和后生白云岩。

primary magma [ˈpraɪməri ˈmæɡmə]

Definition: a magma that originates below the earth's crust. 原生岩浆 yuán shēng yán jiāng

Extended Terms:
anomalous magma 异常岩浆
bench magma 台阶岩浆

primary magmatic structure [ˈpraɪməri ˈmæɡmətik ˈstrʌktʃə]

Definition: 原生岩浆构造 yuán shēng yán jiāng gòu zào

primary: characteristic of or existing in a rock at the time of its formation. 初级的 chū jí de
magmatic: liquid or molten rock deep in the earth, which on cooling solidifies to produce igneous rock. 岩浆的 yán jiāng de
structure: something composed of interrelated parts forming an organism or an organization. 结构 jié gòu

primary mineral ['praiməri 'minərəl]

Definition: 原生矿物 yuán shēng kuàng wù

primary: the Primary or Palaeozoic era. 古生代 gǔ shēng dài
mineral: a solid inorganic substance of natural occurrence. 矿 kuàng; 矿物 kuàng wù

Example:

The mainly orebody is typical arrangement assemblage feature of sulphide of sediment orogenic copper ore because primary mineral has obviously zonation.
主矿体中原生矿石矿物具有明显的分带性,为典型的沉积成因铜的硫化物排列组合特征。

---

primitive magma ['primitiv 'mæɡmə]

Definition: is the "first melt" produced by partial melting within the mantle, and which has not yet undergone any differentiation. A primary magma may therefore evolve into a parental magma by differentiation. For a melt to qualify for the definition of primary magma, it must fulfill the following conditions: (1) have a higher liquidus T compared to its differentiation products, (2) be richer in minerals removed by fractional crystallization compared to its differentiation products, (3) have a composition in equilibrium with the mantle phases from which it was produced by partial melting at high pressure. All primary magmas must have >10% MgO by weight. 原始岩浆 yuán shǐ yán jiāng

Extended Terms:

magma crystallization 岩浆结晶
magma degassing 岩浆去气作用
mother magma 母岩浆

---

primordial mantle [prai'mɔːdjəl 'mæntl]

Definition: 原始地幔 yuán shǐ dì màn

primordial: existing at or from the beginning of time; primeval. 最早的 zuì zǎo de; 原始的 yuán shǐ de
mantle: the region of the earth's interior between the crust and the core, believed to consist of hot, dense silicate rocks (mainly peridotite). [地质]地幔 dì màn

Example:

This type of carbon isotopic distributions is difficult to explain by the current hypotheses, such as phase transformation of carboniferous matters, subduction of sedimentary carbon or mantle degassing, and so it is possible that the primordial mantle is heterogeneous in carbon isotopes.
目前的地幔碳的物相转变、沉积碳的俯冲和地幔去气等假说都很难解释这种碳同位素分布,原始地幔可能是碳同位素不均一的。

prodelta facies [ˌprəuˈdeltə ˈfeiʃiːz]

Definition: 前三角洲相 qián sān jiǎo zhōu xiàng

prodelta: the part of a delta lying beyond the delta front, and sloping gently down to the basin floor of the delta; it is entirely below the water level. 前三角洲 qián sān jiǎo zhōu

facies: the character of a rock expressed by its formation, composition, and fossil content. 相 xiàng

Example:
Braid delta systems mainly consist of braid river deposits in delta plain facies, rapid accumulation deposits, slump deposits and turbidites in delta front and prodelta facies.
辫状三角洲体系的平原相基本上由辫状河道沉积组成,三角洲前缘及前三角洲相带则以快速堆积、滑塌及浊流沉积为主。

prograde metamorphism [ˌprəuˈgreid ˌmetəˈmɔːfizəm]

Definition: 前进变质作用 qián jìn biàn zhì zuò yòng

prograde: (of a metamorphic change) resulting from an increase in temperature or pressure. [地质]前进变质作用的 qián jìn biàn zhì zuò yòng de; 进化变质作用的 Jìn huà biàn zhì zuò yòng de

metamorphism: alteration of the composition or structure of a rock by heat, pressure, or other natural agency. [岩石]变质作用 biàn zhì zuò yòng

Example:
Evolution of mineral assemblages of khondalite series in the process of prograde metamorphism, southeastern Inner Mongolia —Evidence from mineral inclusions in zircons.
内蒙古东南部孔兹岩系进变质过程矿物组合演化——来自锆石中矿物包裹体的证据。

Extended Term:
progressive retrograde metamorphism 逐渐退化变质作用

propylite [ˈprɔpilait]

Definition: a dark-colored form of andesite altered by the action of hot springs and consisting of such minerals as calcite, chlorite, etc. 青磐岩 qīng pán yán

Origin: 20th century: from propylon (see propylaeum) + -ite 1; so named because it is associated with the start of the Tertiary volcanic era.

Extended Term:
quartz-propylite 石英青盘岩

propylitic alteration [prɔpilait ɔːltə'reiʃən]

Definition: 青磐岩化 qīng pán yán huà

Example:
Early propylitic alteration (chlorite-sericite±clinozoisite, sphene) was concentrated above the NW-SE trending, residually warm pluton core, while later intermediate argillic alteration (kaolinite, illite-tourmaline) developed in the vicinity of roof contacts due to trapping and condensation of acid volatiles.
早期青磐岩化(绿泥石-绢云母±斜黝帘石、榍石)集中于北西-南东走向的、当时尚有余温的岩体核部上方,而其后的中性泥质蚀变(高岭石、伊利石-电气石)发育于顶板接触带附近,这是由捕获和凝聚酸性挥发造成的。

Extended Terms:
propylitic facies 青盘岩相
propylitic zone 磐岩化带

protolith [prɔtəlith]

Definition: protolith refers to the precursor lithology of a metamorphic rock. 原岩 yuán yán

Origin: Greek, combining form representing prôtos first, superlative formed from pró.

Example:
Kezlag rock formation whose mother rocks are greywacke is basinal deposit, Karakez rock formation's protolith is island arc volcanics.
克孜拉格岩组原岩为杂砂岩,属盆地沉积;卡拉格兹岩组原岩为岛弧火山岩。

Extended Terms:
protokatagenesis 早后生作用
protometamorphic 初期变质的

pseudobinary system [(p)sjudə'bainəri 'sistəm]

Definition: 假二元体系 jiǎ èr yuán tǐ xì

pseudobinary: not genuine; sham. 伪的 wěi de; 假的 jiǎ de; 冒充的 mào chōng de; 仿真 fǎng zhēn de
system: a set of things working together as parts of a mechanism or an interconnecting network.

系统 xì tǒng；体系 tǐ xì

Example:
Phase equilibria of the Tl 2S Ag 2S pseudobinary system in the thermal variation section at 250℃ were investigated by means of evacuated quartz glass capsules in dry system.
用抽真空石英管在干体系中对 Ag2S-Tl2S 假二元系在 250℃下变温切面相平衡进行了实验研究。

Extended Terms:
pseudocarburizing 假渗炭
pseudobinary alloy 伪二元合金

pseudo concretion ['sju: dəu kən'kri:ʃən]

Definition: 假结核 jiǎ jié hé

pseudo：not genuine；sham. 伪的 wěi de；假的 jiǎ de；冒充的 mào chōng de；仿真 fǎng zhēn de

concretion：a hard solid mass formed by the local accumulation of matter, especially within the body or within a mass of sediment. 固结物 gù jié wù；凝结物 níng jié wù

Example:
Concretions and pseudo-concretions dredged from the sea bottom around Japan.
日本周边海底疏浚的结核与假结核。

Extended Terms:
pseudo conformity 假整合
pseudo conglomerate 假砾岩

pseudo fluidal structure ['sju: dəu f'lu:idl 'strʌktʃə]

Definition: 假流状构造 jiǎ liú zhuàng gòu zào

pseudo：false or counterfeit；fake. 假的 jiǎ de
fluidal：pertaining to a fluid, or to its flowing motion. 流体的 liú tǐ de
structure：something composed of interrelated parts forming an organism or an organization. 结构 jié gòu

pseudomorph ['sju: dəumɔ:f]

Definition: a mineral that has replaced another and taken its shape. 假像 jiǎ xiàng

Origin: mid-19th century：from pseudo- "false" +Greek morphē "form".

Example:

Peridotite has undergone strong serpentinization and was completely replaced by serpentinite, but the pseudomorph of olivine and orthopyroxene still are preserved.
橄榄岩虽然发生了强烈蛇纹岩化并完全被蛇纹岩替代，但蛇纹岩中保留了斜方辉石和橄榄石的假象。

Extended Terms:

incrustation pseudomorph 结壳假象
pseudomorph by alteration 蚀变假象
substitution pseudomorph 交换假象；替换假象

pseudo oolite [ˈsjuːdəu ˈəuəlait]

Definition: 假鲕 jiǎ ér

pseudo: not genuine; sham. 伪的 wěi de；假的 jiǎ de；冒充的 mào chōng de；仿真的 fǎng zhēn de

oolite: (Geology) limestone consisting of a mass of rounded grains (ooliths) made up of concentric layers. [地质]鲕粒岩 ér lì yán

Example:

The pseudo-oolite structures call attention to the occurrence of "pseudo-piezolitic" forms.
假鲕结构使我们注意到"假豆状"形态的出现。

Extended Term:

Millepore oolite Beds 多孔螅鲕状岩层

pseudoplastic rheology [ˌpsjuː dəuˈplæstik riːˈɔlədʒi]

Definition: 假塑性流变 jiǎ sù xìng liú biàn

pseudoplastic: Shear thinning is an effect where viscosity decreases with increasing rate of shear stress. Materials that exhibit shear thinning are called pseudoplastic. 假塑性体 jiǎ sù xìng tǐ
rheology: the study and mathematical representation of the deformation of different fluids in response to surface forces, or stress. 流变学 liú biàn xué

Example:

The paste properties of SSDS were tested, showing that SSDS had a good emulsion ability and stability and the SSDS paste displayed the character of pseudoplastic rheology.
研究了 SSDS 的性质，结果表明 SSDS 乳化能力强，而且乳化稳定性比较好，SSDS 糊符合假塑性流体特征。

Extended Terms:
pseudoplastic behavior 假塑性特性
pseudoplastic consistence 假塑性稠度
pseudoplastic flow 假塑性流
pseudoplastic fluid 拟塑性流体
pseudoplastic hysteresis loop 假塑性滞后回线

pseudo spar [ˈsjuː dəu spɑː]

Definition: 假亮晶 jiǎ liàng jīng; 重结晶方解石(recrystallization calcite) chóng jié jīng fāng jiě shí

pseudo: not genuine; sham. 伪的 wěi de; 假的 jiǎ de; 冒充的 mào chōng de; 仿真的 fǎng zhēn de

spar: a usually metal pole used as part of a crane or derrick. 亮晶 liàng jīng

pseudotachylite [sjuː dəuˈtækilait]

Definition: Pseudotachylite is a fault rock that has the appearance of the basaltic glass, tachylyte. It is dark in color and has a glassy appearance. 假玄武玻璃 jiǎ xuán wǔ bō lí

Origin: Greek, combining form of pseudḗs false, pseûdos falsehood + 19th century: from German Tachylit, from tachy- + Greek lutos.

Example:
The detachment fault dips gently at 10°-15°, in which there occur such tectonites as yellowish white or egg-green microbreccia, pseudotachylite and fault gouges.
拆离断层倾角 10°~15°，其间发育黄白色或蛋青色微角砾岩、假玄武玻璃、断层泥等构造岩。

Extended Terms:
carbonate-pseudotachylite 碳酸盐假熔岩
pseudotachylite brittle structure 脆性构造

ptygmatic migmatite [ˈmigmətait]

Definition: 肠状混合岩 cháng zhuàng hùn hé yán

ptygmatic: generally represent conditions where the folded material is of a much greater viscosity than the surrounding medium. 肠状的 cháng zhuàng de

migmatite: a rock composed of two intermingled but distinguishable components, typically a granitic rock within a metamorphic host rock. [地质]混合岩 hùn hé yán

Example:

The field geological survey across the granite, shows that there are continuously transitional structural-lithofacie zones: Lengjiaxi Group (adjacent rock) without fault dynamic metamorphism→mica quartz tectonic schist bearing garnet→migmatitic quartz-mica schist→banded and ptygmatic migmatite including small rock masses of lenticular granite→gneissic and massive biotite monzonitic granite (host rock).
野外地质剖面观测结果表明，该岩区存在如下连续过渡的构造岩相分带：未卷入断裂变形变质的冷家溪群（围岩）—含石榴子石云母石英构造片岩—混合岩化石英云母片岩—条带状、肠状混合岩夹透镜状花岗质小块岩—片麻状、块状黑云二长花岗岩（主体）。

pulaskite [pəˈlæskait]

Definition: a light-colored, feldspathoid-bearing, granular or trachytoid alkali syenite composed chiefly of orthoclase, soda pyroxene, arfvedsonite, and nepheline. 斑霞正长细晶岩 bān xiá zhèng cháng xì jīng yán

Example:

The pulaskite of the Kangerdlugssuaq alkaline intrusion have been analysed for years.
已对康格尔隆萨克冰川碱性岩体中的斑霞正长细晶岩进行了数年的研究。

Extended Terms:

pulaskite aplite 斑霞正长细晶岩
pulaskite pegmatite 斑霞正长伟晶岩

pumice [ˈpʌmis]

Definition: also called: pumice stone, a light porous acid volcanic rock having the composition of rhyolite, used for scouring and, in powdered form, as an abrasive and for polishing. 浮岩 fú yán

Origin: late Middle English: from Old French pomis, from Latin dialect variant of pumex, pumic-. Compare with pounce.

Example:

The paper points out that 1 GPa/800℃ is the point of thermodynamic phase transformation Okinawa Trough pumice and vicinity andesite, and the point is deeper than 18km.
1 GPa/800℃是浮岩样品和安山岩样品的热动力相变点，推测该点的深度大于18千米。

Extended Terms:

pumice bed 浮石层

pumice concrete 浮石混凝土；浮石水泥
pumice flow 浮石流；浮岩流
pumice slab 浮石
pumice soil 轻石层土壤

pumpellyite [pʌmˈpeliˌait]

Definition: Pumpellyite is a group of closely related sorosilicate minerals. 绿纤石 lǜ xiān shí

Example:
The metamorphic grades of Minleh and Changma areas are prehnite-pumpellyite and prehnite-actinolite facies, respectively.
民乐与昌马地区变质火山岩的变质度分别为葡萄石-绿纤石相和葡萄石-阳起石相。

Extended Term:
prehnite-pumpellyite facies 葡萄石-绿纤石相

P waves [piː weivs]

Definition: a body wave that can pass through all layers of the earth. It is fastest of all seismic waves, traveling at a velocity of 3-4 miles (5-7 kilometers) per second in the crust and 5-6miles (8-9 kilometers) per second in the upper mantle. Also known as compressional wave; longitudinal wave; primary wave. 地震纵波 dì zhèn zòng bō

Origin: 1935-1940.

Examples:
①Seismic surveying is largely concerned with the primary P waves.
地震勘探主要研究一次纵波。
②Secondary waves travel more slowly and arrive after the P waves.
中级的地震波运动缓慢，并且是在初级地震波结束之后到来。

Extended Term:
converted P-SV waves 转换 P-SV 波

pyrite [ˈpairait]

Definition: The mineral pyrite, or iron pyrite, is an iron sulfide with the formula FeS_2. This mineral's metallic luster and pale-to-normal, brass-yellow hue have earned it the nickname fool's gold due to its resemblance to gold. 黄铁矿 huáng tiě kuàng

Origin: late Middle English (denoting a mineral used for kindling fire): via Latin from Greek puritēs "of fire", from pur "fire".

Example:

The minerals are likely to be from the suite pyrite chalcopyrite, sphalerite, galena, fluorite and barite, and to be deposited at temperatures of between 50℃ and 250℃.
这些矿物可能由黄铁矿、黄铜矿、闪锌矿、方铅矿、萤石和重晶石组成，并且可能是在 50℃~250℃之间的温度下沉积出来的。

Extended Terms:

pyrite cinder 黄铁矿烬滓
pyrite furnace 硫铁矿焙烧炉

---

pyroclast tephra [ˈpairəu ˌklæst ˈtefrə]

Definition: solid matter that is ejected into the air by an erupting volcano. 火山碎屑 huǒ shān suì xiè

Origin: 1960-1965; < Greek téphra (singular) ashes.

Example:

Black columns of the particles of ash and cinder that scientists call tephra rose hundreds of feet.
由火山灰和火山岩烬的微粒——科学们称之为火山碎屑形成的黑烟柱冲向数百英尺的高空。

Extended Term:

pyroclastic tephra 火山碎屑物

---

pyroclastic [ˌpairəuˈklæstik]

Definition: describes sedimentary rock that is composed of fragments of volcanic rock produced by the explosion of a volcanic eruption. 火山碎屑物的 huǒ shān suì xiè wù de

Origin: 1870-1875; from Greek klastós broken in pieces (klas-variant stem of klân to break + -tos verbal adjective suffix).

Example:

As bad as the pyroclastic flows are, the ash injected into the atmosphere can have even more far-reaching consequences.
如同火山碎屑流般凶恶，喷发进入大气层的火山灰甚至会造成更深远的影响。

Extended Terms:

pyroclastic cone 集块火山锥

pyroclastic flow 火成碎屑流，火山灰流
pyroclastic ground surge 火成碎屑岩涌
pyroclastic rock 火成碎屑岩
pyroclastic tephra 火山碎屑物

pyroclastic deposit [ˌpaɪrəʊˈklæstɪk dɪˈpɒzɪt]

Definition: 火山碎屑沉积 huǒ shān suì xiè chén jī

pyroclastic：Pyroclastic rocks or pyroclastics are clastic rocks composed solely or primarily of volcanic materials. 火成碎屑物 huǒ chéng suì xiè wù

deposit：something put or left in a place. 沉积物 chén jī wù

Example:

According to the geological features of the mineral deposits and minerogenetic conditions, the author grouped the bentonite deposits in Hunan into four genetic types, i. e., Littoral pyroclastic deposit type, inland lake deposit type, neritic deposit type and weathering residue type.
根据成矿地质条件及矿床特征，笔者将湖南省膨润土矿床归纳为四种成因类型：滨海火山碎屑沉积矿床、内陆湖泊沉积矿床、浅海沉积矿床及风化残余矿床。

Extended Terms:

pyroclastic ground surge 火成碎屑岩涌
pyroclastic rock 火成碎屑岩
pyroclastic tephra 火山碎屑物

pyroclastic flow [ˌpaɪrəʊˈklæstɪk fləʊ]

Definition: A pyroclastic flow (also known scientifically as a pyroclastic density current) is a common and devastating result of certain explosive volcanic eruptions. 火山碎屑流 huǒ shān suì xiè liú

Example:

This paper demonstrates the results of grain-size sieve analyses of the millennium pyroclastic flow deposits from Tianchi volcano.
火山碎屑流堆积因其巨大的危害性而成为火山学研究的重要课题之一。

Extended Terms:

pyroclastic flow plateau 火成碎屑流台地
subaqueous pyroclastic flow 水底火山碎屑流

pyroclastic lava [ˈpairəuˈklæstik ˈlɑːvə]

Definition: 碎屑熔岩 suì xiè róng yán

pyroclastic: relating to, consisting of, or denoting fragments of rock erupted by a volcano. 火山碎屑的 huǒ shān suì xiè de

lava: hot molten or semi-fluid rock erupted from a volcano or fissure, or solid rock resulting from cooling of this. 熔岩 róng yán; 火山岩 huǒ shān yán

Example:
Beijing West Hill was a continental rift basin having pyroclastic lava, pyroclastic rock and coal detrital rock with a total thickness of about 8,000 m in Mesozoic era.
北京西山是中生代的一个大陆裂谷盆地，盆地中堆积了厚达 8000 余米的火山熔岩、火山碎屑岩和含煤碎屑岩。

pyroclastic sedimentary rock [ˈpairəuˈklæstik ˈsediˈmentri rɔk]

Definition: 火山碎屑沉积岩 huǒ shān suì xiè chén jī yán

pyroclastic: relating to, consisting of, or denoting fragments of rock erupted by a volcano. 火山碎屑的 huǒ shān suì xiè de

sedimentary: (of rock) that has formed from sediment deposited by water or air. [地质][岩石]沉积形成的 chén jī xíng chéng de

rock: the solid mineral material forming part of the surface of the earth and other similar planets, exposed on the surface or underlying the soil. 岩石 yán shí; 岩 yán

Example:
The basic pyroclastic rocks in this area belong to the transitional type between normal pyroclastic rock and normal sedimentary rock including two subtypes: sedimentary pyroclastic rock and pyroclastic sedimentary rock.
该区基性火山碎屑岩建造属正常火山碎屑岩与正常沉积岩之间的过渡性岩类，包含沉积火山碎屑岩和火山碎屑沉积岩两个亚类。

pyroclastic texture [ˌpairəuˈklæstik ˈtekstʃə]

Definition: 火山碎屑结构 huǒ shān suì xiè jié gòu

pyroclastic: ash flow not involving high-temperature conditions. 火成碎屑的 huǒ chéng suì xiè de

texture: the arrangement of the particles or constituent parts of any material, as wood, metal, etc., as it affects the appearance or feel of the surface; structure, composition, grain, etc. 结构

jié gòu

> Example:

The residual pyroclastic textures are still preserved to some extent in the seams.
矿层中仍部分保留了残余火山碎屑结构。

pyrolite [ˈpirəlait]

> Definition: Pyrolite is a theoretical rock considered to be the best approximation of the composition of Earth's upper mantle. 地幔岩 dì màn yán

> Extended Term:

host-pyrolite 主火成岩

pyrope [ˈpairəup]

> Definition: a deep red garnet containing magnesium and aluminum. Use：gems. 镁铝榴石 měi lǚ liú shí

> Origin: early 19th century：from German Pyrop, via Latin from Greek purōpos "gold-bronze", literally "fiery-eyed", from pur "fire"+ōps "eye".

> Example:

The variety of garnet includes almandine, pyrope, magnesium-iron garnet, manganese-aluminum garnet, andradite and calcium-chromium garnet.
石榴石的种类包括铁铝榴石、镁铝榴石、镁铁榴石、锰铝榴石、钙铁榴石、钙铬榴石。

> Extended Terms:

Arizona pyrope 亚利桑那镁铝榴石

pyrope garnet 红榴石

pyrrhotite [ˈpirətait]

> Definition: a common yellow-brown lustrous iron sulfide mineral. Source：igneous rocks. Use：source of iron. 磁黄铁矿 cí huáng tiě kuàng

> Origin: mid-19th century：from Greek purrhotēs "redness" + -ite.

> Example:

Pyrrhotite is the most common and abundant iron sulphide mineral in mine wastes worldwide.
磁黄铁矿是矿山尾矿堆中最为常见且分布很广的一种硫化铁矿物。

Extended Terms:

pyrite pyrrhotite 硫化铁，黄铁矿

pyrrhotite peridotite 磁黄铁橄榄岩

pyroxene andesite [paiˈrɒksiːn ˈændizait]

Definition: 辉石安山岩 huī shí ān shān yán

pyroxene：any of a large class of rock-forming silicate minerals, generally containing calcium, magnesium, and iron and typically occurring as prismatic crystals. 辉石 huī shí

andesite：(Geology) a dark, fine-grained, brown or greyish intermediate volcanic rock which is a common constituent of lavas in some areas. [地质]安山岩 ān shān yán

Example:

The Mg-rich volcanic rock (pyroxeneandesite) is also found in the Beitashan Formation of Middle Devonian series of the Saerbulake area, which is a typical boninite formed in the forearc setting.

沙尔布拉克中泥盆统北塔山组中也含有富镁火岩-辉石安山岩，它是典型的玻镁安山岩，其形成环境为前弧。

Extended Terms:

alkali pyroxene 碱性辉石

pyroxene-amphibole-andesite 辉闪安山岩

pyroxene group 辉石组；辉石类

two pyroxene andesite 二辉安山岩

pyroxene hornfels facies [paiˈrɒksiːn ˈhɔrnfls ˈfeiʃiːz]

Definition: 辉石角岩相 huī shí jiǎo yán xiàng

pyroxene：any of a large class of rock-forming silicate minerals, generally containing calcium, magnesium, and iron and typically occurring as prismatic crystals. 辉石 huī shí

hornfels facies：Rock formed at depths in the earth's crust not exceeding 6.2 miles (10 kilometers) at temperatures of 250-800℃; includes albite-epidote hornfels facies, pyroxene-hornfels facies, and hornblende-hornfels facies. 角岩相 jiǎo yán xiàng

Example:

There are three metamorphic facies: epidotealbite hornfels facies, amphibole hornfels facies and pyroxene hornfels facies.

有三个变质相:钠长石角岩、角闪石角岩、辉石角岩。

pyroxenite [paiəˈrɔksinait]

Definition: an igneous rock consisting chiefly of pyroxenes. 辉石岩 huī shí yán

Origin: 1790-1800；from French；see pyro-, xeno-；origin. supposed to be a foreign substance when found in igneous rocks.

Example:
The main rocks in the complex are gabbro, gabbro-diabase, pyroxenite and olivine-pyroxenite etc.
杂岩体主要岩石类型为：辉长岩、辉长辉绿岩、辉石岩、橄榄辉石岩等。

Extended Terms:
magnetite pyroxenite 磁铁辉岩
soda pyroxenite 钠辉岩

pyroxenolite [pairɒkˈsenəlait]

Definition: a heavy, dark igneous rock consisting mostly of pyroxene minerals with smaller amounts of olivine and hornblende. 辉岩 huī yán

Origin: 1860-1865；pyroxene + -ite.

Extended Term:
biotite pyroxenolite 黑云辉石岩

pyrrhotite [ˈpirətait]

Definition: a common yellow-brown lustrous iron sulfide mineral. Source：igneous rocks. Use：source of iron. 磁黄铁矿 cí huáng tiě kuàng

Origin: mid-19th century：from Greek purrhotēs "redness" + -ite.

Example:
Pyrrhotite is the most common and abundant iron sulphide mineral in mine wastes worldwide.
磁黄铁矿是矿山尾矿堆中最为常见且分布很广的一种硫化铁矿物。

Extended Terms:
pyrite pyrrhotite 硫化铁，黄铁矿
pyrrhotite peridotite 磁黄铁橄榄岩

quartile [ˈkwɔːtail]

Definition: one of the values of a variable that divides the distribution of the variable into four groups having equal frequencies. 四分位数 sì fēn wèi shù

Origin: late 19th century: from medieval Latin quartilis, from Latin quartus "fourth".

Example:

The upper quartile is that value above which 25 percent of the observations lie.
上四分位数是25%的观察值位于其上的值。

Extended Terms:

quartile deviation 四分位偏差
quartile division 四分位分法
quartile measurement 四分位量度
sample quartiles 样本四分位数

quartzarenite [ˈkwɔːtzərəˌnait]

Definition: a sandstone composed of greater than 90% detrital quartz, with limited amounts of other framework grains and matri. 石英砂屑岩 shí yīng shā xiè yán

Origin: 1750-1760; from German quarz.

Example:

In south of Sichuan, the lithology of Xujiahe Formation reservoir main is feldspathic-lithic quartz arenite with high compositional maturity and texture maturity. The reservoir physical property is bad because it is low-porosity, low-permeability and special low-porosity, special low-permeability.
蜀南地区须家河组储层岩性主要为结构成熟度和成分成熟度较高的长石岩屑石英砂岩，储层物性差，为低孔、低渗和特低孔、特低渗储层。

Extended Terms:

lithic quartzarenite 岩屑石英砂岩
quartz disc 石英晶片
quartzitic grit 石英岩质粗砂岩
quartz holder 石英试杆
quartzification 石英化

quartz diorite [kwɔːts daiərait]

Definition: 石英闪长岩 shí yīng shǎn cháng yán

quartz: a very hard mineral composed of silica, SiO_2, found worldwide in many different types of rocks, including sandstone and granite. Varieties of quartz include agate, chalcedony, chert, flint, opal, and rock crystal. 石英 shí yīng

diorite: any of various dark, granite-textured, crystalline rocks rich in plagioclase and having little quartz. 闪长岩 shǎn cháng yán

> Extended Terms:

quartz diorite aplite 石英闪长细晶岩
quartz diorite line 石英闪长岩线
quartz diorite porphyrite 石英闪长玢岩
quartz-diorite-porphyry 石英闪长斑岩
quartz gabbro diorite 石英辉长闪长岩

quartzite [ˈkwɔːtsait]

> Definition: Also called feldsparthic quartzite, it is a granular metamorphic rock consisting essentially of quartz in interlocking grains. 石英岩 shí yīng yán

> Origin: 1840-1850; quartz + -ite.

> Example:

There are types of non-metallic phosphorus, limestone, dolomite, quartzite, gypsum, clay, bentonite, and other minerals.
非金属类有磷、石灰岩、白云岩、石英岩、石膏、黏土、膨润土等矿种。

> Extended Terms:

arkose quartzite 长石石英岩
colour quartzite 彩色石英岩
para quartzite 副石英岩

quartz keratophyre [kwɔːts kerətəˈfair]

> Definition: 石英角斑岩 shí yīng jiǎo bān yán

quartz: a hard mineral consisting of silica, found widely in igneous and metamorphic rocks and typically occurring as colourless or white hexagonal prisms. It is often coloured by impurities (as in amethyst, citrine, and cairngorm). 石英 shí yīng

keratophyre: any dike rock or salic lava that is characterized by the presence of albite or albite oligoclase, chlorite, epidote, and calcite. 角斑岩 jiǎo bān yán

> Example:

Spilite-keratophyre sequence is generally divided into spilite, keratophyre and quartz keratophyre.
细碧-角斑岩系一般分为细碧岩、角斑岩和石英角斑岩三类。

Extended Terms:

quartz-keratophyre tuff 石英角斑凝灰岩
riebeckite-quartz keratophyre 钠闪石英角斑岩

quartz monzonite [kwɔːts ˈmɒnzənait]

Definition: an intrusive igneous rock that has an approximately equal proportion of orthoclase and plagioclase feldspars. 石英二长岩 shí yīng èr cháng yán

Example:

The wollastonite ore body occure in the exo-contact zone of quartz monzonite, the host rocks Yanguan Formation siliceous limestone of Lower Carboniferous.
硅灰石矿体产出在石英二长岩外接触带下石炭统岩关阶硅质灰岩特定的地层中。

Extended Terms:

quartz-bearing monzonite 含石英二长岩
quartz labradorite monzonite 石英拉长岩
quartz monzonite aplite 石英二长细晶岩
quartz monzonite porphyry 石英二长斑岩

quartz porphyry [kwɔːts ˈpɔːfiri]

Definition: in petrology, the name given to a group of hemi-crystalline acid rocks containing porphyritic crystals of quartz in a fine-grained matrix, usually of micro-crystalline or felsiticstructure. 石英斑岩 shí yīng bān yán

Example:

The ore bodies, as veins and lenses, occur in the cryptoexplosion breccia on he margin of quartz porphyry.
矿体呈脉状、透镜状产于石英斑岩体边部的隐爆角砾岩中。

Extended Terms:

biotite quartz porphyry 云英斑岩
pyroxene quartz porphyry 辉英斑岩
quartz-free porphyry 无英斑岩
quartz porphyry tuff 石英斑状凝灰岩

quartz sandstone [kwɔːts ˈsændstəun]

Definition: sandstone composed largely of quartz sand grains. 石英砂岩 shí yīng shā yán,

又称正石英岩 zhèng shí yīng yán

Example:

Cuttings sandstone has smaller fractional value than quartz sandstone.
岩屑砂岩较石英砂岩具有较小的分数维值。

Extended Terms:

gray quartz sandstone 灰色石英砂岩
white quartz sandstone 白色石英砂岩
oligomictic quartz sandstone 陆海石英砂岩陆海石英砂岩
quartz sand 石英砂

quartz schist [kwɔːts ʃist]

Definition: (Petrology) a schist whose foliation is due mainly to streaks and lenticles of nongranular quartz. 石英片岩 shí yīng piàn yán

Example:

The mineralizing types are classified into the scheelite-quartz vein in granite and scheelite-quartz vein in amphibolite, mica quartz schist and andalusite-schist.
矿化类型可被分为花岗岩中的白钨矿-石英脉和斜长角闪片岩、云母石英片岩、红柱石片岩中的白钨矿-石英脉。

Extended Term:

quartz-sericite schist 石英绢云片岩

quartz trachyte [kwɔːts 'treikait]

Definition: 石英粗面岩 shí yīng cū miàn yán

quartz: a hard mineral consisting of silica, found widely in igneous and metamorphic rocks and typically occurring as colourless or white hexagonal prisms. It is often coloured by impurities (as in amethyst, citrine, and cairngorm). 石英 shí yīng

trachyte: (Geology) a grey fine-grained volcanic rock consisting largely of alkali feldspar. [地质] 粗面岩 cū miàn yán

Example:

From early to final cycle, the volcanic rock shows the following regular evolution of rock types: andesite tuff, andesite → dacite, Na rhyolith → quartz trachyte → andesite; element abundance (V, Cr, Mn, Co, Ni, Zr, Sr): high→low→lowest (in the quartz trachyte) → high. These regularities form a more perfect evolutionary cycle.
火山喷发旋回由早到晚，火山岩岩性为安山质凝灰岩、安山岩—英安岩、钠流纹岩—石英粗面岩—安山岩，火山岩中铁族元素的丰度由高到低再到最低(在石英粗面岩中)再到高，形成一个较完整的演化周期。

quaternary system [kwəˈtɜːnəri ˈsistəm]

Definition: a major chronostratigraphic subdivision; the latest of the two period forming the Cenozoic era (the other is the Tertiary), itself composed of two epochs, the Pleistocene and the Holocene. 四元体系 sì yuán tǐ xì

Example:
Gravel cemented geological body, called solid formation, has been found in the fluvial sediment of Quaternary system in the exploration and mining of alluvial gold deposit in recent years.
近年来在砂金矿床的勘探和开采中，发现在第四系河流冲积物中存在砂砾石胶结地质体，被称为胶结层。

Extended Terms:
quaternary groundwater system 第四系地下水系统
quaternary mutual system 四元交互体系
quaternary number system 四进制数系
quaternary reciprocal system 四元交互体系
quaternary stratigraphic system 第四纪地层系统

quicksand [ˈkwiksænd]

Definition: a deep, semifluid deposit or bed of sand in which an animal, heavy object. 流沙 liú shā

Origin: middle English quyksond, living sand; quick, quyk, living; see quick+sand, sond.

Example:
The article simply introduces the experience of the way used great excavates and pressing overall support through the quicksand bed at Wenjiazhuang Mine.
本文简要介绍了温家庄煤矿副斜井，采用大开挖与压入整体支架法通过流沙层的经验。

Extended Terms:
cemented quicksand 胶结纱
quicksand coast 流沙海岸
quicksand layer 流沙层
strong quicksand 强流沙层
tackling quicksand 固定流沙

radial blastic texture [ˈreidiːəl ˈblɑːstik ˈtekstʃə]

Definition: 放射状变晶结构 fàng shè zhuàng biàn jīng jié gòu

radial: of or arranged like rays or the radii of a circle; diverging in lines from a common centre. 放射状的 fang shè zhuàng de

blastic: having a specified number or kind of formative elements such as buds, germs, cells, or cell layers. 再结晶的 zài jié jīng de

texture: the arrangement of the particles or constituent parts of any material, as wood, metal, etc., as it affects the appearance or feel of the surface; structure, composition, grain, etc. 结构 jié gòu

Example:
The ore texture is mainly radial blastic texture.
矿石结构主要是放射状变晶结构。

radiation [ˌreidiˈeiʃən]

Definition: the act or process of radiating; specifically, the process in which energy in the form of rays of light, heat, etc. is sent out through space from atoms and molecules as they undergo internal change. 放射物 fang shè wù; 辐射能 fú shè néng; 辐射 fú shè

Origin: 1545-1555; from Latin radiātiōn-(stem of radiātiō).

Example:
Designers should recognize that burner flame length shortens and intensity of radiation increases as pressure increases.
设计时必须认识到当压力增加时火焰的长度缩短而辐射强度增加。

Extended Terms:
radiation accident 放(辐)射事故
radiation effects 辐射效应
radiation hazard 辐射危害性
radiating surface 辐射面

radioactive decay [ˌreidiəuˈæktiv diˈkei]

Definition: 放射性衰变 fàng shè xìng shuāi biàn

radioactive: giving off, or capable of giving off, radiant energy in the form of particles or rays, as alpha, beta, and gamma rays, by the spontaneous disintegration of atomic nuclei: said of certain elements, as plutonium, radium, thorium, and uranium, and their products. 放射性的

fàng shè xìng de
decay: to lose strength, soundness, health, beauty, prosperity, etc. gradually; waste away; deteriorate. 衰退 shuāi tuì

Extended Terms:

radioactive decay series 放射性衰变系
radioactive decay with emission of carbon nuclei 碳核发射放射性衰变
law of radioactive decay 放射性蜕变定律
rate of radioactive decay 放射衰变率

radioactive heat sources [ˌreidiəuˈæktiv hiːt sɔːsis]

Definition: 放射性热源 fàng shè xìng rè yuán

Radioactive: giving off or capable of giving off, radiant energy in the form of particles or rays, as alpha, beta, and gamma rays, by the spontaneous disintegration of atomic nuclei: said of certain elements, as plutonium, radium, thorium, and uranium, and their products. 放射性的 fàng shè xìng de

heat: the quality of being hot; hotness: in physics, heat is considered a form of energy existing as the result of the random motion of molecules and is the form of energy that is transferred between bodies as a result of their temperature difference. 高温 gāo wēn

source: that from which something comes into existence, develops, or derives. 来源 lái yuán

Example:

In the Earth's core, there exists not only huge thermal energy, but huge rotation energy and radioactive heat source.
在重力分异和热对流过程中，地核不仅有巨量热能，而且有巨量的旋转能和放射性热源。

radioactive isotopes [ˌreidiəuˈæktiv ˈaisətəups]

Definition: an isotope having an unstable nucleus that decomposes spontaneously by emission of a nuclear electron or helium nucleus and radiation, thus achieving a stable nuclear composition. 放射性同位素 fàng shè xìng tóng wèi sù

Example:

She had carried test tubes containing radioactive isotopes in her pocket and stored them in her desk drawer, remarking on the pretty blue-green light that the substances gave off in the dark.
她把装有放射性同位素的试管放进口袋，将它们存放在办公桌的抽屉内，这些元素在夜晚总是会发出美丽的蓝绿色光芒。

Extended Terms:

daughter radioactive isotopes 蜕嬗放射性同位素

isotopes that emit radioactive particles 放出放射性粒子的同位素
stable and radioactive isotopes. 稳定和放射性同位素

radiogenic heat production [ˌreidiəuˈdʒenik hiːt prəuˈdʌkʃən]

Definition: 放射性产热 fàng shè xìng chǎn rè

radiogenic: produced by ionizing radiation. 放射产生的 fàng shè chǎn shēng de
heat: the quality of being hot; hotness; in physics, heat is considered a form of energy existing as the result of the random motion of molecules and is the form of energy that is transferred between bodies as a result of their temperature difference. 高温 gāo wēn
production: the act or process of producing. 生产 shēng chǎn

Example:

Usually the radiogenic heat production is determined from the potassium(K), uranium(U) and thorium(Th) content and the rock density.
通常，放射性热产量由 K、U、Th 的含量和岩石密度来确定。

radiogenic isotopes [ˌreidiəuˈdʒenik ˈaisəutəups]

Definition: 放射成因同位素 fàng shè chéng yīn tóng wèi sù

radiogenic: produced by ionizing radiation. 放射产生的 fàng shè chǎn shēng de
isotopes: any of two or more forms of an element having the same or very closely related chemical properties and the same atomic number but different atomic weights (or mass numbers): U-235, U-238, and U-239 are three isotopes of uranium. 同位素 tóng wèi sù

Example:

Studies in geochemistry, stable and radiogenic isotopes of ores, gangues and fluid inclusions suggest that mineralization in Lanping basin is characterized by multiple sources for metals, from either the strata in the basins or deep in the Earth's interior.
对矿石矿物和脉石矿物的地球化学、稳定同位素、放射性同位素和流体包裹体研究显示，兰坪盆地成矿物质具有多来源的特点，既有盆地内浅源地层物质，又有深部物质的参与。

radiolaria [ˌreidiəuˈlɛəriə]

Definition: are amoeboid protozoa that produce intricate mineral skeletons, typically with a central capsule dividing the cell into inner and outer portions, called endoplasm and ectoplasm. 放射虫 fàng shè chóng; 放射虫目 fàng shè chóng mù

Origin: 1875-1880; from New Latin Radiolari(a) name of the group (Latin radiol(us) a small beam, equivalent to radi(us) radius + -olus-ole1 + -aria-aria) + -an.

> Extended Terms:

radiolarian chert 放射虫燧石
radiolaria clay 放射虫黏土
radiolarian fossil 放射虫化石
radiolarian ooze 放射虫软泥

radiometric dating [ˌreɪdiəʊˈmɛtrɪk ˈdeɪtɪŋ]

> Definition: a method for determining the age of an object based on the concentration of a particular radioactive isotope contained within it. 同位素年龄(代)测定 tóng wèi sù nián líng (dài) cè dìng

> Example:

Radiometric dating methods, measuring rods of "deep time," indicate how old the earth really is.
放射性元素定年法是"深层时间"的测量杆，可以告诉我们地球的年龄究竟有多大。

> Extended Term:

radiometric age dating 放射性年龄测定

radium [ˈreɪdiəm]

> Definition: a rare, bright-white, highly radioactive element of the alkaline-earth group. It occurs naturally in very small amounts in ores and minerals containing uranium, and it is naturally luminescent. 镭 léi

> Origin: 1895-1900; from New Latin, equivalent to Latin rad(ius) ray (see radius) + -ium -ium.

> Example:

Five-year survival rates of the X-ray bodycavity tube group, the rotatory radiotherapy group and the radium therapy group were 50.5%(95/188 cases), 57.3%(43/75 cases) and 67.9%(1413/2081 cases) respectively and the curative effect of the radium therapy group was the best.
体腔管组，旋转组和镭疗组的5年存活率分别为50.5%(95/188例)，57.3%(43/75例)和67.9%(1413/2081例)；以上3组的疗效以镭疗组为最佳。

> Extended Terms:

radium age 镭龄
radium appliance 镭装置
radium cell 镭管
radium compounds 镭化合物

radium hill 镭定山

raindrop imprints [ˈreindrɔp imˈprints]

Definition: small, shallow depressions formed in soft sediment or mud by the impact of falling raindrops. Also known as raindrop impressions. 雨痕 yǔ hén

ramification migmatite [ˌræmifiˈkeiʃən ˈmigmətait]

Definition: 分枝状混合岩 fēn zhī zhuàng hùn hé yán

ramification：(formal or technical) the action or state of ramifying or being ramified. (正式的或技术性的)分枝(或分支)形成 fēn zhī xíng chéng；分枝(或分支)排列 fēn zhī pái liè
migmatite：(Geology) a rock composed of two intermingled but distinguishable components, typically a granitic rock within a metamorphic host rock. [地质]混合岩 hùn hé yán

rapakivi granite [ˌrɑːpəˈkiːviː ˈgrænit]

Definition: a hornblende-biotite granite containing large rounded crystals of orthoclase mantled with oligoclase. 更长环斑花岗岩 gèng cháng huán bān huā gāng yán

Origin: Rapakivi was first described by Finnish Jakob Sederholm in 1891. Since then southern Finland's Rapakivi formations have been the type locality of this type of granite.

Example:
This research shows that Yingfeng rapakivi granite is a typical Proterozoic rapakivi granite that has rapakivi texture and features of A-type granite, and belongs to A1 subtype, and the magma assemblage has two-apex characteristics.
初步研究表明，鹰峰环斑花岗岩是具环斑结构和 A 型花岗岩特征的元古宙环斑花岗岩体，且属于 A1 亚型，岩浆组合具双峰式特征.

Extended Terms:
rapakivi aplite 奥长环斑细晶岩
rapakivi granite 奥长环斑花岗岩
rapakivi syenite 奥长环斑正长岩
rapakivi texture 环斑结构，奥环状花岗岩结构

rapakivi texture [ˌrɑːpəˈkiːviː ˈtekstʃə]

Definition: 奥环状花岗岩结构 ào huán zhuàng huā gāng yán jié gòu rapakivi: a form of granite found in southeastern Finland. 奥环斑花岗岩 ào huán bān huā gǎng yán

texture: the arrangement of the particles or constituent parts of any material, as wood, metal, etc., as it affects the appearance or feel of the surface; structure, composition, grain, etc. 结构 jié gòu

Example:

Rapakivi texture is the distinctive texture of Shachang rapakivi granite massif.
奥环状花岗岩结构是沙厂斜长环球斑花岗岩的特征结构。

rare earth elements (REE) [reə ɜːθ ˈelimənts]

Definition: 稀土元素 xī tǔ yuán sù

rare: not frequently encountered; scarce; unusual. 稀有的 xī yǒu de

earth: the soft, granular or crumbly part of land; soil. 泥土 ní tǔ

element: any of the four substances (earth, air, fire, and water) formerly believed to constitute all physical matter. 元素 yuán sù

Example:

The researches of Rare Earth Elements (REE) in life field include two parts: biological effects and their applications as a tool.
稀土元素在生物领域内的研究可以分为两个方面: 稀土的生物效应和稀土在其他研究中的应用。

rate of metamorphic reactions [reit ɔv metəˈmɔːfik riˈækʃəns]

Definition: 变质反应率 biàn zhì fǎn yìng lǜ

rate: the amount, degree, etc. of anything in relation to units of something else. 比率 bǐ lǜ

metamorphic: of, characterized by, causing, or formed by metamorphism or metamorphosis. 变质的 biàn zhì de

reaction: a return or opposing action, force, influence, etc. 反应 fǎn yìng

Example:

The calculations yielded estimates of the effective rates of metamorphic reactions and the effective diffusion coefficient.
计算得出有效变质反应率和有效扩散系数的估计值。

Rayleigh-Taylor instability [ˈreili ˈteilə instəˈbiliti]

Definition: is an instability of an interface between two fluids of different densities, which occurs when the lighter fluid is pushing the heavier fluid. 瑞利-泰勒不稳定性 ruì lì tài lè bù wěn dìng xìng

Example:

The Rayleigh-Taylor instability caused by a density gradient is very important in the inertial confinement fusion.
瑞利-泰勒不稳定性是一种由于密度梯度引起的界面不稳定性，在惯性约束聚变中具有重要的意义。

Extended Terms:

Rayleigh-Sommerfeld formula 瑞立-索末菲公式
Rayleigh-wing scattering 瑞立翼散射

Rayleigh fractionation ['reili ˌfrækʃən'eʃən]

Definition: 瑞利分馏 ruì lì fēn liú

Rayleigh: Rayleigh, 3d Baron (John William Strutt) 1842-1919; physicist. 瑞利 ruì lì fractionation: term used in the energy industry to describe the separation of saturated hydrocarbons from natural gas into distinct parts or fractions (such as propane, butane, ethane, or other elements). 分馏法 fēn liú fǎ

Example:

These patterns cannot be modeled using Rayleigh fractionation accompanied by mineral breakdown reactions.
这些模型无法用伴随矿物分解反应的瑞利分馏构建。

Rayleigh number ['reili 'nʌmbə]

Definition: is a dimensionless number associated with buoyancy driven flow (also known as free convection or natural convection). 瑞利系数 ruì lì xì shù

Example:

Rayleigh Bénard (RB) convective system is simulated numerically using Direct Simulation Monte Carlo (DSMC) method which origins from micro-molecular dynamics. On the basis of DSMC simulation results, the critical Rayleigh number is obtained, which is consistent with the linear stability theory.
应用发端于微观分子动力学的蒙特卡罗直接模拟（DSMC）方法实现了 RB（Rayleigh Bénard）对流的数值模拟，并在此基础上得到了与线性稳定性理论分析结果相一致的 RB 系统第一次分叉的临界瑞利数。

Extended Terms:

critic rayleigh number 临界瑞利数
critical rayleigh number 临界瑞利数
magnetic rayleigh number 磁瑞利数
mean rayleigh number 平均瑞利数
modified rayleigh number 修正瑞利数

Rb (rubidium) [ruːˈbidiəm]

Definition: a soft, silvery-white, metallic chemical element, one of the alkali metals, that ignites spontaneously in air and reacts violently in water: used in photocells and in filaments of vacuum tubes: symbol, Rb; at. no., 37. 元素铷(rubidium)的符号 yuán sù rú de fú hào

Origin: so named (1861) by Robet Wilhelm Bunsen & Gustay Robert Kirchhoff, from Latin rubidus, red (from the red lines in its spectrum) + -ium.

Example:

The measurement of the decay of radioactive isotopes, especially uranium, strontium, rubidium, argon and carbon, has allowed geologists to more precisely determine the age of rock formations.
对放射性同位素衰变的测量，尤其是铀、锶、铷、氩和碳，可以帮助地质学家更精确地确定岩层的年龄。

Extended Terms:

rubidium atoms 铷原子
rubidium chloride 氯化铷
rubidium clock 铷钟
rubidium magnetometer 铷磁力仪
rubidium nitrate 硝酸铷

reaction curve [riˈækʃən kɜːv]

Definition: a curve showing the best price or quantity for one competitor, for each possible price or quantity of another. An equilibrium is where the reaction curves of two competitors intersect. 反应曲线 fǎn yìng qū xiàn

Example:

The concentration-time curve of glucan was hyperbola and its time-reaction curve had low reaction rate.
葡聚糖的浓度时间曲线为双曲线，时间反应曲线显示反应速率较低。

Extended Terms:

nuclear reaction excitation curve 核反应激发曲线
process reaction curve 制程反应曲线

reaction infiltration instability [riˈækʃən ˌinfilˈtreiʃən instəˈbiliti]

Definition: 反应渗透不稳定 fǎn yìng shèn tòu bù wěn dìng

reaction: a return or opposing action, force, influence, etc. 反应 fǎn yìng
infiltration: the act or process of infiltrating. 渗透 shèn tòu
instability: lack of stability; unstableness. 不稳定 bù wěn dìng

Example:

Reaction infiltration instabilities in experiments on partially molten mantle rocks.
部分熔融表层眼的反应渗透不稳定性实验

reaction mechanism [riˈækʃən ˈmekənizəm]

Definition: is the step by step sequence of elementary reactions by which overall chemical change occurs. 反应机制 fǎn yìng jī zhì

Example:

This paper introduced a method prepared nanometersized pareicles-microemulsion method, introduced the constitute and reaction mechanism of microemulsion.
本文介绍了一种制备纳米材料的方法——微乳液法,介绍了微乳液的组成、反应机理。

Extended Terms:

enzyme reaction mechanism 酶反应机理
organic reaction mechanism 有机反应机理
reaction mechanism for isotope exchange 同位素交换机制
rearrangement reaction mechanism 重排反应机理
reaction servo-mechanism 反作用伺服机构

reaction point [riˈækʃən pɔint]

Definition: 反应点 fǎn yìng diǎn

reaction: a return or opposing action, force, influence, etc. 反应 fǎn yìng
point: a stage, condition, level, or degree reached or indicated. 点 diǎn

Example:

The porous conductive polymer nanowire electrodes can be modified to provide many of the

electrochemical reaction load point and catalyst for the reaction of places.
多孔的导电聚合物纳米线修饰电极能够为众多的电化学反应提供负载催化剂的位点和反应场所。

Extended Terms:

individual reaction point 独立反应点
near point reaction 近点反应
point reaction force 接触点反力
reaction of single support point 单支点反力

reaction progress variable [riˈækʃən ˈprəugres veəriəbl]

Definition: 反应过程变量 fǎn yìng guò chéng biàn liàng

reaction: a return or opposing action, force, influence, etc. 反应 fǎn yìng
progress: a moving forward or onward. 过程 guò chéng
variable: anything changeable; especially, a quality or quantity that varies or may vary. 变量 biàn liàng

Example:

Many combustion models that are based on the flamelet paradigm employ a reaction progress variable.
很多燃烧模型是基于小火焰单元使用反应过程变量。

reaction rim texture [riˈækʃən rim ˈtekstʃə]

Definition: 反应边结构 fǎn yìng biān jié gòu

reaction: a response to a stimulus. 反应 fǎn yìng
rim: an edge, border, or margin, esp. of something circular; often, a raised or projecting edge or border. 边 biān
texture: the arrangement of the particles or constituent parts of any material, as wood, metal, etc., as it affects the appearance or feel of the surface; structure, composition, grain, etc. 结构 jié gòu

reaction [riˈækʃən]

Definition: a return or opposing action, force, influence, a movement back to a former or less advanced condition, stage, etc.; especially, such a movement or tendency in economics or politics; extreme conservatism. 反应 fǎn yìng

Origin: 1635-1645; re- + action, modeled on react.

> Example:

The most common reactions include skin rashes, itching, breathing problems and swelling in areas such as the face.
最常见的反应包括皮肤出疹、瘙痒、呼吸问题以及诸如脸部等部位的肿胀。

> Extended Terms:

chain reactions 链反应
dark reactions 暗反应
elementary reactions 基元反应
photochemical reactions 光化反应
transfer reactions 转变反应

reactivation surfaces structure [ri(ː)ˌæktiˈveiʃən ˈsɔːfisis ˈstrʌktʃə]

> Definition: 再作用面构造 zài zuò yòng miàn gòu zào

reactivation: restoring (something) to a state of activity; bringing (something) back into action. 恢复活力 huī fù huó lì; 重新起作用 chóng xīn qǐ zuò yòng
surface: the outside part or uppermost layer of something. 表面 biǎo miàn
structure: the arrangement of and relations between the parts or elements of something complex. 结构 jié gòu; 构造 gòu zào

reaction path [riˈækʃən pɑːθ]

> Definition: 反应路径 fǎn yìng lù jìn

reaction: a return or opposing action, force, influence, etc. 反应 fǎn yìng
path: a particular course of action or way of achieving something. 途径 tú jìng; 路径 lù jìng

> Example:

Three examples are cited for illustration of theoretical methods used by the present authors for calculating the intermediate, transition state and reaction path of organic reaction.
本文以三个例子,分别介绍了用量子化学方法求得有机反应的中间体、过渡态和反应路径的方法。

reaction rind [riˈækʃən raind]

> Definition: 果皮反应 guǒ pí fǎn yìng

reaction: a return or opposing action, force, influence, etc. 反应 fǎn yìng
rind: a thick, hard or tough natural outer covering, as of a watermelon, grapefruit, or orange. 外皮 wài pí

> **Example:**

Reaction rinds between differing lithologies are commonly considered as the result of fluid-assisted metasomatic alteration.
不同岩性间的果皮反应通常被视作流体辅助的硅化蚀变结果。

reconstructive phase transition [ˌriːkənˈstrʌktɪv feɪz trænˈsɪʃən]

> **Definition:** 重建性相变 chóng jiàn xìng xiàng biàn

reconstructive: serving to rebuild, restore, or correct the appearance and function of defective, damaged, or misshaped body structures or parts. 重建的 chóng jiàn de

phase: any of the stages or forms in any series or cycle of changes, as in development. 相 xiàng

transition: a passing from one condition, form, stage, activity, place, etc. to another. 转变 zhuǎn biàn

> **Example:**

Three possible mechanisms for the reconstructive phase transition from the B1 to the B2 structure types are examined.
仔细调查从 B1 到 B2 结构的重建性相变的三个可能机制。

recrystallization [riːˌkrɪstəlaɪˈzeɪʃən]

> **Definition:** In geology, solid-state recrystallization is a metamorphic process that occursunder situations of intense temperature and pressure where grains, atoms or molecules of a rock or mineral are packed closer together, creating a new crystal structure. 变质重结晶作用 biàn zhì chóng jié jīng zuò yòng

> **Origin:** 1790-1800; re- + crystallize.

> **Example:**

The longer the heat treatment time, the easier the recrystallization is.
热处理时间越长, 则再结晶越容易发生。

> **Extended Terms:**

allotropic recrystallization 同素变态再结晶
paratectonic recrystallization 同造山再结晶作用
pretectonic recrystallization 构造变形前再结晶
primary recrystallization 一次再结晶；初级再结晶
secondary recrystallization 二次再结晶；次生重结晶；次级再结晶

recycled sediment [ˌriːˈsaikld ˈsedimənt]

Definition: 再生沉积物 zài shēng chén jī wù

recycled: to use again and again, as a single supply of water in cooling, washing, diluting, etc. 可循环再造的 kě xún huán zài zào de

sediment: matter deposited by water or wind. 沉淀物 chén diàn wù

Example:

This corresponds to the Paleoproterozoic Qinling Group and its recycled sediments in the North Qinling belt.
这相当于北秦岭构造单元中古元古代的秦岭群及秦岭群组分的再生沉积物部分。

recycling of sediment [ˌriːˈsaikliŋ ɔv ˈsedimənt]

Definition: 沉积物再循环 chén jī wù zài xún huán

recycling: use again and again, as a single supply of water in cooling, washing, diluting, etc. 再循环利用 zài xún huán lì yòng

sediment: particulate matter that is carried by water or wind and deposited on the surface of the land or the seabed, and may in time become consolidated into rock. 沉积物 chén jī wù

Example:

Recycling of Harbor Sediment as Lightweight Aggregate
轻砂石的海港沉积物再循环

Red Sea Rift [red siː rift]

Definition: is a spreading center between two tectonic plates, the African Plate and the Arabian Plate. It extends down the length of the Red Sea, stretching from the southern end of the Dead Sea Transform to a triple junction with the Aden Ridge and the East African Rift (the Afar Triple Junction) in the Afar Depression of eastern Africa. 红海裂谷 hóng hǎi liè gǔ

Example:

Great Rift Valley is one of the world's longest Rift Valley, South East Branch from Shire River, head north through the Red Sea in the northern Dead Sea, is about 6400 kilometers.
东非大裂谷是世界上最长的裂谷，东南支起希雷河河口，向北越过红海至死海北部，全长约 6400 公里。

Red Sea [red siː]

Definition: sea between NE Africa & W Arabia, connected with the Mediterranean Sea by the Suez Canal & with the Indian Ocean by the Gulf of Aden; c. 1,400 mi (2,253 km) long; c. 174,900 sq mi (452,989 sq km). 红海 hóng hǎi

Example:

Indo-Pacific region including Red Sea, Persian Gulf and East Africa, India, Sri Lanka, China and Japan.
印度太平洋地区，包括红海、波斯湾和东非、印度、斯里兰卡、中国及日本。

Redlich-Kwong equation [ˈredlitʃ-kwɔŋ iˈkweiʒen]

Definition: is an equation that is derived from the van der Waals equation. It is generally more accurate than the van der Waals equation and the ideal gas equation, but not used asfrequently because the increased difficulty in its derivatives and overall use. It was formulated by Otto Redlich and J. N. S. Kwong in 1949. 瑞德利奇广方程 ruì dé lì qí guǎng fāng chéng

redox states [ˈriːdɒks steits]

Definition: 氧化还原状态 yǎng huà huán yuán zhuàng tài

redox: oxidation-reduction. 氧化还原反应 yǎng huà huán yuán fǎn yìng
states: a condition or mode of being, as with regard to circumstance. 状态 zhuàng tài

Example:

Chromium is a redox active metal that persists as either Cr(Ⅲ) or Cr(Ⅵ) in the environment. These two oxidation states have opposing toxicities and mobilities.
铬酸盐在环境中属活泼、容易发生氧化还原反应的金属，其普遍存在三价或六价的型态，这两种氧化态彼此之间能够互相转换。

reduced heat flow [riˈdjuː st hiːt fləu]

Definition: 简化热流 jiǎn huà rè liú

reduced: made smaller or less, resulting from reduction. 简化的 jiǎn huà de
heat: the quality of being hot; hotness; in physics, heat is considered a form of energy existing as the result of the random motion of molecules and is the form of energy that is transferred between bodies as a result of their temperature difference. 高温 gāo wēn
flow: to change in shape under pressure without breaking or splitting, as ice in a glacier or rocks deep in the earth. 流量 liú liàng

Example:

Results of measurements in the United States are presented in the form of a map of reduced heat flow.

美国的测量结果以简化热流地图的方式呈现。

REE [riː] (rare earth element) [rɛə əːθ ˈelimənt]

Definition: any of the series of metallic chemical elements, with consecutive atomic numbers of 57 (lanthanum) through 71 (lutetium) inclusive. 稀土元素 xī tǔ yuán sù

Example:

The chondrite-normalized distribution of the rare earth element (REE) shows light REE enrichment for the majority of samples.
磷灰石中稀土元素含量对于球粒状陨石的标绘表明，绝大多数磷灰石样品对于轻稀土元素有富集现象。

reef limestone [riːf ˈlaimstəun]

Definition: Limestone composed of the remains of sedentary organisms such as sponges, and of sediment-binding organic constituents such as calcareous algae. Also known as coral rock.
礁灰岩 jiāo huī yán, 又称生物骨架灰岩 (organic framework limestone) shēng wù gǔ jià huī yán

Example:

This paper introduces the geological features and the unhomogeneous weathering phenomenon of coral reef limestone of Sudan Port Area, and the bearing capacity of this layer.
本文介绍了苏丹港区珊瑚礁灰岩的地质特征及不均匀风化现象，以及该层的地基承载力。

reef of platform margin [riːf ɒv ˈplætfɔːm ˈmɑːdʒin]

Definition: 台地边缘生物礁 tái dì biān yuán shēng wù jiāo

reef: a ridge of jagged rock, coral, or sand just above or below the surface of the sea. 生物礁 shēng wù jiāo
platform: a raised level surface on which people or things can stand. 台 tái; 平台 píng tái
margin: the edge or border of something. 边 biān; 边沿 biān yán; 边缘 biān yuán

reflux [ˈriːflʌks]

Definition: a flowing back; ebb; specif., regurgitation of food from the stomach to the esophagus. 逆流 nì liú

Origin: medieval Latin reflūxus: Latin re-, re- + Latin flūxus, flow, from past participle of fluere, to flow; see bhleu-in Indo-European roots.

Example:

The paper presents a new iterative calculation method of minimum reflux ratio of multiple-side streams complex rectification for binary non-ideal system.
对二元非理想系多侧线精馏过程，本文介绍了一种迭代计算最小回流比的新的解析方法。

Extended Terms:

reflux drum 回流罐

reflux duty 回流负荷

reflux ratio 回流比

reflux stabilizer 回流稳定塔

reflux valve 回流阀

refluxed magma chambers ['riːflʌksd 'mægmə 'tʃeimbəs]

Definition: 逆流岩浆房 nì liú yán jiāng fáng

refluxed: a flowing back; ebb; specif., regurgitation of food from the stomach to the esophagus. 逆流 nì liú

chamber: a room in a house, especially a bedroom. 室 shì; 膛 táng

magma: the molten rock material under the earth's crust, from which igneous rock is formed by cooling. 岩浆 yán jiāng

regenerated granite [ri'dʒenəritid'grænit]

Definition: 再生花岗岩 zài shēng huā gāng yán

regenerated: (of a living organism) regrow (new tissue) to replace lost or injured tissue. (活生物体)再生的 zài shēng de; 重新生长出的(取代失去或受损组织) chóng xīn shēng zhǎng chū de

granite: a very hard, granular, crystalline, igneous rock consisting mainly of quartz, mica, andfeldspar and often used as a building stone. 花岗岩 huā gāng yán

Example:

The geochemical types of granitic rocks belonging to the regenerated granite of calc-alkali series and granite of latite series are good for latent ore-forming of gold.
花岗质岩石的地球化学类型属钙碱系列的改造型花岗岩和安粗岩系列的花岗岩，具有金的潜在含矿性。

regional dynamo thermal flow metamorphism
[ˈriːdʒənl ˈdainəməu ˈθəːməl fləu metəˈmɔːfizəm]

Definition: 区域动力热流变质作用 qū yù dòng lì rè liú biàn zhì zuò yòng

regional: of, relating to, or characteristic of a region. (与)地区(有关)的 yǔ dì qū yǒu guān de；区域性的 qū yù xìng de

dynamo: a generator, especially one for producing direct current. 动力 dòng lì

thermal flow: flow of or relating to heat. 热流 rè liú

metamorphism: alteration of the composition or structure of a rock by heat, pressure, or other natural agency. [地质][岩石]变质作用 biàn zhì zuò yòng

Extended Term:

regional low temperature dynamo metamorphism 区域低温动力变质作用

regional metamorphism [ˈriːdʒənl metəˈmɔːfizəm]

Definition: a type of metamorphism in which the mineralogy and texture of rocks are changed over a wide area by deep burial and heating associated with the large-scale forces of plate tectonics. In regional metamorphism, rocks that form closer to the margin of the tectonic plates, where the heat and pressure are greatest, often differ in their minerals and texture from those that form farther away. 区域变质作用 qū yù biàn zhì zuò yòng

Example:

The bedded metamorphite series in Foping area had been formed by regional metamorphism of amphibolite facies.
经过角闪岩相区域变质作用，形成了佛坪地区较为完整的孔兹岩系。

Extended Terms:

dynamo regional metamorphism 动力区域变质作用
regional dynamothermal metamorphism 区域热动力变质作用
regional dynamic metamorphism 区域动力变质作用

regional migmatization [ˈriːdʒənl ˌmigmətaiˈzeiʃən]

Definition: 区域混合岩化作用 qū yù hùn hé yán huà zuò yòng

regional: of, relating to, or characteristic of a region. (与)地区(有关)的 yǔ dì qū yǒu guān de；区域性的 qū yù xìng de

migmatization: formation of migmatite; involves either injection or in-place melting. [岩化]混合作用 hùn hé zuò yòng

> **Example:**

Regional migmatization by anatexis took place at the later stage of granulite facies.
麻粒岩相晚期发生了重熔型区域性混合岩化作用。

regional petrology [ˈriːdʒənl pɪˈtrɒlədʒi]

> **Definition:** 区域岩石学 qū yù yán shí xué

regional: of, relating to, or characteristic of a region. (与)地区(有关)的 yǔ dì qū yǒu guān de；区域性的 qū yù xìng de

petrology: [mass noun] the branch of science concerned with the origin, structure, and composition of rocks. 岩石学 yán shí xué

> **Example:**

Resources in this category cover sedimentology, stratigraphy, hydrogeology, ore geology, structural geology, regional geology, and petrology.
资源种类包括沉积学、地层学、水文地质学、矿石地质学、结构地质学、区域地质学、岩石学。

> **Extended Terms:**

compositional petrology 岩石组成学
reservoir petrology 储层岩石学
structural petrology 构造岩石学

regular solution model [ˈregjulə səˈluːʃen ˈmɒdəl]

> **Definition:** 常规溶液模型 cháng guī róng yè mó xíng

regular: conforming in form, build, or arrangement to a rule, principle, type, standard, etc.; orderly; symmetrical. 定期的 dìng qī de

solution: the act or process of dispersing one or more liquid, gaseous, or solid substances in another, usually a liquid, so as to form a homogeneous mixture; a dissolving. 溶液 róng yè

model: a small copy or imitation of an existing object, as a ship, building, etc., made to scale. 模型 mó xíng

> **Example:**

So the fitting effect of MIVM is better than that of the regular solution model in binary solid alloys.
分子相互作用体积模型在二元固态合金中的活度拟合效果优于正规溶液模型。

relict [ˈrelikt]

> **Definition:** a physical feature, mineral, structure, etc. remaining after other components

have wasted away or been altered. 残留体 cán liú tǐ

Origin: 1525-1535; from Medieval Latin relicta widow, noun use of feminine of Latin relictus.

Example:
A relict fauna and anomalously coarse grains are common in the upper few feet of some shelf sands.
残余化石和异常粗的颗粒通常出现在某些陆棚砂上部的几英尺内。

Extended Terms:
relict ecotype 残遗生态型
relict landforms 残余地形
relict relief 残余地形
relict structure 残留结构

remnant arc ['remnənt ɑːk]

Definition: 残余弧 cán yú hú

remnan: what is left over; remainder; residue. 剩余的 shèng yú de
arc: any part of a curve, especially. of a circle. 弧 hú

Example:
The model of the inter-arc spreading and the remnant arc development are verified by the deep-sea drilling in the Philippine Sea.
菲律宾海的深海钻探证实了弧间扩张和残留弧形成模式。

remote sensing [ri'məut 'sensiŋ]

Definition: the use of satellites to gather data, images, etc., as to study the earth or other bodies of the solar system. 遥感 yáo gǎn; 遥测 yáo cè; 远距离读出 yuǎn jù lí dú chū

Example:
It is of significance to dynamically monitor the mine environment via remote sensing for protecting mine ecology and developing mineral resources.
运用遥感技术动态地监测矿区环境,对保护矿区生态、合理开发矿产资源有着重要的意义。

Extended Terms:
remote error sensing 远程误差检测(技术,方法)
remote-sensing regulator 遥感调节器

remote sensing sounding 遥感测深
remote sensing with audio frequency pulse 音频脉冲遥感
remote sensing with radio frequency pulse 射频脉冲遥感

replacement texture [riˈpleismənt ˈtekstʃə]

Definition: it occurs where a mineral or mineral aggregate has the external crystal form of a preexisting different mineral (pseudomorphism) or where the juxtaposition of two minerals indicates that one was formed at the expense of the other. 交代结构 jiāo dài jié gòu

residual dilution [riˈzidjuəl daiˈluːʃən]

Definition: 残余稀释 cán yú xī shì

residual: of, or having the nature of, a residue or residuum; left over after part or most is taken away; remainin. 剩余的 shèng yú de

dilution: a diluting or being diluted. 稀释 xī shì

residual enrichment [riˈzidjuəl inˈritʃmənt]

Definition: 残余富集 cán yú fù jí

residual: of, or having the nature of, a residue or residuum; left over after part or most is taken away; remainin. 剩余的 shèng yú de

enrichment: something added to another for embellishment or completion. 丰富 fēng fù

Example:

Mass balance calculations suggest that upgrading of the manganese ore can be attributed to leaching from the sedimentary ore with residual enrichment of Mn.
质量守恒运算表示锰矿的叠积可归因为锰残余富集从沉积矿中浸出。

residual magma [riˈzidjuəl ˈmægmə]

Definition: 残余岩浆 cán yú yán jiāng

residual: remaining after the greater part or quantity has gone. 剩余的 shèng yú de; 残留的 cán liú de

magma: hot fluid or semi-fluid material below or within the earth's crust from which lava and other igneous rock is formed by cooling. 岩浆 yán jiāng

Example:

The poor content of Fe in residual magma of Huangmeijian body favored mineralization of W,

Sn、Nb and Ta; the rich content in Fe in residual magma of Dalongshan favored Fe mineralization.
其中黄梅尖岩体残余岩浆贫铁，有利于W、Sn、Nb、Ta矿化，而大龙山岩体残余岩浆富铁，有利于铁矿化。

Extended Terms:

residual loss 残余损失；剩余损耗
residual magnesium content 残余镁量

residuum composition [riˈzidjuəlm ˌkɒmpəˈziʃən]

Definition: 剩余组分 shèng yú zǔ fēn

residuum: something remaining after removal of a part; a residue. 残留物 cán liú wù
composition: the makeup of a thing or person; aggregate of ingredients or qualities and manner of their combination; constitution. 合成物 hé chéng wù

restite [ˈrestait]

Definition: is the residual material left at the site of melting during the in place production of granite through intense metamorphism. 残余体 cán yú tǐ

Example:
Based on the quantitative modeling calculation Rongcheng gneiss suite was formed from the 42% partial melting of amphibolite source rock and the corresponding restite iseclogite.
根据稀土元素定量模拟计算，荣成片麻岩套是由斜长角闪岩源岩经42%的部分熔融形成，相应的残留体是榴辉岩。

restricted platform facies [riˈstriktid ˈplætfɔːm ˈfeiʃiːz]

Definition: 局限台地相 jú xiàn tái dì xiàng

restricted: limited in extent, number, scope, or action. 受限制的 shòu xiàn zhì de；有限的 yǒu xiàn de
platform: a raised level surface on which people or things can stand. 台 tái；平台 píng tái
facies: the character of a rock expressed by its formation, composition, and fossil content. 相 xiàng

Example:
The lower Triassic Jialingjiang formation in Nanzhang, Hubei province, is composed of carbonate rocks and can be divided into four different sedimentary facies zones: shoal facies on

edge of platform, open platform facies, restricted platform facies and platform evaporate facies. 湖北南漳地区下三叠统嘉陵江组为一套碳酸盐岩沉积，由台地边缘浅滩相、开阔台地相、局限台地相和蒸发台地相4种不同的相带组成。

resurgent boiling [rɪˈsɜːdʒənt ˈbɔɪlɪŋ]

Definition: 复活沸腾 fù huó fèi téng

resurgent: rising or tending to rise again; resurging. 复活的 fù huó de
boiling: stewing, steeping, percolating, steaming, bubbling, seething, simmering, evaporating, distilling, boiling over. 沸腾的 fèi téng de

Example:

The high salinity, high temperature stage is interpreted to be the result of resurgent boiling. 高盐度高温阶段被解释为复活沸腾的结果。

resurgent dome [rɪˈsɜːdʒənt dəum]

Definition: is a dome formed by swelling or rising of a caldera floor due to movement in the magma chamber beneath it. 再生穹窿 zài shēng qióng lóng

retro arc basin [ˈretrəu ɑːk beisən]

Definition: 弧后盆地 hú hòu pén dì

retroarc: is the area behind a volcanic arc. 弧后的 hú hòu de
basin: a wide, depressed area in which the rock layers all incline toward a central area. 盆地 pén dì

Example:

The middle upper Jurassic Jienu Group, distributed widely along Gerze Baingoin area of northern Tibet, is terrigenous deposit of remnant retroarc basin with a complete transgressive regressive cycle.
西藏北部改则班戈地区出露的中上侏罗统接奴群是残余弧后盆地的滨浅海相陆缘碎屑沉积，具有完整的水进水退沉积旋回。

Extended Term:

retroarc foreland basin 弧后陆前盆地

retrograde boiling [ˈretrəgreid ˈbɔɪlɪŋ]

Definition: 逆行沸腾 nì xíng fèi téng

retrograde: moving or directed backward; retiring or retreating. 倒退的 dào tuì de
boiling: stewing, steeping, percolating, steaming, bubbling, seething, simmering, evaporating, distilling, boiling over. 沸腾的 fèi téng de

Example:

The volatile components are sometimes separated from the silicate molten mass due to the retrograde boiling.
挥发相的组成及其与硅酸盐熔融体由于逆行沸腾导致不相溶性。

retrogressive metamorphism [ˌretrəˈɡresiv ˌmetəˈmɔːfizəm]

Definition: 退化变质作用 tuì huà biàn zhì zuò yòng

retrogressive: of or relatinf with the process of returning to an earlier state, typically a worse one. 倒退 dào tuì; 退化 tuì huà; 衰退 shuāi tuì; 恶化 è huà
metamorphism: alteration of the composition or structure of a rock by heat, pressure, or other natural agency. [地质][岩石]变质作用 biàn zhì zuò yòng

Example:

In addition, also a great number of silicon can be separated from weathered basalt during retrogressive metamorphism process.
另外,在风化过程中玄武岩的退变质过程亦能排出大量硅质。

reversible process [riˈvɜːsəbl prəˈses]

Definition: any process in which a system can be made to pass through the same states in the reverse order when the process is reversed. 可逆过程 kě nì guò chéng

Example:

Chemical additions to the DNA, or to the histone proteins around which it coils, somehow silence important genes, but in a reversible process quite different from mutation.
DNA 或 DNA 所盘绕的组织蛋白上出现了其他化学物质,使得重要的基因失去了功能,不过此属可逆过程,与突变大不相同。

Extended Terms:

adiabatic reversible process 绝热可逆过程
external reversible process 外可逆过程
reversible Markov process 可逆马尔可夫过程
reversible saturation adiabatic process 可逆饱和绝热过程

reversible thermodynamic process
[riˈvɜːsəbl ˌθɜːməudaiˈnæmik prəˈses]

Definition: 可逆的热力学过程 kě nì de rè lì xué guò chéng

reversible: that can reverse; specifically, that can change and then go back to the original condition by a reversal of the change; said of a chemical reaction, etc. 可逆的 kě nì de
thermodynamic: of, or relating to the conversion of heat into other forms of energy. 热力学的 rè lì xué de
process: a particular method of doing something, generally involving a number of steps or operations. 过程 guò chéng

Example:
A study is focused on the reversible thermodynamic process for desalination.
一项研究聚焦于可逆的热力学过程在脱盐过程中的运用。

Reynolds Number [ˈrenəlds ˈnʌmbə]

Definition: a dimensionless number used in fluid mechanics to indicate whether fluid flow past a body or in a duct is steady or turbulent. 雷诺数 léi nuò shù

Origin: early 20th century: named after Osborne Reynolds (1842-1912), English physicist.

Example:
The friction drag coefficient increases with increasing afterbody fineness ratio and decreasing Reynolds Number.
后体摩擦阻力系数分别随后体的长细比的增加和雷诺数的减小而增加。

rhenium isotope [ˈriːniəm ˈaisətəup]

Definition: 铼同位素 lái tóng wèi sù

rhenium: a rare, metallic chemical element that is a silver-white solid or a gray-to-black powder, used in thermocouples, electrodes, etc.; symbol, Re; at. no., 75. 铼 lái
isotopes: any of two or more forms of an element having the same or very closely related chemical properties and the same atomic number but different atomic weights (or mass numbers): U-235, U-238, and U-239 are three isotopes of uranium. 同位素 tóng wèi sù

Rheology of magma [riːˈɒlədʒi ɔv ˈmægmə]

Definition: 岩浆流变学 yán jiāng liú biàn xué

rheology: the study of the change in form and the flow of matter, embracing elasticity, viscosity, and plasticity. 流变学 liú biàn xué

magma: the molten rock material under the earth's crust, from which igneous rock is formed by cooling. 岩浆 yán jiāng

rheomorphism [ˌriːəuˈmɔːfizəm]

Definition: (Geology) the liquefaction of rock, which results in its flowing and intruding into surrounding rocks. 流变作用 liú biàn zuò yòng

Origin: Greek-morphos, *adj.* derivative of morphḗ form.

Example:

It forms four tectonic defor-mation facies, i. e. brittle shear, flexural slope, solid-state rheomorphism and plastic flow or fusion.
形成了脆裂剪切、弯曲滑动、固态流变和柔流或熔融四种构造变形相。

Rhine Graben [rainˈɡrɑːbən]

Definition: is a major rift, straddling the border between France and Germany. It forms part of the European Cenozoic Rift System, which extends across central Europe. 莱茵地堑 lái yīn dì qiàn

Example:

The Eisenberg Basin is situated in the southwestern part of the Mainz Basin and is part of the western shoulder of the Rhinegraben.
艾森贝格盆地位于梅因茨盆地西南部，为莱茵地堑的一部分。

Rhodolith [ˈrədəliθ]

Definition: are colorful, unattached, branching, crustose benthic marine red algae that resemble coral. Rhodolith beds create biogenic habitat for diverse benthic communities. Common rhodolith species include Lithophyllum margaritae, Lithothamnion muellerii, and Neogoniolithon trichotomum. 红藻石 hóng zǎo shí

Origin: Greek, combining form of rhódon rose.

Example:

Abundant rhodolith like grains, which are branching, spheroidal and irregular in shape and 0.5-1.2mm in size, occur in the upper part of the Neoproterozoic Doushantuo Formation at Weng'an, Guizhou Province, southwest China.

贵州瓮安上震旦统陡山沱组上段产出大量似红藻石，呈枝状、球状和不规则团块状，大小一般为 0.5mm~1.2 mm。

rhomb porphyry [rɒmbˈpɔːfiri]

Definition: 菱长斑岩 líng cháng bān yán

rhomb：菱形 líng xíng

porphyry：rock containing relatively large conspicuous crystals, especially feldspar, in a fine-grained igneous matrix. 斑岩 bān yán

Example:

Monzonitic rhomb porphyry flows predominate, with basaltic flows in-between.
二长岩的菱长斑岩占主导位置，少量玄武岩在中间流动。

rhyodacite [raiˈɔdəsait]

Definition: is an extrusive volcanic rock intermediate in composition between dacite and rhyolite. It is the extrusive equivalent of granodiorite. 流纹英安岩 liú wén yīng ān yán

Example:

In this region, the parent magma for the lower series of volcanic rocks was similar to dacite, but for the upper series of volcanic rocks, it was relative more acid (similar to rhyodacite).
区内下火山岩系的母岩浆与石英闪长质相似，而上火山岩系的母岩浆则与流纹英安质相似。

Extended Term:

rhyodiabasic texture 流纹辉绿结构

rhyolite [ˈraiəlait]

Definition: a light-colored igneous rock with a fine-grained, granitelike texture. 流纹岩 liú wén yán

Origin: Greek rhuāx, stream (from rhein, to flow; see sreu-in Indo-European roots) + -lite.

Example:

The volcanic rocks belong to high kalium calc alkalic series and can be divided into andesite, dacite, trachydacite and rhyolite by petrochemical composition.
据岩石化学成分，金刚台组火山岩可划分为安山质、英安质、粗面英安质和流纹质几种类型，并属于高钾钙碱性系列。

Extended Terms:

plagioclase rhyolite 斜长流纹岩

porphyritic rhyolite 斑状流纹岩
potash rhyolite 钾疗岩
riebeckite rhyolite 钠闪流纹岩
soda rhyolite 钠疗岩

rhyolite porphyry ['raiəlait'pɔːfiri]

Definition: 流纹斑岩 liú wén bān yán

rhyolite：(Geology) a pale fine-grained volcanic rock of granitic composition, typically porphyritic in texture. [地质]流纹岩

porphyry：a hard igneous rock containing crystals of feldspar in a fine-grained, typically reddish groundmass. 斑岩 bān yán

Example:

The Cretaceous rhyolite-porphyry is the main parent rock of the deposit, extending in NE-orientation.
白垩纪流纹斑岩为其主要的成矿母岩，可能为东北向展布。

rhyolitic structure ['raiəlait'strʌktʃə]

Definition: 流纹构造 liú wén gòu zào

rhyolitic：a fine-grained extrusive volcanic rock, similar to granite in composition and usually exhibiting flow lines. 流纹 liú wén

structure：something composed of interrelated parts forming an organism or an organization. 结构 jié gòu

rhythmic bedding ['riðmik'bediŋ]

Definition: 韵律层理 yùn lǜ céng lǐ

rhythmic：having or relating to rhythm. 有节奏的 yǒu jié zòu de；与节奏有关的 yǔ jié zòu yǒu guān de

bedding：a base or bottom layer. 底层 dǐ céng；基层 jī céng

Extended Term:

rhythmic graded bedding 韵律性递变层理

ribbon cementation ['ribən ˌsiːmen'teiʃən]

Definition: 带状胶结 dài zhuàng jiāo jié

ribbon: a long, narrow strip of fabric. 带状 dài zhuàng

cementation: (chiefly Geology) the binding together of particles or other things by cement. [主地质] 黏结 nián jié; 胶接作用 jiāo jiē zuò yòng

richterite [ˈriktərait]

Definition: a sodium calcium magnesium silicate mineral belonging to the amphibole group. 锰闪石 měng shǎn shí

Origin: combining form of Greek chlōrós light green, greenish yellow.

Example:

the origin of phlogopite and potassic richterite bearing peridotite xenoliths from South Africa
南非金云母和含橄榄捕虏岩的钾锰闪石的起源

Extended Term:

fluor-richterite 氟钠透闪石

riebeckite [ˈriːbekait]

Definition: is a sodium-rich member of the amphibole group of silicate minerals, chemical formula $Na_2(Fe, Mg)_5Si_8O_{22}(OH)_2$. It forms a series with magnesioriebeckite. It crystallizes in the monoclinic. 钠闪石 nà shǎn shí

Origin: 1885-1890; named after Emil Riebeck (died 1885), German explorer.

Extended Terms:

riebeckite aplite 钠闪细晶岩
riebeckite aplite-granit 钠闪细晶花岗岩
riebeckite rhyolite 钠闪流纹岩

rift basin [rift ˈbeisən]

Definition: a basin formed when a continental plate breaks apart. 裂谷盆地 liè gǔ pén dì

Example:

The coupling of rift basin and extensional mountain system is the effect of evolution of the interaction of crust-mantle boundary.
裂陷盆地与伸展山岭的耦合关系主要是深部壳幔作用在浅层的响应。

Extended Term:

protooceanic rift basin 原始大洋裂谷盆地

rifting [ˈrɪftɪŋ]

Definition: an opening caused by or as if by splitting; cleft; fissure. 分裂 fēn liè; 裂开 liè kāi

Origin: 1250-1300; Middle English, from Old Norse ript breaking of an agreement (compare Danish, Norwegian rift cleavage), derivative of rīfa to tear (cognate with rive).

Example:
Analogue modeling of rifting provides direct and effective ways for the study of the relation between geometry and dynamics of continental rift systems.
裂谷作用的物理模拟是研究大陆裂谷系统的几何结构与动力学机制之间关系的一种直观有效的手段。

Extended Terms:
arc rifting 弧的张裂作用
global rifting system 全球裂谷系
rifting phase 裂谷作用带，裂谷作用期
rifting rod 钩杆

rill mark [rɪl mɑːk]

Definition: a small, dendritic channel formed on beach mud or sand by a rill, especially if on the lee side of a partially buried obstruction. 流痕 liú hén

Example:
The channel wall often has slump (deformation) structures, rill marks and silt stalactite-like structures.
沟槽壁上常发育滑塌、流痕或类钟乳等次级构造。

Extended Term:
lobate rill mark 朵状细流痕

rimmed platform facies [rɪmd ˈplætfɔːm ˈfeɪʃiːz]

Definition: 镶边台地相 xiāng biān tái dì xiàng

rimmed: of the upper or outer edge of an object, typically something circular or approximately circular. 有边缘的 yǒu biān yuán de

platform: a raised level surface on which people or things can stand. 台 tái; 平台 píng tái

facies: the character of a rock expressed by its formation, composition, and fossil content. 相 xiàng

Example:

the model of steep-sloped enclosed rimmed platform facies, northern Sichuan basin
四川北部盆地的陡坡封闭型镶边台地相模型

rimmed texture [rimd 'tekstʃə]

Definition: 镶边结构 xiāng biān jié gòu

rimmed: of the upper or outer edge of an object, typically something circular or approximately circular. 有边缘的 yǒu biān yuán de

texture: the feel, appearance, or consistency of a surface or a substance. (物质、表面的)结构 jié gòu; 构造 gòu zào; 纹理 wén lǐ; 肌理 jī lǐ

Example:

The CL images reveal that most zircons have core rim texture. The cores show igneous zircons characteristics.
锆石 CL 图像显示其有明显的镶边结构，核部具岩浆锆石，边部是变质增生锆石的特征。

ring complex [rɪŋ 'kɒmpleks]

Definition: is a set of circular igneous intrusions around an igneous centre. 环状杂岩 huán zhuàng zá yán

Extended Terms:

chelate ring complex 螯合环状络合物
complex cobordism ring 复配边环
gamma tubulin ring complexy 微管蛋白环形复合体

ring dyke [rɪŋ daik]

Definition: 环状岩墙 huán zhuàng yán qiáng

ring: a circumscribing object, (roughly) circular and hollow, looking like an annual ring, earring, finger ring etc. 环形物 huán xíng wù

dyke: used as a disparaging term for a lesbian. 堤 dī

ring silicates [rɪŋ 'silikeits]

Definition: 环状硅酸盐 huán zhuàng guī suān yán

ring: a circumscribing object, (roughly) circular and hollow, looking like an annual ring, earring, finger ring etc. 环形物 huán xíng wù

silicates: any of numerous compounds containing silicon, oxygen, and one or more metals; a salt of silicic acid. 硅酸盐 guī suān yán

Example:

The chemical bonding in the ring silicate mineral dioptase is investigated.
调查环状硅酸盐绿铜矿的化学胶合。

ripped up plastic shard [ript ʌp ˈplæstik ʃɑːd]

Definition: 塑变撕裂状玻屑 sù biàn sī liè zhuàng bō xiè

ripped up: to tear something into small pieces. 撕裂 sī liè

plastic: a synthetic material made from a wide range of organic polymers such as polyethylene, PVC, nylon, etc., that can be moulded into shape while soft, and then set into a rigid or slightly elastic form. 塑料 sù liào

shard: a fragment of a brittle substance, as of glass or metal. 碎片 suì piàn

ripped up shard [ript ʌp ʃɑːd]

Definition: 撕裂状玻屑 sī liè zhuàng bō xiè

ripped up: to tear something into small pieces. 撕裂 sī liè

shard: a fragment of a brittle substance, as of glass or metal. 碎片 suì piàn

ripple index (RI) [ˈripl ˈindeks]

Definition: on a rippled surface, the ratio of the crest-to-crest distance to the crest-to-trough distance. 波痕指数 bō hén zhǐ shù

Example:

vertical ordering variation type in ripple index
波痕指数垂向有序变化型

ripple mark [ˈripl mɑːk]

Definition: In geology, ripple marks are sedimentary structures (i.e. bedforms of the lower flow regime) and indicate agitation by water (current or waves) or wind. 波痕 bō hén

Example:

These trace fossils yield at the surface of a sandstone with current ripple mark and cross bedding. The sediments belonged to the shallow marine ami subtidal low-energy and Cruziana

ichnofacies.
这些遗迹化石产于具波痕和斜层理的砂岩下层面上，为浅海潮下低能带的克鲁斯迹遗迹相。

rock [rɒk]

Definition: any consolidated material consisting of more than one mineral and, sometimes, organic material, e. g. granite or limestone. 岩石 yán shí

Origin: from Old French rocque, from medieval Latin rocca, of unknown ultimate origin.

Example:
Comets are conglomerations of ice, dust and rock from the early solar system.
彗星是早期太阳系中冰、灰尘和岩石的聚合物。

Extended Terms:
rock leather 一种石棉
sedimentary rocks 沉积岩

rock-forming minerals [rɒk ˈfɔːmɪŋ ˈmɪnərəls]

Definition: 造岩矿物 zào yán kuàng wù

rock：a large mass of stone forming a peak or cliff. 岩石 yán shí
forming：the shape, outline, or configuration of anything; structure as apart from color, material, etc. 形成 xíng chéng
mineral：an inorganic substance occurring naturally in the earth and having a consistent and distinctive set of physical properties (e. g., a usually crystalline structure, hardness, color, etc.) and a composition that can be expressed by a chemical formula; sometimes applied to substances in the earth of organic origin, such as coal. 矿物 kuàng wù

rock fall deposit [rɒk fɔːl dɪˈpɒzɪt]

Definition: 岩崩沉积 yán bēng chén jī

rock fall：refers to quantities of rock falling freely from a cliff face. 岩崩 yán bēng；岩石崩落 yán shí bēng luò
deposit：a layer or body of accumulated matter. 沉积物 chén jī wù；沉淀 chén diàn

rock forming element [rɒkˈfɔːmɪŋˈelɪmənt]

Definition: 造岩元素 zào yán yuán sù

rock: the solid mineral material forming part of the surface of the earth and other similar planets, exposed on the surface or underlying the soil. 岩石 yán shí；岩 yán

forming: the act or process of giving form or shape to anything; as, in shipbuilding, the exact shaping of partially shaped timbers. 形成 xíng chéng；生成 shēng chéng

element: a part or aspect of something, especially one that is essential or characteristic. 基本组成部分 jī běn zǔ chéng bù fēn；要素 yào sù

Example:

In the Huachangshan fault zone, the change of rock-forming element contents has certain relation to different types of structural rock.
在华昌山断裂带内，造岩元素的含量变化与不同的构造岩石类型有一定的关系。

roof foundering [ruːf ˈfaundəriŋ]

Definition: 岩浆坍顶作用 yán jiāng tān dǐng zuò yòng；顶落作用 dǐng luò zuò yòng

roof: the top or peak of anything. 顶部 dǐng bù

foundering: to stumble, fall, or go lame. 陷落 xiàn luò

roof pendant [ruːf ˈpendənt]

Definition: downward extension of the surrounding rock that protrudes into the upper surface of an igneous intrusive body. 顶垂体 dǐng chuí tǐ

Example:

Other types such as those associated with mafic-ultramafic rocks and in the contact between granites and dolomitic marbles or enclaves and roof pendants in granites are less significant.
其他类型的滑石，如与镁铁质-超镁铁质岩共生的、在花岗岩与白云质大理岩接触带的、在花岗岩体中呈包体或顶垂体状的，则相对不重要。

rotary texture [ˈrəutəri ˈtekstʃə]

Definition: 旋转结构 xuán zhuǎn jié gòu

rotary: (of motion) revolving around a centre or axis; rotational. 旋转的 xuán zhuǎn de；轮转的 lún zhuàn de；转动的 zhuàn dòng de

texture: the feel, appearance, or consistency of a surface or a substance. (物质、表面的)结构 jié gòu；构造 gòu zào；纹理 wén lǐ；肌理 jī lǐ

rubble ['rʌbl]

Definition: waste or rough fragments of stone, brick, concrete, etc., especially as the debris from the demolition of buildings. 毛石 máo shí; 碎石 suì shí

Origin: late Middle English: perhaps from an Anglo-Norman French alteration of Old French robe "spoils"; compare with rubbish.

Example:

Therefore, the air-cooling embankment with rubble slope protection, as a kind of conveniently and widely used construction measure, was a positive frozen soil protection measure for its effectively decreasing ground temperature and protecting permafrost.
因此，片石气冷护坡能够有效发挥降低地温、保护多年冻土的作用，是一种施作方便、适用条件较广泛的主动保护多年冻土措施。

Extended Terms:

rubble concrete 毛石混凝土
rubble disposal 毛石处置
rubble mound 石堆
rubble wall 毛石墙

Rubidium-strontium dating [ruːˈbidiəm ˈstrɔntiəm ˈdeitiŋ]

Definition: is a radiometric dating technique that geologists use to determine the age of rocks. 铷锶定年 rú sī dìng nián

Example:

the data process for rubidium-strontium dating method
铷锶定年方法的实验数据处理

rubidium isotope [ruː ˈbidiəm ˈaisətəup]

Definition: 铷同位素 rú tóng wèi sù

rubidium: a soft, silvery-white, metallic chemical element, one of the alkali metals, that ignites spontaneously in air and reacts violently in water: used in photocells and in filaments of vacuum tubes: symbol, Rb; at. no., 37. 铷 rú

isotopes: any of two or more forms of an element having the same or very closely related chemical properties and the same atomic number but different atomic weights (or mass numbers): U-235, U-238, and U-239 are three isotopes of uranium. 同位素 tóng wèi sù

> Example:

study of faraday resonant effect of a natural mixture of two rubidium isotopes
同位素铷原子的超精细法拉第共振效应研究

rudite ['ruːdait]

> Definition: is any sedimentary clastic rock with a grain size exceeding 2 mm (0.08 in) such as conglomerates and breccias. 砾屑岩 lì xiè yán; 砾状岩 lì zhuàng yán

> Extended Terms:

cataclastic rudite 碎裂砾屑岩
silico-rudite 硅砾岩

rutile ['ruːtail]

> Definition: a hard, reddish, tetragonal form of titanium dioxide, an ore of titanium. 金红石 jīn hóng shí

> Origin: French, from German Rutil, from Latin rutilus, red; see rutilant.

> Example:

Rutile resource is rich in our country, but rutile ore grade is low and the product is poor competitive because of its high production cost.
我国天然金红石资源储量较大，但绝大部分为低品位的原生矿石，所以加工成本高，产品缺乏市场竞争力。

> Extended Terms:

artificial rutile 人造金红石
rutile Structure 金红石结构
rutile powder 金红石粉末
synthetic rutile 合成金红石

salic mineral ['sælik 'minərəl]

> Definition: 硅铝质矿物 guī lǚ zhì kuàng wù

salic: of or relating to certain minerals, such as quartz and the feldspars, that commonly occur in igneous rocks and contain large amounts of silica and alumina. 硅铝质的 guī lǚ zhì de
Mineral: a solid inorganic substance of natural occurrence 矿 kuàng; 矿物 kuàng wù

> Example:

Their source rocks are salic mineral and the granites belong to S-type granites.

其源岩为硅铝质矿物，属 S 型花岗岩。

saline lake ['seilain leik]

Definition: a landlocked body of water which has a concentration of salts (mostly sodium chloride) and other minerals significantly higher than most lakes (often defined as at least three grams of salt per liter). 盐湖 yán hú

Example:
The saline lake is one of the most favorable environments for developing hydrocarbon source rocks.
盐湖沉积环境是烃源岩发育的最重要的地质环境之一。

Extended Term:
saline lake deposit 盐湖矿床

salt dome [sɔːlt dəum]

Definition: a domelike structure produced in stratified rocks by the intrusion of a mass of salt in a plastic state and frequently containing oil, gas, etc. 盐丘 yán qiū

Extended Terms:
salt dome boundary 盐丘界面
salt dome coast 盐丘海岸
salt dome reservoir 盐丘油藏
salt dome storage 盐丘地下储库
salt dome well 盐丘井

salt pans [sɔːlt pænz]

Definition: a flat expanse of ground covered with salt and other minerals, usually found in deserts. 盐田 yán tián

Example:
From March, 1986 to September, 1990, samples of mainly planktonic algae are collected from 3 salt pans and 4 saline lakes (salinity 61.0-320.0) in northern China.
从 1986 年 3 月到 1990 年 9 月，对中国北方一些盐田和盐湖的高盐水域(盐度 61.0~320.0)中的藻类(主要是浮游藻)进行调查。

Extended Term:
Makgadikgadi Salt Pans 马卡迪卡迪盐沼

salt rock [sɔːlt rɒk]

Definition: Salt Rock is a small town just north of Ballito and Shaka's Rock on the Dolphin Coast of KwaZulu-Natal, South Africa. It is a favorite holiday destination for many local South Africans. It is about 40 minutes north of Durban. 石盐岩 shí yán yán，又称蒸发岩（evaporite）zhēng fā yán

Example:
Salt rock sequence is characterized by salt rock and salt rock interbedded with mudstones of deep lake.
盐岩层序主要发育盐岩以及盐岩夹深湖泥岩为主。

Extended Terms:
crude salt 粗盐
salt-roof rock 盐盖层

sand pillow structure [sænd'piləu'strʌktʃə] = sand ball structure

Definition: 砂枕构造 shā zhěn gòu zào，又称砂球构造 shā qiú gòu zào
sand：a stratum of sandstone or compacted sand 沙石层 shā shí céng；沙层 shā céng
pillow：a piece of wood or metal used as a support；a block or bearing like a pillow. 枕状的 zhěn zhuàng de
structure：the arrangement of and relations between the parts or elements of something complex 结构 jié gòu；构造 gòu zào

Example:
Pillow structure consists of a suit of deformed concave upward bodies made of silt and sand whose original layering bent round parallel the basal surface and it truncated at the top surface.
枕状构造是砂层中一组呈"凹"形弯曲的变形沉积体，它的原始层平行于枕状体的底面，顶面则是一个平直的截切面。

Samarium-neodymium dating [sə'meəriəm 'niːəu'dimiəm 'deitiŋ]

Definition: is useful for determining the age relationships of rocks and meteorites, based on decay of a long-lived samarium (Sm) isotope to a radiogenic neodymium (Nd) isotope.
钐-钕(Sm-Nd)shān nǚ；同位素定年 tóng wèi sù dìng nián

sand [sænd]

Definition: a loose granular substance, typically pale yellowish brown, resulting from the erosion of siliceous and other rocks and forming a major constituent of beaches, river beds, the seabed, and deserts. 砂 shā

Origin: Old English, of Germanic origin; related to Dutch zand and German Sand.

Extended Terms:

abrasive sand 研磨砂
adhering sand 粘砂
antiquated sand 废砂，粘砂
sand grass 沙草
sand hill 沙丘

sandstone [ˈsændstəun]

Definition: a sedimentary rock made up of particles of sand bound together with a mineral cement. 砂岩 shā yán

Origin: Old English, of Germanic origin; related to Dutch zand and German sand.

Example:

The types of reservoirs main include base rock weathering crust, sandstone, tuffaceous sandstone and conglomerate.
储层类型主要包括基岩风化壳、岩、灰质砂岩和砾岩。

Extended Terms:

argillaceous sandstone 泥质砂岩
arkosic sandstone 长石砂岩
asphaltic sandstone 沥青砂岩
crystal sandstone 晶屑砂岩
micaceous sandstone 云母砂岩

sanidine facies [ˈsænidiːn ˈfeiʃiːz]

Definition: 透长石相 tòu cháng shí xiàng

sanidine: a glassy mineral of the alkali feldspar group, typically occurring as tabular crystals. 透长石 tòu cháng shí

facies: the character of a rock expressed by its formation, composition, and fossil content. [地质]相 xiàng

saponite [ˈsæpənait]

Definition: a complex hydrous silicate of aluminum and magnesium, a light-colored, soft clay mineral, often found in veins and cavities in serpentine and basaltic rocks. 皂石 zào shí

Origin: Latin sāpō, sāpōn-, hair dye; see saponin+ -ite.

Example:

Saponite arts are the representative works of Zimbabwe in Africa.
皂石雕刻艺术是非洲地区津巴布韦具有代表性的艺术创造。

Extended Terms:

magnesium-saponite 镁皂石
nickel saponite 镍皂
pillared saponite 交联皂石
synthetic saponite 合成皂石
zinc-saponite 锌蒙脱石锌皂石

sapphirine [ˈsæfəriːn]

Definition: a rare, blue or green, very hard, monoclinic mineral, (Mg, Al)(Al, Si)O, found in some metamorphic rocks. 假蓝宝石 jiǎ lán bǎo shí

Origin: 1375-1425; late Middle English saphyryn (from Old French), Greek sappheírinos like lapis lazuli.

Extended Term:

Sapphirine Cool 蓝宝石酷乐

satellite interferometry [ˈsætəlait ˌintəfəˈrɔmitri]

Definition: a remote sensing technique recently adopted by the earth science community to study various active earth processes. 卫星干涉仪 wèi xīng gān shè yí

Example:

Based on interferometry measurement theory of satellite carrier phase, the technology of land vehicle heading determination through Beidou Bi-Satellite System was studied.
基于卫星载波相位干涉测量原理，研究了利用北斗双星系统确定车辆航向的相关技术。

Extended Term:

satellite radio interferometry 卫星射电干涉测量

saturated mineral [ˈsætʃəreitid ˈminərəl]

Definition: a mineral that forms in the presence of free silica. 饱和矿物 bǎo hé kuàng wù

Example:

Water-saturated mineral oil was layered above the fluid to prevent fluid evaporation.
包水矿油层积在液体之上以防止液体蒸发。

Extended Terms:

saturated temperature 饱和温度
saturated infiltration 饱和入渗

saturated rock ['sætʃəreitid rɔk]

Definition: an igneous rock composed principally of saturated minerals. 饱和岩 bǎo hé yán

Example:

The conclusions are derived that the saturated rock's averaged energy is less than that of natural rock, but its averaged deformation is lager, and the conclusions are explained according to the mechanism of rock breach.
结论是：浸水饱和岩石较天然含水岩石的平均能量小、平均形变大，这是根据岩石破裂损伤机理得出的。

Extended Terms:

oil-saturated reservoir rock 储油岩
saturated soil paste 饱和土浆
Si-saturated rock 硅饱和岩

saturation [sætʃə'reiʃən]

Definition: the condition of a magnetic substance that has been magnetized to the maximum. 饱和(状态) bǎo hé (zhuàng tài)

Origin: 1545-1555; from Late Latin saturātiōn-(stem of saturātiō) a filling, equivalent to saturāt(us).

Example:

The results show that the new method by new ratio is the best way to calculate the oil-saturation, and the calculated oil-saturation values agree with the real values.
比较所得结果得出，用新碳氧比和新的求含油饱和度方法的计算结果与实际测量值一致，是目前结合最好的求含油饱和度的方法。

Extended Terms:

saturation magnetization 饱和磁化
saturation point 饱和点

Sc (scandium)

saturation ratio 饱和系数
saturation state 饱和状态
saturated solution 饱和溶液

Sc (scandium) [ˈskændiəm]

Definition: a rare, silvery-white, metallic chemical element occurring with various rare-earth elements, used to produce high-intensity light sources. 元素钪(scandium)的符号 yuán sù kàng de fú hào

Example:
Scandium in these alloys exhibits obvious grain refining, improved strength and serves to inhibit recrystallization in Al alloys, strengthen welds and eliminate hot cracking from welds.
钪对铝合金有明显的晶粒细化作用, 提高了铝合金强度, 抑制了再结晶, 强化了焊缝, 消除了焊缝出现的热裂。

Extended Terms:
metallic scandium 金属钪
scandium carbide 碳化钪
scandium concentrate 钪精矿
scandium nitride 氮化钪
scandium oxide 氧化钪

scaly structure [ˈskeili ˈstrʌktʃə]

Definition: 鳞片构造 lín piàn gòu zào

scaly: covered in scales. 鳞片覆盖的 lín piàn fù gài de
structure: the arrangement of and relations between the parts or elements of something complex. 结构 jié gòu; 构造 gòu zào

Example:
The strongly deformed mylonite or ultramylonite zone in the down-side of "scaly structure belt" has an obvious characteristic of permeability, so that it provides passages for Au-bearing fluid.
鳞片状构造带下盘由糜棱岩和超糜棱岩组成的强变形带可透性好, 为含金流体来源通道。

scatter diagram of size parameters
[ˈskætə ˈdaiəgræm ɔv saiz pəˈræmitəs]

Definition: 粒度参数离散图解 lì dù cān shù lí sàn tú jiě

scatter: occur or be found at intervals rather than all together. 散布 sàn bù; 分散 fēn sàn diagram: a simplified drawing showing the appearance, structure, or workings of something; a schematic representation. 图解 tú jiě; 简图 jiǎn tú; 示意图 shì yì tú

size parameter: the ratio of the size of a spherical scattering particle to the wavelength of the radiation being scattered, that is, $\alpha = \pi d/\lambda$, where d is the diameter of the particle, and λ is the wavelength of the incident radiation. 尺寸参数 chǐ cùn cān shù

Example:

Such parameters as standard deviation, skewness, kurtosis, etc. revealed by scatter diagram of quartz grain size indicate that each profile basically plots in different area.
石英粒度的标准差、偏度、尖度等参数也显示，各剖面样品基本上分布于不同的区域。

schiller structure [ˈʃilə ˈstrʌktʃə]

Definition: 席勒构造 xí lè gòu zào

schiller: a peculiar bronzelike luster in certain minerals, often iridescent, caused by the diffraction of light in embedded crystals. 青铜色光泽 qīng tóng sè guāng zé

structure: manner of building, constructing, or organizing. 构造 gòu zào

schist [ʃist]

Definition: any of a group of metamorphic rocks containing parallel layers of flaky minerals, as mica or talc, and splitting easily into thin, parallel leaves. 片岩 piàn yán

Origin: French schiste, from Latin (lapis) schistos, fissile (stone), a kind of iron ore, from Greek skhistos, split, divisible, from skhizein, to split; see skei-in Indo-European roots.

Example:

The emerald deposits in Brazil belong to the classic schist-hosted emerald deposit type.
巴西的祖母绿矿床主要属于所谓的片岩型祖母绿矿床。

Extended Terms:

chlorite schist 泥石片岩
green schist 绿片岩
gas schist 含天然气片岩
schistoil 页岩油
schist forming process 片理

schistose structure [ˈʃistəus ˈstrʌktʃə]

Definition: 片状构造 piàn zhuàng gòu zào

schistose: (of metamorphic rock) having a laminar structure like that of schist. (变形岩石)片岩结构的 piàn yán jié gòu de

structure: the arrangement of and relations between the parts or elements of something complex. 结构 jié gòu; 构造 gòu zào

> Example:

The latter two rock types have a slaty to schistose structure, they are green in fresh state and show a silvery lustre on foliation planes.

后两种岩类有着从板状构造向片状构造的发展趋势，在清新状态下呈绿色，并在叶理面上显示一种银色光泽。

> Extended Term:

schistose granular structure 片粒状构造

schistose texture [ˈʃistəus ˈtekstʃə]

> Definition: is caused by the parallel orientation of microscopic grains. 片状结构 piàn zhuàng jié gòu

> Example:

The other evidences are as follows: the development of fault gouge with new stria and schistose texture and clastic materials.

其他证据有带有新条痕、片状结构和碎屑状的断层泥的发展。

schistosity [ʃisˈtɔsiti]

> Definition: mode of foliation that occurs in certain metamorphic rocks as a consequence of the parallel alignment of platy and lath-shaped mineral constituents. 片理 piàn lǐ

> Example:

In Xuefeng arcuate structural belt of Hunan, Miaoershan ductile shear zone includes two types of tectonites, mylonite and schistosity rocks, which have different deformed characteristics.

湖南雪峰弧形构造带中的苗儿山韧性剪切带存在糜棱岩和片理化岩石两大构造岩类，各具不同的变形特征。

> Extended Terms:

crystallization schistosity 结晶片理
gneissic schistosity 片麻理
linear schistosity 线状片理
marginal schistosity 边缘片理
plane schistosity 面状片理

schlieren [ˈʃliərən]

Definition: irregular dark or light streaks in plutonic igneous rock that differ in composition from the principal mass. 析离体 xī lí tǐ

Origin: German, lit., streaks; akin to slur.

Example:
Schlieren device is an important method for the measurement of transient flow field.
纹影法是进行流场测量的重要手段之一。

Extended Terms:
schlieren chamber 超快扫描照相机
schlieren optic 纹影光学
schlieren pattern 纹影图样
schlieren projector 纹影投影机
underwater schlieren 水下纹影

scoria [ˈskɔːriə]

Definition: the slag or refuse left after metal has been smelted from ore. 矿渣 kuàng zhā；火山渣 huǒ shān zhā

Origin: Middle English, dross, from Latin scōria, from Greek skōriā, from skōr, excrement, dung; see sker-in Indo-European roots.

Example:
Rare earth original mine, final mine, tailing mine and scoria were formed in the course of produce and letting.
稀土在生产和排放过程中形成了原矿、精矿、尾矿和矿渣。

Extended Terms:
lead scoria 铅析渣(粗铅氧化除杂质时产出的浮渣)
scoria cone 火山渣锥
scoria slag 矿渣
volcanic scoria 符山石

scoriaceous basalt [ˌskɔːriˈeiʃəs ˈbæsɔːlt]

Definition: 铁渣玄武岩 tiě zhā xuán wǔ yán

scoriaceous：the slag or refuse left after metal has been smelted from ore. 铁渣子的 tiě zhā zǐ de basalt: a dark, fine-grained, usually extrusive igneous rock that is more basic than andesite, consisting chiefly of plagioclase feldspars and pyroxene: often found in vast sheets, it is the most common extrusive igneous rock. 玄武岩 xuán wǔ yán

Scottish Highlands [ˈskɔtiʃ ˈhailəndz]

Definition: include the rugged and mountainous regions of Scotland north and west of the Highland Boundary Fault, although the exact boundaries are not clearly defined, particularly to the east. 苏格兰高地 sū gé lán gāo dì

Example:
Most of Northern Scotland is a mountainous area known as the Scottish Highlands.
苏格兰的大部分地方是多山地区，被称为苏格兰高地。

sea-floor spreading [ˈsiː flɔː ˈspredin]

Definition: In the theory of plate tectonics, the process by which new oceanic crust is formed by the convective upwelling of magma at mid-ocean ridges, resulting in the continuous lateral displacement of existing oceanic crust. 海底扩张 hǎi dǐ kuò zhāng

Example:
Both sea-floor spreading and oceanization may lead to the formation of oceanic crust.
海底扩张和陆壳大洋化均能形成洋壳。

Extended Term:
plate tectonics and sea-floor spreading 板块结构与海床扩展

seamount [ˈsiːmaunt]

Definition: a mountain rising from the sea floor but not reaching the surface. 海山 hǎi shān

Example:
Seamounts are valuable real estate for deep-sea creatures because they provide rocky surfaces to which the animals can affix themselves.
海山对深海生物而言是宝贵真实的资产，因为它们提供让动物依附的岩石表面。

Extended Terms:
emperor seamounts 北潍平洋海岭
mid-Pacific seamounts 中太平洋海底山群，中太平洋海底山脉

seawater ['siːˌwɔːtə]

Definition: the salty water of the ocean. 海水 hǎi shuǐ

Example:

Evidently, the sulfate normally in seawater are converted to hydrogen sulfide as the water passes through the warm crust.
显而易见，海水中的硫酸盐在海水通过暖地壳时转变成了硫化氢。

Extended Terms:

concentrated seawater 浓缩海水
seawater deaeration 海水脱氧
seawater intrusion 海水入侵
seawater encroachment 海水侵蚀

secondary enlargement ['sekəndəri in'lɑːdʒmənt]

Definition: overgrowth by chemical deposition on a mineral grain of additional material of identical composition in optical and crystallographic continuity with the original grain; crystal faces characteristic of the original mineral often result. Also known as secondary growth. 次生加大作用 cì shēng jiā dà zuò yòng

Example:

Main effects of diagenesis on porosity formation and evolution are compaction during burying, quartz secondary enlargement, carbonate cementation and dissolution.
研究结果表明：对孔隙形成与演化起主要控制作用的成岩作用是埋藏压实、石英的次生加大、碳酸盐的胶结及溶蚀溶解。

secondary growth cementation ['sekəndəri grəuθ ˌsiːmen'teiʃən]

Definition: 再生生长胶结 zài shēng shēng zhǎng jiāo jié

secondary growth: growth in vascular plants from production of secondary tissues by alateral meristem, usually resulting in wider branches and stems. 再生生长 zài shēng shēng zhǎng cementation: (chiefly Geology) the binding together of particles or other things by cement. [主地质] 黏结 nián jié; 胶接作用 jiāo jiē zuò yòng

second boiling ['sekənd 'bɔiliŋ]

Definition: 再次沸腾 zài cì fèi téng

second: another; other; additional; supplementary. 第二 dì èr

boiling: stewing, steeping, percolating, steaming, bubbling, seething, simmering, evaporating, distilling, boiling over. 沸腾的 fèi téng de

Example:

The produced second boiling water is very pure, can be used as special analysis water, widely fits to produce and prepare pure analysis water in colleges.

生产的再沸水纯度高,可以满足各种特殊分析用水,广泛用于各大中院校制备高纯度分析用水。

Extended Term:

second boiling point 第二沸点

sector zoning in crystal [ˈsektə ˈzəuniŋ in ˈkristəl]

Definition: 晶体的扇形分带 jīng tǐ de shàn xíng fēn dài

sector: part of a circle bounded by any two radii and the arc included between them. 扇形 shàn xíng

zone: a region or stratum distinguished by composition or content. 地带 dì dài

crystal: a clear, transparent mineral. 晶体 jīng tǐ

sediment [ˈsedimənt]

Definition: particulate matter that is carried by water or wind and deposited on the surface of the land or the seabed, and may in time become consolidated into rock. 沉积物 chén jī wù

Origin: mid-16th century: from French sédiment or Latin sedimentum "settling", from sedere "sit".

Example:

The underground sediment of the Ganges Delta contains arsenic.

恒河三角洲的地下沉积物中含有砷。

Extended Terms:

lacustrine sediments 湖成沉积物
land sediment 陆相沉积物
marine sediment 海底沉积物
oligomictic sediment 陆海沉积
residual sediment 残余沉积物
terrigenous sediment 陆源沉积物

sedimentary basin [ˌsedi'mentəri 'beisən]

Definition: The term sedimentary basin is used to refer to any geographical feature exhibiting subsidence and consequent infilling by sedimentation. 沉积盆地 chén jī pén dì

Example:

Gold and zinc deposits in Lixian-Taibai area occur in Devonian sedimentary basin.
礼县-太白地区的金、铅锌矿床产于泥盆纪沉积盆地。

Extended Terms:

antecedent sedimentary basin 先成沉积盆地
continental sedimentary basin 陆相沉积盆地
peripheral sedimentary basin 边缘沉积盆地

sedimentary contact [ˌsedi'mentəri 'kɒntækt]

Definition: 沉积接触 chén jī jiē chù

sedimentary: of or relating to rocks formed by the deposition of sediment. 沉淀的 chén diàn de
contact: the act or state of touching or meeting. 接触 jiē chù

sedimentary differentiation [ˌsedi'mentəri ˌdifərenʃi'eiʃən]

Definition: the progressive separation (by erosion and transportation) of a well-defined rock mass into physically and chemically unlike products that are resorted and deposited as sediments in more or less separate areas. 沉积分异作用 chén jī fēn yì zuò yòng

Example:

During the sedimentary period of Shanxi Formation, the sedimentary facies pattern was the meandering fluvial-delta deposit, which showed a sedimentary differentiation from north to south.
山西期沉积相格局呈南北向相分异的曲流河-三角洲沉积。

Extended Term:

mechanical sedimentary differentiation 机械沉积分异作用

sedimentary environment [ˌsedi'mentəri in'vaiərənmənt]

Definition: 沉积环境 chén jī huán jìng

sedimentary: (of rock) that has formed from sediment deposited by water or air. [地质][岩

石]沉积形成的 chén jī xíng chéng de

environment: the setting or conditions in which a particular activity is carried on. 环境 huán jìng; 条件 tiáo jiàn

Example:

It is a research on the sedimentary environments and characteristics of the coralreef of the Yongxing Island.
这是关于永兴岛珊瑚礁的沉积环境和沉积特征的研究。

Extended Terms:

ancient oceanic sedimentary environments 古海洋沉积环境
modern sedimentary environments 现代沉积环境
saline lake sedimentary environments 盐湖沉积环境

sedimentary facies [ˌsedi'mentəri 'feiʃiːz]

Definition: 沉积相 chén jī xiàng

sedimentary: (of rock) that has formed from sediment deposited by water or air. [地质][岩石]沉积形成的 chén jī xíng chéng de

facies: the character of a rock expressed by its formation, composition, and fossil content. [地质]相 xiàng

Example:

There is an analysis on the middle-late carboniferous sedimentary facies and model in the South of China.
这是关于中国南方中晚石炭世沉积相及沉积模式的研究。

Extended Terms:

sedimentary facies forecast 沉积相预测
sedimentary facies region 沉积相区
sedimentary facies sequence 沉积相序

sedimentary geochemical facies [ˌsedi'mentəri ˌdʒiːə'kemikəl 'feiʃiːz]

Definition: 沉积地球化学相 chén jī dì qiú huà xué xiàng

sedimentary: (of rock) that has formed from sediment deposited by water or air. [地质][岩石]沉积形成的 chén jī xíng chéng de

geochemical: the study of the chemical composition of the earth and its rocks and minerals. 地球化学 dì qiú huà xué

facies: the character of a rock expressed by its formation, composition, and fossil content. [地

质]相 xiàng

Example:

sulphur isotopic compositions of hydrogen sulphides in natural gases and the sedimentary geochemical facies.
天然气中硫化氢的硫黄同位素组成和沉积地球化学相

sedimentary model [ˌsedi'mentəri 'mɒdl]

Definition: 沉积模式 chén jī mó shì

sedimentary: (of rock) that has formed from sediment deposited by water or air. [地质][岩石]沉积形成的 chén jī xíng chéng de

model: a three-dimensional representation of a person or thing or of a proposed structure, typically on a smaller scale than the original. 模型 mó xíng

Example:

Ten sedimentary Facies were divided and four sedimentary models were gained in Western Shandong Block, while, the strata in the Taeksan Basin was divided into 6 sedimentary facies and 4 sedimentary models were set up.
在鲁西地块上寒武统中共划分出了 10 个不同的沉积相，根据沉积相组合形式，识别出了 4 个沉积模式。同时，在韩国 Taebaeksan 盆地中，晚寒武世地层划分为 6 个沉积相，并在相组合的基础上建立了 4 个沉积模式。

Extended Term:

sedimentary micro-facies models 沉积微相模式

sedimentary quartzite [ˌsedi'mentəri 'kwɔːtsait]

Definition: 沉积石英岩 chén jī shí yīng yán

sedimentary: (of rock) that has formed from sediment deposited by water or air. [地质][岩石]沉积形成的 chén jī xíng chéng de

quartzite: (Geology) an extremely compact, hard, granular rock consisting essentially of quartz. It often occurs as silicified sandstone, as in sarsen stones. [地质]石英岩 shí yīng yán

Example:

Some sandstone with siliceous cement and poor matrix have become sedimentary quartzite.
当胶结物为硅质时，一些砂岩成为沉积石英岩。

sedimentary recrystallization [ˌsedi'mentəri riːˌkristəlai'zeiʃən]

Definition: 沉积重结晶作用 chén jī chóng jié jīng zuò yòng

sedimentary：(of rock) that has formed from sediment deposited by water or air.［地质］［岩石］沉积形成的 chén jī xíng chéng de

recrystallization：forming or cause to form crystals again.（使）再结晶 zài jié jīng

Example:

chemical and isotopic constraints for sedimentary recrystallization from the Western Canada sedimentary basin
加拿大西部沉积盆地沉积重结晶作用的化学和同位素限制

sedimentary replacement [ˌsediˈmentəri riˈpleismənt]

Definition: 沉积交代作用 chén jī jiāo dài zuò yòng

sedimentary：(of rock) that has formed from sediment deposited by water or air.［地质］［岩石］沉积形成的 chén jī xíng chéng de

replacement：the action or process of replacing someone or something. 替代 tì dài；更换 gēng huàn；复位 fù wèi

sedimentary rhythm [ˌsediˈmentəri ˈriðəm]

Definition: 沉积韵律 chén jī yùn lǜ

sedimentary：(of rock) that has formed from sediment deposited by water or air.［地质］［岩石］沉积形成的 chén jī xíng chéng de

rhythm：a strong, regular repeated pattern of movement or sound. 韵律 yùn lǜ；节奏 jié zòu

Extended Terms:

annual sedimentary rhythm 年沉积节律
sedimentary rhythm characteristics 沉积韵律特征

sedimentary rock [ˌsediˈmentəri rɔk]

Definition: a rock type that has been created by the deposit and compression of sediment. This type of rock is created over millions of years while igneous rock can be created overnight. Sandstone is a good example of a sedimentary rock. 沉积岩 chén jī yán

Example:

Limestone is a sedimentary rock.
石灰石是一种沉积岩。

sedimentary sequence of transgression
[ˌsedi'mentəri 'siːkwəns ɔv træns'greʃn]

Definition: 海侵沉积层序 hǎi qīn chén jī céng xù

sedimentary: (of rock) that has formed from sediment deposited by water or air. [地质][岩石]沉积形成的 chén jī xíng chéng de
transgression: (of the sea) spread over (an area of land). (海水)侵入(陆地)(hǎ shuǐ) qīn rù (lù dì)

sedimentary structure [ˌsedi'mentəri 'strʌktʃə]

Definition: Sedimentary structures are those structures formed during sediment deposition. 沉积构造 chén jī gòu zào

Example:

The generation mechanism of orientating sedimentary structure is very important to the recognition of the internal-waves sediments.
明确指向沉积构造的形成机理对于内波沉积的识别具有重要意义。

sediment gravity flow ['sedimənt 'grævəti fləu]

Definition: 沉积物重力流 chén jī wù zhòng lì liú

sediment: particulate matter that is carried by water or wind and deposited on the surface of the land or the seabed, and may in time become consolidated into rock. [地质]沉积物 chén jī wù
gravity flow: a form of glacier movement in which the flow of the ice results from the downslope gravitational component in an ice mass resting on a sloping floor. 重力流 zhòng lì liú

Example:

Starting with core description and integrating with regional geological setting, core analysis and logging data, we make a thorough study on the origin and evolution of F4 layer in Devonian in Zarzaitine Oilfield based on the integrated analysis of depositional features with the theories of sedimentary petrology, sediment gravity flow geology, logging geology and lithofacies palaeogeography.
从岩心描述入手，结合区域地质背景、岩心分析和测井等资料，应用沉积岩石学、沉积物重力流地质学、测井地质学和岩相古地理学等理论，在对沉积特征进行综合分析的基础上，对扎尔则油田泥盆系F4层沉积成因及其演化过程进行了研究。

Extended Terms:

sediment gravity flows 沉积物重力流
sediment gravity underflow 沉积物重力流

sedimentology [ˌsedimenˈtɔlədʒi]

Definition: the branch of geology concerned with the nature and formation of sedimentary rock. 沉积学 chén jī xué

Origin: 1935-1940; sediment + -o- + -logy.

Example:
The actual analysis shows that the applied seismic sedimentology has unique advantages in sedimentary facies analysis and in lithologic trap forecasting.
实例分析表明，实用地震沉积学在沉积相分析和岩性圈闭预测方面具有独特的优势。

Extended Terms:
biologic sedimentology 生物沉积学
experimental sedimentology 实验沉积学
historical sedimentology 历史沉积学
reservoir sedimentology 储层沉积学
sedimentology related to mineral deposits 矿床沉积学

sediment transport [ˈsedimənt ˈtrænspɔːt]

Definition: Sediment transport is the movement of solid particles (sediment), typically due to a combination of the force of gravity acting on the sediment, and/or the movement of the fluid in which the sediment is entrained. 沉积物运移 chén jī wù yùn yí

Example:
Density currents generated by saline and fresh water interactions are also important in the transport of sediment in estuaries.
由盐水和淡水相互作用所产生的异重流，在河口泥沙输移中也是很重要的。

Extended Terms:
hydraulic transport of sedimen 水力输沙
sediment transport capacity of flow 水流挟沙能力

seepage reflux [ˈsiːpidʒ ˈriːflʌks]

Definition: 盐水渗透回流作用 yán shuǐ shèn tòu huí liú zuò yòng

seepage: the slow escape of a liquid or gas through porous material or small holes.（液体或气体的）渗出 shèn chū，渗漏 shèn lòu

reflux: the process of boiling a liquid so that any vapour is liquefied and returned to the stock. 回流 huí liú

Example:

Crystalline dolostone is formed by seepage-reflux dolomitization and burial dolomitization.
晶粒白云岩有两种成因：一为回流渗透白云石化；二为深埋藏白云石化。

Extended Term:

seepage reflux mechanism 渗透率

segregation of magma [ˌsegriˈgeiʃən ɔv ˈmægmə]

Definition: 岩浆的分离作用 yán jiāng de fēn lí zuò yòng

segregation: the act or process of segregating or the condition of being segregated. 分离 fēn lí
magma: The molten rock material under the earth's crust, from which igneous rock is formed by cooling. 岩浆 yán jiāng

Example:

The segregation of magma from residual crystals is strongly dependent on the viscosity of the melt.
残余晶体的岩浆分离作用高度依赖于融化物的黏度。

seismic [ˈsaizmik]

Definition: of, having to do with, or caused by an earthquake or earthquakes or by man-made earth tremors. 地震的 dì zhèn de

Origin: 1880-1885; from Greek seismós, equivalent to seis-, stem of seíein to shake, quake + -mos noun suffix.

Example:

Every link of source shot, wave spread, geophone receiving and seismic recording all has influence to the data fidelity.
震源的激发、传播、波器的接收和仪器记录各个环节对地震信息的保真度都有影响。

Extended Terms:

seismic force 地震力
seismic load 地震荷载；地震载重
seismic surface wave 地震表面波
seismic survey 地震测量；地震调查；地震探查；反射法勘探
seismic travel time 地震波传播时间

seismic discontinuity [ˈsaizmik disˌkɔntiˈnjuːiti]

Definition: 地震不连续面 dì zhèn bù lián xù miàn

seismic: of, having to do with, or caused by an earthquake or earthquakes or by man-made earth tremors. 地震的 dì zhèn de
discontinuity: a break or change in a continuous process. 中断 zhōng duàn

seismic meteorite [ˈsaizmik ˈmiːtiərait]

Definition: 地震陨石 dì zhèn yǔn shí

seismic: of, having to do with, or caused by an earthquake or earthquakes or by man-made earth tremors. 地震的 dì zhèn de
meteorite: that part of a relatively large meteoroid that survives passage through the atmosphere and falls to the surface of a planet or moon as a mass of metal or stone. 陨星 yǔn xīng

semimature stage [semiməˈtjuə steidʒ]

Definition: 半成熟阶段 bàn chéng shú jiē duàn

semimature: half developed physically; half-grown. 半成熟的 bàn chéng shú de
stage: a point, period, or step in a process or development. 阶段 jiē duàn; 时期 shí qī; 步骤 bù zhòu

Example:

The study shows that the sandstone of the third member of the Shahejie Formation have undergone immature stage, semimature stage and mature stage.
研究表明沙河街组的三段上部砂岩在埋藏过程中经历了未成熟期、半成熟期和成熟期三个阶段。

sepiolite clay [ˈsiːpiəlait klei]

Definition: 海泡石黏土 hǎi pào shí nián tǔ

sepiolite: also called meerschaum, a white, yellowish, or pink compact earthy mineral consisting of hydrated magnesium silicate: used to make tobacco pipes and as a building stone. 海泡石 hǎi pào shí
clay: a stiff, sticky fine-grained earth, typically yellow, red, or bluish-grey in colour and often forming an impermeable layer in the soil. It can be moulded when wet, and is dried and baked to make bricks, pottery, and ceramics. 黏土 nián tǔ; 泥土 ní tǔ; 陶土 táo tǔ

Example:

There are many species and types of plastic clay in China. According to their mineral compositions, they are divided into Kaolinitic clay, Montmorillonite clay, Sepiolite clay and Attapulgite clay.

我国塑性黏土种类繁多，按矿物成分可分为高岭石质黏土、蒙脱石黏土、海泡石黏土、凹凸棒石黏土等。

septarium [sepˈteəriəm]

Definition: a nodule of mineral containing cracks filled with crystalline material. 龟背石 guī bèi shí

Origin: late 18th century. From modern Latin; from Latin septum (see septum).

Example:
Calculations at 600 Hz show that for thicknesses of 1 m the septarium can be detected more easily.
600 赫兹的计算表明厚度 1 米的龟背石能够更容易被察觉。

Extended Term:
septarian structure 龟甲构造

seriate porphyritic texture [ˈsiəriət ˌpɔːfiˈritik ˈtekstʃə]

Definition: 连续不等粒斑状结构 lán xù bù děng lì bān zhuàng jié gòu

seriate：arranged or occurring in a series or in rows. 连续的 lián xù de

porphyritic：containing relatively large isolated crystals in a mass of fine texture. 斑岩的 bān yán de

texture：the arrangement of the particles or constituent parts of any material, as wood, metal, etc., as it affects the appearance or feel of the surface; structure, composition, grain, etc. 结构 jié gòu

Example:
These dikes show the characteristic seriate-porphyritic texture of lamprophyres.
这些堤坝表现出煌斑岩的连续不等粒斑状结构特征。

serpentinite [ˌsɜːpənˈtiːnait]

Definition: a metamorphic rock consisting almost entirely of minerals in the serpentine group. Serpentinite forms from the alteration of ferromagnesian silicate materials, such as olivine and pyroxene, during metamorphism. 蛇纹岩 shé wén yán

Origin: 1350-1400; Middle English (adj.); from Latin serpentīnus snakelike, equivalent to serpent-serpent + -īnus ine.

> Example:

Water changes the rock to serpentinite, named for its green snakeskin pattern, which, as it weathers, breaks down into nutrient-poor soil, heavy with metals.
水力把岩石侵蚀成为蛇纹岩，因其绿色的如同蛇皮的图案而得名。

> Extended Terms:

Alpine peridotite serpentinite 阿尔卑斯型橄榄蛇纹岩
serpentinite mylonite 蛇纹糜棱岩

serrate granoblastic texture [ˈsereit ɡrænəuˈblæstik ˈtekstʃə]

> Definition: 齿形粒状变晶结构 chǐ xíng lì zhuàng biàn jīng jié gòu

serrate: notched on the edge like a saw 锯齿状的 jù chǐ zhuàng de
granoblastic: is an anhedral phaneritic equi-granular metamorphic rock texture. Granoblastic texture is typical of quartzite, marble and other non-foliated metamorphic rocks without porphyroblasts. 花岗变晶状 huā gāng biàn jīng zhuàng
texture: the arrangement of the particles or constituent parts of any material, as wood, metal, etc., as it affects the appearance or feel of the surface; structure, composition, grain, etc. 结构 jié gòu

settling of crystal [ˈsetliŋ ɔv ˈkristəl]

> Definition: 晶体沉淀 jīng tǐ chén diàn

settling: dregs; sediment. 沉淀物 chén diàn wù
crystal: a clear, transparent mineral. 晶体 jīng tǐ

> Example:

According to the observation, the settling of crystal takes place in 15 minutes.
据观测，晶体沉淀在15分钟内发生。

settling velocity [ˈsetliŋ viˈlɒsəti]

> Definition: the rate at which suspended solids subside and are deposited. Also known as fall velocity. (mechanics) The velocity reached by a particle as it falls through a fluid, dependent on its size and shape, and the difference between its specific gravity and that of the settling medium; used to sort particles by grain size. 沉速 chén sù

> Extended Terms:

group settling velocity 群体沉速

hydrostatic settling velocity 静水沉降速度
terminal settling velocity 终端沉降速度

shale [ʃeil]

Definition: Shale is a fine-grained, clastic sedimentary rock composed of mud, which is a mix of flakes of clay minerals and tiny fragments (silt-sized particles) of other minerals, especially quartz and calcite. 页岩 yè yán

Origin: mid-18th century. Ultimately from Germanic, "split".

Example:
North America has vast deposits of shale.
北美有大量油页岩沉积。

Extended Terms:
bituminous shale 沥青页岩
carbonaceous shale 碳质页岩
oil shale 油母页岩
shale oil 页岩油

shatter cones ['ʃætə kəunz]

Definition: are rare geological features that are only known to form in the bedrock beneath meteorite impact craters or underground nuclear explosions. 震裂锥 zhèn liè zhuī

Example:
All these observations imply a restricted range of pressure allowing the formation of shatter cones.
所有的观测显示震裂锥形成的压力在一个有限的范围内。

shear [ʃiə]

Definition: to cut with shears or a similar sharp-edged instrument. 剪切 jiǎn qiē

Origin: before 900; (v.) Middle English sheren, Old English sceran, cognate with Dutch, German scheren, Old Norse skera; (noun) (in sense "tool for shearing") Middle English sheres (plural), continuing Old English scērero, scēar.

Example:
This linear, narrow, highly strained belt is a ductile shear zone. The rocks within this zone have

been suffered intense deformation which is accomplished entirely by ductile flow.
这个线状的狭窄高应变带是一条韧性剪切带，带内的岩石曾遭受了强烈的变形，而这种变形则是由韧性流动所完成的。

> Extended Terms:

shear deformation/strain 剪切变形
shear reinforcement 抗剪钢筋
shear strength 剪切强度

shear band cleavages [ʃiə bænd ˈkliːvidʒz]

> Definition: 剪切带劈理 jiǎn qiē dài pī lǐ

shear: to remove the fleece from a sheep etc by clipping. 剪切 jiǎn qiē
band: a thin layer of distinctive rock, ore, etc. 带 dài
cleavages: a cleaving, splitting, or dividing. 劈理 pī lǐ

> Example:

Sets of synthetic and antithetic shear band cleavages (SBC) developed in strongly anisotropic zones.
合成的和反向剪切带劈理集合在各强向异性区域中形成。

shear sense indicator [ʃiə sens ˈindikeitə]

> Definition: 剪切指向标志 jiǎn qiē zhǐ xiàng biāo zhì

shear: to remove the fleece from a sheep etc by clipping. 剪切 jiǎn qiē
sense: an intuitive or acquired perception or ability to estimate. 感觉 gǎn jué
indicator: a person or thing that indicates. 指示符 zhǐ shì fú

> Example:

In the SS domain lots of shear sense indicators show sinistral strike-slip, features, and pure shear strain indicators are seen locally in outcrops perpendicular to foliations.
在 SS 区域内，大量的标志显示剪切作用为左旋走滑，在垂向剖面露头上可看到一些纯剪现象。

shear zone [ʃiə zəun]

> Definition: 剪切带 jiǎn qiē dài

shear: to remove the fleece from a sheep etc by clipping. 剪切 jiǎn qiē
zone: a region or stratum distinguished by composition or content. 地带 dì dài

Example:

The mantle shear zones not only control distribution of the chromites, but also provide heat souce and tectonic dynamics.

地幔剪切带不仅控制了铬铁矿的分布，而且为铬铁矿的形成提供了热动力和构造动力。

sheeted dyke complex [ˌʃiːtid daik ˈkɒmpleks]

Definition: is a normal component of an ophiolite, a piece of oceanic crust that has been emplaced within a sequence of continental rocks. 席状岩墙复合体 xí zhuàng yán qiáng fù hé tǐ

Example:

The sheeted dyke complex consists of a series of diabase dykes with single-side chilled borders, one against the other and without any fillings between these dykes.

席状岩墙杂岩由一系列具单向冷凝边的辉绿岩墙组成，以一墙挨一墙的形式产出，岩墙间无任何填充物。

shelf facies [ʃelf ˈfeiʃiːz]

Definition: a sedimentary facies characterized by carbonate rocks and fossil shells and produced in the neritic environments of marginal shelf seas. Also known as foreland facies; platform facies. 陆架相 lù jià xiàng

Example:

In Middle-Late Ordovician, there mainly were carbonate-classic mixed rock shelf facies in the west part and deep-water slope facies in the east part.

中-晚奥陶世，塔中地区西部以混积陆架相为主，东部以深水斜坡相为主。

shell limestone [ʃel ˈlaimstəun]

Definition: 贝壳灰岩 bèi ké huī yán

shell: the hard protective outer case of a mollusc or crustacean(软体或甲壳纲动物的). 壳 ké limestone: a hard sedimentary rock, composed mainly of calcium carbonate or dolomite, used as building material and in the making of cement. 灰岩 huī yán; 石灰岩 shí huī yán

Example:

The shell limestone is thin bed, compact and anisotropism, so the reservoir is of low porosity, low permeability and pore fracture character.

介壳灰岩层薄、致密、非均质性强，属低孔、低渗孔隙的裂缝性储层。

shield volcano [ʃiːld vɒlˈkeinəu]

Definition: 盾状火山 dùn zhuàng huǒ shān, 循形火山 xún xíng huǒ shān
shield: anything shaped like a shield, as a plaque, trophy, badge, or emble. 盾 dùn
volcano: a vent in the earth's crust through which molten rock (lava), rock fragments, gases, ashes, etc. are ejected from the earth's interior; a volcano is active while erupting, dormant during a long period of inactivity, or extinct when all activity has finally ceased. 火山 huǒ shān

Example:
Petrography, petrology, paleomagnetism and variations of thickness of the basalt andsedimentary interlayers indicate that the oil reservoir basalt came from a shield volcano whose vent is situated within 1.5-2km to the northeast of the oil area.
岩相岩石学、古地磁学以及玄武岩和其中沉积岩夹层的厚度变化显示储油玄武岩来自同一座盾火山, 火口位置在油区东北方1.5~2km 范围内。

shoal of platform margin facies [ʃəul ɔv ˈplætfɔːm ˈmɑːdʒin ˈfeiʃiːz]

Definition: 台地边缘浅滩相 tái dì biān yuán qiǎn tān xiàng
platform: a raised level surface on which people or things can stand. 台 tái; 平台 píng tái
shoal: a submerged sandbank visible at low water. 浅滩 qiǎn tān; 沙洲 shā zhōu
facies: the character of a rock expressed by its formation, composition, and fossil content. [地质]相 xiàng

Example:
reservoir characteristics of middle triassic shoal of platform margin facies at Qingyan in Guiyang, Guizhou
贵州贵阳青岩的中三叠世台地边缘浅滩相的储层特性

shock metamorphism [ʃɔk ˌmetəˈmɔːfizəm]

Definition: Shock metamorphism or impact metamorphism describes the effects of shock-wave related deformation and heating during impact events. The formation of similar features during explosive volcanism is generally discounted due to the lack of metamorphic effects unequivocally associated with explosions and the difficulty in reaching sufficient pressures during such an event. 冲击变质作用 chōng jī biàn zhì zuò yòng

Example:
Since the neocatastrophism has rapidly developed from the beginning of 1980's, the study of meteorite crater and shock metamorphism has become one of the most important topics in the earth sciences all over the world.

自 20 世纪 80 年代初新突变学说兴起以来，陨击坑及冲击变质作用研究已成为国际地学界的重大课题之一。

shoe string sand [ʃuː strɪŋ sænd]

Definition: a shoestring composed of sand and usually buried in mud or shale, usually a sandbar or channel fill. 鞋带砂堆 xié dài shā duī；鞋带砂岩（shoe string sandstone）xié dài shā yán

Example:
The buried intervein areas are similar to the "shoe-string" sands of Kansas and Oklahoma.
掩埋的交错区域和堪萨斯以及俄克拉荷马周的鞋带砂岩类似。

Extended Term:
shoe string deposit 鞋带状油藏

shonkinite ['ʃɒŋkinait]

Definition: a very dark (often black) nepheline syenite with over 50% dark minerals, mostly Aegirine and Augite. 富辉正长岩 fù huī zhèng cháng yán

Example:
Syenite and monzonite are intruded by phlogopite and shonkinite.
正长岩和二长岩被金云母和富辉正长岩侵蚀。

Extended Terms:
mica-shonkinite 云母等色岩
nepheline shonkinite 霞石等色岩
shonkinite-porphyry 等色斑岩；暗辉正长斑岩

shore conglomerate [ʃɔː kənˈɡlɒmərət]

Definition: 滨岸砾岩 bīn àn lì yán

shore：the land along the edge of a sea, lake, or other large body of water.（海、湖或其他大水域的）岸 àn；滨 bīn
conglomerate：(Geology) a coarse-grained sedimentary rock composed of rounded fragments embedded in a matrix of cementing material such as silica. [地质]砾岩 lì yán

Example:
The sediments vary from coarse conglomerate to fine silt and clay.
沉积物各不相同，有粗糙的砾岩，也有精细的泥沙和黏土。

shoshonite [ʃəuːˈʃəunait]

Definition: is a basaltic rock, properly a potassic trachyandesite, composed of olivine, augite and plagioclase phenocrysts in a groundmass with calcic plagioclase and sanidine and some dark-colored volcanic glass. 钾玄岩 jiǎ xuán yán

Example:
The rocks include shoshonite, latite, trachyte, mugearite, dacite and rhyolite.
岩石类型主要有钾玄岩、安粗岩、粗面岩、橄榄粗安岩、英安岩和流纹岩。

shoshonite series [ʃəuːˈʃəunaitˈsiəriːz]

Definition: 钾玄岩系列 jiǎ xuán yán xì liè

shoshonite: a basaltic rock composed of olivine and augite phenocrysts in a groundmass of labradorite with orthoclase rims, olivine, augite, a small amount of leucite, and some dark-colored glass. 橄榄玄粗岩 gǎn lǎn xuán cū yán

series: a number of things, events, or people of a similar kind or related nature coming one after another. 系列 xì liè

Example:
High-K and shoshonite series calcic pyroxenes extend into the diopside and salite compositional fields.
高钾和钾玄岩系列含钙辉石延伸到透辉石和次透辉石组合地域。

siderite [ˈsaidərait]

Definition: a yellowish to brownish, semihard mineral, iron carbonate, $FeCO$, that is a valuable ore of iron. 菱铁矿 líng tiě kuàng

Example:
Under supergene process influencing, kaolinites inverted dickites, siderite were been ferritization, part of carbonate mineral have been dissolute producing calcite.
在表生作用的影响下，高岭石向地开石转化，菱铁矿褐铁矿化，部分碳酸盐矿物溶解生成方解石。

Extended Terms:
aerolite aerolith cloud stone meteorolite 陨石
aerosiderite siderite 陨铁
magnesian siderite 镁菱铁矿

siderite content 菱铁矿含量
siderite concentrate 菱铁精矿

sideromelane [ˌsidərəˈmelein]

Definition: is a vitreous basaltic volcanic glass, usually occurring in palagonite tuff, for which it is characteristic. 铁镁矿物 tiě měi kuàng wù

Example:

A rind consists of sideromelane, tachylyte, and trachylytic basalt.
由铁镁矿物、玄武玻璃和粗面玄武岩构成的壳。

siderophile element [saidˈəʊfail ˈelimənt]

Definition: 亲铁元素 qīn tiě yuán sù

siderophile: in the Goldschmidt classification, an element that forms alloys easily with iron and is concentrated in the Earth's core. 亲铁的 qīn tiě de

element: any of the four substances (earth, air, fire, and water) formerly believed to constitute all physical matter. 元素 yuán sù

Example:

The lithophile elements in ordinary and carbonaceous chondrite groups may display either chalcophile or siderophile elements in enstatite chondrites.
普通球粒陨石和碳质球粒陨石中的亲石元素在顽火辉石球粒陨石内显示亲铜或亲铁的元素。

sieve sediment [siv ˈsedimənt]

Definition: 筛积物 shāi jī wù

sieve: a utensil consisting of a wire or plastic mesh held in a frame, used for straining solids from liquids, for separating coarser from finer particles, or for reducing soft solids to a pulp 筛 (子) shāi zǐ; 筛网 shāi wǎng; 滤器 lǜ qì

sediment: particulate matter that is carried by water or wind and deposited on the surface of the land or the seabed, and may in time become consolidated into rock. [地质]沉积物 chén jī wù

sieve texture [siv ˈtekstʃə]

Definition: 筛状结构 shāi zhuàng jié gòu

sieve: a utensil having many small meshed or perforated openings, used to strain solids from liquids, to separate fine particles of loose matter from coarser ones, etc.; sifter; strainer. 筛子 shāi zǐ

texture: the arrangement of the particles or constituent parts of any material, as wood, metal, etc., as it affects the appearance or feel of the surface; structure, composition, grain, etc. 结构 jié gòu

Example:

Research results show that the special texture—sieve texture of andalusite in Zhoukoudian ore is a chief cause of difficult to mineral processing of ore.

结果表明，其特殊结构——周口店矿石的筛孔状结构是红柱石矿难选的主要原因。

silcrete ['silkriːt]

Definition: is an indurated soil duricrust formed when silica is dissolved and resolidifies as a cement. It is a hard and resistant material, and though different in origin and nature, appears similar to quartzite. 硅质壳层 guī zhì qiào céng

Example:

The silcrete found at archaeological sites was glossy with a fine grain and a reddish color.

在南非的考古地点发现的硅结砾岩很光滑，带有细致的颗粒，呈红色。

silica activity in magma ['silikə æk'tivəti in'mægmə]

Definition: 岩浆中二氧化硅的活动性 yán jiāng zhōng èr yǎng huà guī de huó dòng xìng

silica: a glassy, very hard mineral, silicon dioxide, SiO, found in a variety of forms, as in quartz, opal, chalcedony, sand, or chert. 二氧化硅 èr yǎng huà guī

magma: the molten rock material under the earth's crust, from which igneous rock is formed by cooling. 岩浆 yán jiāng

silica alkalic index ['silikə æl'kælik 'indeks]

Definition: 硅碱指数 guī jiǎn zhǐ shù

silica: a hard, unreactive, colourless compound which occurs as the mineral quartz and as a principal constituent of sandstone and other rocks. 硅石 guī shí

alkalic: (of a rock or mineral) richer in sodium and/or potassium than is usual for its type. [地质](岩石或矿物)碱性的 jiǎn xìng de

index: a figure in a system or scale representing the average value of specified prices, shares, or other items as compared with some reference figure. 指数 zhǐ shù

Example:

Silica alkalic index is also known as Rittmann index.
硅碱指数又称里特曼指数。

silica fire clay [ˈsilikə ˈfaiə klei]

Definition: 硅质耐火黏土 guī zhì nài huǒ nián tǔ

silica: a hard, unreactive, colourless compound which occurs as the mineral quartz and as a principal constituent of sandstone and other rocks. 硅石 guī shí

fire clay: a type of clay that is able to withstand intense heat, used to make firebricks, crucibles, and other objects that are exposed to high temperatures. 耐火黏土 nài huǒ nián tǔ

Example:

A study was made of the factors affecting the pyrometric cone equivalent (P. C. E.) of silica fire clay made of ganister, clay and silica-brick.
一项研究针对由致密硅岩、黏土和硅砖构成的硅质耐火黏土是温熔锥当量的影响因素。

silica mineral [ˈsilikə ˈminərəl]

Definition: any of the forms of silicon dioxide (SiO_2), including quartz, tridymite, cristobalite, coesite, stishovite, lechatelierite, and chalcedony. Various kinds of silica minerals have been produced synthetically; one is keatite. 二氧化硅矿物 èr yǎng huà guī kuàng wù

Example:

Translucent, semiprecious variety of the silica mineral chalcedony that owes its red to reddish brown color to the incorporation of small amounts of iron oxide.
二氧化硅矿物玉髓的半透明、次宝石变种由于存在分散的胶体赤铁矿（氧化铁）故呈红色到红褐色或肉红色。

silica saturation/undersaturation

[ˈsilikə sætʃəˈreiʃən/ ˌʌndəˌsætʃəˈreiʃən]

Definition: 二氧化硅饱和/未饱和 èr yǎng huà guī bǎo hé/wèi bǎo hé

silica: a glassy, very hard mineral, silicon dioxide, SiO, found in a variety of forms, as in quartz, opal, chalcedony, sand, or chert. 二氧化硅 èr yǎng huà guī

saturation: a saturating or being saturated. 饱和 bǎo hé

Example:

It aims at analyzing soil color, mechanical composition, pH, base saturation, charge

properties, ba value, silica-alumina molar ratio of silt and clay, clay mineralogy, iron oxide forms, the index of silica reactivity and the index of silica saturation.
旨在对土壤颜色、颗粒组成、土壤酸度、电荷特性、土壤风化淋溶系数、粘粒和粉砂的硅铝率、铁形态组成、自然粘粒率、硅反应指标和二氧化硅饱和度指标等做出分析。

silicate liquid [ˈsilikit ˈlikwid]

Definition: 硅酸盐流体 guī suān yán liú tǐ

silicate: a compound of silica which does not dissolve. There are many different kinds of silicate. 硅酸盐 guī suān yán

liquid: a substance which is not solid but which flows and can be poured, for example, water. 流体 liú tǐ

Extended Terms:

liquid for silicate cement 硅粘固粉液，瓷（粘固）粉液
liquid sodium silicate 液态硅酸钠

siliceous rock [siˈliʃəs rɔk]

Definition: Siliceous rocks are sedimentary rocks that have silica (SiO_2) as the principal constituent. The most common siliceous rock is Chert other types include Diatomite. They commonly form from silica-secreting organisms such as radiolarians, diatoms, or some types of sponges. 硅质岩 guī zhì yán

Example:

The existence of phengites in the metamorphic siliceous rock indicates that it was formed through middle-high pressure metamorphism.
岩石中多硅白云母的存在显示其曾经受了中-高压的变质作用。

Extended Terms:

silicatization 硅化作用
silicic acid 硅酸

siliceous shale [siˈliʃəs ʃeil]

Definition: 硅质页岩 guī zhì yè yán

siliceous: of a hard, unreactive, colourless compound which occurs as the mineral quartz and as a principal constituent of sandstone and other rocks 硅石的 guī shí de

shale: soft finely stratified sedimentary rock that formed from consolidated mud or clay and can be split easily into fragile plates. 页岩 yè yán

Example:

By the experiments, the effects of several mineral admixtures, such as silica fume, superfine slag and siliceous shale on the High Strength Concrete (HSC) is researched.

通过实验,研究了硅灰、超细矿渣、硅质页岩几种矿物质掺合料对高强混凝土强度的影响。

Extended Terms:

siliceous sediments 硅质沉积

siliceous sinter 硅华

---

sill [sil]

Definition: an approximately horizontal sheet of igneous rock intruded between older rock beds. 岩床 yán chuáng

Origin: before 900; Middle English sille, Old English syl, sylle; cognate with Low German süll, Old Norse syll; akin to German Schwelle sill.

Example:

In the northern part of mine area, stock of granite porphyry played a dominate role in coal metamorphism, and intrusive sill in coal seam was found locally.

在矿区北部,花岗斑岩岩株对煤变质起了主导作用,侵入煤层的岩床的影响是局部的。

Extended Term:

laccolithic sill 岩盖岩床

---

sillimanite [ˈsilimənait]

Definition: a very hard, mineral, aluminum silicate, AlSiO, usually found in metamorphic rock as long, slender crystals of various colors. 硅线石 guī xiàn shí

Origin: after B. Silliman (1779-1864), U. S. chemist and geologist+ -ite.

Example:

By adding brown corundum powder and sillimanite, the clay bonded castable used in the car surface of ceramic tunnel kiln will increase in hot strength, thermal shock rest resistance, use life.

通过在黏土结合浇注料中加入棕刚玉、硅线石,将其用于陶瓷隧道窑车面,具有高温强度大、抗剥落性好、使用寿命长的特点。

Extended Terms:

kyanite-sillimanite type facies series 蓝晶石-硅线石型相系

sillimanite brick 硅线石砖

sillimanite gneiss 硅线片麻岩

sillimanite zone 硅线石带

silt [silt]

Definition: sediment whose particles are between clay and sand in size (typically 0.002-0.06 mm). 粉砂 fěn shā

Origin: late Middle English: probably originally denoting a salty deposit and of Scandinavian origin, related to Danish and Norwegian sylt "salt marsh".

Example:
After many years of repeated analysis of the floods caused by the Yellow River, experts found that the flow of the river is fully capable of washing down the silt into the sea, as long as balanced amount of silt and water is controlled.
通过多年来对黄河洪水过程的大量分析，专家们终于发现，只要找到一个合适的水沙平衡关系，黄河水流是完全有能力输沙入海的。

Extended Terms:
abrasive silt 腐蚀淤泥
clayey silt 黏质粉土
coal silt 煤泥滓
siltpan 粉砂磐
silt-seam 粉土(淤泥)层

siltstone [ˈsiltstəun]

Definition: Siltstone is a sedimentary rock which has a grain size in the silt range, finer than sandstone and coarser than claystones. 粉砂岩 fěn shā yán

Example:
These folded rocks are composed of medium to thick bedded limestone of Ordovician and of thin bedded shale and siltstone of Silurian.
这些褶皱岩层是由奥陶系的中层至厚层石灰岩以及志留系的薄层页岩和粉砂岩组成。

Extended Terms:
tuffaceous siltstone 凝灰质粉砂岩
dolomitic siltstone 白云石质粉砂岩

sinuosity index [ˌsinjuˈɒsiti ˈindeks]

Definition: Sinuousity or sinuosity index is a measure of deviation of a path length from the shortest possible path. 弯度指数 wān dù zhǐ shù

Example:

This study aims at developing a new sinuosity index for cyclone tracks in the tropical South Pacific.

该研究旨在在热带南太平洋建立一个新的气旋路线弯度指数。

skarn [skɑːn]

Definition:
a coarse-grained metamorphic rock formed by the contact metamorphism of carbonate rocks. Skarn typically contains garnet, pyroxene, epidote, and wollastonite. 矽卡岩 xī kǎ yán

Example:
The assemblage of garnet-diopside in the silicate alteration rocks is different from that in the skarn of traditional idea.

硅酸盐蚀变岩中的"石榴子石-透辉石"组合与传统理解的"矽卡岩"内的相同组合名同实异。

Extended Terms:
garnet skarn 石榴矽卡岩
ore skarn 矿石矽卡岩
primary skarn 原生矽卡岩
reaction skarn 反应矽卡岩
zoned skarn 带状矽卡岩

skeletal limestone [ˈskelitəlˈlaimstəun]

Definition:
骨粒灰岩 gǔ lì huī yán

skeletal: of, relating to, or functioning as a skeleton. 骨骼的 gǔ gé de
limestone: a hard sedimentary rock, composed mainly of calcium carbonate or dolomite, used as building material and in the making of cement. 灰岩 huī yán

Example:
The fossil, preserved in 418-million-year-old limestone, is around 33 centimeters long and shows the skeletal anatomy of a small sarcopterygian.

这块化石保存在距今大约 4.18 亿年前的石灰岩中，长约 33 厘米，带有小型肉鳍鱼的解剖学特征。

Extended Terms:
non-skeletal limestone 非骨骼灰岩
skeletal grain 骨粒
skeletal remains 骨骼残骸

skeletal sand 骨屑砂

skeletal particle [ˈskelitəlˈpɑːtikl] = skeleton grain

Definition: 骨粒 gǔ lì，又称骨骼颗粒 gǔ gé kē lì

skeletal：of, relating to, or functioning as a skeleton. 骨骼的 gǔ gé de
particle：a minute portion of matter. 颗粒 kē lì

Example:

The results were as follows：invariegated horizon, the skeleton grain, with porphritic distribution, was dominated byquartz.
结果表明：其骨骼颗粒以石英为主，呈斑晶状分布。

Extended Term:

grain skeleton 颗粒骨架

slate [sleit]

Definition: a hard, fine-grained, metamorphic rock, typically formed from shale, that cleaves naturally into thin, smooth-surfaced layers. 板岩 bǎn yán

Extended Terms:

slate blue 石蓝色
slate cement 板岩水泥
slate grey 灰石色，鼠灰色
clay slate 粘板岩
slate switch 石板闸

slaty cleavage [ˈsleiti ˈkliːvidʒ]

Definition: is a type of foliation expressed by the tendancy of a rock to split along parallel planes. 板劈理 bǎn pī lǐ

Example:

At the late Palaeozoic the study area entered Jinlin stage at which consolidated Tarim platform basement broke apart partially, resulting in a scrips of east-west fault basins The Xingdi Fault accepted the second-phase deformation, characterized by plastic deformation mainly of slaty cleavage and phyllitic cleavage.
晚元古代本区进入晋宁期发展阶段，团结的塔里木地台基底局部裂开，形成一系列近东西向的断陷盆地，发生了第二期变形，形成以塑性变形为主的板劈理和千枚理为其特征。

slide sediments facies [slaid 'sedimənts 'feiʃiːz]

Definition: 滑坡堆积相 huá pō duī jī xiàng

slide: the descent of a large mass of earth or rocks or snow etc. 崩塌 bēng tā; 滑坡 huá pō

sediment: sand, stones, mud, etc. carried by water or wind and left, for example, on the bottom of a lake, river, etc. 沉积物 chén jī wù

facies: The character of a rock expressed by its formation, composition, and fossil content. [地质]相 xiàng

slide deposit ['slaid di'pɔzit]

Definition: 滑移沉积 huá yí chén jī

slide: move along a smooth surface while maintaining continuous contact with it. 滑动 huá dòng; 移动 yí dòng

deposit: a natural layer of sand, rock, coal, or other material. (沙、岩石、煤等的)天然沉积层 tiān rán chén jī céng

Example:

Sedimentary processes identified include deep-water original position deposit, deep-water contour-culrent deposit, gravity-flow deposit and gravity-slide deposit.

已鉴别出的沉积作用包括深水原地沉积、深水等深流沉积、深水重力流沉积及重力滑移沉积。

slip direction [slip di'rekʃən]

Definition: 滑移方向 huá yí fāng xiàng

slip: to go, move, pass, etc., smoothly, quickly, or easily. 滑动 huá dòng

direction: the point toward which something faces or the line along which something moves or lies. 方向 fāng xiàng

Example:

In mylonitized saxonite, the mantle-formed forsterite shows a typical multilayed structure and an obviously slip direction, the trace direction is 211, and a number of exsolved materials further generated out.

在糜棱岩化斜辉橄榄岩内的幔成镁橄榄石表现出典型的多层结构，而且滑移方向明显，测定的速线方向为211，进一步还形成了许多出溶物。

slip plane [slip plein]

Definition: 滑动面 huá dòng miàn

slip: to go, move, pass, etc, smoothly, quickly, or easily. 滑动 huá dòng

plane: a surface containing all the straight lines that connect any two points on it. 面 miàn

Example:

The distribution of carbides depends on the deformed temperature, random distribution at higher deformed temperature and directional distribution along slip plane F111G_γ at lower deformed temperature.

碳化物的分布与形变温度有关，高温形变呈随机分布，而低温下则趋于沿 F111G_γ 滑移面析出呈方向性分布。

Extended Terms:

bedding-plane slip 曲滑，层面滑动
crystal-plane slip 晶面滑移
non-slip plane 非滑移面
slip plane westbank 滑面

slope basin facies [sləup ˈbeisən ˈfeiʃiːz]

Definition: 斜坡盆地相 xié pō pén dì xiàng

slope: a surface of which one end or side is at a higher level than another; a rising or falling surface. 斜坡 xié pō；坡地 pō dì

basin: (Geology) a circumscribed rock formation where the strata dip towards the centre. [地质] 盆地 pén dì

facies: the character of a rock expressed by its formation, composition, and fossil content. [地质] 相 xiàng

Example:

The Namgyaixoi Group consists dominantly of basic volcanic rocks in the slope-basin facies in which abundant thin-shelled bivalves occur.

朗杰学群中含大量基性火山岩，总体上为斜坡盆地相，产大量薄壳型双壳，各岩组、岩性段间多呈断层接触。

slope wash facies [sləup wɔːʃ ˈfeiʃiːz]

Definition: 坡积相 pō jī xiàng

slope: a surface of which one end or side is at a higher level than another; a rising or falling surface. 斜坡 xié pō;坡地 pō dì

wash: of water to flow or carry something/somebody in a particular direction. (向着某一方向)流动 liú xiàng; 冲向 chōng xiàng

facies: the character of a rock expressed by its formation, composition, and fossil content. [地质]相 xiàng

Example:

Slope wash facies sedimentation refers to a type of band sedimentation of weathering products.
坡积相沉淀指风化物的一种带状沉淀。

slow-spreading ridge [sləu'spredɪŋ rɪdʒ]

Definition: 慢速扩张脊 màn sù kuò zhāng jǐ

slow: not quick or clever in understanding; dull; obtuse. 慢的 màn de

spreading: extended, extensive, spread out, growing, widening, ever-widening, parasitic, radial. 扩张 kuò zhāng

ridge: a long, narrow elevation of land or a similar range of hills or mountains. 山脊 shān jǐ

Example:

The thickness of oceanic crust calculated with the main elements of spilites is 3.2km, implying that the Shangzhou ophiolite suite was derived from a slow-spreading ridge.
用细碧岩主要元素估算的洋壳厚度为 3.2km,证明商州蛇绿岩带形成于慢速扩张脊。

slump structure [slʌmp'strʌktʃə]

Definition: 滑塌构造 huá tā gòu zào, 又称滑陷构造 huá xiàn gòu zào

slump: to sit or fall down heavily. 坐下或倒下 zuò xià huò dǎo xià

structure: the arrangement of and relations between the parts or elements of something complex. 结构 jié gòu;构造 gòu zào

Example:

Ice-induced slump structure is an unusual structure that is formed in such a state that the sediment is transformed from the being frozen to being melt, and its motion is featured by sliding, mainly with horizontal motion.
冰成滑塌构造是在沉积物从冰冻状态转向冰成状态发育的一种特殊构造,其运动学特点是以滑,即以水平运动为主,冰成滑塌构造多呈鳞片状或皱褶状。

Extended Term:

syndepositional slump structure 同生滑移构造;同生积堆移构造

Sm(samarium) [səˈmeəriəm]

Definition: a chemical element, one of the rare-earth elements: symbol, Sm; at. no., 62. 元素钐(samarium)的符号 yuán sù shān de fú hào

Origin: named (1879) by Boisbaudran (see gallium), from French samarskite (see samarskite), in which it occurs + -ium.

Example: This paper reports reductive reaction of the thiocarbamic esters with samarium diiodide to give disulfidesand carboxamides in good results.
在二碘化钐作用下，硫代氨基甲酸酯发生还原反应，得到较好产率的二硫醚和酰胺化合物。

Extended Terms:
samarium boride 硼化钐
samarium carbide 碳化钐
samarium nitride 氮化钐
samarium oxide 氧化钐
samarium silicate 硅酸钐

smectite [ˈsmektait]

Definition: a type of clay mineral (e.g., montmorillonite) that undergoes reversible expansion on absorbing water. 蒙脱石 méng tuō shí

Origin: 1805-1815; from Greek smēkt(ós) smeared + -ite.

Example: Wind was the main agent for transporting illite from land to ocean, and also ocean currents that influenced the distribution of the clays, especially the smectite.
风力的吹扬是研究区伊利石物质来源的主要输送方式，洋流作用对黏土矿物，尤其是蒙皂石矿物的分布具有一定影响。

Extended Term:
dioctahedral smectite 蒙脱石

sodalite [ˈsəudəlait]

Definition: a usually blue, hard mineral, Na(AlSiO)Cl, used in making jewelry; sodium aluminum silicate chloride. 方钠矿 fāng nà kuàng；方钠石 fāng nà shí；钠沸石 nà fèi shí；苏打石 sū dá shí

Origin: 1550-1560.

Example:

The family of zeolites that can be synthesized by the new dry powder method was extended from pentasil zeolites to sodalite having the structure composed of 4- and 6-member rings.
用新颖的干粉方法合成系列沸石的研究已从五元环沸石延伸到具有四元环和六元环的方钠石。

Extended Terms:

sodalite Braun tube 方钠石显像管
sodalite nephelinite 方钠霞石岩
sodalite phonelite 方钠响岩
sodalite syenite 方钠正长岩
transparent sodalite 透明方钠石

sodic ['səudik]

Definition: relating to or containing sodium. 钠的 nà de，含钠的 hán nà de

Origin: 1800-1810; from New Latin; see soda, -ium.

Example:

PAM application through mixing obviously decreased steady hydraulic conductivity of the two soils, but its effect on sodic soil was greater than on non sodic soil.
PAM 混合施用显著地减小了土壤的稳定水力传导度，但它对碱土的影响程度大于对非碱土的影响。

Extended Terms:

sodic albite-magnetite alteration 钠长石-磁铁矿蚀变作用
sodic metasomatism 钠质交代作用
sodic-metasomatism type uranium deposit 钠交代型铀矿
sodic soil 苏打土

soft sediment deformation [sɔft'sedimənt ˌdiːfɔː'meiʃən]

Definition: 软沉积物变形 ruǎn chén jī wù biàn xíng

soft: easy to mould, cut, compress, or fold; not hard or firm to the touch. 易弯曲的 yì wān qū de；柔软的 róu ruǎn de

sediment: sand, stones, mud, etc. carried by water or wind and left, for example, on the bottom of a lake, river, etc. 沉积物 chén jī wù

deformation: the action or process of changing in shape or distorting, especially through the application of pressure. (尤指受压)变形 biàn xíng

> Example:

Soft sediment deformation is one of the key indicators identifying the existence of paleo-earthquake.
软沉积物变形构造是确定古地震存在的关键证据之一。

solid-solid reaction ['sɔlid'sɔlid ri'ækʃən]

> Definition: 固-固反应 gù gù fǎn yìng

solid: a substance that is solid, not a liquid or gas. 固体的 gù tǐ de
reaction: a return or opposing action, force, influence, etc. 反应 fǎn yìng

> Example:

This study analyzes the process for preparing organic smectite by solid-solid reaction.
该研究分析了通过固-固反应制备有机蒙脱石的过程。

solidification index (SI) [sə,lidəfə'keʃən 'indeks]

> Definition: 固结指数 gù jié zhǐ shù

solidification: the process of making or becoming hard or solid. 凝固 níng gù；固结 gù jié
index: a figure in a system or scale representing the average value of specified prices, shares, or other items as compared with some reference figure. 指数 zhǐ shù

> Extended Terms:

directional solidification 定向凝固
exogenous solidification 外生长凝固
grease solidification 润滑脂固化
monotectic solidification 偏晶凝固

solid solution ['sɔlid sə'luːʃən]

> Definition: a homogeneous crystalline structure in which one or more types of atoms or molecules may be partly substituted for the original atoms and molecules without changing the structure. 固溶体 gù róng tǐ

> Example:

Hardening and strengthening of metals result from alloying in which a solid solution is formed.
合金化形成固溶体导致材料硬化和强化。

> Extended Terms:

addition solid solution 加成固溶体

continuous solid-solution series 连续固溶体系列
solid solution hardening 固溶体硬化
solid solution range 固溶体区域
solid solution strengthening 固溶强化

solidus [ˈsɔlidəs]

Definition: the highest temperature at which a metal is completely solid. 固相线 gù xiàng xiàn

Origin: 1350-1400; Middle English; from Late Latin solidus (nummus) a solid (coin), a gold (coin).

Example:
And it is discovered that the liquidus temperature hardly has effects on the solidification rate of alloy if solidus temperature is fixed.
对于合金凝固在固相线温度一定的情况下，液相线温度的高低对凝固速度几乎没有影响。

Extended Terms:
solidus line 固相线
solidus tempreture 固线温度

solubility of minerals in metamorphic fluid
[ˌsɔljuˈbiləti ɔvˈminərəl inmetəˈmɔːfik ˈflu(ː)id]

Definition: 变质流体中矿物的溶解度 biàn zhì liú tǐ zhōng kuàng wù de róng jiě dù
solubility: the quality of being soluble and easily dissolved in liquid. 溶解度 róng jiě dù
mineral: solid homogeneous inorganic substances occurring in nature having a definite chemical composition. 矿物 kuàng wù
metamorphic: of or relating to metamorphosis (especially of rocks). 变质的 biàn zhì de
fluid: a substance that is fluid at room temperature and pressure. 流体 liú tǐ

solution transfer [səˈluː ʃən transfə]

Definition: 溶移 róng yí；溶解转移 róng jiě zhuǎn yí
solution: the process or state of being dissolved in a solvent. 溶解过程 róng jiě guò chéng
transfer: move (someone or something) from one place to another. 转移 zhuǎn yí

Example:
Spatial variability of soil water retention function is a primary prerequisite for quantitative

research on water movement and solution transfer in unsaturated zone.
土壤水分特征函数空间异质性是定量研究土壤非饱和带水分运动以及溶质运移的先决条件。

solvus [ˈsɔlvəs]

Definition: In a phase or equilibrium diagram, the locus of points representing the solid-solubility temperatures of various compositions of the solid phase. 不混溶线 bù hùn róng xiàn

Example: However, at high temperature, when the matrix is in the state of low saturation (near solvus), precipitations in grain and on grain boundary are controlled by driving force (thermodynamics).
然而，高温时，当基体出于低饱和状态时(接近固溶相线)，颗粒中和晶界的沉淀由热力学控制。

Extended Terms:
solvus tempreture 溶线温度
solvus line 溶解度曲线

sorting [ˈsɔːtiŋ]

Definition: the process by which sedimentary particles become separated according to some particular characteristic, as size or shape. 分选 fēn xuǎn

Origin: 1200-1250; (noun) Middle English from Middle French sorte, from Medieval Latin sort-(stem of sors) kind, allotted status or portion, lot, Latin: orig., voter's lot; (v.) Middle English sorten toallot, arrange, assort (Middle French sortir), from Latin sortīrī to draw lots, derivative of sors.

Example: Because deposition is rapid, sorting and stratification are generally not well developed.
因为沉积作用是快速的，分选作用和层理就得不到很好的发展。

Extended Terms:
machine sorting 机器选别
manual sorting 人工分拣, 人工整理
marble sorting 选球, 球分等级
numeric sorting 数字分拣(整理)
ore sorting (拣)选矿(石)

source contamination [sɔːs kən ˌtæmiˈneiʃən]

Definition: 源污染 yuán wū rǎn

source: a place, person, or thing from which something comes or can be obtained. 来源 lái yuán

contamination: impure by exposure to or addition of a poisonous or polluting substance. 使不纯 shǐ bù chún, 污染 wū rǎn

Example:

Our country non-point source contamination concern day by day is also serious. How the science know and the active control non-point source pollution has become anurgent research topic.
我国非点源污染问题也日益严重，如何科学认识并有效控制非点源污染已成为一个紧迫的研究课题。

sparry [ˈspɑːri]

Definition: any light-colored lustrous mineral that cleaves easily. 亮晶 liàng jīng, 又称淀晶 (cumulus crystal) diàn jīng

Origin: late 16th century: from Middle Low German; related to Old English spærstān "gypsum".

Example:

The dolostone includes sparry dolostone, residual-grained dolostone, residual limy dolostone and algae dolostone.
白云岩主要包括结晶白云岩、残余颗粒白云岩、残余灰质白云岩和隐藻白云岩。

Extended Terms:

sparry allochemical rock 亮晶异化灰岩
sparry calcite 亮晶方解石
sparry intraclastic calcarenite 亮晶内碎屑砂屑灰岩
sparry limestone 颗粒大理石

spatter cone [ˈspætə kəun]

Definition: 寄生熔岩锥 jì shēng róng yán zhuī

spatter: cover with drops or spots of something. 溅污 jiàn wū
cone: a solid or hollow object which tapers from a circular or roughly circular base to a point. 圆锥体 yuán zhuī tǐ

Example:

Based on the composition and texture differences of the cones, the cones have been subdivided into scoria cone, spatter cone and mixed cone.

在火山锥中,依据锥体组成与结构的差异又可进一步分为岩渣锥、寄生熔岩锥和混合锥等碎屑锥。

spessartite [ˈspesətait]

Definition: a usually dark-red type of garnet, MnAl(SiO); manganese aluminum silicate, also called spessartine. 斜煌岩 xié huáng yán

Origin: French, after Spessart, mountain range in Bavaria + -ite, -ite.

Example:
Inspired by Swarovski's expertise in crystal and Signity's brilliance in gems, the Luna Watch is made of crystal, stainless steel and a Spessartite Garnet.
施华洛世奇处理水晶的技术以及升丽缇处理宝石的技术十分精湛,受此启发,这款月亮手表用水晶、不锈钢和深红色榴石打造而成。

Extended Terms:
diabase spessartite 辉绿闪斜煌斑岩
proterobase spessartite 角闪斜煌斑岩

sphene [sfiːn]

Definition: a variously colored mineral, CaTiSiO, that is an ore of titanium; calcium titanium silicate. 榍石 xiè shí

Origin: French sphène; from Greer sphēn, a wedge: so named because of its crystal form.

Example:
The sphene is a kind of stable mineral which is widely used in solidifying actinides separated from high level radioactive waste.
榍石是一种稳定矿物,是人造岩石固化高放射性废弃物较理想的基材之一。

Extended Terms:
natural sphene 变彩榍石;天然榍石
tin sphene 铬锡红色料

sphericity [ˌsfiəˈrisəti]

Definition: the roundness of a 3-dimensional object. 球度 qiú dù

Origin: 1615-1625.

Example:

As the eutectic time increases, the graphite morphology departs from sphericity to the abnormal morphology, and the roundness of the graphite particles decreases as well.
随着共晶时间的延长,铸件的石墨形态逐渐偏离球形,向异常化趋势发展,石墨颗粒的圆整度随之降低。

Extended Terms:

spherical washer 球面垫圈
sphericity test 球形检验
spheroidal graphite 球状石墨
spheroidized cementite 球状雪明碳铁
wetted sphericity 湿球形度

spherulitic texture [sfeˈrulitik ˈtekstʃə]

Definition: 球粒结构 qiú lì jié gòu

spherulitic: relating to the texture of a rock composed of numerous spherulites. Also known as globular; sphaerolitic. 球粒状的 qiú lì zhuàng de

texture: the arrangement of the particles or constituent parts of any material, as wood, metal, etc., as it affects the appearance or feel of the surface; structure, composition, grain, etc. 结构 jié gòu

Example:

The petrological textures of lava flows consist of porphyritic texture, glomeroporphyritic texture, spherulitic texture, and the groundmass typically consists of microlites and glass.
熔岩流岩石呈斑状结构、聚斑结构,发育球粒结构,基质为玻璃和微晶。

spilite [ˈspailait]

Definition: is a very-fine-grained igneous rock, resulting particularly from alteration of oceanic basalt. 细碧岩 xì bì yán

Example:

Spilite-keratophyre sequence is generally divided into spilite, keratophyre and quartz keratophyre.
细碧-角斑岩系一般分为细碧岩、角斑岩和石英角斑岩三类。

Extended Terms:

kalikeratophyre spilite 钾角斑细碧岩
keratophyre spilite 角斑细碧岩

soda keratophyre spilite 钠角斑细碧岩
spilite magma 细碧岩岩浆
spilitic lava 细碧岩质熔岩

spilite texture [ˈspailait ˈtekstʃə]

Definition: 细碧结构 xì bì jié gòu

spilite: an altered basalt containing albitized feldspar accompanied by low-temperature, hydrous crystallization products such as chlorite, calcite, and epidote. 细碧岩 xì bì yán

texture: the arrangement of the particles or constituent parts of any material, as wood, metal, etc., as it affects the appearance or feel of the surface; structure, composition, grain, etc. 结构 jié gòu

spinel [spiˈnel]

Definition: a variously colored, crystalline mineral, MgAlO, used as a gem; magnesium aluminum oxide. 尖晶石 jiān jīng shí

Origin: Italian spinella, diminutive of spina, thorn (from its sharply pointed crystals), from Latin spīna.

Extended Terms:
ferromagnetic spinel 铁磁性尖晶石
ruby spinel 红尖晶石
spinel structure 尖晶石型结构
synthetic spinel 合成尖晶石
white spinel 白色晶玉

spinel lherzolite [spiˈnel lˈheəzəlait]

Definition: 尖晶石二辉橄榄岩 jiān jīng shí èr huī gǎn lǎn yán

spinel: a hard glassy mineral consisting of an oxide of magnesium and aluminum; occurs in various colors that are used as gemstones. 尖晶石 jiān jīng shí

lherzolite: 二辉橄榄岩 èr huī gǎn lǎn yán

Example:

On the basis of the study on the petrology, trace element, and isotopic geochemistry, the primary magma of volcanic rocks can be divided into two series according to their originated rocks and degree of partial melting. One is the basanite-alkaline basalt-Olivine tholeiite magma series, which are generated by partial melting from spinel lherzolite.

岩石化学、微量元素、同位素地球化学证据都表明，根据源岩和局部熔融程度的不同，可以分为两个原生岩浆系列：一个是源岩为尖晶石二辉橄榄岩，随着局部熔融程度的增加，形成的碧玄岩岩浆-碱性玄武岩岩浆-橄榄拉斑玄武岩岩浆。

spinifex texture ['spinifeks 'tekstʃə]

Definition: 鬣刺结构 liè cì jié gòu

spinifex: any of various clump-forming, perennial Australian grasses, chiefly of the genus Triodia, growing in arid regions and having awl-shaped, pointed leaves. 鬣刺 liè cì

texture: the arrangement of the particles or constituent parts of any material, as wood, metal, etc., as it affects the appearance or feel of the surface; structure, composition, grain, etc. 结构 jié gòu

Example:

The skeleton crystals of clinoaugites are 0.1-0.5 mm in length, 0.025-0.05mm in width. That is micro spinifex texture.

单斜辉石骸晶长 0.1~0.5mm，宽 0.025~0.05mm，属显微鬣刺结构。

spongolite ['spɒŋgə‚lait] =spongolith

Definition: spongolite is a stone made almost entirely from fossilised sponges. It is light and porous. 海绵岩 hǎi mián yán

Origin: before 1000; (n.) Middle English, Old English, from Latin spongia, spongea, from Greek spongiā; (v.) Middle English spongen.

Example:

Spongolites and spiculites are common in the rock record, and represent environments that were dominated by sponges.

海绵岩和锤锥晶在岩石记录中常见且代表了海绵材质主导的环境。

spotted cementation ['spɒtid ‚si:men'teiʃən]

Definition: 斑点状胶结 bān diǎn zhuàng jiāo jié

spotted: marked or decorated with spots. 有斑点的 yǒu bān diǎn de

cementation: (chiefly Geology) the binding together of particles or other things by cement. [主地质] 黏结 nián jié; 胶接作用 jiāo jiē zuò yòng

spotted slate [ˈspɔtid sleit]

Definition: (Petrology) a type of slate containing dark spots that represent the beginning of porphyroblast development. 斑点板岩 bān diǎn bǎn yán

Example:

anisotropy in the flow deformation and failure of spotted slate
斑点板岩流动变形破坏的各向异性

spotted structure [ˈspɔtidˈstrʌktʃə]

Definition: 斑点状构造 bān diǎn zhuàng gòu zào

spotted: marked or decorated with spots. 有斑点的 yǒu bān diǎn de
structure: the arrangement of and relations between the parts or elements of something complex. 结构 jié gòu; 构造 gòu zào

Example:

the spotted structure and gold mineralization in devonian rocks in West Qinling Cai Yunhua and Li Wanhua
西秦岭泥盆系岩石中斑点状构造与金矿化

spotted texture [ˈspɔtidˈtekstʃə]

Definition: 斑状黏土结构 bān zhuàng nián tǔ jié gòu

spotted: marked or decorated with spots. 有斑点的 yǒu bān diǎn de
texture: the feel, appearance, or consistency of a surface or a substance. (物质、表面的)结构 jié gòu; 构造 gòu zào

Extended Term:

strongly spotted 多斑点钻石

stable isotope [ˈsteibl ˈaisəutəup]

Definition: an isotope of an element that shows no tendency to undergo radioactive breakdown. 稳定同位素 wěn dìng tóng wèi sù

Example:

The stable isotope labeling technique provided an ideal method for accurately quantitating the low abundance proteins which play a key role in cells or tissues.
稳定同位素标记技术的提出，为准确定量在细胞或组织体系中发挥重要功能的低丰度蛋白质提供了一个理想的方法。

Extended Terms:

stable isotope of lithium 锂稳定同位素
stable isotope of oxygen 氧稳定同位素
stable isotope standard 稳定同位素标准
stable isotope stratigraphy 稳定同位素地层学
stable tracer isotope 稳定示踪同位素

stable mineral [ˈsteibl ˈminərəl]

Definition: 稳定矿物 wěn dìng kuàng wù

stable: (of an object or structure) not likely to give way or overturn; firmly fixed. 稳定的 wěn dìng de

mineral: a solid inorganic substance of natural occurrence. 矿物 kuàng wù

Example:

The sphene is a kind of stable mineral which is widely used in solidifying actinides separated from high level radioactive waste.
榍石是一种稳定矿物，是人造岩石固化高放射性废弃物较理想的基材之一。

Extended Term:

stable relict mineral 稳定残余矿物；稳定残存矿物

standard deviation [ˈstændəd diːviˈeiʃən]

Definition: a statistic used as a measure of the dispersion or variation in a distribution, equal to the square root of the arithmetic mean of the squares of the deviations from the arithmetic mean. 标准偏差 biāo zhǔn piān chā

Example:

A portfolio that gives the highest expected return for a given standard deviation, or the lowest standard deviation for a given expected return, is known as an efficient portfolio.
对一个给定的标准差提供最高的期望回报率，或者对一个给定的期望回报率提供最低的标准差的投资组合，就被称之为一个有效投资组合。

Extended Terms:

polled standard deviation 合并标准差
residual standard deviation 剩余标准差
standard frequency deviation oscillator 标准频偏振荡器
standard geometric deviation 标准几何离差

standard normal deviation 标准正态偏离

standard state [ˈstændəd steit]

Definition: 标准态 biāo zhǔn tài

standard: a level of quality or attainment. 标准 biāo zhǔn; 水平 shuǐ píng
state: the particular condition that someone or something is in at a specific time. 状态 zhuàng tài

Example:

Then the concept of thickness in standard state was introduced for the absorption of CO_2, and the calculation methods of spectrum transmittance of H_2O, CO_2 were deduced.
针对二氧化碳的吸收引入了"标准状态下的厚度"的概念,给出了水、二氧化碳光谱透过率的计算方法.

Extended Terms:

standard atmospheric state 标准大气状态
standard basal state 标准基础状态

state property [steitˈprɔpəti]

Definition: 态性 tài xìng

state: the particular condition that someone or something is in at a specific time. 状态 zhuàng tài
property: an attribute, quality, or characteristic of something. 特性 tè xìng; 性质 xìng zhì; 性能 xìng néng

Example:

That soft mode is a normal state property that has nothing to do with superfluidity.
这种软模是一种通常态性,和超流动性无关。

staurolite [ˈstɔːrəlait]

Definition: a dark-colored mineral, (Fe, Mg) AlSiO (OH), a silicate of iron and aluminum: the crystals are often found twinned in the form of a cross. 十字石 shí zì shí; 交沸石 jiāo fèi shí

Origin: Greek stauros, cross; see stā-in Indo-European roots + -lite.

Example:

A thermal anticline—formed progressive metamorphic belt comprises chlorite-biotite, garnet, staurolite and sihlimanite zones.

变质岩石由绿泥石、黑云母带、石榴石带、十字石带和矽线石带组成。

Extended Terms:

staurolite zone 十字石带
staurolite kyanite subfacies 十字蓝晶分相

steady-state flow ['stedi steit fləu]

Definition: 稳态流 wěn tài liú

steady-state: permanent, not changing. 不变的 bù biàn de; 永恒的 yǒng héng de
flow: move along or out steadily and continuously in a current or stream. 流动 liú dòng

Example:

Certain hydronamical systems exhibit steady-state flow patterns, while others oscillate in a regular periodic fashion.
某些水动力学系统存在着稳定状态的流型,另一些则以某种规则的周期作往复运动。

Extended Terms:

pseudo-steady state fluid flow through porous medium 拟稳定渗流
radial steady-state flow equation 径向稳定流动方程

Stefan-Boltzmann law ['stævn 'bɔːltsmɑːn lɔː]

Definition: 斯特藩-玻尔兹曼定律 sī tè fān bō ěr zī màn dìng lǜ

law: a statement of fact, deduced from observation, to the effect that a particular natural or scientific phenomenon always occurs if certain conditions are present. 定律 dìng lǜ; 定理 dìng lǐ; 定则 dìng zé

Example:

The law of Maxwell speed distribution, n-dimensional heat capacity of solid and Stefan-Boltzmann law are related on dimensionality and the density of states.
麦克斯韦速度分布律、n 维情况固体热容量及斯特藩-玻尔兹曼定律都与维度及态密度有关。

Stereology [ˌsteriˈɒlədʒi]

Definition: the study of three-dimensional properties of objects or matter usually observed two-dimensionally. 体视学 tǐ shì xué; 立体测量学 lì tǐ cè liáng xué

Origin: Greek stereós + Middle English-logie, from Latin-logia.

Example:

With the development of computer technology and the improvement of morphometry, biological

stereology and image analytical technique progress deeply.
随着计算机技术的发展和形态结构测试手段的改进，生物体视学和图像分析技术得到了深化和发展。

Extended Terms:

Chinese Journal of Stereology and Image Analysis 中国体视学与图像分析杂志
mathematical stereology 数学体视学，数学立体测量学
principles and applications of stereology 体视学原理及应用

stishovite ['stiʃəvait]

Definition: a very dense, tetragonal form of silica that is colorless and transparent. 超石英（比重为4.2的高压变体石英）chāo shí yīng

Origin: after S. Stishov, Soviet mineralogist who discovered it + -ite.

Example:

The silicon, stishovite and β-quartz crystallites were found in some specimens.
某些试样中还出现硅、超石英和β石英晶相。

stock [stɔk]

Definition: A body of intrusive igneous rock of which less than 100 square kilometers (40 square miles) is exposed. 岩株 yán zhū

Origin: before 900; (n.) Middle English; Old English stoc(c) stump, stake, post, log; cognate with German Stock, Old Norse stokkr tree-trunk; (v.) derivative of the noun.

Example:

In order to study the change of tempersture field in the coalbeds after the stock and sillsintruding, thermadynamic models are set up in given hypothetic conditions for an imitative calculation around the intrusions by computer.
为了探讨岩株和岩床侵入后造成煤层内的温度场变化，在一定假设条件下，建立热力学模型，使用电子计算机进行模拟计算。

Extended Term:

stock intruding 岩株侵入

stone [stəun]

Definition: the hard, solid, nonmetallic mineral matter of which rock is composed. 石 shí;

石块 shí kuài

> Example:

The sun rises over gray stone.
太阳从灰色的岩石上升起。

> Extended Terms:

almond stone 杏仁石
limbra stone 白沙石
paving stone 铺路石

stony iron meteorite [ˈstəuni ˈaiən ˈmiːtiərait]

> Definition: 石质铁陨石 shí zhì tiě yǔn shí

stony：made of or resembling stone. 石制的 shí zhì de
iron：a strong, hard magnetic silvery-grey metal. 铁 tiě
meteorite：a piece of rock or metal that has fallen to the earth's surface from outer space as a meteor. 陨石 yǔn shí

> Example:

A Stony-Iron Meteorite Fallen to Dongujimqin, Inner Mongolia, China
中国内蒙东乌旗石铁陨石

stony meteorite [ˈstəuni ˈmiːtiərait]

> Definition: 石质陨石 shí zhì yǔn shí

stony：made of or resembling stone. 石制的 shí zhì de
meteorite：a piece of rock or metal that has fallen to the earth's surface from outer space as a meteor. 陨石 yǔn shí

> Example:

Lujiang meteorite is a stony meteorite fallen in Lujiang County, Anhui Province.
庐江陨石是一块陨落在安徽省庐江县的球粒石陨石。

stratified magma [ˈstrætifaid ˈmægmə]

> Definition: 分层岩浆 fēn céng yán jiāng

stratified：arranged in a sequence of grades or ranks. 分层的 fēn céng de
magma：hot fluid or semi-fluid material below or within the earth's crust from which lava and other igneous rock is formed by cooling. 岩浆 yán jiāng

> Example:

The isotopic and other chemical change are interpreted as the result of tapping a stratified magma chamber.
同位素和其他化学变化是挖潜分层岩浆室的结果。

strato volcano [ˌstretə vɔlˈkeinəu]

> Definition: also known as a composite volcano, is a tall, conical volcano built up by many layers (strata) of hardened lava, tephra, pumice, and volcanic ash. 层状火山 céng zhuàng huǒ shān

> Example:

El Misti is a 5,822 m strato volcano in southern Peru.
埃尔米斯蒂火山是位于南秘鲁的一座5822米高的层状火山。

stream competence [striːmˈkɒmpitəns]

> Definition: 流能力 liú néng lì

stream: a continuous flow of liquid, air, or gas. 流 liú; 股 gǔ; 缕 lǚ
competence: the ability to do something successfully or efficiently. 能力 néng lì

strength [streŋθ]

> Definition: the state or quality of being strong; force; power; vigor. 强度 qiáng dù

> Origin: Middle English, from Old English strengthu.

> Example:

The wind force is a directional force with velocity and strength. It generates a force that speeds up particles or objects to a target velocity.
风力是一个有速度和强度的方向力。它使一个使粒子或者物体在一个目标方向上产生速度变化。

> Extended Terms:

breaking strength 断裂强度;抗断强度
competitive strength 竞争优势/能力
compressive strength 抗压强度
impact strength 冲击强度
ultimate strength 极限强度

stress [stres]

Definition: strain or straining force; specif., emphasis; importance; significance. 应力 yìng lì; 压力 yā lì

Origin: 1275-1325; (n.) Middle English stresse, aphetic variant of distresse distress; (v.) derivative of the noun.

Example:
The optimized die is subject to less stress during forging process, which can ensure the service life of dies and guarantee the quality of forging part.
优化后的模具在锻造过程中承受较小的应力分布，从而可以确保模具的使用寿命以及锻件的质量不受影响。

Extended Terms:
breaking stress 断裂应力
nominal stress 名义应力，公称应力
shear stress 剪切应力
stress relief 应力释放

stromatolitic limestone [strəumætə'laitik 'laimstəun]

Definition: 叠层石灰岩 dié céng shí huī yán

stromatolitic: a widely distributed sedimentary structure consisting of laminated carbonate or silicate rocks, produced over geologic time by the trapping, binding, or precipitating of sediment by groups of microorganisms, primarily cyanobacteria. 叠层石 dié céng shí
limestone: a hard sedimentary rock, composed mainly of calcium carbonate or dolomite, used as building material and in the making of cement. 灰岩 huī yán; 石灰岩 shí huī yán

Example:
The Neoproterozoic Jiuliqiao Formation in the Huainan region consists of sandy, silty, stromatolitic and dolomitic limestones.
安徽淮南地区新元古代九里桥组主要由砂质、泥质灰岩、叠层石灰岩和白云质灰岩组成。

Extended Terms:
stromatolitic bioherm 叠层石礁
stromatolitic biostrome 叠层石层
stromatolitic structure 叠层构造

stromatolitic structure [strəumætə'laitik 'strʌktʃə]

Definition: 叠层石构造 dié céng shí gòu zào

stromatolitic: a widely distributed sedimentary structure consisting of laminated carbonate or silicate rocks, produced over geologic time by the trapping, binding, or precipitating of sediment by groups of microorganisms, primarily cyanobacteria. 叠层石 dié céng shí

structure: the arrangement of and relations between the parts or elements of something complex. 结构 jié gòu；构造 gòu zào

> Example:

development of recent stromatolitic structures and phosphatic enrichment in Precambrian marble of Sri Lanka
斯里兰卡前寒武纪大理石的近代叠层石构造和磷酸盐富集

> Extended Terms:

stromatolitic bioherm 叠层石礁
stromatolitic biostrome 叠层石层

Strombolian eruption [strɒmˈbəuljən iˈrʌpʃən]

> Definition: are relatively low-level volcanic eruptions, named after the Italian volcano Stromboli, where such eruptions consist of ejection of incandescent cinder, lapilli and lava bombs to altitudes of tens to hundreds of meters. 斯特隆博利型喷发 sī tè lóng bó lì xíng pēn fā

Strombolian volcanic activity [strɒmˈbəuljən vɒlˈkeinəu ækˈtivəti]

> Definition: 斯特隆博利火山活动 sī tè lóng bó lì huǒ shān huó dòng

Strombolian: denoting volcanic activity of the kind typified by Stromboli, with continual mild eruptions in which lava fragments are ejected. 斯特隆博利火山式的 sī tè lóng bó lì huǒ shān shì de

volcanic: of, relating to, or produced by a volcano or volcanoes. 火山的 huǒ shān de

activity: any specific behavior. 活动 huó dòng

> Example:

Strombolian volcanic activity is an intermittent phenomenon.
斯特隆博利火山活动是一种间断性现象。

Stromboli volcano [ˈstrɒmbəliːvɒlˈkeinəu]

> Definition: 斯特隆博利火山 sī tè lóng bó lì huǒ shān

Stromboli: a volcanic island in the Mediterranean, the most north-easterly of the Lipari Islands. 斯特隆博利岛 sī tè lóng bó lì dǎo

volcano: a mountain or hill, typically conical, having a crater or vent through which lava, rock fragments, hot vapour, and gas are or have been erupted from the earth's crust. 火山 huǒ shān

Example:

A seismic survey was carried out at Stromboli volcano during August 1973.
1973年8月进行了斯特隆博利火山的地震勘测。

strontium [ˈstrɒntiəm]

Definition: a pale-yellow, metallic chemical element, one of the alkaline-earth metals, resembling calcium in properties and found only in combination: strontium compounds burn with a red flame and are used in fireworks: symbol, Sr; at. no., 38: a deadly radioactive isotope (strontium 90) is present in the fallout of nuclear explosions. 锶(元素符号为：Sr) sī

Origin: Middle English: so named (1808) by Sir Humphry Davy, who first isolated it, from strontia + -ium.

Example:

A method for the determination of barium in the optical material barium strontium titanate by flame emission spectrometry is described.
本文介绍了火焰发射光谱法测定光学材料钛酸钡锶中钡的方法。

Extended Terms:

strontium phosphate 磷酸锶
strontium ranelate 雷尼酸锶
strontium sulfide 硫化锶
strontium unit 锶单位
strontium zirconate 锆酸锶

strontium isotopes [ˈstrɒntiəm ˈaisəutəups]

Definition: 锶同位素 sī tóng wèi sù

strontium: a soft silver-white or yellowish metallic element of the alkali metal group; turns yellow in air; occurs in celestite and strontianite. 锶 sī

isotope: each of two or more forms of the same element that contain equal numbers of protons but different numbers of neutrons in their nuclei, and hence differ in relative atomic mass but not in chemical properties; in particular, a radioactive form of an element. 同位素 tóng wèi sù

structural diapir [ˈstrʌktʃərəl ˈdaiəpiə]

Definition: 结构底辟 jié gòu dǐ pì

structural: of or relating to the arrangement of and relations between the parts or elements of a complex whole. 结构的 jié gòu de diapirs: a domed rock formation in which a core of rock has moved upward to pierce the overlying strata. 底辟 dǐ pì; 挤入构造 jǐ rù gòu zào

structure of rocks [ˈstrʌktʃə ɔv rɔks]

Definition: 岩石构造 yán shí gòu zào

structure: something composed of interrelated parts forming an organism or an organization. 结构 jié gòu

rocks: a large mass of stone forming a peak or cliff. 岩石 yán shí

Example:

The texture and structure of rocks are often inhomogeneous and thus display anisotropy of physical and mechanical properties.
岩石的结构、构造往往具有不均一性，因而表现出物理力学性质的各向异性。

stylolite [ˈstailəlait]

Definition: (Geology) an irregular surface or seam within a limestone or other sedimentary rock, characterized by irregular interlocking pegs and sockets around 1 cm in depth and a concentration of insoluble minerals. 沉积缝合线 chén jī féng hé xiàn

Origin: combining form representing Latin stilus.

Example:

Three diagenesis types are help to differentiate the middle Caledonian and early Hercynian karst, which are stylolite, dolomitization and dedolomitization.
有三种成岩作用类型有助于区分加里东中期和海西早期岩溶，它们分别是缝合线、与缝合线相关的埋藏白云石化及其相关的去白云石化。

Extended Terms:

bed-normal stylolite 立层缝合线
bed-parallel stylolite 顺层

subduction [səbˈdʌkʃən]

Definition: a geologic process in which one edge of one crustal plate is forced below the edge of another. 潜没（指地壳的板块沉到另一板块之下）qián méi

Origin: French, from Latin subductus, past participle of subdūcere, to draw away from below: sub-, sub- + dūcere, to lead; see deuk-in Indo-European roots.

> Example:

South China Sea, which locates within the Pacific-Oceanic Plate, the India Plate and Eurasia Plate, is the result of Pacific-Oceanic Plate subduction and the collision of India Plate and Eurasia Plate.
南海处于欧亚、太平洋、印-澳三大板块的交汇处,是在太平洋板块俯冲和印度板块与欧亚板块碰撞共同作用下经过扩张形成的。

> Extended Terms:

sag subduction 沉陷俯冲消亡作用
subduction earthquake 隐没式地震
subduction erosion 隐没侵蚀
subduction plate 俯冲板块

subduction zones [səbˈdʌkʃən zəuns]

> Definition: 俯冲带 fǔ chōng dài

subduction: the sideways and downward movement of the edge of a plate of the earth's crust into the mantle beneath another plate. 潜没 qián mò
zone: an area or stretch of land having a particular characteristic, purpose, or use, or subject to particular restrictions. 地带 dì dài; 区域 qū yù

> Example:

Deep seated earthquakes, often severe in magnitude, generally occur beneath subduction zones, as one plate is forced deep into the mantle beneath another.
当一个板块被迫深入另一个板块下面的地幔中时,强烈的深源地震一般会发生在俯冲带的下部。

subhedral crystal [ˌsʌbˈhiːdrəl ˈkristəl]

> Definition: 半自形晶 bàn zì xíng jīng

subhedral: pertaining to an individual mineral crystal that is partly bounded by its own crystal faces and partly bounded by surfaces formed against preexisting crystals. Descriptive of a crystal having partially developed crystal faces. 半形的 bàn xíng de
crystal: a mineral, especially a transparent form of quartz, having a crystalline structure, often characterized by external planar faces. 晶体 jīng tǐ

> Example:

The mineral occurs as granular subhedral crystals on the edges of iridosmine and native platinum.
矿样在天然铂和铱锇矿的边缘呈颗粒状半自形晶的铱矿物。

subhedral equigranular texture [ˌsʌbˈhiːdrəl iːkwiˈɡrænjulə ˈtekstʃəs]

Definition: 半自纹理 bàn zì wén lǐ
subhedral：半形的 bàn xíng de
equigranular：等粒度的 děng lì dù de
texture：the feel, appearance, or consistency of a surface or a substance. 结构 jié gòu；纹理 wén lǐ

subsidence [səbˈsaidəns]

Definition: sinking or settling in a bone, as of a prosthetic component of a total joint implant. 沉淀 chén diàn；下沉 xià chén；陷没 xiàn mò
Origin: 1640-1650；< Latin subsī dere, equivalent to sub-sub- + sī dere to sit, settle; akin to sedēre to be seated.
Example:
Differential subsidence and uplift in later period are the main factors for a large set of up dip pinchout sand body trap (reservoir).
差异沉降和后期不均衡抬升是南部断超带形成大套砂岩上倾尖灭圈闭（油藏）的主要因素。
Extended Terms:
block subsidence 块体沉陷；块状断陷
land subsidence 土地沉陷；地陷；地面沉降
secular subsidence 缓慢沉陷
subsidence curve 沉降曲线
subsidence rate 沉降速率

subsolvus [sʌbˈzɒlvəs]

Definition: a range of conditions in which two or more solid phases can form by exsolution from an original homogeneous phase. 次固溶线 cì gù róng xiàn
Example:
In contrast, the non-alkalinehypersolvus and subsolvus granites do not show the same degree of enrichment.
相反，非碱性的超熔花岗岩和次固溶花岗岩未表现出相同程度的富集。

subsolvus granite [sʌbˈzɔlvəs ˈɡrænit]

Definition: 次固溶线花岗岩 cì gù róng xiàn huā gāng yán

subsolvus: a range of conditions in which two or more solid phases can form by exsolution from an original homogeneous phase. 次固溶线 cì gù róng xiàn

granite: a very hard, granular, crystalline, igneous rock consisting mainly of quartz, mica, and feldspar and often used as a building stone. 花岗岩 huā gāng yán

subsolvus texture [ˌsʌbˈzɔlvəs ˈtekstʃə]

Definition: 次固溶线纹理 cì gù róng xiàn wén lǐ

subsolvus: a range of conditions in which two or more solid phases can form by exsolution from an original homogeneous phase. 次固溶线 cì gù róng xiàn

texture: the feel, appearance, or consistency of a surface or a substance 结构 jié gòu；纹理 wén lǐ

subtidal zone facies [suˈtaidl zəun ˈfeʃiːz]

Definition: 潮下带相 cháo xià dài xiàng

subtidal zone: the zone of the shoreline that is below low tide and is always covered by water. 次潮间带 cì cháo jiān dài

facies: the character of a rock expressed by its formation, composition, and fossil content. [地质]相 xiàng

Example:

Based on the detailed analysis of these microfacies, the hydrodynamic conditions and the trend of relative change in water depth during deposition of carbonate are determined, and four types of high-frequency cycles, types A1 and A2 in the low-energy subtidal zone facies and types B1 and B2 in high-energy shoals, are distinguished.

通过详细的微相特征分析，确定了碳酸盐岩沉积过程中的水动力状况和相对水深变化趋势，识别出低能潮下带相碳酸盐旋回 A1、A2 及高能浅滩型碳酸盐旋回 B1、B2 两类 4 种高频旋回。

Sudbury intrusion [ˈsʌdbəri inˈtruː ʒən]

Definition: 萨德伯里侵入 sà dé bó lǐ qīn rù

Sudbury: a city in SW central Ontario; pop. 110,670 (1991). It lies at the centre of Canada's largest mining region. 萨德伯里 sà dé bó lǐ

intrusion: the action or process of forcing a body of igneous rock between or through existing formations, without reaching the surface. 侵入 qīn rù

suite [swiːt]

Definition: a group of minerals, rocks, or fossils occurring together and characteristic of a location or period. 岩套(或岩石,化石) yán tào

Origin: late 17th century: from French, from Anglo-Norman French siwte (see suit).

Example:
In SE China the Fuzhou plutonic complex is made up of two major suites: a) A calc-alkaline suite (Danyang, Fuzhou) emplaced about 105-100 Ma ago (Rb-Sr ages).
中国东南部福州复式岩体主要由两个岩套构成:(1)钙碱性岩套(丹阳岩体、福州岩体)侵位于约105—100Ma前(Rb-Sr 年龄)。

Extended Terms:
Atlantic suite [地质] 大西洋套
mineral suite 矿物系列

sulfate [ˈsʌlfeit]

Definition: a salt of sulfuric acid containing the divalent, negative radical SO. 硫酸盐 liú suān yán

Example:
The sulfate process is similar to the soda process except that the alkali which is lost in the process is replaced with sodium sulfate instead of sodium carbonate.
硫酸盐法同烧碱法很相似,但在蒸煮过程中损失的碱在硫酸盐法中是由硫酸钠来补充的而不是碳酸钠。

Extended Terms:
cupric sulfate 硫酸铜;胆矾;蓝矾
dimethyl sulfate 硫酸二甲酯
ferrous sulfate 硫酸亚铁,硫酸低铁
protamine sulfate 硫酸鱼精蛋白
sodium sulfate 硫酸钠

sulfide [ˈsʌlfaid]

Definition: a compound of bivalent sulfur with an electropositive element or group, especially a binary compound of sulfur with a metal. 硫化物 liú huà wù

Origin: 1300-1350; Middle English sulphur < Latin sulpur, sulphur, sulfur brimstone, sulfur.

> **Example:**

Sulfide in the wastewater was determined by iodimetry via dissociated to sulfureted hydrogen under the conditions of acidity, adsorbed and created precipitated sulfide.
废水中硫化物在酸性条件下游离成硫化氢气体,被吸收液吸收,生成硫化物沉淀,用碘量法测定硫化物含量。

> **Extended Terms:**

barium sulfide 硫化钡
calcium sulfide 硫化钙
carbonyl sulfide 硫化羰
cyclical sulfide 环状硫化物
zinc sulfide 硫化锌

sulfur-isotope fractionation ['sʌlfə'aisəutəup ˌfrækʃən'eʃən]

> **Definition:** 硫同位素分馏 liú tóng wèi sù fēn liú

sulfur: the chemical element of atomic number 16, a yellow combustible nonmetal. (Symbol: S) 硫元素 liú yuán sù

isotope: (Chemistry) each of two or more forms of the same element that contain equal numbers of protons but different numbers of neutrons in their nuclei, and hence differ in relative atomic mass but not in chemical properties; in particular, a radioactive form of an element. [化]同位素 tóng wèi sù

fractionation: the act of separating (a mixture) by fractional distillation. 分馏 fēn liú;分级(混合物) fēn jí

> **Example:**

Improved increment method was used to calculate sulfur isotope fractionation factors in sulfides.
用改进的增量方法计算硫化物中硫的同位素分馏系数。

sulfur ['sʌlfə]

> **Definition:** a pale-yellow, nonmetallic chemical element found in crystalline or amorphous form; it burns with a blue flame and a stifling odor and is used in vulcanizing rubber and in making matches, paper, gunpowder, insecticides, sulfuric acid, etc.; symbol, S; at. no., 16. 硫 liú

> **Origin:** Middle English, from Anglo-Norman sulfre, from Latin sulfur.

> **Example:**

Combined with water vapor, sulfur dioxide forms sulfate aerosols, which can spread around the

globe, blocking solar radiation and chilling the air before becoming acid rain and snow.
二氧化硫跟水蒸气结合形成硫酸盐气溶胶，这种气溶胶可以遍布全球，在变成酸雨和酸雪之前阻隔太阳辐射，降低气温。

Extended Terms:

free sulfur 单体硫
sulfur dioxide 二氧化硫
sulfur ether 硫醚
sublimed sulfur 硫黄华
sulfur trioxide 三氧化硫

sulphide [ˈsʌlfaid]

Definition: is an anion of sulfur in its lowest oxidation number of -2. Sulfide is also a slightly archaic term for thioethers, a common type of organosulfur compound that are well known for their bad odors. 硫化物 liú huà wù; 硫化物类 liú huà wù lèi

Example:

Three sulphide zones with platinum group elements(PGE) mineralization have been discovered in the upper part of the ultramafic rock formations of the Great Dyke, Zimbabwe.
大岩墙超铁镁质岩层上部主要有三个硫化物带，铂族元素(PGE)主要分布在这些矿带中，尤其富集在硫化物带的底部。

Extended Terms:

carbonyl sulphide 氧硫化碳；碳酰硫
copper sulphide 硫化铜
dimethyl sulphide 甲硫醚
sodium sulphide 硫化钠
sulphide ores 硫化矿

sunda arc [ˈsʌndə ɑːk]

Definition: is a volcanic arc that has produced the islands of Sumatra and Java, the Sunda Strait and the Lesser Sunda Islands. A chain of volcanoes forms the topographic spine of these islands. 巽他弧 xùn tā hú

Example:

The thrust earthquakes are mainly distributed on the circum-pacific seismic belt, the Himalaya collision boundary and the Sunda arc.
逆冲型地震主要发生在环太平洋地震带上，在喜马拉雅碰撞边界和巽他弧也有发生。

supercontinent [ˈsjuːpəkɒntinənt]

Definition: a large hypothetical continent, especially Pangaea, that is thought to have split into smaller ones in the geologic past. Also called protocontinent. 超大陆 chāo dà lù

Origin: 1350-1400; Middle English from Latin continent-(stem of continēns, present participle of continēre to contain), equivalent to con-con- + -tin-, combining form of ten-hold + -ent--ent.

Example:
The Columbia Supercontinent proposed by Rogers and Santosh (2002) was composed of the Nena, Ur and Atlantic continents from 1.9 Ga to 1.5 Ga.
由 Rogers 和 Santosh 于 2002 年提出的哥伦比亚超大陆，是约从 1.9~1.5Ga 由 Nena, Ur 和 Atlantic 等 3 个大陆块体群，通过逐步汇聚而形成的一个超级大陆。

supercooling [ˌsjuːpəˈkuːliŋ]

Definition: to cool (a liquid) below a transition temperature without the transition occurring, especially to cool below the freezing point without solidification. 过冷现象(过度冷却现象) guò lěng xiàn xiàng (guò dù lěng què xiàn xiàng)

Origin: Latin super (preposition and v. prefix) above, beyond, in addition, to an especially high degree; akin to Greek hypér (see hyper-), Sanskrit upari.

Example:
The pressure drop in the tube should be controlled because the degree of supercooling of liquid refrigerant at the condenser outlet decreases with increasing pressure drop in the tube.
冷凝器出口制冷剂的过冷度随着管内压降的增大而减小，所以必须控制过冷段换热器的管内压降。

Extended Terms:
supercooling degree 过冷度
supercooling field 过冷磁场
supercooling melt 过冷液体
supercooling on contact 让液体出现超冷状态的一种方法
supercooling point 过冷却点

supercritical fluids [ˌsjuːpəˈkritikəl ˈfluiz]

Definition: is any substance at a temperature and pressure above its critical point. It can effuse through solids like a gas, and dissolve materials like a liquid. 超临界流体 chāo lín jiè liú tǐ

Example:

Compressed or supercritical fluids are excellent extraction solvents. They have got wide applications in food processing and storage, medicine making, as well as biomaterials processing.
压缩或超临界流体作为良好的萃取溶剂，在食品加工与保藏、制药和生物材料加工等领域有广阔的应用前景。

supercritical temperature [ˌsjuːpəˈkritikəl ˈtempəritʃə]

Definition: 超临界温度 chāo lín jiè wēn dù

supercritical: of, relating to, or denoting a fluid at a temperature and pressure greater than its critical temperature and pressur. (流体)超临界温度的 chāo lín jiè wēn dù de
temperature: the degree or intensity of heat present in a substance or object, especially as expressed according to a comparative scale and shown by a thermometer or perceived by touch. 温度 wēn dù; 气温 qì wēn

Example:

Almost all hydrogen bonding of water is broken down at the supercritical temperature of 400℃.
水在约400℃的超临界温度下，几乎所有的氢键都裂解了。

superheat [ˌsjuːpəˈhiːt]

Definition: the number of degrees by which the temperature of superheated steam exceeds the temperature of the steam at its saturation point. 过热化 guò rè huà; 过热 guò rè

Origin: Latin super (preposition and v. prefix) above, beyond, in addition, to an especially high degree; akin to Greek hypér (see hyper-), Sanskrit upari.

Example:

The droplet size, initial velocity of gas flow, superheat degree and spray distance all have great influences on the heat transfer behavior of atomized droplets.
熔滴尺寸、雾化气流初始速度、熔滴过热度及喷涂距离对雾化熔滴的热传输行为均有很大的影响。

Extended Terms:

nuclear superheat 核过热
superheat region 过热区
superheat steam 过热蒸汽
superheat steam controller 过热蒸汽控制器
superheat working 过热运转

superheat in magma [ˌsjuːpəˈhiːt in ˈmæɡmə]

Definition: 岩浆中的过热 yán jiāng zhōng de guò rè

superheat: the excess of temperature of a vapour above its temperature of saturation. (气体)过热 guò rè

in: expressing the situation of something that is or appears to be enclosed or surrounded by something else. 在……里 zài... lǐ

magma: hot fluid or semi-fluid material below or within the earth's crust from which lava and other igneous rock is formed by cooling. 岩浆 yán jiāng

---

supersaturation [ˈsjuːpəˌsætʃəˈreɪʃən]

Definition: refers to a solution that contains more of the dissolved material than could be dissolved by the solvent under normal circumstances. 过度饱和 guò dù bǎo hé

Origin: 1545-1555；< Late Latin saturātiōn-(stem of saturātiō) a filling, equivalent to saturāt(us) (see saturate) + -iōn-ion.

Example:

In some certain condition, when the scale assume supersaturation state, they are likely to precipitate on the heater surface.
当这些盐类呈过饱和状态时，在一定条件下，就可能在换热器表面以水垢的形式析出。

Extended Terms:

critical supersaturation 临界过饱和
critical supersaturation ratio 临界过饱和比率
limited supersaturation 极限过饱和
relative supersaturation 相对过度饱和度
supersaturation solution 过饱和溶液

---

supracrustal rock [ˌsjuːprəˈkrʌstəl rɔk]

Definition: Supracrustal rocks are rocks that were deposited on the existing basement rocks of the crust, hence the name. They may be further metamorphosed from both sedimentary and volcanic rocks. 表壳岩 biǎo ké yán

Example:

The REE patterns for bitumen is similar to those for supracrustal sedimentary rocks in areas with no basement faults but stabilized structural settings.
在基底断裂不发育且构造稳定的区域原油沥青，其稀土模式具有与上地壳沉积岩相似的特征。

supratidal zone facies [ˌsjuːprəˈtaidəl zəun ˈfeiʃiːz]

Definition: 潮上带相 cháo shàng dài xiāng

supratidal: describing the marginal zone above the level of high tide. 潮上 cháo shàng
zone: an area or a region with a particular feature or use. 地带 dì dài
facies: the character of a rock expressed by its formation, composition, and fossil content. 相 xiàng

Example:

We observed fifteen samples collected from supratidal zone facies to subtidal zone facies of these two sand coasts.
我们观测了采自这两个沙岸潮上带相到潮下带相的 15 个样本。

surface free energy [ˈsɜːfis friː ˈenədʒi]

Definition: is the amount of increase of free energy when the area of surface increases by every unit area. It can be calculated using Stefan's formula. 表面自由能 biǎo miàn zì yóu néng

Example:

The dispersive component of the surface free energy of HTPB, as well as the total surface free energy decreases as the temperature increases.
结果表明，HTPB 的色散分量随着温度的升高而降低，总表面自由能表现出与色散分量相一致的规律。

Extended Term:

surface tension and surface free energy 表面张力与表面自由能

surface tension [ˈsɜːfis ˈtenʃən]

Definition: a property of liquids in which the exposed surface tends to contract to the smallest possible area because of unequal molecular cohesive forces near the surface; measured by the force per unit of length. 表面张力 biǎo miàn zhāng lì

Example:

Since different liquids have different surface tension, so when liquids flow across the small orifice, if surface tension is smaller, the flow speed will be faster.
不同的液体有着不同的表面张力，而液体流过小孔时，表面张力越小，所通过的速率就越快。

Extended Terms:

equilibrium surface tension 平衡表面张力
skin surface oxygen tension 皮肤表面氧张力

surface tension balance 表面张力平衡
surface-tension energy 表面张力能
surface-tension property 表面张力特性

surface uplift [ˈsɜːfis ʌpˈlift]

Definition: 地面抬升 dì miàn tái shēng

surface: the outside part or uppermost layer of something (often used when describing its texture, form, or extent). 表面(常用来描述质地,形状,大小)biǎo miàn;地面 dì miàn
uplift: the upward movement of part of the earth's surface. 抬升 tái shēng;隆起 lóng qǐ

Example:

Except for shortage in research area and degree, the different of evidences for plateau uplift are responsible for the rise debate. After analyzing recording mechanism of every evidence, we think that planation surfaces, river terraces, conglomerate and depositional rate in the surroundings of the Plateau are more important factors for surface uplift.

在分析了各种证据对高原地面上升的记录机理后,我们认为夷平面、河流阶地、山麓相砾岩和高原周围盆地物质堆积速率等证据较准确地记录了高原的地面抬升,今后必须加强这类证据的研究。

surge deposit [sɜːdʒ diˈpɔzit]

Definition: 涌浪堆积 yǒng làng duī jī

surge: a sudden powerful forward or upward movement, especially by a crowd or by a natural force such as the waves or tide. 汹涌 xiōng yǒng
deposit: an accumulation of sand, sediment, minerals, or other substances that has built up over a period of time through a natural process. 堆积 duī jī

Example:

Grain size analysis shows that the ejecta are composed mainly of fall deposit with an interbed of surge deposit, and the type of eruption is phreatomagmatic.

粒度分析表明,天池火山最近一次喷发物以空降堆积为主,夹一层薄层涌浪堆积,火山喷发类型为射气岩浆型。

survival potential [səˈvaivəl pəˈtenʃəl]

Definition: 生存潜能 shēng cún qián néng

survival: the state or fact of continuing to live or exist, typically in spite of an accident, ordeal, or difficult circumstances. 幸存 xìng cún potential: latent qualities or abilities that may be developed and lead to future success or usefulness. 潜力 qián lì; 潜能 qián néng

suspended loads [səˈspendid ləuds]

Definition: is the term for the sediment particles which settle slowly enough to be carried in flowing water (such as a stream or coastal area) either without touching the bed or while only intermittently touching it. 悬移质 xuán yí zhì

Example:

The various hydrodynamic factor functions on lakes and reservoirs was researched systematically in consideration of bed friction under effect of wave and current, radiational stress, wind shear stress, wind current, suspended load.
系统地研究湖泊、水库中各类水动力外负荷作用，考虑了波、流共同作用下的底摩阻、波浪剩余动量流、风作用剪应力、风吹流及悬移质扩散分布。

Extended Terms:

nonuniform suspended load 非均匀悬移质
suspended load flux 悬沙通量
suspended load gradation 悬沙级配
suspended load model 悬沙模型
suspended load transport 悬移质输

suture basin [ˈsuː tʃə ˈbeisən]

Definition: 缝合盆地 féng hé pén dì

suture: a stitch or stitches made when sewing up a wound, especially after an operation. 缝合 féng hé
basin: a place where the earth's surface is lower than in other areas of the world. 盆地 pén dì

sutured contact [ˈsuː tʃəd ˈkɒntækt]

Definition: 锯齿状接触 jù chǐ zhuàng jiē chù, 缝合接触 féng hé jiē chù

sutured: relating to a line of junction formed by two crustal plates which have collided. (板块碰撞形成的)接缝的 jiē fèng de
contact: the state or condition of physical touching. 接触 jiē chù

swamp facies [swɔmp ˈfeiʃiːz]

Definition: 沼泽相 zhǎo zé xiàng

swamp: an area of ground that is very wet or covered with water and in which plants, trees, etc. are growing. 沼泽 zhǎo zé

facies: The character of a rock expressed by its formation, composition, and fossil content. 相 xiàng

Example:

Ore beds are located in two horizons of middle Carboniferous shallow and littoral facies strata and early Permian continental lacustrine and swamp facies strata.
它在本区有两个层位，一个是中石炭世的浅海和滨海相地层，另一个是早二叠世的陆相湖沼泽沉积地层。

Swave [sweiv]

Definition: wave motion in a solid medium in which the particles of the medium oscillate in a direction perpendicular to the direction of travel of the wave; it cannot be transmitted through a fluid. S 波 S bō; 横波 héng bō

Origin: s(econdary) wave.

Example:

The crack porosity and closed pressure obtained by the experiment provided the basic data for understanding anisotropy of rocks, S wave splitting and crust dynamics.
实验得到的裂纹孔隙率和岩石裂纹的闭合压力，为了解岩石各向异性、S 波分裂以及地壳动力学提供了基础数据。

Extended Terms:

polarized S wave 极化 S 波
S wave method S 波法
S wave splitting S 波分离
S wave interval velocity 横波层速度
S wave velocity S 波速度

syenite [ˈsaiənait]

Definition: is a coarse-grained intrusive igneous rock of the same general composition as granite but with the quartz either absent or present in relatively small amounts (<5%). 黑花岗石 hēi huā gāng shí

Example:

Generally, the intrusions is small and lithologic compositions mainly include diorite, adamellite or syenite.

侵入岩体规模一般较小，岩性主要为闪长岩类、石英(二长)斑岩类或正长岩类的中酸性岩石。

Extended Terms:

biotite syenite 黑云正长岩
calcite syenite 方解石正长岩
nepheline syenite 霞石正长岩
riebeckite syenite 钠闪正长岩
sodalite syenite 方钠正长岩

sylvinite [ˈsilvinait]

Definition: Sylvinite is the most important ore for the production of potash in North America. It is a mechanical mixture of sylvite (KCl, or potassium chloride) and halite (NaCl, or sodium chloride). 钾石盐岩 jiǎ shí yán yán

Origin: late 19th century: from sylvine + -ite.

Example:

Sylvinite beds carry 35%-45% KCl and <1% either $MgCl_2$ or insoluble residues. The salt-bearing basin is separated from the open sea by the surrounding reefs and old-lands. The potassium-bearing depression is the deepest region on the platform.

钾石盐层中氯化钾含量达35%~45%，氯化镁和水不溶物含量均小于1%，含盐盆地四周以堡礁带和陆岛与广海相隔，含钾凹陷处于地台上凹陷最深部。

Extended Terms:

sylvinite fertilizer 钾盐肥
symbiotic 钾石膏

sylvite [ˈsilvait]

Definition: a white or colorless mineral, KCl, that is the chief ore of the potassium used to make fertilizers. 钾盐(天然氯化钾) jiǎ yán (tiān rán lǜ huà jiǎ)

Origin: alteration of sylvine, from French, from New Latin (sal digestivus) Sylvii, (digestive salt) of sylvius, probably after Franz de la Boë, or Franciscus Sylvius (1614-1672), German-born Dutch physician.

Example:

The results demonstrated that the organic sylvite can reduce tar and a relationship between tar and

the kind and content of organic sylvite exists to a certain extent.
研究结果表明，所加的有机酸钾能够降低焦油量，且在一定范围内添加钾盐种类及添加量与焦油量具有密切的相关性。

Extended Terms:

halo-sylvite 钾盐

sylvite deposit 钾盐矿床

syneresis [siˈniərisis]

Definition: contraction of a gel so that liquid is exuded at the surface, as in the separation of serum from a blood clot. 脱水收缩作用 tuō shuǐ shōu suō zuò yòng

Origin: Late Latin synaeresis, from Greek sunairesis, from sunairein, to contract: sun-, syn- + hairein, to take, grasp.

Example:

Syneresis of gel under certain conditions is the inherent behavior of gel. Slight syneresis has no effect on the application of gel in the porous media.
凝胶在一定条件下发生脱水是其固有的性质，轻微的脱水并不影响凝胶在孔隙介质中的作用。

Extended Terms:

syneresis crack 脱水收缩裂隙

synaeresis syneresis 脱水收缩

vitreous syneresis 玻璃体凝缩

syneresis crack [siˈniərisis kræk]

Definition: 脱水收缩裂缝 tuō shuǐ shōu suō liè fèng

Syneresis: (Chemistry) the contraction of a gel accompanied by the separating out of liquid. (化)脱水收缩 tuō shuǐ shōu suō; (胶体)凝缩(作用) níng suō

crack: a line on the surface of something along which it has split without breaking into separate parts. 裂缝 liè fèng

Example:

Syneresis cracks: subaqueous shrinkage in argillaceous sediments was caused by earthquake-induced dewatering.
脱水收缩裂缝——地震诱发的脱水作用引起泥质沉积水下收缩。

syngenesis [sinˈdʒenisis]

Definition: sexual reproduction. 同生作用 tóng shēng zuò yòng

Origin: 1830-1840.

Example:

This paper shows that the economic tin deposits hardly formed by syngenesis, but preliminary enrichment of tin by it widely exists in many metallogetic regious or belts.
本文提出同生作用很难生成工业性锡矿床，但由其所形成的初始富集在诸多成矿区、带中普遍存在。

Extended Terms:

syngenetic anomaly 同生异常
syngenetic deposit 同生矿床
syngenetic evaporation 同生蒸发
syngenetic relationship 同生关系

syngenetic concretion [ˌsirdʒiˈnetik kənˈkriːʃən]

Definition: 同生结核 tóng shēng jié hé

syngenetic: relating to or denoting a mineral deposit or formation produced at the same time as the enclosing or surrounding rock. 同生 tóng shēng

concretion: a hard solid mass formed by the local accumulation of matter, esp. within the body or within a mass of sediment. 结核 jié hé

Extended Terms:

syngenetic deposit 同生矿床
syngenetic structure 同生构造

syntectonic crystallization [ˌsintekˈtɔnik ˌkristəlaiˈzeiʃən]

Definition: 同构造期结晶作用 tóng gòu zào qī jié jīng zuò yòng

syntectonic: a geologic event, process, rock or feature formed contemporary to tectonism. 同构造期 tóng gòu zào qī

crystallization: the process of forming or causing to form crystals. 结晶 jié jīng; 晶化 jīng huà

Example:

Strontium is the signature element of the calcite resulting from syntectonic crystallization and zinc is the signature element of the calcite of fracture filling type.

Sr 是高应力环境同构造重结晶方解石带的特征元素，Zn 是低应力环境裂隙充填型方解石脉的特征元素。

syntectonic rock [ˌsintekˈtɔnik rɔk]

Definition: 同构造岩 tóng gòu zào yán

syntectonic: a geologic event, process, rock or feature formed contemporary to tectonism. 同构造期 tóng gòu zào qī

tectonic: of or relating to the structure of the earth's crust and the large-scale processes which take place within it. 地壳构造上的 dì qiào gòu zào shàng de

rock: the solid mineral material forming part of the surface of the earth and other similar planets, exposed on the surface or underlying the soil. 岩 yán

Example:
A synthetical study of various rocks in a rock suite and of syntectonic rock suites is proposed.
提出采用岩套内岩类配套及同构造岩套综合分析途径的建议。

Syros [ˈsairɔs]

Definition: an island of Greece in the north-central Cyclades. It is the richest and most populous of the Cyclades. 锡罗斯 xī luó sī

Origin: Greek Síros.

Example:
Pyroxene samples from the Greek island of Syros were investigated by chemical and XRD analyses.
用化学和 XRD 分析观测采自希腊锡罗斯岛的辉石样本。

tachylite [ˈtækilait]

Definition: a kind of dark-colored, basaltic volcanic glass. 玄武玻璃 xuán wǔ bō lí

Origin: Ger tachylit, from Greek tachys, swift (see tachy-) + lytos, soluble, from lyein, to dissolve (see lysis): from its rapid decomposition in acids.

Example:
the Tachylite of the Cleveland Dyke
克利夫兰岩脉中的玄武玻璃

Extended Term:
tachylite basalt glass 玄武玻璃

taconite [ˈtækənait]

Definition: an iron-bearing chert containing from 25 to 35 percent hematite and magnetite; it is a low-grade iron ore that is pelletized for blast-furnace reduction. 角岩 jiǎo yán

Origin: from Taconic, old name for rock formations first identified in the Taconic Range+ -ite.

Example:
the affirmance of three types of taconite limestone in Sutong bridge and it's engineering significance
苏通大桥区三种角砾灰岩的成因及其工程意义

Extended Terms:
magnetite-taconite 磁铁矿-铁燧岩
taconite limestone 角砾灰岩
taconite pellets 铁燧岩球团

Tahitian alkaline series [tɑːˈhiːʃən ˈælkəlain ˈsiəriːz]

Definition: 大溪地碱性系列 dà xī dì jiǎn xìng xì liè

Tahitian: 大溪地 dà xī dì

alkaline: having the properties of an alkali, or containing alkali; having a pH greater than 7. 碱性的 jiǎn xìng de

series: a number of things, events, or people of a similar kind or related nature coming one after another. 系列 xì liè

talc [tælk]

Definition: a soft, light-colored, monoclinic mineral, $MgSiO(OH)$, with a greasy feel, used to make talcum powder, lubricants, etc.; magnesium silicate. 滑石 huá shí; 云母 yún mǔ

Origin: French, from Medieval Latin talcum and Old Spanish talco, both from Arabic talq, from Persian talk.

Example:
The results showed that the property indexes of ethylene-propylene rubber added powdery quartz was better than the property indexes of ethylene-propylene rubber added talc.
结果表明添加了粉石英改性粉体的乙丙绝缘胶的性能指标优于添加滑石粉的。

Extended Terms:
disseminated talc 浸染状滑石

fired talc 焙烧滑石
grey talc 灰滑石
talc schist 滑石片岩
white talc 白滑石

Talcott basalt ['tælkət 'bæsɔ:lt]

Definition: 托尔克特玄武岩 tuō ěr kè tè xuán wǔ yán

Talcott：托尔克特 tuō ěr kè tè

basalt：a dark fine-grained volcanic rock that sometimes displays a columnar structure. It is typically composed largely of plagioclase with pyroxene and olivine. 玄武岩 xuán wǔ yán

talc schist [tælk ʃist]

Definition: a regional metamorphic rock composed predominantly of talc, and displaying a schistosity. The rock forms by the metamorphism and deformation of ultrabasic igneous rocks in regional terranes. 滑石片岩 huá shí piàn yán

Example:
The talc ore shows schisteous texture (talc schist) or in compact mass (steatite).
滑石矿石显示片理化结构(滑石片岩)或压实块状结构(块滑石)。

talus facies ['teiləs'feiʃii:z]

Definition: 塌积相 tā jī xiàng

talus：a sloping mass of loose rocks at the base of a cliff. [地]碎石堆 suì shí duī
facies：The character of a rock expressed by its formation, composition, and fossil content. [地]相 xiàng

Example:
The top of explosion talus facies were the best favorable facies zones for accumulation. And they are the main exploration targets in searching for the oil and gas pools of volcanic rocks in the future.
火山碎屑流相中上部及爆发塌积相顶部为有利的储集相带，是今后寻找火山岩油藏的主要勘探目标区。

Tanganyika shield [ˌtæŋgən'ji:kə ʃi:ld]

Definition: 坦噶尼喀地盾 tǎn gá ní kā dì dùn

Tanganyika: former country in East Africa, constituting the mainland part of what is now Tanzania. 坦噶尼喀 tǎn gá ní kā

shield: a large rigid area of the earth's crust, typically of Precambrian rock, which has been unaffected by later orogenic episodes, e. g. the Canadian Shield. 地盾 dì dùn

Example:

An Archaean greenstone belt rests in the central part of the Tanganyika shield.
一个太古代的绿岩带位于坦噶尼喀地盾的中心。

Taupo caldera [ˈtaupə kɒlˈdeərə]

Definition: 陶波火山口 táo bō huǒ shān kǒu

Taupo: city in the center of the North Island, New Zealand, on the northern shore of Lake Taupo. It is a commercial center and tourist resort. Population: 22,100 (2005 estimate). 陶波 táo bō

caldera: a large volcanic crater, especially one formed by a major eruption leading to the collapse of the mouth of the volcano. 破火山口 pò huǒ shān kǒu

Example:

These processes may relate to rim-collapse of the Taupo caldera.
这些过程也许和陶波火山口的边缘坍塌相关。

taxitic structure [tækˈsitik ˈstrʌktʃə]

Definition: 斑杂构造 bān zá gòu zào

structure: something composed of interrelated parts forming an organism or an organization. 结构 jié gòu

Example:

The metasomatite assumes massive form with local taxitic structure.
换质岩与当地斑杂构造形成巨大的形态。

Taylor series expansion [ˈteilə siəriːz ikˈspænʃən]

Definition: 泰勒级数展开 tài lè jí shù zhǎn kāi

Taylor: 泰勒 tài lè

series: a set of quantities constituting a progression or having the several values determined by a common relation. (数)级数 jí shù; 序列项的总和 xù liè xiàng de zǒng hé

expansion: the action of becoming larger or more extensive. 扩大 kuò dà; 扩展 kuò zhǎn

technical petrology ['teknikəl pi'trɒlədʒi]

Definition: 工艺岩石学 gōng yì yán shí xué，又称工业岩石学(industrial petrology)

technical: of or relating to a particular subject, art, or craft, or its techniques. 专门的 zhuān mén de；工艺的 gōng yì de；技巧上的 jì qiǎo shàng de；技术上的 jì shù shàng de

petrology: [mass noun] the branch of science concerned with the origin, structure, and composition of rocks. 岩石学 yán shí xué

Example:

From the point of view of technical petrology, in order to improve the quality of cement clinker, we must increase the quantity of alite, brownmillerite, and decrease the quantity of free lime.

从工艺岩石学的观点来看，提高水泥熟料的质量，必须增加硅酸三钙、铁铝酸四钙的数量，减少自由氧化钙的数量。

tectonic arkose [tek'tɒnik 'ɑːkəus]

Definition: 构造长石砂岩 gòu zào cháng shí shā yán

tectonic: connected with the structure of the earth's surface. 地壳构造的 dì qiào gòu zào de

arkose: a quartzose sandstone with greater than 25% feldspar grains. Usually interpreted to indicate a provenance with a granitic composition (granite, gneiss, others), and commonly interpreted as being deposited fairly close to the provenance region (ie., an immature sandstone). 长石砂岩 cháng shí shā yán

Example:

This is a tectonic arkose brought up under humid climate.
这是一个在湿度气候下形成的构造长石砂岩。

Extended Term:

tectonic layer 构造层

tectonic classification [tek'tɒnik ˌklæsifi'keiʃən]

Definition: 构造分类 gòu zào fēn lèi

tectonic: of or relating to the structure of the earth's crust and the large-scale processes which take place within it. 地壳构造上的 dì qiào gòu zào shàng de

classification: the action or process of classifying something according to shared qualities orcharacteristics. 分类 fēn lèi

Example:

Concrete suggestions are put forward about deepening the study of the problems about the petrological tectonic classification and emplacement mechanism of granitoids of China.
对深化我国花岗岩类岩石成因、构造分类及定位机制的研究提出了具体建议。

tectonic overpressure [tekˈtɒnikˈəuvəˌpreʃə]

Definition: 构造超压 gòu zào chāo yā

tectonic: connected with the structure of the earth's surface. 地壳构造的 dì qiào gòu zào de
overpressure: the pressure increase or accumulation above the set pressure when the valve is discharging flow. 超压 chāo yā

Example:

The tectonic overpressure is an important condition for gold metallogenesis in the study area.
构造附加压力是促进该区金成矿的重要动力条件。

tectonic pressure on magma [tekˈtɒnik ˈpreʃə ɒn ˈmægmə]

Definition: 构造岩浆压力 gòu zào yán jiāng yā lì

tectonic: of or relating to the structure of the earth's crust and the large-scale processes which take place within it. 地壳构造上的 dì qiào gòu zào shàng de
pressure: the force per unit area exerted by a fluid against a surface with which it is in contact. 压力 yā lì; 压强 yā qiáng
on: physically in contact with and supported by (a surface). 在……上 zài... shàng
magma: hot fluid or semi-fluid material below or within the earth's crust from which lava and other igneous rock is formed by cooling. 岩浆 yán jiāng

tectonic setting [tekˈtɒnikˈsetiŋ]

Definition: 构造环境 gòu zào huán jìng

tectonic: of or relating to the structure of the earth's crust and the large-scale processes which take place within it. 地壳构造上的 dì qiào gòu zào shàng de
setting: the place or type of surroundings where something is positioned or where an event takes place. 环境 huán jìng

Example:

Magmatic activities are developed under special tectonic setting, and granite masses of different types and properties are widely formed in different orogenic stages and tectonic settings.

独特的大地构造环境使岩浆活动异常发育，广泛出露面积巨大的花岗岩体，不同造山阶段、不同构造环境均有不同类型、不同特征的花岗岩出露。

tektite [ˈtektait]

Definition: any of certain small, yellowish-green to black glassy bodies of various shapes, found in isolated locations around the world and thought to have originated as meteorites or from meteorite impacts. 玻陨石 bō yǔn shí

Origin: Greek tēktos, molten (from tēkein, to melt) + -ite.

Example:
The tektite was formed from the upper crust rocks by impact melting.
玻陨石是地壳岩石受撞击熔融形成的。

telogenetic stage [tiːləˈdʒiˈnetik steidʒ]

Definition: 晚期表成岩阶段 wǎn qī biǎo chéng yán jiē duàn

telogenetic: (geology) related to or arising during telogenesis, the process of erosion and oxidation that occurs when sedimentary rocks are subject to uplift. 后期 hòu qī

stage: a period or state that something/somebody passes through while developing or making progress. (发展或进展的)阶段 jiē duàn

temperature [ˈtemprətʃə]

Definition: a measure of the quantity of heat in an object, usually as measured on a thermometer; specif. 温度 wēn dù; 体温 tǐ wēn; 热度 rè dù; 发烧 fā shāo

Origin: Middle English, temperate weather, Latin temperātūra, due measure, from temperātus, past participle of temperāre, to mix.

Example:
The soil water and the land surface evaporation decreased with temperature rising and increased with precipitation increasing.
陆面蒸发和土壤储水量随着温度的升高而减少，随着降水的增加而增加。

Extended Terms:
temperature coefficient 温度系数
temperature inversion 逆温温度系数
temperature range 温度范围

temperature buffering

temperature regulator 温度调节器
temperature scale 温标

temperature buffering [ˈtemprətʃə ˈbʌfəriŋ]

Definition: 温度缓冲 wēn dù huǎn chōng

temperature: the degree or intensity of heat present in a substance or object, especially as expressed according to a comparative scale and shown by a thermometer or perceived by touch. 温度 wēn dù; 气温 qì wēn

buffering: an effect that prevents incompatible or antagonistic people or things from coming into contact with or harming each other. 缓冲作用 huǎn chōng zuò yòng

Example:
A reason for this is that RH buffering by materials is feeble and short lived compared with temperature buffering when the air change rate is large.
原因是当空气交换量大时，和热缓冲材料相比，湿度缓冲材料的功能较弱，而且寿命不长。

temperature variation [ˈtemprəʃə veəriˈeiʃən]

Definition: 温度变动 wēn dù biàn dòng

temperature: the degree or intensity of heat present in a substance or object, especially as expressed according to a comparative scale and shown by a thermometer or perceived by touch. 温度 wēn dù; 气温 qì wēn

variation: a change or slight difference in condition, amount, or level, typically with certain limits. 变动 biàn dòng

Example:
Quartz crystal has a biggish temperature expansion coefficient, therefore temperature variation will cause use errors for quartz crystal devices.
石英晶体有着较大的温度膨胀系数，因而温度变化会引起石英晶体制作器件的使用误差。

Extended Terms:
ace temperature variation 区域温差
curve temperature variation 温度变化曲线
diurnal temperature variation 日温度变化
interdiurnal temperature variation 温度日际变化
TVR Temperature Variation of Resistance 阻温度变化

tempestite deposit [tempisˈtait diˈpɒzit]

Definition: 风暴沉积 fēng bào chén jī

tempestite: a storm deposit. 风暴岩 fēng bào yán

deposit: a layer or body of accumulated matter. 沉积物 chén jī wù; 沉淀 chén diàn

Example:

The types of ancient shallow sea tempestites deposit in North China are given as follows.
中国北部古代浅海的风暴沉积类型如下。

tensile strength ['tensail streŋθ]

Definition: resistance to lengthwise stress, measured (in force per unit of cross-sectional area) by the greatest load pulling in the direction of length that a given substance can bear without tearing apart. 抗张强度 kàng zhāng qiáng dù; 拉伸强度 lā shēn qiáng dù

Example:

The result shows that the elastic modulus, the tensile strength and yield strength of the alloy increase, but the elongation of the alloy decreases as content of beryllium increasing.
结果表明，随着铍含量的增加，铝含量的降低，铍铝合金的弹性模量、抗拉强度、屈服强度显著升高，延伸率降低。

Extended Terms:

tensile impact strength 拉伸冲击强度
tensile splitting strength 拉裂强度极限
tensile-strength safety factor 抗拉安全系数
tensile yield strength 抗拉屈服强度
wet tensile breaking strength 湿法拉断劲

tepee structure ['tiːpiː 'strʌktʃə]

Definition: 帐篷构造 zhàng péng gòu zào

tepee: a type of tall tent shaped like a cone, used by Native Americans in the past. (美洲印第安人旧时使用的)圆锥形帐篷 yuán zhuī xíng zhàng péng

structure: the way in which the parts of something are connected together, arranged or organized; a particular arrangement of parts. 构造 gòu zào

Example:

Tepee structure, which is defined for its inverted "V" shape, is a special structure that is mainly found in submarine hardgrounds, peritidal, lacustrine and caliche carbonate sediments.
帐篷构造是碳酸盐岩中的一种特殊沉积组构，因其倒"V"字形形态类似于帐篷(tepee)而得名。

tephrite [ˈtefrait]

Definition: a group of basaltic extrusive rocks composed chiefly of calcic plagioclase, augite, and nepheline or leucite, with some sodic sanidine. 碱玄岩 jiǎn xuán yán

Origin: 1875-1880; from Greek tephr (ós) ash-colored + -ite.

Example:
Leucite tephrite was derived from the more enriched mantle source with lower degree of partial melting of mantle source than alkaline olivine basalt.
相对于碱性橄榄玄武岩，白榴石碱玄岩来自于更加富集的地幔源区，其地幔的部分熔融程度相对较低。

Extended Terms:
haüyne-tephrite 蓝方碱玄岩
leucite-tephrite 白榴石碱玄岩
nepheline tephrite 霞石碱玄岩
nosean tephrite 黝方碱玄岩
sodalite tephrite 方钠碱玄岩

tephrochronology [ˌtefrəukrəuˈnɔlədʒi]

Definition: is a geochronological technique that uses discrete layers of tephra—volcanic ash from a single eruption—to create a chronological framework in which paleoenvironmental or archaeological records can be placed. 火山灰年代学
huǒ shān huī nián dài xué

Origin: 1940-1945; from Swedish tefrokronologi, equivalent to tefr (a) volcanic ejecta (from Greek téphra ash) + -o--o- + kronologi chronology.

Example:
The cores span approximately the last 1000 years, based on 210Pb dating, AMS 14C dating and tephrochronology.
基于210Pb年代测定，AMS 14C年代测定和火山灰年代学，该核心跨度约为过去的1000年。

ternary eutectic point [ˈtɜːnəri juː ˈtektik pɔint]

Definition: 三元共晶点 sān yuán gòng jīng diǎn
ternary: composed of three parts. 三个组成的 sān gè zǔ chéng de
eutectic: relating to or denoting a mixture of substances (in fixed proportions) that melts and solidifies at a single temperature that is lower than the melting points of the separate

constituentsor of any other mixture of them. 低共熔的 dī gòng róng de; 易熔的 yì róng de
point: a particular spot, place, or position in an area or on a map, object, or surface. (地区、地图、物体或表面上的)点 diǎn; 处 chù; 位置 wèi zhì

ternary systems [ˈtɜːnəri sistəms]

Definition: 三元系 sān yuán xì

ternary: using three as a base. 三元的 sān yuán de

system: a set of things working together as parts of a mechanism or an interconnecting network. 系统 xì tǒng

Extended Terms:

ternary liquid systems 三元液体系统
ternary polyelectrolyte systems 三元多电解质系统

terrigenous clast [teˈridʒinəs klæst]

Definition: 陆源碎屑 lù yuán suì xiè

terrigenous: (of a marine deposit) Made of material eroded from the land. [地]陆生的 lù shēng de

clast: a constituent fragment of a clastic rock. 碎屑 suì xiè

terrigenous constituents [teˈridʒinəs kənˈstituənts]

Definition: 陆源组分 lù yuán zǔ fēn

terrigenous: (of a marine deposit) Made of material eroded from the land. [地]陆生的 lù shēng de

constituent: one of the parts of something that combine to form the whole. 成分 chéng fèn; 构成要素 gòu chéng yào sù

Example:

In the sea area of between south of Taiwan Province of China and 17°N, the sediments consist mainly of terrigenous constituents.
台湾省以南到 17°N 以北海区沉积物以陆源沉积物分布为主。

terrigenous detritus [teˈridʒinəs diˈtraitəs]

Definition: 陆源碎屑 lù yuán suì xiè

terrigenous: (of a marine deposit) made of material eroded from the land. (海洋沉积)陆源的

607

lù yuán de；陆地的 lù dì de

detritus：gravel, sand, silt, or other material produced by erosion. 碎屑 suì xiè；瓦砾 wǎ lì

Example:

The clay-size terrigenous detritus supplied by Yangtze River and East Asia winter monsoon is also higher in the last glaciation maximum (LGM) while terrigenous flux is high and changes smoothly.

在末次盛冰期时，陆源碎屑物质通量高而稳定，长江和东亚冬季风输运的黏土粒级碎屑物质增多。

terrigenous mineral [teˈridʒinəs ˈminərəl]

Definition: 陆源矿物 lù yuán kuàng wù

terrigenous：(of a marine deposit) Made of material eroded from the land. [地] 陆生的 lù shēng de

mineral：a substance that is naturally present in the earth and is not formed from animal or vegetable matter, for example gold and salt. Some minerals are also present in food and drink and in the human body and are essential for good health. 矿物 kuàng wù

Example:

The results of our study showed that the mineral assemblages in the sediments of the area investigated can also be divided into three types：terrigenous minerals, authigenic minerals and volcanic minerals from oceanic volcanoes.

分析资料表明，调查区内共有三种类型的矿物：陆源矿物、自生矿物和海底火山喷发形成的火山型矿物。

terrigenous province [teˈridʒinəsˈprɔvins]

Definition: 陆源区 lù yuán qū

terrigenous：(of a marine deposit) Made of material eroded from the land. [地] 陆生的 lù shēng de

province：an area of the world with respect to its flora, fauna, or physical characteristics. 区域 qū yù

Example:

Controlled by the distance to the terrigenous province, basement structure and main faults, these lake floor fans can be subdivided into three subfacies, i. e. inner fan, middle fan and outer fan.

受物源区的距离、基底构造和主干断层活动的控制，湖底扇平面上具有分带性，可划分为内扇、中扇和外扇 3 个亚相带。

terrigenous sedimentary rock [teˈridʒinəs sediˈmentəri rɒk]

Definition: 陆源沉积岩 lù yuán chén jī yán

terrigenous: (of a marine deposit) made of material eroded from the land. [地] 陆生的 lù shēng de

sedimentary rock: (geology) one of the major groups of rock that makes up the crust of the Earth; formed by the deposition of either the weathered remains of other rocks, the results of biological activity, or precipitation from solution. 沉积岩 chén jī yán

Example:

Terrigenous sedimentary rocks are the dominant rock types in a continental sedimentary basin, whose components of these rocks are mainly controlled by source rocks.
陆源碎屑岩是陆相沉积盆地的主要充填物，其成分主要受物源区母岩成分控制。

tertiary granite [ˈtɜːnəri ˈgrænits]

Definition: 第三花岗岩 dì sān huā gāng yán

tertiary: of, relating to, or denoting the first period of the Cenozoic era, between the Cretaceous and Quaternary periods, and comprising the Palaeogene and Neogene sub-periods. 第三纪(或系)的 dì sān jì de

granite: a very hard, granular, crystalline, igneous rock consisting mainly of quartz, mica, and feldspar and often used as a building stone. 花岗岩 huā gāng yán

Example:

study on the composition and structure of feldspars in tertiary granites of Himalayas Area, China
我国境内喜马拉雅地区第三纪花岗岩类中的长石成分和结构状态研究

teschenite [ˈteʃənait]

Definition: coarse-to fine-grained, rather dark-coloured, intrusive igneous rock that occurs in sills (tabular bodies inserted while molten between other rocks), dikes (tabular bodies injected in fissures), and irregular masses and is always altered to some extent. 沸方绿岩 fèi fāng lǜ yán

Origin: 1865-1870; Teschen (German name of Czech Těšín, town in the Czech Republic) + -ite.

Example:

study on the petrology and geochemistry of differentiated teschenite intrusions from the Hunter Valley, New South Wales, Australia

澳大利亚新南威尔士猎人谷的分化型沸方绿岩侵蚀的岩石学和地球化学探究

Extended Term:

nepheline teschenite 霞石方沸徨绿岩

Tethys Sea [ˈtiːθis siː]

Definition: former tropical body of salt water that separated the supercontinent of Laurasia in the north from Gondwana in the south during much of the Mesozoic Era (251 to 65.5 million years ago). 特提斯海 tè tí sī hǎi

Example:

The Meso-Tethys was a sea lying between Laurasia and Gondwana during the Middle and Late Mesozoic.
中特提斯是中生代中晚期存在于南、北大陆之间的海洋。

tetrahedron [ˌtetrəˈhiːdrən]

Definition: a solid figure with four triangular faces. 四面体 sì miàn tǐ

Origin: 1560-1570; tetra- + -hedron, modeled on Late Greek tetráedron, noun use of neuter of tetráedros four-sided.

Example:

The pattern of microspores in the tetrad period is of tetrahedron type.
小孢子四分体时期其排列为四面体型。

Extended Terms:

elementary tetrahedron 元四面体
orthocentric tetrahedron 垂心四面体
pyramidal tetrahedron 锥形四面体
wind tetrahedron 风标

textural inversion [ˈtekstʃərəl inˈvɜːʃən]

Definition: 质地反演 zhì dì fǎn yǎn

Textural: relating to the feel, appearance, or consistency of a surface or a substance. （物质、表面的）结构的 jié gòu de; 纹理的 wén lǐ de

inversion: (Chemistry) a reaction causing a change from one optically active configuration to the opposite configuration, especially the hydrolysis of dextrose to give a laevorotatory solution of fructose and glucose. （化）倒反 dǎo fǎn; 反转 fǎn zhuǎn

Example:

The sedimentary history of Devonian alluvial fan—a study of textural inversion
泥盆纪冲积扇的沉积史——质地反演研究

textural maturity [ˈtekstʃərəl məˈtjuərəti]

Definition: 结构成熟度 jié gòu chéng shú dù

textural: relating to texture. 质地的 zhì dì de；纹理的 wén lǐ de
maturity: the state, fact, or period of being mature. 成熟度 chéng shú dù

Example:

The Jurassic reservoirs in the Moxizhuang region, Junggar Basin, Xinjiang are characterized by low compositional maturity, abundant debris, low matrix contents and high textural maturity.
准噶尔盆地莫西庄地区侏罗系储层具有低成分成熟度、富含岩屑、低杂基含量和高结构成熟度的岩石学特征。

texture [ˈtekstʃə]

Definition: a structure of interwoven fibers or other elements. 质地 zhì dì；(材料等的)结构 (cái liào děng de) jié gòu

Origin: Middle English, from Old French, from Latin textūra, from textus, past participle of texere, to weave.

Example:

According to the spectrum of fabric texture and defects, a new method for defect detection was presented.
根据织物纹理和疵点的频谱不同分布，提出了织物疵点检测的新方法。

Extended Terms:

soil texture 土壤质地
texture map 纹理贴图 材质贴图
texture modes 材质模式
texture swapping 纹理交换

texture of rocks [ˈtekstʃə ɔv rɒks]

Definition: 岩石结构 yán shí jié gòu

texture: the arrangement of the particles or constituent parts of any material, as wood, metal, etc., as it affects the appearance or feel of the surface; structure, composition, grain, etc. 结构

jié gòu

rocks: a large mass of stone forming a peak or cliff. 岩石 yán shí

> Example:

All the units are composed with biotite-bearing admellite, but they are obviously varied in their texture of rocks, composition of minerals, chemical composition of rocks, the geochemical characteristics of trace elements and rare earth elements and their evolution trend etc.
岩性均为黑云母二长花岗岩，但岩石结构、矿物成分、岩石化学成分、微量元素和稀土元素地球化学有明显差异和演化趋势。

textures of igneous rock ['tekstʃəs ɔv 'igniəs rɒk]

> Definition: 火成岩纹理 huǒ chéng yán wén lǐ

textures: the feel, appearance, or consistency of a surface or a substance.（物质、表面的）结构 jié gòu; 纹理 wén lǐ

of: indicating an association between two entities, typically one of belonging, in which the first is the head of the phrase and the second is something associated with it. [表示所属关系] 属于……的 shǔ yú ... de

igneous: (of rock) having solidified from lava or magma. 火成的 huǒ chéng de

rock: the solid mineral material forming part of the surface of the earth and other similar planets, exposed on the surface or underlying the soil. 岩 yán

Th (thorium) ['θɔːriəm]

> Definition: thorium-a soft silvery-white tetravalent radioactive metallic element; isotope 232 is used as a power source in nuclear reactors; occurs in thorite and in monazite sands. 元素钍 (thorium) 的符号 yuán sù tǔ de fú hào

> Origin: before 1050; Old English Thōr, from Old Norse Thōrr +New Latin, Latin, neuter suffix.

Theory of the Earth ['θiəri ɔv ði]

> Definition: 地球理论 dì qiú lǐ lùn

theory: a set of principles on which the practice of an activity is based. 理论 lǐ lùn; 原理 yuán lǐ

earth: the planet on which we live; the world. 地球 dì qiú

> Example:

It is a complete update of Anderson's Theory of the Earth (1989).
这是安德森的地球理论的完整更新(1989)。

Extended Term:

tetrahedral theory of the earth 地球四面体说

theralite [ˈθiərəˌlait]

Definition: a coarse-grained, phaneritic rock composed of labradorite, nepheline, and augite. 霞斜岩 xiá xié yán

Origin: Greek thḗra hunting + -lite; said to be so called because success in hunting it down was thought to be certain.

Example:

The paper presents electron microprobe analyses of nephelines and analcimes in alkaline igneous rocks ranging from theralite and basanite to mugearite and tinguaite.
本文提供碱性火成岩中的霞石和方沸石的电子显微探针分析，取材范围包括霞斜岩、碧玄岩、橄榄粗安岩和丁古岩。

Extended Terms:

olivine theralite gabbro 橄榄霞斜辉长岩
theralite diabase 霞斜辉绿岩
theralite porphyrite 霞斜玢岩

thermal Rayleigh number [ˈθɜːməlˈreiliˈnʌmbə]

Definition: 热瑞利数 rè ruì lì shù

thermal: connected with heat. 热的；热量的
Rayleigh number: In fluid mechanics, the Rayleigh number for a fluid is a dimensionless number associated with buoyancy driven flow (also known as free convection or natural convection). When the Rayleigh number is below the critical value for that fluid, heat transfer is primarily in the form of conduction; when it exceeds the critical value, heat transfer is primarily in the form of convection. The Rayleigh number is named after Lord Rayleigh and is defined as the product of the Grashof number, which describes the relationship between buoyancy and viscosity within a fluid, and the Prandtl number, which describes the relationship between momentum diffusivity and thermal diffusivity. Hence the Rayleigh number itself may also be viewed as the ratio of buoyancy forces and (the product of) thermal and momentum diffusivities. 瑞利数 ruì lì shù

Example:

The mechanism of bifurcations for the thermal convection flow of porousmedia is studied by analytical method. It is pointed out that the Rayleigh number Ra in thenonlinear flow equations

is an important parameter to determine the fiow characteristics.
本文利用解析分析方法研究了数值模拟发现的多孔介质层中出现的对流分叉机理，指出控制方程中的瑞利数，是决定流动的特征参数。

thermal [ˈθɜːməl]

Definition: having to do with heat, hot springs, etc. 热的 rè de；由热造成的 yóu rè zào chéng de；保暖的 bǎo nuǎn de；上升的热气流 shàng shēng de rè qì liú

Origin: from Greek, combining form of thermós hot, thérmē heat.

Example:
Shale pottery thermal insulation surface mortar has good heat preservation and heat insulation performances.
页岩陶砂保温抹面砂浆具有良好的保温隔热等性能。

Extended Terms:
thermal analyzer 热分析仪
thermal conductivity 导热性
thermal neutron 热中子
thermal reactor 热反应堆
thermal slug 散热片

thermal boundary layer [ˈθɜːməl ˈbaʊndəri ˈleɪə]

Definition: (TBL) 热边界层 (TBL) rè biān jiè céng de

thermal: of or relating to heat. 热量的 rè liàng de；由热造成的 yóu rè zào chéng de
boundary: line which marks the limits of an area; a dividing line. 分界线 fēn jiè xiàn；边界 biān jiè
layer: a sheet, quantity, or thickness of material, typically one of several, covering a surface or body (figurative). 层 céng；覆盖层 fù gài céng

Example:
It is shown that there exists a thermal boundary layer for the convective heat transfer of ideal fluids also, similar to that for viscous fluid.
结果表明理想流体的对流传热与黏性流体同样存在着热边界层。

thermal conductivity [ˈθɜːməl kɒndʌkˈtiviti]

Definition: a measure of the ability of a material to transfer heat. Given two surfaces on either side of the material with a temperature difference between them, the thermal conductivity is the heat energy transferred per unit time and per unit surface area, divided by the temperature difference. It is measured in watts per degree Kelvin. 导热性 dǎo rè xìng; 热导率 rè dǎo lǜ

Example:

Thermal conductivity of hydrocarbon fiber is much higher compare to other analogues used in heating systems. This characteristic allows achieving more thermal efficiency with less energy consumption.
碳纤化合物用于供暖系统的电导率比其他类似物要高的多。这一特点使热能更有效地发挥以减少能源的消耗。

Extended Terms:

coefficient of thermal conductivity 导热系数
effective thermal conductivity 有效热传导率
thermal conductivity detector 热导探测器
thermal conductivity gas analyzer 热导式气体分析器
thermal semi conductivity sensor 热半传导传感器

thermal contact metamorphism [ˈθɜːməl ˈkɒntækt metəˈmɔːfɪzəm]

Definition: 热接触变质作用 rè jiē chù biàn zhì zuò yòng

thermal contact: a state in which two or more systems can exchange thermal energy. 热接触 rè jiē chù

metamorphism: alteration of the composition or structure of a rock by heat, pressure, or other natural agency. 变质作用 biàn zhì zuò yòng

Example:

On the condition of thermal contact metamorphism, clay minerals in coal are changed into andalusite.
无烟煤中红柱石是煤中黏土质矿物经热接触变质作用形成的。

thermal diffusivity [ˈθɜːməl ˌdɪfjuːsɪvəti]

Definition: is the thermal conductivity divided by the volumetric heat capacity. It has the SI unit of m^2/s. 热扩散系数 rè kuò sàn xì shù; 热扩散率 rè kuò sàn lǜ

Example:

Experiment result indicates that the RFHP has greater thermal diffusivity, and can reduce junction-to-case thermal resistance of the module.
结果表明径向平板热管具有更高的热扩散能力，可降低模块的结壳热阻。

Extended Terms:

soil thermal diffusivity 土壤热扩散性，土壤热扩散系数
test method for thermal diffusivity of hardmetal (carbide) 硬质合金热扩散率的测定方法
thermal diffusivity of rocks 岩石热扩散率

thermal divide [ˈθɜːməl diˈvaid]

Definition: 热鸿沟 rè hóng gōu

thermal: of or relating to heat. 热量的 rè liàng de；由热造成的 yóu rè zào chéng de
divide: a boundary between two things.（两事物间的）界限 jiè xiàn

thermal energy [ˈθɜːməl ˈenədʒi]

Definition: 热能 rè néng

thermal: of or relating to heat. 热量的 rè liàng de；由热造成的 yóu rè zào chéng de
energy: power derived from the utilization of physical or chemical resources, especially to provide light and heat or to work machines. 能 néng

Extended Terms:

thermal energy and dynamic engineering 热能与动力工程
thermal excitation energy 热激发能量
thermal jet energy 热喷能
thermal ionization energy 热电离能

thermal expansion coefficient [ˈθɜːməl ikˈspænʃən ˌkəuiˈfiʃənt]

Definition: 热膨胀系数 rè péng zhàng xì shù

thermal expansion: Thermal expansion is the tendency of matter to change in volume in response to a change in temperature. All materials have this tendency. 热膨胀 rè páng zhàng
coefficient: a multiplier or factor that measures some property. 系数 xì shù

Extended Term:

linear thermal expansion coefficient 线热膨胀系数

thermal gradient [ˈθɜːməlˈgreidiənt]

Definition: A temperature gradient is a physical quantity that describes in which direction and at what rate the temperature changes the most rapidly around a particular location. The

temperature gradient is a dimensional quantity expressed in units of degrees (on a particular temperature scale) per unit length. 热梯度 rè tī dù

Example:

The larger the thermal gradient is, the lower the anisothermal ductility is.
温度梯度越大，不等温韧性越低。

Extended Terms:

axial thermal gradient 轴向温度梯度
conductive thermal gradient 传导热梯度
equilibrium thermal gradient 平衡热梯度
thermal gradient kiln 热梯度炉

thermal shock [ˈθɜːməl ʃɒk]

Definition: Thermal shock is the name given to cracking as a result of rapid temperature change. Glass and ceramic objects are particularly vulnerable to this form of failure, due to their low toughness, low thermal conductivity, and high thermal expansion coefficients. However, they are used in many high temperature applications due to their high melting point. 热冲击 rè chōng jī；热休克 rè xiū kè；温度急增 wēn dù jí zēng

Example:

In the process of rapid quenching iron-base alloy, chemical stability, washing resistance, thermal shock resistance at high temperature and easy machining the nozzle are required.
铁基合金的快淬工艺要求陶瓷喷嘴在高温下具有化学稳定性、耐冲刷性、抗热震性，且易机械加工。

Extended Terms:

pressurized thermal shock 承压热冲击
thermal shock stability 抗热震性能
thermal shock tester 冷热冲击仪
thermal shock break 热震破裂

thermal structure [ˈθɜːməl ˈstrʌktʃə]

Definition: 热构造 rè gòu zào，热结构 rè jié gòu

thermal: connected with heat. 热的 rè de；热量的 rè liàng de
structure: the arrangement of and relations between the parts or elements of something complex. 结构 jié gòu

> **Example:**

Efficient heat transfer elements formed by the heat of the thermal structure of a heat interchanger tight have the advantage of both small and pressure reduction, efficient auxiliary power, and without merit.
由高效传热元件热管组成的热管换热器具有结构紧凑、体积小、压降低、效率高以及无辅助动力等优点。

thermodynamics [ˌθəːməudaiˈnæmiks]

> **Definition:** the branch of physics dealing with the transformation of heat to and from other forms of energy, and with the laws governing such conversions of energy. 热力学 rè lì xué

> **Origin:** thermo- + dynamics.

> **Example:**

The idea that energy is conserved is the first law of thermodynamics.
能量守恒这一概念是热力学的第一定律。

> **Extended Terms:**

energetics thermodynamics 热力学
equilibrium thermodynamics 平衡态热力学
irreversible thermodynamics 非平衡态热力学
metallurgical thermodynamics 冶金热力学
statistic thermodynamics 统计热力学

thermometry [θəˈmɔmitri]

> **Definition:** measurement of temperature; the science of making or using thermometers. 计温 jì wēn; 温度测定法 wēn dù cè dìng fǎ

> **Origin:** 1855-1860; thermo- + -metry.

> **Example:**

Thermistors and thermocouples are common types of transducers used in differential thermometry. The choice of transducer depends on the specific microcalorimetry application.
热敏电阻和热偶是各种测温学中常用的传感器类型。传感器的选择依赖于特定的微热量测量应用。

> **Extended Terms:**

geological thermometry 地温测定法
infrared thermometry 红外线测温
magnetic thermometry 磁温度量测术

radiation thermometry 辐射测温法
thermistor thermometry 热敏计温学

thermos bottle effect ['θəːmɔs 'bɔtl i'fekt]

Definition: 保温瓶效应 bǎo wēn píng xiào yìng

thermos bottle: a particular kind of vacuum flask(= a container like a bottle with double walls with a vacuum between them, used for keeping liquids hot or cold). 瑟姆斯保温瓶 sè mǔ sī bǎo wēn píng

effect: a change that somebody/something causes in somebody/something else; a result. 效应 xiào yìng

Thiel Mountain siderolite ['θil 'mauntən 'sidərəlait]

Definition: 提尔山菱铁矿 tí ěr shān ling tiě kuàng

Thiel: Thiel is a lunar impact crater on the far side of the Moon. It is located to the south of the larger crater Quetelet, and to the north-northwest of Charlier. 提尔 tí ěr

mountain: A mountain is a large landform that stretches above the surrounding land in a limited area usually in the form of a peak. A mountain is generally steeper than a hill. the adjective montane is used to describe mountainous areas and things associated with them. 山 shān

siderolite: the category of meteorites commonly referred to as Stoney-Irons. The three classes of Siderolites are Lodranites, Mesosiderites and Pallasites. 菱铁矿 ling tiě kuàng

thin sections [θin 'sekʃənz]

Definition: a thin, flat piece of material prepared for examination with a microscope, in particular a piece of rock about 0.03 millimeters thick, or, for electron microscopy, a piece of tissue about 30 nanometers thick. 薄片 báo piàn

Example:
Observation of cores and thin sections indicates that primary pores are the basic reservoir space of the volcanic rocks, while secondary pores have important influence.
岩芯和薄片观察表明，原生孔隙是火山岩储集空间形成的基础，次生孔隙是储集空间的重要组成部分。

tholeiite ['θəuliait]

Definition: Tholeiitic basalt is a type of basalt (an igneous rock) which includes very little sodium as compared with other basalts. Chemically, this type of rock has been described as a subalkaline basalt. 拉斑玄武岩 lā bān xuán wǔ yán

> Example:

Basalt series in Baiyinchang ore-field are consist of basalts and alkali-basalts, in which basalts belong to calc-alkalic basalts and tholeiite, alkali basalts belongs to kali-alkali series.
白银厂矿田玄武岩主要由玄武岩和碱性玄武岩组成，其中玄武岩属于钙碱性系列和拉斑系列，碱性玄武岩属于钠质碱性玄武岩系列。

> Extended Terms:

bronzite tholeiite 古铜硅质玄武岩
island arc tholeiite 岛弧型拉斑玄武岩
oceanic ridge tholeiite 洋脊拉斑玄武岩
oceanic tholeiite 海洋性拉斑玄武岩
olivine tholeiite 橄榄多硅玄武岩

tholeiite basalt ['θəuliait 'bæsɔːlt]

> Definition: Tholeiitic basalt is a type of basalt (an igneous rock) which includes very little sodium as compared with other basalts. Chemically, this type of rock has been described as a subalkaline basalt. 拉斑玄武岩 lā bān xuán wǔ yán

> Extended Term:

tholeiite basalt magma 拉斑质玄武质岩浆

tholeiitic series [ˌθəuliːˈitik ˈsiəriːz]

> Definition: 拉斑系列 lā bān xì liè

tholeiitic: of a basaltic rock containing augite and a calcium-poor pyroxene (pigeonite or hypersthene), and with a higher silica content than an alkali basalt. [地质]拉斑玄武岩的 lā bān xuán wǔ yán de

series: (Geology) (in chronostratigraphy) a range of strata corresponding to an epoch in time, being a subdivision of a system and itself subdivided into stages. [地质]（年代地层学）统 tǒng, （岩系的）段 duàn

> Example:

Their original rocks are volcanic rocks, epicontinental clastic sedimentary rocks, and ophiolite of volcanic and island arc tholeiitic series or calcium alkali series.
其原岩岩石类型有火山岩，陆缘碎屑沉积岩，火山、岛弧拉斑系列或钙碱系列火山岩、蛇绿岩等。

Extended Terms:
tholeiitic magma 硅质玄武岩浆
tholeiitic texture 硅质玄武结构

thorium [ˈθɔːriəm]

Definition: a rare, grayish, radioactive, metallic chemical element, one of the actinides, found in monazite and thorite: it is used in magnesium alloys, the making of gas mantles, electronic equipment, etc., and as a nuclear fuel: symbol, Th; at. no., 90. 钍(符号 Th，原子序 90) tǔ

Origin: named (1829) by Baron Jöns Jakob Berzelius.

Example:
Chemical compounds and mixtures containing uranium and thorium are active in direct proportion to the amount of these metals contained in them.
含铀和钍的化合物和混合物活跃的程度与这些化合物和混合物中这两种金属的含量直接相关。

Extended Terms:
thorium chloride 氯化钍
thorium phosphate 磷酸钍
thorium reactor 钍堆
thorium sulfate 硫酸钍
unalloy thorium 金属钍

three-component system [θriːkəmˈpəunənt sistəm]

Definition: 三组分体系 sān zǔ fēn tǐ xì

three: equivalent to the sum of one and two; one more than two. 三 sān

component: one of several parts of which something is made. 组成部分 zǔ chéng bù fèn；成分 chéng fèn

system: a set of things working together as parts of a mechanism or an interconnecting network. 体系 tǐ xì

Thulean igneous province [ˈθjuːlin ˈigniəs ˈprɒvins]

Definition: 杜林桥火成岩区 dù lín qiáo huǒ chéng yán qū

igneous: of rocks formed when magma (= melted or liquid material lying below the earth's surface) becomes solid, especially after it has poured out of a volcano. 火成的(尤指火山喷出

的)huǒ chéng de
province：a biogeographical area within a region that is defined by the plants and animals that inhabit it. (生物地理)区域 qū yù

Ti [tiː]

Definition: symbol of chemical elements titanium：a light strong grey lustrous corrosion-resistant metallic element used in strong lightweight alloys (as for airplane parts); the main sources are rutile and ilmenite 元素钛的符号 yuán sù tài de fú hào

Origin: late 18th century：from Titan, on the pattern of uranium.

Example:

The poison is a mixture of Alcoho, Calcium, Titanium, Red Phosphorus, Xenon, Silica. 毒药是酒精、钙、钛、红磷、氙气、二氧化硅组成的混合物。

Extended Terms:

low titanium 低钛
titanium powders 钛粉

Tibesti dome [təˈbesti dəum]

Definition: 提贝斯提穹顶 tí bèi sī tí qióng dǐng

Tibesti：The Tibesti Mountains are a volcanic group of inactive volcanoes with one potentially active volcano in the central Sahara desert in the Bourkou-Ennedi-Tibesti Region of northern Chad. The northern slopes extend a short distance into southern Libya. 提贝斯提 tí bèi sī tí dome：a round roof with a circular base. 穹顶 qióng dǐng；圆屋顶 yuán wū dǐng

tidal friction [ˈtaidəl ˈfrikʃən]

Definition: Tidal acceleration is an effect of the tidal forces between an orbiting natural satellite (e.g. the Moon), and the primary planet that it orbits (e.g. the Earth). The "acceleration" is usually negative, as it causes a gradual slowing and recession of a satellite in a prograde orbit away from the primary, and a corresponding slowdown of the primary's rotation. The process eventually leads to tidal locking of first the smaller, and later the larger body. The Earth-Moon system is the best studied case. 潮汐摩擦 cháo xī mó cā

Example:

The theory of tidal friction is the concern of dynamic oceanography.

潮夕摩擦理论是海洋动力学的课题。

tie line [tai lain]

Definition: a direct telephone line between extensions in two or more PBX systems; a line used to connect one electric power or transportation system with another. 两平衡相成分点间连线 liǎng píng héng xiāng chéng fēn diǎn jiān lián xiàn; 接入线 jiē rù xiàn; 专用线路 zhuān yòng xiàn lù; 连接线 lián jié xiàn; 联络线 lián luò xiàn

Example:
The research on the fault level of tie line is of significance for the security and stability evaluation of power system as well as the preventive control of transient stability.
研究联络线故障水平对于电力系统的安全稳定评估以及暂态稳定的预防控制都有着重要意义。

Extended Terms:
area tie line 电力系统地区联络线
automatic tie line 自动联络线
power system tie line 电力系统联络线
tie line access 专线接入
weak tie line 弱联络线

tie triangle [tai 'traiæŋgl]

Definition: 联系三角 lián xì sān jiǎo
tie: a strong connection between people or organizations. 联系 lián xì; 关系 guān xì
triangle: a flat shape with three straight sides and three angles; a thing in the shape of a triangle. 三角形 sān jiǎo xíng; 三角形物体 sān jiǎo xíng wù tǐ

Example:
According to the comparison results of different survey programs, the tying triangle directional survey program is presented to directionally survey the vertical shaft and is successfully applied to this pipeline project.
通过对各种测量方案的比选，提出了采用联系三角形定向测量法进行竖井定向测量的技术方案。

tillite ['tilait]

Definition: sedimentary rock consisting of compacted rock fragments of various sizes that have been formed by glacial action 冰碛岩 bīng qì yán，又称冰川砾岩 (glacial conglomerate)

bīng chuān lì yán

Example:

In addition, the fabric characteristics of modern till are also of importance to the study of the ancient till and tillite.

此外，现代冰碛物的组构特征对于研究古冰碛和冰碛岩也有着重要的意义。

Extended Term:

para-tillite 准冰碛岩

time-integrated fluid flux [taim 'intigreitid 'fluːid flʌks]

Definition: 时间集成流体流量 shí jiān jí chéng liú tǐ liú liàng

time: what is measured in minutes, hours, days, etc. 时间 shí jiān

integrated: with various parts or aspects linked or coordinated. 综合的 zōng hé de;完整统一的 wán zhěng tǒng yī de

fluid: a liquid; a substance that can flow. 液体 yè tǐ;流体 liú tǐ

flux: a flow; an act of flowing. 通量 tōng liàng;流动 liú dòng

Tinos ['tiːnəus]

Definition: an island of southeast Greece in the Cyclades Islands east-southeast of Athens. 蒂诺斯 dì nuò sī

Example:

Pigeon houses dot the hillsides on the Greek island of Tinos.

鸽子屋遍布于希腊蒂诺斯岛的山坡上。

Extended Terms:

Tinos green 希腊绿

Tinos marble 绿色大理石

titanite ['taitənait]

Definition: Titanite, or sphene (from the Greek sphenos (σφην ώ), meaning wedge), is a calcium titanium nesosilicate mineral, $CaTiSiO_5$. Trace impurities of iron and aluminium are typically present. also commonly present are rare earth metals including cerium and yttrium; calcium may be partly replaced by thorium. 榍石 xiè shí

Origin: 1790-1800; from German Titanit. See titanium, -ite.

> Example:

Biotite, titanite, cassiterite, rutile, wolframite, scheelite and wolframoixiolite are essentially useful minerals in clarifying the ore-forming potential of granites.
黑云母、榍石、锡石、金红石、黑钨矿、白钨矿和钨铁铌矿是讨论的重点矿物，它们可用于判别花岗岩的成矿能力。

> Extended Terms:

sphene, titanite 榍石
zinc titanite 钛酸锌
zirconium titanite 钛酸锆

titanium in quartz geothermometer
[taiˈteiniəm in kwɔːts ˌdʒiːəuθəˈmɔmitə]

> Definition: 石英温度计中的钛 shí yīng wēn dù jì zhōng de tài

titanium: the chemical element of atomic number 22, a hard silver-gray metal of the transition series, used in strong, light, corrosion-resistant alloys. 钛 tài

quartz: a hard white or colorless mineral consisting of silicon dioxide, found widely in igneous, metamorphic, and sedimentary rocks. It is often colored by impurities (as in amethyst, citrine, and cairngorm). 石英 shí yīng

geothermometer: (also called geologic thermometer) a thermometer designed to measure temperatures in deep-sea deposits or in bore holes deep below the surface of the earth. 温度计 wēn dù jì

titanomagnetite [taitənəuˈmægnitait]

> Definition: a mineral containing oxides of titanium and iron. 钛磁铁矿 tài cí tiě kuàng

tomography X [təˈmɒgrəfi eks]

> Definition: X 射线断层摄影 X shè xiàn duàn céng shè yǐng

tomography: technique for displaying a representation of a cross section through a human body or other solid object using X-rays or ultrasound. [医] X 线断层照相术 X xiàn duàn céng zhào xiàng shù

> Example:

X-ray fluorescence tomography X 射线荧光层析

tonalite [ˈtəunəlait]

Definition: a coarse-grained plutonic rock consisting chiefly of sodic plagioclase, quartz, and hornblende or other mafic minerals. 英云闪长岩 yīng yún shǎn cháng yán

Origin: late 19th century: from Tonale Pass, northern Italy, + -ite.

Example:
We usually don't care whether a rock is classified as a granodiorite or diorite or tonalite.
我们通常并不注意某种岩石是属于花岗闪光岩、闪长岩还是英云闪长岩。

Extended Terms:
tonalite-pegmatite 英云闪长伟晶岩
tonalite-porphyrite 英云闪长玢岩

Topology [təˈpɒlədʒi]

Definition: the study of changes in topography that occur over time and, especially, of how such changes taking place in an area affect the history of that area. 地志学 dì zhì xué

Origin: late 19th century: via German from Greek topos "place" + -logy.

Example:
Topology is a crucial sub-discipline of Geoscience.
地志学是地球科学的一个重要分支学科。

tortuosity factor [ˌtɔːtjuˈɒsəti ˈfæktə]

Definition: 扭曲系数 niǔ qū xì shù

tortuosity: the state of being twisted or crooked. 曲折 qū zhé
factor: a quantity or level of something. 系数 xì shù

Example:
The tortuosity factor of A301 ammonia synthesis catalyst is measured using dynamic state method single pellet string reactor method (SPSRM).
采用动态法，在单颗粒线反应器中对 A301 氨合成催化剂的曲折因子进行了测定。

Extended Term:
hydraulic tortuosity factor 水力弯曲系数

tourmaline [ˈtuəməliːn]

Definition: Tourmaline is a crystal silicate mineral compounded with elements such as aluminium, iron, magnesium, sodium, lithium, or potassium. 电气石 diàn qì shí

Origin: mid-18th century: from French, based on Sinhalese tōramalli "carnelian".

Example:

Tourmaline is a semi-precious stone and mineral that emits negative ions and far-infrared rays (FIR).

电气石在矿产中是一种次级的宝石，它可以放射负离子和远红外线。

Extended Terms:

aquamarine tourmaline 海蓝宝石碧玺
Brazilian tourmaline 巴西碧硒
iron tourmaline 铁电气石
Siberian tourmaline 西伯利亚碧
tourmaline gauge 电气石计

trace element [treisˈelimənt]

Definition: a chemical element present only in minute amounts in a particular sample orenvironment. 痕量元素 hén liàng yuán sù

Extended Term:

trace element deficiency 微量元素缺乏症

tracer [ˈtreisə]

Definition: a substance introduced into a biological organism or other system so that its subsequent distribution can be readily followed from its color, fluorescence, radioactivity, or other distinctive property. 示踪剂 shì zōng jì

Origin: 1250-1300; late Middle English tracen, Middle English: to make one's way, proceed, from Middle French tracier; from Vulgar Latin *tractiāre, derivative of Latin tractus, past participle of trahere to draw, drag; (noun) Middle English: orig., way, course, line of footprints, from Old French, derivative of tracier.

Example:

At present, the methods of isotope tracer and turbine flowmeter are universally used for testing annulus well in home oilfield.

同位素示踪法和涡轮流量计法是目前国内各油田普遍采用的环空测井方法。

Extended Term:

tracer method 示踪法

trachyandesite [trəkaiæn'desait]

Definition: an extrusive igneous rock. It has little or no free quartz, but is dominated by alkali feldspar and sodic plagioclase along with one or more of the following mafic minerals: amphibole, biotite or pyroxene. Small amounts of nepheline may be present and apatite is a common accessory mineral. 粗安岩 cū ān yán

Example:
The rocks fall into four groups, namely Cenozoic basalt, Late Cretaceous basaltic trachyandesite, Late Creatceous trachydacite and liparite, and Early Triassic dacite.
主要岩石类型有新生代老第三纪玄武岩、中生代晚白垩世玄武粗安岩、中生代晚白垩世粗面英安岩和流纹岩和中生代早三叠世英安岩。

Extended Term:
trachyandesite bassalt 粗安玄武岩

trachybasalt ['treikibə'sɔːlt]

Definition: an extrusive rock characterized by calcic plagioclase and sanidine, with augite, olivine, and possibly minor analcime or leucite. 粗面玄武岩 cū miàn xuán wǔ yán

Example:
We regard Dahongyu volcanic rocks as trachybasalt and trachyte, Calc-alkalic series, potassic.
作者认为大红峪组火山岩应为钾质的粗面岩和粗面玄武岩，属钙碱系列。

Extended Term:
vesicular olivine trachybasalt 多孔质橄榄石粗面玄武岩

trachyte ['treikait]

Definition: a fine-grained volcanic rock, characterized by the presence of alkaline feldspar minerals. 粗面岩 cū miàn yán

Origin: early 19th century (denoting a volcanic rock with a rough or gritty surface): from Greek trakhus "rough" or trakhutēs "roughness".

Extended Terms:
quartz trachyte 石英粗面岩
soda trachyte 钠性粗面岩
trachyte andesite 粗面安山岩

trachytic [trəˈkitik]

Definition: describes igneous rocks in which the crystals are arranged in parallel and show the flow of the molten lava from which they were formed. 粗面岩状的 cū miàn yán zhuàng de

Origin: 1815-1825；< French < Greek trāchýtēs roughness, equivalent to trāchý(s) rough + -tēs noun suffix.

Example:
Petrochemistry analysis of pumice component indicates that the magma can be assigned to trachytic magma.
火山碎屑堆积物中的浮岩岩石化学分析结果表明岩浆成分为粗面质。

Extended Terms:
trachytic porphyry 粗面斑岩
trachytic texture 粗面结构

trachytic texture [trəˈkitikˈtekstʃə]

Definition: 粗面结构 cū miàn jié gòu

trachytic: describes igneous rocks in which the crystals are arranged in parallel and show the flow of the molten lava from which they were formed. 粗面状 cū miàn zhuàng

texture: the structure of a substance or material such as soil or food, especially how it feels when touched or chewed. 结构 jié gòu

Example:
Petrochemistry analysis of pumice component indicates that the magma can be assigned totrachytic magma.
火山碎屑堆积物中的浮岩岩石化学分析结果表明岩浆成分为粗面质。

traction [ˈtrækʃən]

Definition: the act or process of pulling something, especially by means of a motor, or the fact or state of being pulled along. 牵引作用 qiān yǐn zuò yòng

Origin: 1610s. "a drawing or pulling," from M. L. tractionem (nom. tractio) "a drawing" (mid-13c.), noun of action from stem of L. trahere "to pull, draw" (see tract (1)). Sense of "rolling friction of a vehicle" first appears 1825.

Example:
This is a research on the Electromagnic Performance of Permanent Magnet Brushless Traction Motors.

这是一个关于永磁无刷牵引电机电磁性能的研究。

Extended Terms:

anterior traction 前方牵引
elastic traction 弹性牵引
slow traction 慢速牵引
traction load 推移质
traction network 牵引网
traction process 牵引法

tractive current [ˈtræktiv ˈkʌrənt]

Definition: 牵引流 qiān yǐn liú

tractive: relating to or denoting the power exerted in pulling, especially by a vehicle or other machine. (尤指车辆等)牵引的 qiān yǐn de
current: a body of water or air moving in a definite direction, especially through a surrounding body of water or air in which there is less movement. 水流 shuǐ liú; 气流 qì liú

Example:

The tractive current deposition has been a new field of sedimentological research since the 60's of 20th century.
深水牵引流沉积是20世纪60年代以来沉积学迅速发展的一个新的研究领域。

Extended Terms:

deep water tractive current deposits 深水牵引流沉积
direct current tractive electric machinery 直流牵引电机
interference of tractive current 牵引电流干扰

transform fault [trænsˈfɔːm fɔːlt]

Definition: a fault which runs along the boundary of a tectonic plate. The relative motion of such plates is horizontal in either sinistral or dextral direction. 转换断层 zhuǎn huàn duàn céng

Example:

Transform fault earthquakes are most noticeable where the fault passes through a continent, rather than the sea floor.
转换断层地震是最值得注意的，在那里断层是通过陆地，而不是海底。

Extended Term:

arc transform fault 左旋脊弧转换断层

transform margin [træns'fɔːm 'mɑːdʒin]

Definition: 变换边缘 biàn huàn biān yuán

transform: to change somebody or something completely, especially improving their appearance or usefulness. 转换 zhuǎn huàn

margin: the edge of something, especially the outer edge, or the area close to it. 边缘 biān yuán

Example:

It has experienced three geological stages as: the transform margin during the Mcsozoic time, the passive margin of Paleocene-Miocene, and the active margin in the late Eocene-Quaternary. 经历了中生代期间的变换边缘、古新至中新世的被动边缘和晚始新世至第四纪的主动边缘三个发展阶段。

transform plate boundary [træns'fɔːm pleit 'baundəri]

Definition: an edge between two plates with motion parallel to the boundary. 转换板块边缘 zhuǎn huàn bǎn kuài biān yuán

Example:

crustal structure of a transform plate boundary: San Francisco Bay and the central California continental margin
转换板块边缘的地壳结构——旧金山湾和加利福尼亚州中部大陆边缘

transitional basalt [træn'zɪʃənəl 'bæsɔːlt]

Definition: 过渡玄武岩 guò dù xuán wǔ yán

transitional: relating to or characteristic of a process or period of transition. 过渡的 guò dù de
basalt: a hard, black, often glassy, volcanic rock. It was produced by the partial melting of the Earth's mantle. 玄武岩 xuán wǔ yán

transitional contact [træn'zɪʃənəl 'kɒntækt]

Definition: 过渡接触 guò dù jiē chù

transitional: of or pertaining to transition; involving or denoting transition; as, transitional changes; transitional stage. 过渡期的 guò dù qī de
contact: the act or state of touching or meeting. 接触 jiē chù

Example:

The transitional contact zones fall into two types.
过渡接触区域分为两类。

transition element [trænˈsiʒən ˈelimənt]

Definition: any chemical element with valence electrons in two shells instead of only one. 过渡元素 guò dù yuán sù

Extended Terms:

inner transition element 内过渡元素
solid-shell transition element 实心壳转换元
post-transition element 过渡后元素

transition flow regime [trænˈsiʒən fləu riˈdʒiːm]

Definition: 过渡流态 guò dù liú tài

transition：[mass noun] the process or a period of changing from one state or condition to another. 过渡 guò dù

flow regime：combinations of river discharge and corresponding water levels and their respective (yearly or seasonally) averaged values and characteristic fluctuations around these values. 流态 liú tài

Example:

When the second derivative of the bubble number in unit bed volume with respect to gas velocity equals zero, the flow regime transition occurs. The model fits fairly well with the experimental data.
导出了当单位床层体积中气泡数目对气速的二阶导数为零时发生流型转变，该模型的计算机结果与实验数据吻合良好。

transition zone [trænˈsiʒən zəun]

Definition: The transition zone is part of the Earth's mantle, and is located between the lower mantle and the upper mantle, between a depth of 410 and 660 km. 过渡区域 guò dù qū yù

Example:

This mixing of fresh and salt water createsaunique environment filled with life of all kinds a transition zone between the land and sea.

淡水和盐水的混合创造了一个富含各种生物的奇特的环境陆地向海洋的过渡区域。

Extended Terms:

arid transition life zone 旱转换生物带
density transition zone 密度变化区
gas-oil transition zone 油-气过渡带
ramp-transition zone 斜坡过渡层
transition finger zone 指状过渡带

translation gliding [trænsˈleiʃən ˈglaidiŋ]

Definition: 平移滑动 píng yí huá dòng

translation: (formal or technical) the process of moving something from one place to another. (正式或技术) 移动 yí dòng

glide: to move smoothly and quietly, especially as though it takes no effort. 滑行 huá xíng; 滑动 huá dòng

Example:

The western margin of the Yangtze platform was dismembered into a number of displaced structural blocks (fault blocks) owing to subsequent thrust napping and translation gliding.
扬子台地西缘由于构造的逆冲推覆与平移滑动而受到严重破坏。

transpositional structure [trænspəˈziʃənəl ˈstrʌktʃə]

Definition: 转位构造 zhuǎn wèi gòu zào

transpositional: [mass noun] the action of transposing something. 互换 hù huàn; 换位 huàn wèi

structure: the arrangement of and relations between the parts or elements of something complex. 构造 gòu zào

Example:

After sediment deposition or just after consolidation, water plastic flow or fluid flow disturbance forms transpositional structure.
沉积物沉积后或刚固结后,水塑性流动或流体流动形成转位构造。

trapp [træp]

Definition: any of several dark, fine-grained igneous rocks often used in making roads. 暗色岩 àn sè yán

Origin: before 1000; Middle English trappe (noun), trappen, Old English træppe (n.), cognate with Middle Dutch trappe (Dutch trap) trap, step, staircase; akin to Old English treppan to tread, German Treppe staircase.

> Example:

This paper gives an effective method for the accurate description and studying of the low-amplitude structure and the lithology trapp, it improved the accuracy of the structural mapping. 将该方法应用于塔中地区低幅度构造、暗色岩岩性精细描述和研究，提高了构造成图的精度。

trap rock [træp rɒk]

> Definition: Trap rock is a form of igneous rock that tends to form polygonal vertical fractures, most typically hexagonal, but also four to eight sided. 暗色岩 àn sè yán

> Example:

Through 1∶50000 scale unit—superunit geologic mapping of Qitianling granitic bathlith, the author proceeded from studies on petrological, petrochemical, trap rock inclusion, trace element, rare earth element and accessory mineral features.
通过1∶50000骑田岭花岗岩基单元——超单元填图，笔者从岩石学、岩石化学、暗色岩包体、微量元素、稀土元素和副矿物等特征研究入手。

tremolite [ˈtreməlait]

> Definition: a white, gray, or pale green hydrated silicate mineral containing calcium, magnesium, and some iron. Source：metamorphic rocks. Use：substitute for asbestos. 透闪石 tòu shǎn shí

> Origin: late 18th century：from Tremola Valley, Switzerland, + -ite.

> Example:

The results show that the samples are almost exclusively composed of tremolite, and are quite similar to Hetian jade in chemical composition.
结果表明，样品的主要矿物组成为透闪石，其化学组成特征与和田玉相似。

> Extended Terms:

asbestos tremolite 阳起石
epidote-tremolite schist 绿帘石-透闪石片岩
fibrous tremolite 纤维状透闪石
soda tremolite 钠透闪石

trench [trentʃ]

Definition: valley on ocean floor: a long narrow valley on an ocean or sea floor. 堑壕 qiàn háo

Origin: late Middle English (in the senses "track cut through a wood" and "sever by cutting"): from Old Frenchtrenche (noun), trenchier (verb), based on Latintruncare (see truncate).

Example:
Elsewhere, two plates may shift toward each other and converge along a deepsea trench and island arc.
在别的地方，两块板块又可能相向移动而沿着深海沟和岛弧逐渐聚合。

Extended Terms:
dovetail trench 斜槽位；鱼尾槽位
trench-trench transform fault 沟-沟转换断层

triangular diagram [traiˈæŋgjulə ˈdaiəgræm]

Definition: 三角形图解 sān jiǎo xíng tú jiě

triangular: shaped like a triangle. 三角形的 sān jiǎo xíng de
diagram: a simple drawing using lines to explain where something is, how something works, etc. 图解 tú jiě

Extended Terms:
gle triangular diagram 直角三角形相图
triangular phase diagram 相三角图

tridymite [ˈtridimait]

Definition: a high-temperature form of quartz found as thin hexagonal crystals in some igneous rocks and stony meteorites. 鳞石英 lín shí yīng

Origin: mid-19th century: from German Tridymit, from Greek tridumos "threefold", from tri-"three" + -dumos (as in didumos "twin"), because of its occurrence in groups of three crystals.

Example:
The notes describe the common minerals such as quartz and its polymorphs, coesite, tridymite and cristobalite, feldspar, plagioclase, feldspathoids, pyroxenes, amphiboles, etc.
讲稿详细地介绍了火成岩矿物，如石英及它的同构异形体、柯石英、鳞石英、方石英、长石、斜长石、似长石、辉石、闪石等矿物。

Extended Terms:

tridymite alboranite 鳞英暗玄岩

tridymite latite 鳞英二长安岩

triple junction [ˈtripl ˈdʒʌŋkʃən]

Definition: A triple junction is the point where the boundaries of three tectonic plates meet. 三联点 sān lián diǎn

Example:

Spreading or rifting along the other arms of the triple junction can form new oceanic basins, whereas the aulacogen can become a sediment-filled graben.
沿三联点的其他臂发生扩张作用或裂谷作用时，可形成新的大洋盆地，而断陷槽则成为被沉积物充填的地堑。

Extended Term:

recent triple junction 现代三合点

triple point [ˈtripl pɔint]

Definition: A temperature point at which a substance can be either a solid, liquid, or gas. 三相点 sān xiàng diǎn

Example:

At temperatures and pressures lower than the triple point, carbon dioxide may be either a solid or a vapor, again depending on conditions.
如果温度和压力低于三态点，依据当时的状况，二氧化碳可能以固态或气态存在。

Extended Term:

water triple-point 水三相点

troctolite [ˈtrɔktəlait]

Definition: gabbro made up mainly of olivine and calcic plagioclase, often having a spotted appearance likened to a trout's back. 橄长岩 gǎn cháng yán

Origin: late 19th century: from German Troklotit, from Greek trōktēs, a marine fish (taken to be "trout").

Example:

The olivine grains in the troctolite of the Upper Zone are rather short and thick.
上岩带橄长岩中的橄榄石颗粒相当短粗。

troilite [ˈtrɔuilait]

Definition: a variety of iron sulfide found in some meteorites. 陨硫铁 yǔn liú tiě

Origin: mid-19th century. After Domenico Troili, 18th century Italian scientist.

Example:

Its mineral modal composition is olivin(55vol%), pyroxene(37.5vol%, pigeonite and augite), a few maskelynite (6vol%), chromite (1.5vol%), and trace whitlockite, troilite and so on.
矿物模式组成以橄榄石(55%)、辉石(37.5%)为主，有少量熔长石(6%)、铬铁矿(1.5%)以及微量白磷钙石、陨硫铁等。

Extended Terms:

metal-troilite 金属-陨硫铁
paragenetic troilite 中共生的陨硫
synthetic troilite 合成陨硫铁

trona [ˈtrəunə]

Definition: a grayish white or yellowish hydrated sodium carbonate mineral. Source：salt deposits. 天然碱 tiān rán jiǎn

Origin: late 18th century：from Swedish, from Arabic natrūn (see natron).

Extended Terms:

trona deposit 天然碱矿床
trona process 天然碱流程

trondhjemite [ˈtrɔnjeˌmait]

Definition: Trondhjemite is a leucocratic (light-colored) intrusive igneous rock. It is a variety of tonalite in which the plagioclase is mostly in the form of oligoclase. 奥长花岗岩 ào cháng huā gāng yán

Origin: Trondhjem (now Trondheim) + -ite.

Example:

The adakite in Heishishan of Ganshu Province is of O-type composed of trondhjemite.

甘肃黑石山埃达克岩属 O 型埃达克岩，其岩石类型为奥长花岗岩。

Extended Terms:

trondhjemite pegmatite 奥长花岗伟晶岩
trondhjemite aplite 奥长花岗细晶岩

trough cross bedding [trɒf krɒs ˈbedɪŋ]

Definition: 槽状交错层理 cáo zhuàng jiāo cuò céng lǐ

trough: a long narrow open container for animals to eat or drink from. 槽 cáo
cross bedding: In geology, cross-bedding refers to inclined sedimentary structures in a horizontal unit of rock. 交错层理 jiāo cuò céng lǐ

Example:

Sedimentary structures are mainly massive bedding, large trough cross bedding and oblique bedding with few parallel bedding, wave bedding and horizotal bedding.
沉积构造以块状层理、大型槽状交错层理和斜层理为主，可见平行层理、波状层理和极少量的水平层理。

Extended Terms:

trough cross set 槽形交错层组
trough cross-stratification 槽状交错层理

true solution [truː səˈluːʃən]

Definition: 溶液 róng yè

solution: a liquid in which something is dissolved. 溶液 róng yè

Example:

Certain organic substances and compounds that are considered soluble in water do not form true solution.
某些被认为溶于水的有机物和化合物不形成真溶液。

tuff [tʌf]

Definition: a rock formed by the fusing together on the ground of small rock fragments (less than 2 mm across) ejected from a volcano. 凝灰岩 níng huī yán

Origin: 16th century: from Old French tuff, from Italian tufo; see tufa.

Example:

Typical features of subaqueous volcanic rocks include perlite with glassy structure and pillow

structure, lamellar tuff, bentonite.
水下喷发火山岩典型标志为玻璃质结构的珍珠岩、枕状构造、纹层状凝灰岩和膨润土。

Extended Terms:

crystal tuff 晶屑凝灰岩
lapilli tuff 火山砾凝灰岩
volcanic tuff 凝灰岩；火山凝灰岩
welded tuff 熔结凝灰岩；流纹岩状凝灰岩

tuffaceous shale [tʌˈfeiʃəs ʃeil]

Definition: 凝灰质页岩 níng huī zhì yè yán

tuffaceous: Tuff (from the Italian "tufo") is a type of rock consisting of consolidated volcanic ash ejected from vents during a volcanic eruption. Tuff is sometimes called tufa, particularly when used as construction material, although tufa also refers to a quite different rock. 凝灰岩 níng huī yán

shale: soft, finely stratified sedimentary rock that formed from consolidated mud or clay and can be split easily into fragile slabs. 页岩 yè yán

Example:

The tuffaceous blanketed the entire basin.
凝灰岩充填着整个盆地。

Extended Terms:

tuffaceous concretio 凝灰质结块
tuffaceous facies 凝灰岩相
tuffaceous sandstone 凝灰质砂岩

tuffaceous texture [tʌˈfeiʃəs ˈtekstʃə]

Definition: 凝灰结构 níng huī jié gòu

tuffaceous: pertaining to sediments which contain up to 50% tuff. 凝灰质的 níng huī zhì de
texture: the arrangement of the particles or constituent parts of any material, as wood, metal, etc., as it affects the appearance or feel of the surface; structure, composition, grain, etc. 结构 jié gòu

Example:

They are of obvious palimpsest tuffaceous texture and pseudofluidal structure.
其具有明显的变余凝灰结构和假流纹构造。

turbidite facies ['tɜːbidait 'feiʃiːz]

Definition: 浊积相 zhuó jī xiàng

turbidite: a sediment or rock deposited by a turbidity current. 浊积岩 zhuó jī yán
facies: the character of a rock expressed by its formation, composition, and fossil content. 相 xiàng

Example:

Petroleum reservoirs are obviously controlled by sedimentary facies, and are mainly distributed in fan delta facies, delta facies and turbidite fan facies.
沉积相带控制油气作用明显，油气藏主要分布在扇三角洲、三角洲和浊积扇相。

turbidite fan deposit ['tɜːbidait fæn di'pɒzit]

Definition: 浊积扇沉积 zhuó jī shàn chén jī

turbidite: a sediment or rock deposited by a turbidity current. 浊积岩 zhuó jī yán
fan: a thing that you hold in your hand and wave to create a current of cool air. 扇子 shàn zi
deposit: a layer of a substance that has been left somewhere, especially by a river, flood, etc., or is found at the bottom of a liquid 沉积物. chén jī wù

Example:

Jishan sand bed, a delta-packed turbidite fan-deposit distributing along Jishan trough, has been attracted our attention in recent years.
基山砂岩体是近年来引起关注的一种沿基山槽展布的三角洲充填浊积扇沉积。

turbidity current [tɜː'bidəti 'kʌrənt]

Definition: a turbid, dense current of sediments in suspension moving along the slope and bottom of a lake or ocean. 浊流 zhuó liú

Example:

This is a preliminary knowledge on turbidity current sedimentary in ore-bearing rocks of Houpaao Silver Mine.
这是一个对厚婆坳银矿区含矿岩系浊流沉积的初步认识。

Extended Terms:

dilute turbidity current 稀释浊流
spasmodic turbidity current 突发浊流
turbidity current deposition 浊流沉积

turbidity current sediment 浊流沉积
under flow type turbidity current 底流性浊流

turbulence [ˈtɜːbjuləns]

Definition: violent or unsteady movement of air or water, or of some other fluid. 紊流 wěn liú

Origin: late Middle English: from Old French, or from late Latin turbulentia, from turbulentus "full of commotion" (see turbulent).

Example:
Result indicates that the ring fins could effectively enhance the turbulence of the mixed gas in the combustor, improve combustion status, and increase the combustion efficiency.
结果表明，双环形翅片能增强微燃烧室内混合气体的湍流扰动，改善燃烧状况，有效地提高燃烧效率。

Extended Terms:
hydraulic turbulence 水力干扰
turbulence transfer 湍流输运
turbulence balance 湍动能平衡

turbulent flow [ˈtɜːbjulənt fləu]

Definition: flow in which the velocity at any point varies erratically. 紊流 wěn liú

Extended Terms:
turbulent flow burner 湍流喷灯
turbulent flow conditions 紊流状态
turbulent flow domain 紊流场
turbulent flow motion 紊流动
dimension turbulent flow 三维紊流流动

tuya [ˈtʌjə]

Definition: a tuya is a type of distinctive, flat-topped, steep-sided volcano formed when lava erupts through a thick glacier or ice sheet. 平顶火山 píng dǐng huǒ shān

Example:
the relative importance of supraglacial versus subglacial meltwater escape in tuya eruptions
平顶火山的冰川上对比冰川下的融水漏出相对重要性研究

twin gliding [twinˈglaidiŋ]

Definition: 双晶滑动 shuāng jīng huá dòng

twin: used to describe two things that are used as a pair. 成对的 chéng duì de；成双的 chéng shuāng de

gliding: move along a smooth surface while maintaining continuous contact with it. 滑动 huá dòng

Example:

Twin gliding and dislocation creep are mutually enhanced during twinning nucleation.
双晶滑动和位错蠕变在双晶成核过程中共同增强。

Extended Terms:

twin gliding plane 双晶滑移面
twinning crystal 双晶

twinning [ˈtwiniŋ]

Definition: a compound crystal consisting of two mirror-image crystals that share a common plane. 双晶形成 shuāng jīng xíng chéng

Origin: Old English twinn, from Indo-European, "two by two".

Extended Terms:

calcite twinning 方解石双晶
crossed twinning 十字双晶
deformation twinning 塑变双晶；形变孪生
twinning crystal 双晶

two-feldspar geothermometry [tuː ˈfeldspɑː ˌdʒiːəʊθəˈmɒmitri]

Definition: 二长石地热测温术 èr cháng shí dì rè cè wēn shù

feldspar: an abundant rock-forming mineral typically occurring as colorless or pale-colored crystals and consisting of aluminosilicates of potassium, sodium, and calcium. 长石 cháng shí
geothermometry: 地热测温术 dì rè cè wēn shù

Example:

Recent improvements in the experimental and thermodynamic basis of two-feldspar geothermometry allow one to recover temperatures of coexistence more reliably.
实验和热力学的二长石地热测温术原理在近期的改进促成更可靠的共存温度。

type of cementation [taip ɔv ˌsiːmenˈteiʃən]

Definition: 胶结类型 jiāo jié lèi xíng

cementation: the binding together of particles or other things by cement. 胶结 jiāo jié

Example:

The thin sand bed is composed of tight mixed type cementation facies and untight mixed type cementation facies.
其中，薄砂层由致密混合型胶结相和非致密混合型胶结相组成。

---

ultra acid rock [ˈʌltrə ˈæsid rɒk]

Definition: 超酸性岩 chāo suān xìng yán

ultra: going beyond what is usual or ordinary; excessive; extreme. 很 hěn; 非常 fēi cháng
acid rock: is a form of psychedelic rock, which is characterized with long instrumental solos, few (if any) lyrics and musical improvisation. 酸性岩 suān xìng yán

---

ultracataclasite [ˌʌltrəˌkætəˈklesait]

Definition: Ultracataclasite is a type of cataclasite characterized by a matrix occupying greater than 90% of the total volume. 超碎裂岩 chāo suì liè yán

Origin: from Greek katáklasis refraction, equivalent to katakla-, stem of kataklân to break off, refract, break down (kata-cata- + klân to break) + -sis-sis; compare Norwegian kataklasstruktur (1885).

Example:

There was some discussion of the difference between gouge and ultracataclasite.
断层泥和超碎裂岩的区别引发了一些讨论。

---

ultrahigh-temperature metamorphism
[ˌʌltrəˈhaiˈtemprətʃə metəˈmɔːfizəm]

Definition: 超高温变质作用 chāo gāo wēn biàn zhì zuò yòng

ultrahigh-temperature: 超高温 chāo gāo wēn
metamorphism: change in the structure of rock by natural agencies such as pressure or heat or introduction of new chemical substances. 变质 biàn zhì

Example:

Ultrahigh-temperature metamorphism was followed by two-stage decompression of granulites.

在超高温变质作用之后的是麻粒岩的两阶段减压。

ultrahigh pressure [ˌʌltrəˈhaɪ ˈpreʃə]

Definition: High pressure science and engineering is studying the effects of high pressure on materials and the design and construction of devices. 超高压 chāo gāo yā

Example:
The eclogites in northern Jiangsu have at least three types: high pressure crustal type (Htype), ultrahigh pressure crustal type (U-type) and mantle type (M-type).
苏北榴辉岩至少存在高压壳源型(H 型)、超高压壳源型(U 型)和幔源型(M 型)三类。

Extended Terms:
ultrahigh pressure metamorphism 超高压变质作用
ultrahigh pressure minerals 超高压矿物
ultrahigh pressure phase 超高压相超高压相
ultrahigh pressure treatment 超高压处理
ultrahigh pressure water jet 超高压水射流

ultrahigh pressure metamorphism [ˌʌltrəˈhaɪ ˈpreʃə metəˈmɔːfizəm]

Definition: 超高压变质作用 chāo gāo yā biàn zhì zuò yòng

ultrahigh pressure: High pressure science and engineering is studying the effects of high pressure on materials and the design and construction of devices. 超高压 chāo gāo yā

metamorphism: change in the structure of rock by natural agencies such as pressure or heat or introduction of new chemical substances. 变质 biàn zhì

ultramafic [ˌʌltrəˈmæfɪk]

Definition: relating to or denoting igneous rocks composed chiefly of mafic minerals. 超镁铁质的 chāo měi tiě zhì de

Origin: 1910-1915; ma(gnesium)+Latin f(errum) iron + -ic.

Example:
With regard to the oxide minerals, chromite is almost entirely contained in the olivine-richultramafic rocks in the lower part of the intrusion.
关于氧化矿物,铬铁矿几乎全部包含在侵入体下部的富橄榄石的超镁铁质岩石中。

Extended Terms:
mafic ultramafic 镁铁质-超镁铁质

ultramafic enclaves 超镁铁岩包体
ultramafic lavas 超镁铁质熔岩
ultramafic rock 超铁镁岩
ultramafic xenoliths 超镁铁质岩

ultramafic rock [ˌʌltrəˈmæfik rɒk]

Definition: rocks are igneous and meta-igneous rocks with very low silica content (less than 45%), generally >18% MgO, high FeO, low potassium, and are composed of usually greater than 90% mafic minerals (dark colored, high magnesium and iron content). The Earth's mantle is composed of ultramafic rocks. 超镁铁质岩 chāo měi tiě zhì yán

Example:
Paragenesis of ultramafic rock and eclogite inclusions show that the formed environment of the diamond is very complex.
超镁铁岩与榴辉岩包裹体共生显示了该金刚石生成环境的复杂性。

Extended Terms:
mafic ultramafic rock 镁铁-超镁铁岩
meta-ultramafic rock 变超镁铁质岩
ultramafic rock block 超镁铁岩块

ultramafic rock classification [ˌʌltrəˈmæfik rɒk ˌklæsifiˈkeiʃən]

Definition: 超镁铁质岩分类 chāo měi tiě zhì yán fēn lèi

ultramafic: (Geology) relating to or denoting igneous rocks composed chiefly of mafic minerals. [地质]超镁铁质的 chāo měi tiě zhì de

rock: the solid mineral material forming part of the surface of the earth and other similar planets, exposed on the surface or underlying the soil. 岩石 yán shí; 岩 yán

classification: the action or process of classifying something according to shared qualities or characteristics. 分类 fēn lèi

ultramylonite [ˈʌltrəmilənait]

Definition: A more thoroughly deformed and fine-grained rock containing more than 90% matrix and less than 10% relict grains. Ultramylonites are mylonites taken to the edge of recognition. 超糜棱岩 chāo mí léng yán

Origin: 1885-1890; mylon-(representing Greek mýlos mill) + -ite.

> Example:

From altered mylonite to altered ultramylonite, the crystallinity of illite decreases whereas that of chlorite shows no obvious variation with the increasing strain.
由应变较弱的蚀变糜棱岩带到应变较强的蚀变超糜棱岩带，伊利石的结晶度有变小的趋势，而绿泥石的结晶度则基本保持不变。

ultrapotassic magmatism [ˌʌltrəˈpotasik ˈmæɡmətizəm]

> Definition: 超钾岩浆作用 chāo jiǎ yán jiāng zuò yòng

magmatism：[mass noun] (Geology) the motion or activity of magma. （地质）岩浆活动 yán jiāng huó dòng

> Example:

subduction-related shoshonitic and ultrapotassic magmatism：a study of Ordovician syenites from the Scottish Caledonides
隐没相关的钾玄质岩浆和超钾岩浆作用——苏格兰加里东山带的奥陶纪正长岩研究

ultrapotassic series [ˌʌltrəˈpotasik ˈsiəriːz]

> Definition: 超钾系列 chāo jiǎ xì liè

series：Series are subdivisions of rock layers made based on the age of the rock and corresponding to the dating system unit called an epoch. 系列 xì liè

umber [ˈʌmbə]

> Definition: a natural pigment resembling but darker than ochre, normally dark yellowish-brown in colour (raw umber) or dark brown when roasted. 赭土 zhě tǔ

> Origin: mid-16th century.

> Example:

The contrasting green and umber of the Egyptian landscape forms a sharp line between fertile earth and parched desert.
这明显的绿色与埃及领地的焦茶色在肥沃的土地与沙漠间构成了明显的分割线。

> Extended Terms:

burnt umber 烧褐土，熟褐（颜料）
cyprus umber 绿褐色棕土
raw umber 富锰棕土，生褐（颜料）
sepia umber 乌贼褐

Turkey umber 土耳其褐

unary systems [ˈjuːnəriˈsistəms]

Definition: 一元体系 yī yuán tǐ xì

ynary: consisting of or involving a single component or element. 一元的 yī yuán de
system: a major range of strata that corresponds to a period in time, subdivided into series. 地层的系 dì céng de xì

Example:

Unary system is a chemical system, which is called unicomponent system.
一元体系是一种化学体系，又称作单元系。

unconformity [ˌʌnkənˈfɔːmiti]

Definition: a surface of contact between two groups of unconformable strata. 不整合接触 bù zhěng hé jiē chù

Origin: 1375-1425; late Middle English conformite, from Middle French, from Late Latin confōrmitās. See conform, -ity.

Example:

In fact, the available sand is insufficient to form a continuous blanket overlying the unconformity.
事实上，所补入的砂质物不足以在不整合面上形成连续的盖层。

Extended Terms:

concealed unconformity 隐蔽不整合
contemporaneous unconformity 同生不整合
depositional unconformity 沉积不整合
erosional unconformity 侵蚀不整合
geographical unconformity 地理不整合
graded unconformity 深风化不整合

underplating [ˌʌndəˈpleitiŋ]

Definition: Underplating is the result of partial melts being produced in the mantle wedge above a subducting plate. 底侵作用 dǐ qīn zuò yòng

Example:

The crust mantle transition in the west is considerably thicker than in the east, probably due to

different extents of magmatic underplating.
西部壳幔过渡带较厚而东部较薄，反映两地不同的岩浆底侵作用程度。

Extended Terms:

magma underplating 岩浆底侵作用
magmatic underplating 深部岩浆的底贴作用

undulatory extinction [ˈʌndjulətəri ikˈstiŋkʃən]

Definition: 波状消光 bō zhuàng xiāo guāng

undulatory: resembling waves in form or outline or motion. 波动的 bō dòng de
extinction: the state or process of ceasing or causing something to cease to exist. （使）消亡 xiāo wáng

Example:

It is found that the horn-blendes show no micro-structure while their coexisting quartz displays apparent undulatory extinction.
笔者发现，当石英呈明显波状消光时，普通角闪石未显示变形结构。

uniformitarianism [ˌjuːnifɔːmiˈteəriənizəm]

Definition: the theory that changes in the earth's crust during geological history have resulted from the action of continuous and uniform processes. 均变说 jūn biàn shuō

Origin: 1400-1450；late Middle English uniformite < Middle French uniformite < Late Latin ūnifōrmitās, equivalent to Latin ūnifōrm(is) uniform + -itās-ity + <Latin-āri(us) or-ary+ -an.

Example:

Wernerians thought strata were deposits from shrinking seas, but James Hutton proposed a self-maintaining infinite cycle, anticipating uniformitarianism.
沃纳里恩斯认为地层是收缩的海洋的沉积物，但詹姆斯·赫顿提出了一个自我维持的无限周期，促进了地质均变论的发展。

Extended Term:

uniformitarianism science 均变学说

unimodal grain size [ˌjuːniˈməudəl ɡrein saiz]

Definition: 单峰晶粒尺寸 dān fēng jīng lì chǐ cùn

unimodal: having one maximum. 单峰的 dān fēng de
grain: a small hard particle of a substance such as salt or sand. 小而硬的颗粒 xiǎo ér yìng de

kē lì

size: the relative extent of something; a thing's overall dimensions or magnitude; how big something is. 尺寸 chǐ cùn

Example:

The calculated results show that, after some fines is washed out, piping failure more easily takes place in the piping-typed soil with a bimodal grain-size-distribution than in the soil with the unimodal grain-size-distribution.

计算结果表明，当有细颗粒流失后，级配单峰晶粒尺寸型土比级配单峰晶粒尺寸型土更易发展成管涌破坏。

univariance [juːniˈvɛərɪəns]

Definition: the condition of being univariant. 单变量 dān biàn liàng

Origin: 1300-1350; Middle English < Latin variantia, equivalent to vari(āre) to vary) + -antia-ance.

Example:

Both waveform and response amplitude versus intensity relationship of the ERG were independent of the wavelength of stimuli, and hence followed the Univariance Principle.

其波形和振幅-强度曲线与刺激波长无关，均符合单变量原理。

univariant assemblage [juː niˈvɛərɪənt əˈsemblɪdʒ]

Definition: 单变集合物 dān biàn jí hé wù

univariant: having a variance of one. 单变的 dān biàn de

assemblage: the action of gathering or fitting things together. 聚集 jù jí

Example:

In this study the solubility of the univariant assemblage has been determined in vapor saturated hydrothermal solutions.

该项研究中，单变集合物的溶解度在蒸汽饱和的热液试剂中测定。

univariant equilibrium [juː niˈvɛərɪənt ˌiːkwɪˈlɪbrɪəm]

Definition: 单变平衡 dān biàn píng héng

univariant: having a variance of one. 单变的 dān biàn de

equilibrium: a state in which opposing forces or influences are balanced. 平衡 píng héng

> Example:

a simple easy and universal method for auto discrimination of the stable univariant equilibrium curves
关于稳定单变平衡曲线自动判别的一种简便的普适性方法

univariant reaction [juːniˈvɛəriənt riˈækʃən]

> Definition: 单变反应 dān biàn fǎn yìng

univariant: having a variance of one. 单变的 dān biàn de
reaction: an action performed or a feeling experienced in response to a situation or event. 反应 fǎn yìng

unsaturated/undersaturated rock [ʌnˈsætʃəˌreitid ˌʌndəˈsætʃəreitid rɒk]

> Definition: are rocks which contain feldspathoids (no quartz). 不饱和岩 bù bǎo hé yán

> Example:

A statistical damage softening constitutive model for saturated and unsaturated rock is developed based on the mixture theory.
在混合物理论基础上，建立了饱和和非饱和岩石损伤软化统计本构模型。

unstable [ʌnˈsteibl]

> Definition: prone to change, fail, or give way; not stable. 不稳定的 bù wěn dìng de

> Origin: Middle English: from Anglo-Norman French, from Latin stabilis, from the base of stare "to stand".

> Example:

Using the survey data of shallow seabed in Chengbei since 1988, we studied the formation and evolution of unstable topography, and analyzed the recent evolutional trend of unstable topography of the old Yellow River subaqueous delta.
根据1988年以来埕北浅海区的海底调查资料，深入探讨了海底不稳定现象的形成及发育机制，分析了老黄河口水下三角洲不稳定地形的近期演变动态。

> Extended Terms:

unstable atherosclerosis 不稳定斑块
unstable field 不稳定场
unstable margin 不稳定裕度

unstable manifolds 不稳定流形
unstable topography 不稳定地形

unstable phase [ˌʌnˈsteibl feiz]

Definition: 不稳定阶段 bù wěn dìng jiē duàn

unstable: prone to change, fail, or give way; not stable. 易变的 yì biàn de; 易动摇的 yì dòng yáo de; 不稳定的 bù wěn dìng de

phase: a distinct period or stage in a process of change or forming part of something's development. 时期 shí qī; 阶段 jiē duàn

unstable state [ʌnˈsteibl steit]

Definition: 不稳定状态 bù wěn dìng zhuàng tài

unstable: prone to change, fail, or give way; not stable. 易变的 yì biàn de; 易动摇的 yì dòng yáo de; 不稳定的 bù wěn dìng de

state: the particular condition that someone or something is in at a specific time. 状态 zhuàng tài, 情况 qíng kuàng

Example:

In ultrafine grinding, minerals exhibit remarkable changes in crystal structure and activity and are in a highly efficient unstable state and their reaction activity is enhanced.
在超细磨过程中，矿物在晶体结构、活性等方面具有显著变化，处于高效失稳状态，反应活性增强。

upper mantle [ˈʌpəˈmæntl]

Definition: the upper part of the mantle. 上地幔 shàng dì màn

Example:

The upheaval zone in upper mantle is the primary factor controlling the distribution of gulch-gold mines and lode gold mines.
上地幔隆起区是控制砂金矿和岩金矿面式分布的主要因素。

Extended Terms:

earth upper mantle 上地幔
oceanic upper mantle 大洋上地幔
omalous upper mantle 异常上地幔
upper mantle discontinuities 上地幔间断面
upper mantle materials 上地幔物质

upper mantle structure 上地幔结构

upper mantle rock [ˈʌpə ˈmæntl rɔk]

Definition: rocks of the portion of the mantle lying above a depth of about 600 miles (1000 kilometers). Also known as outer mantle; peridotite shell. 上地幔岩 shàng dìmàn yán

Example:
The former is not only controled by general composition of melting magma from the upper mantle rock but also by crystalization velocity which is govened by differentiation crystalization of the magma and its emplacement.
岩浆铬铁矿床的形成除决定于上地幔岩熔融岩浆总成分外，还受岩浆分异结晶作用、岩浆就位的地质条件和规模所决定的结晶速度和状态的直接影响。

upwelling [ʌpˈweliŋ]

Definition: a rising up of seawater, magma, or other liquid. 上涌 shàng yǒng

Origin: before 900; Middle English, Old English wel(l) (adj. and adv.); cognate with Dutch wel, German wohl, Old Norse vel, Gothic waila.

Example:
Their upwelling currents bring deep, cold water to the surface and create some of the richest biological areas of the sea.
上升的洋流将深处的冷水带到了海洋表面，创造出一片片生机勃勃的海域。

Extended Terms:
coastal upwelling 近岸上升流
infrastructural upwelling 下部构造层物质上涌
magmatic upwelling 岩浆涌动
oceanic upwelling 大洋上升流
upwelling action 上升流

uralite [ˈjuərəlait]

Definition: an amphibole silicate mineral with the chemical formula. 纤闪石 xiān shǎn shí

Origin: 1825-1835; from German Uralit, named after the Ural Mountains, where found; see -ite.

Example:
Industrial tests calculate the wear resistance of uralite milling bodies.

工业测试计算了纤闪石研磨体的耐磨性。

urtite ['ɜːtait]

Definition: a dark greenish-grey fine to medium grained intrusive rock, occoring in alkaline intrusions. It consists of nepheline (70%-90% with aegirine and feldspar. Urtite has a porphyric structure. The type locality of Urtite is Parga in Lovozero. (Ramsay 1896). 磷霞岩 lín xiá yán

Extended Term:
biotite urtite 云磷霞岩

vadose pisolith ['veidəus 'paisəuliθ]

Definition: 渗流豆石 shèn liú dòu shí

vadose: relating to or denoting underground water above the water table. 渗流的 shèn liú de pisolith: (Geology) any of the component pieces of which pisolite consists. [地质]豆石 dòu shí

Extended Term:
vadose water 渗流水

vadose zone ['veidəus zəun]

Definition: The vadose zone, also termed the unsaturated zone, is the portion of Earth between the land surface and the phreatic zone or zone of saturation. 渗流带 shèn liú dài;包气带 bāo qì dài

Example:
The key to modeling the vadose zone is the ability to predict an accurate surface boundary condition.
对渗流带进行建模分析的关键问题是预测准确的面边界条件。

Extended Terms:
soil moisture of vadose zone 包气带土壤水分
thick vadose zone 厚包气带
vadose zone processes 包气带过程

Valles ['vælis]

Definition: Vallés (variant: Samartín del Vallés) is one of 41 parishes (administrative divisions) in Villaviciosa, a municipality within the province and autonomous community of Asturias, in northern Spain. 巴列斯 bā liè sī

variance [ˈveərɪəns]

Definition: the fact or quality of being different, divergent, or inconsistent. 变异度 biàn yì dù

Origin: Middle English: via Old French from Latin variantia "difference", from the verb variare.

Example:
In contrast, layered variance shadow maps require only a single texture sample per shaded pixel, just like variance shadow maps.
相比较而言，多层方差阴影图对于每个着色像素只需要单个纹理采样，就像方差阴影图算法。

Extended Terms:
average comparable variance 平均可比较的变异性
between-cluster variance 组间变异
between-strata variance 区间变异；层间变异
capacity utilization variance 能量利用差异
component variance 成分变异数
controllable variance 可控制差异

variation diagram [veərɪˈeɪʃən ˈdaɪəɡræms]

Definition: 变化图 biàn huà tú

variation: an instance of change; the rate or magnitude of change. 变异 biàn yì；变化 biàn huà
diagram: a drawing intended to explain how something works; a drawing showing the relation between the parts. 图解 tú jiě

varve [vɑːv]

Definition: A varve is an annual layer of sediment or sedimentary rock. 纹泥 wén ní

Origin: 1920-1925; from Swedish varv a round, (complete) turn.

Example:
The variation of varve thickness may show a decrease of sediment flux during the Little Ice Age, and an increase of sediment flux after 1960s.

但纹泥厚度的变化可能揭示出小冰期沉积通量的减少和 20 世纪 60 年代之后沉积通量的增加。

Extended Terms:

glacial varves 冰川融稳层
varve clay 纹泥

vein [vein]

Definition: a fracture in rock containing a deposit of minerals or ore and typically having an extensive course underground. 矿脉 kuàng mài

Origin: Middle English: from Old French veine, from Latin vena. The earliest senses were "blood vessel" and "small natural underground channel of water".

Example:

For most of independent gold deposits, especially those of vein type, the ore mineral assemblages and mineral sequences are similar.
绝大多数独立金矿床，特别是脉状金矿的矿石矿物组合及其成矿阶段都是十分相似的。

Extended Terms:

accessory subcostal vein 副亚前缘脉
accessory vein 副脉
accretion vein 加填矿脉
ambient vein 围脉
arciform vein 弓状静脉
arcuate vein 弓脉

vein material [vein məˈtiəriəl]

Definition: In geology, a vein is a distinct sheetlike body of crystallized minerals within a rock. Veins form when mineral constituents carried by an aqueous solution within the rock mass are deposited through precipitation. 脉体 mài tǐ

Example:

The effects of shock have eliminated porosity from the vein material.
震动的效果削减了脉体的多孔性。

verite [ˈverait]

Definition: 金橄玻基煌斑岩 jīn gǎn bō jī huáng bān yán

Example:

A bronze vein will give gold ore, a verite vein will give valorite and so on.
青铜矿脉可以挖出金矿，而金橄玻基煌斑岩矿脉可以挖出蓝矿。

Extended Term:

verite elemental 篮矿元素

vesicle [ˈvesikl]

Definition: a small cavity in volcanic rock, produced by gas bubbles. 气泡 qì pào

Origin: late 16th century: from French vésicule or Latin vesicula, diminutive of vesica "bladder".

Extended Terms:

germinal vesicle 生发泡
membrane vesicle 膜泡
polymeric vesicle 高分子微泡
secretory vesicle 分泌泡
synaptic vesicle 突触小泡

vesicular structure [viˈsikjulə ˈstrʌktʃə]

Definition: 气孔构造 qì kǒng gòu zào

vesicular: composed of or containing vesicles. 气孔质的 qì kǒng zhì de
structure: something composed of interrelated parts forming an organism or an organization. 结构 jié gòu

Example:

The present paper has studied magnetic fabrics of Quaternary basalts in Chifeng area, Inner Mongolia and analysed magnetic anisotropic characteristics of such different structures of basalts as ropy flow structure, prismatic joints and vesicular structure.
本文对内蒙古赤峰地区第三纪玄武岩的磁组构进行了研究，分析了流动绳状构造、柱状节理、气孔构造等不同构造的玄武岩的磁各向异性特征。

viscosity [viˈskɔsiti]

Definition: resistance of a liquid to sheer forces (and hence to flow). 黏度 nián dù

Origin: late Middle English: from Old French viscosite or medieval Latin viscositas, from late Latin viscosus.

> Example:

Some improvement in efficiency can be gained at high speed by reducing viscosity and at low speed by increasing viscosity.
在高速时降低黏度，在低速时增大黏度，均可提高效率。

> Extended Terms:

absolute viscosity 绝对黏度
anisotropic viscosity 各向异性黏性
anomalous viscosity 反常黏滞性
apparent shear viscosity 表观切变黏度
apparent viscosity 表观黏度
ash viscosity 灰黏度

vitreous [ˈvitriəs]

> Definition: Of, relating to, resembling, or having the nature of glass; glassy. 玻璃质 bō lí zhì

> Origin: 1640-1650; < Latin vitreus, equivalent to vitr(um) glass + -eus-eous.

> Extended Terms:

vitreous china 玻化瓷器
vitreous fibre 玻璃质纤维
vitreous luster 玻璃状光泽
vitreous state 玻璃态
vitreous solid 玻璃质固体

vitreous texture [ˈvitriəs ˈtekstʃə]

> Definition: 玻璃质结构 bō lí zhì jié gòu

vitreous: Of, relating to, resembling, or having the nature of glass; glassy. 玻璃质 bō lí zhì
texture: the arrangement of the particles or constituent parts of any material, as wood, metal, etc., as it affects the appearance or feel of the surface; structure, composition, grain, etc. 结构 jié gòu

> Example:

vitreous texture and vitreoschisis 玻璃质结构与玻璃体劈裂

vitric fragment/shard [ˈvitrik ˈfrægmənt/ʃɑːd]

> Definition: pieces of a pyroclastic material which is characteristically glassy, that is,

contains more than 75% glass. 玻屑 bō xiè

Example:
Developed in the upper part of the Nanlinhu Formation in the Chaohu area of Anhui is aseries of pyroclastic flow deposits composed largely of dacite breccia, crystal and glass shard, dacite tuff breccia and tuff, 1.8m thick with distinct boundaries.
在安徽巢县下三叠统南陵湖组上部,发育着一套火山碎屑沉积物,主要由英安质角砾岩、英安质晶屑、玻屑凝灰角砾岩和英安质玻屑、晶屑凝灰岩等组成。它厚 1.8m,层序清楚。

Extended Terms:
rhyolitic vitric tuff 流纹质玻屑凝灰岩
vitric tuff 玻屑凝灰岩

vitrinite ['vitrinait]

Definition: Vitrinite is one of the primary components of coals and most sedimentary kerogens. 镜质体 jìng zhì tǐ

Origin: 1875-1880; from French, equivalent to vitre pane of glass + -ine.

Extended Terms:
specularite vitrinite 镜质体
subhydrous vitrinite 低氢镜质体
vitrinite reflectance 镜质体反射率
vitrinite reflectogram 镜质体反射率分布图

volatile ['vɒlətail]

Definition: easily evaporated at normal temperatures. 易挥发的 yì huī fā de

Origin: Middle English (in the sense "creature that flies", also, as a collective, "birds"): from Old French volatil or Latin volatilis, from volare "to fly".

Example:
The flame retardants containing bromine or chlorine also decompose when the polymer decomposes to form volatile combustibles. The halogen escapes with the volatile combustibles and acts in the vapor phase to inhibit flaming combustion.
在聚合物分解时,含有溴或氯的阻燃剂也会分解成可燃性气体,卤素也与可燃性气体一起释放出,它在气态时会阻止燃烧。

Extended Terms:
volatile acid 挥发(性)酸
volatile storage 易失性存储

volatile component ['vɒlətail kəm'pəunənt]

Definition: a component of magma whose vapor pressures are high enough to allow them to be concentrated in any gaseous phase. Also known as volatile flux. 挥发分 huī fā fēn

Example:
The volatile component was not only the activator of melting of the mantle to form initial magma but also one of the main factors to control the trend of magma evolution and the formation of zoned magma chamber under the trough.
岩浆中的挥发性组分不仅是导致地幔熔融产生岩浆的"催化剂"，而且是岩浆结晶演化及形成分层岩浆房的主要控制因素之一。

Extended Terms:
active optical component 有源光学部件
acidic component 酸性成分
semi-volatile component 半挥发性成分
volatile organic component 挥发性有机物

volcanic [vɒl'kænik]

Definition: of, relating to, or produced by a volcano or volcanoes. 火山的 huǒ shān de

Origin: late 18th century: from French volcanique, from volcan.

Example:
A volcanic rock is formed by the welding together of tuff material from an explosive volcanic eruption.
熔结凝灰岩火山喷发后凝灰物质聚集后形成一种火成岩。

Extended Terms:
volcanic ash 火山灰
volcanic rock 火山岩

volcanic agglomerate [vɒl'kænik ə'glɒmərət]

Definition: 火山集块岩 huǒ shān jí kuài yán
volcanic: relating to, or produced by a volcano or volcanoes. 火山的 huǒ shān de
agglomerate: [mass noun] (Geology) a volcanic rock consisting of large fragments bonded together. [地质] (火山)集块岩 jí kuài yán

Example:
There are Cenozoic continental intermediate and acid volcanic rocks in the Tianshui area,

western Qinling. The rocks are mainly composed of rhyolite, parts of rhyolitic ignimbrite (breccia tuff), a little amount of breccia tuff and volcanic agglomerate.
西秦岭甘肃天水地区分布的新生代陆相中酸性火山岩,主要由流纹岩、部分流纹质熔结(角砾)凝灰岩及少量角砾凝灰岩、火山集块岩等组成。

volcanic breccia [vɒlˈkænik ˈbretʃiə]

Definition: a rock composed of broken fragments of minerals or rock cemented together by a fine-grained matrix, that may be similar to or different from the composition of the fragments.
火山角砾岩 huǒ shān jiǎo lì yán

Example:
Jurassic volcanic rocks in Yingshan fault depression are mainly andesite, basalt, dacite, volcanic breccia and tuff.
莺山断陷侏罗系火山岩主要为安山岩、玄武岩、英安岩、火山角砾岩及凝灰岩。

Extended Term:
volcanic friction breccia 火山摩擦角砾岩

volcanic breccia texture [vɒlˈkænik ˈbretʃiə ˈtekstʃə]

Definition: 火山角砾结构 huǒ shān jiǎo lì jié gòu

volcanic: of, thrown from, caused by, or characteristic of a volcano. 火山的 huǒ shān de
breccia: rock consisting of sharp-cornered bits of fragmented rock, cemented together by sand, clay, or lime. 角砾岩 jiǎo lì yán
texture: the arrangement of the particles or constituent parts of any material, as wood, metal, etc., as it affects the appearance or feel of the surface; structure, composition, grain, etc. 结构 jié gòu

Example:
Volcanites in Dawa Oilfield are distributed in Fangshenpao Formation and Mesozoic group with laminar distribution, lithology is mainly consisted of basalts, andesites, volcanic breccia and tuffs, of which volcanic breccia texture is a key reservoir of Mesozoic group.
大洼油田火山岩主要分布于新生界房身泡组和中生界,呈层状分布,岩性主要为玄武岩、安山岩、火山角砾岩和凝灰岩,其中火山角砾结构是中生界的主要储层。

volcanic formation [vɒlˈkænik fɔːˈmeiʃən]

Definition: volcanoes form when hot magma from below rises and leaks into the crust. Magma, rising from lower reaches, gathers in a reservoir, in a weak portion of the overlying rock called the magma chamber. Eventually, but not always, the magma erupts onto the surface. Strong earthquakes accompany rising magma, and the volcanic cone may swell in appearance, just before an eruption. 火山岩建造 huǒ shān yán jiàn zào

Example:

The large scale gas reservoirs have been found in deep volcanic formation in the north of Songliao Basin of China.

在我国松辽盆地北部深层火山岩储层发现了大规模的天然气藏。

Extended Terms:

marine volcanic-sedimentary formation 海相火山-沉积建造区
volcanic tuff formation 凝灰质储层

volcanic mud ball [vɒlˈkænik mʌd bɔːl]

Definition: 火山泥球 huǒ shān ní qiú

volcanic: of, thrown from, caused by, or characteristic of a volcano. 火山的 huǒ shān de
mud: wet, soft, sticky earth. 泥 ní
ball: any round, or spherical, object; sphere; globe. 球 qiú

volcanic rock [vɔlˈkænik rɒk]

Definition: finely crystalline or glassy igneous rock resulting from volcanic activity at or near the surface of the earth. Also known as extrusive rock. 火山岩 huǒ shān yán

Example:

There is a large area of Masozoic acidic volcanic rock in southeastern China, which is an important area of volcanic uranium deposits.

中国东南部分布着大面积的中生代中酸性火山岩,是我国重要的火山岩型铀矿产区。

Extended Terms:

decomposed volcanic rock 风化火山岩
highly decomposed volcanic rock 高度风化火山岩
marine volcanic rock 海相火山岩
volcanic rock distribution 火山岩分布
volcanic rock probe 火山岩探针

volcanic rock QAPF classification [vɔlˈkænik rɒk ˈklæsifiˈkeiʃən]

Definition: 火山岩 QAPF 分类 huǒ shān yán fēn lèi

volcanic: of, relating to, or produced by a volcano or volcanoes. 火山的 huǒ shān de
rock: the solid mineral material forming part of the surface of the earth and other similar planets, exposed on the surface or underlying the soil. 岩石 yán shí; 岩 yán
classification: the action or process of classifying something according to shared qualities or characteristics. 分类 fēn lèi

volcanic rock TAS classification [vɔlˈkænik rɔk ˌklæsifiˈkeiʃən]

Definition: 火山岩 TAS 分类 huǒ shān yán fēn lèi

volcanic: of, relating to, or produced by a volcano or volcanoes. 火山的 huǒ shān de
rock: the solid mineral material forming part of the surface of the earth and other similar planets, exposed on the surface or underlying the soil. 岩石 yán shí; 岩 yán
classification: the action or process of classifying something according to shared qualities or characteristics. 分类 fēn lèi

volcanism [ˈvɔlkənizəm]

Definition: volcanic activity or phenomena. 火山活动 huǒ shān huó dòng

Origin: 1865-1870; volcan(o) + -ism.

Example:
Periods of volcanism are indicated by the presence of pyroclastic material in the formation.
群系中火成碎屑物的存在表明了火山酌的形成年代。

Extended Terms:
bimodal rift volcanism 双向裂谷火山活动
epigenetic volcanism 后期火山作用
intraplate volcanism 板内火山活动
lunar volcanism 月球火山作用
sedimentary volcanism 沉积火山作用

volcano [vɔlˈkeinəu]

Definition: a mountain or hill, typically conical, having a crater or vent through which lava, rock fragments, hot vapour, and gas are or have been erupted from the earth's crust. 火山 huǒ shān

Origin: early 17th century: from Italian, from Latin Volcanus "Vulcan".

> Example:

In fact, Mount Saint Helens appears to be the youngest, most recently active volcano in the Cascade Range.

事实上，圣海伦斯火山是喀斯特山脉中最年轻又是最近期活动的活火山。

> Extended Terms:

composite volcano 复式火山
compound volcano 复合火山
dormant volcano 休眠火山
extinct volcano 死火山
lava volcano 熔岩火山
mud volcano 泥火山
oceanic volcano 海底火山

Von Wolff diagram [vɒn ˈwul ˈdaɪəɡræm]

> Definition: 冯乌尔夫图解 féng wū ěr fū tú jiě

Von Wolff: 冯乌尔夫 féng wū ěr fū

diagram: a drawing intended to explain how something works; a drawing showing the relation between the parts. 图解 tú jiě

vug [vʌg]

> Definition: cavity in rock, lined with mineral crystals. 晶洞 jīng dòng

> Origin: early 19th century: from Cornish vooga.

> Extended Terms:

basaltic rock vug 玄武质岩石晶洞
solution vug 溶洞
vug crystal 窝内结晶
vug porosity 溶洞孔隙度

vulcanian [vʌlˈkeɪnɪən]

> Definition: relating to or denoting a type of volcanic eruption marked by periodic explosive events. 火山(周期性)喷发的 huǒ shān (zhōu qī xìng) pēn fā de

> Origin: early 20th century: from Vulcano, the name of a volcano in the Lipari Islands, Italy, + -ian.

> Example:

From the beginning of activity on 20 May to 26-27 August, Krakatau produced a discontinuous series of vulcanian to sub-plinian eruptions.
自5月20日火山活动开始到8月26—27日，喀拉卡托火山产生了间断性的周期性喷发和亚普林尼式喷发。

> Extended Term:

vulcanian type eruption 乌尔堪型火山喷发

wackestone [wæˈkestən]

> Definition:

Wackestone is a matrix-supported carbonate rock that contains over 10% allochems in a carbonate mud matrix. 粒泥灰岩 lì ní huī yán

> Origin:

1803, from German Wacke, from Middle High German wacke "a large stone," from Old High German wacko "gravel," probably from Old High German wegan "to move." A miner's word, brought into geology by German geologist Abraham Gottlob Werner (1750-1817).

> Example:

Calcarenite silicon-wackestone of the upper and lower silicon members deposit from contour currents and occur on the lower of the carbonate slope.
等深流沉积物形成了上、下硅质层段的砂屑硅质粒泥灰岩，主要分布于斜坡下部。

> Extended Terms:

bioclastic wackestone 生物碎屑粒泥灰岩
wackestone packstone 藻屑粒泥状泥粒状灰岩

wall rocks [wɔːl rɒks]

> Definition:

the rock adjacent to or enclosing a vein, hydrothermal ore deposit, fault, or other geological feature. 围岩 wéi yán

> Example:

Isotope geochemical studies show that sulfur, lead and carbon come from the wall rocks of the Jingshan Group.
矿床地球化学特征表明，矿质主要来自围岩荆山群地层。

wave base [weiv beis]

> Definition:

The wave base is the maximum depth at which a water wave's passage causes significant water motion. 浪基面 làng jī miàn

Example:

Under wave action, zone liquefaction sediment traps and finger like sandbody of wave base above their form, and finger like sandbody can bring isolation slumped turbidites.
波浪作用形成环带状分布的液化沉积区和浪基面之上的指状砂体，指状砂体在重力作用下可产生孤立的滑塌浊积体。

Extended Term:

normal wave base 正常浪基面

wave ripples [weiv'ripls]

Definition: 浪成波痕 làng chéng bō hén

wave: a long body of water curling into an arched form and breaking on the shore. 波浪 bō làng

ripple: ripple: a small wave or series of waves on the surface of water, especially as caused by a slight breeze or an object dropping into it. 波痕 bō hén

Example:

Wave ripple crest trends show that the lake shoreline had an overall east-northeast to west-southwest orientation.
浪成波痕顶表明湖岸线总体呈东北向西南的走向。

wavy bedding ['weivi 'bediŋ]

Definition: 波状层理 bō zhuàng céng lǐ

wavy: (of a line or surface) having or consisting of a series of undulating and wave-like curves. (线，表面)波状的 bō zhuàng de；起伏的 qǐ fú de

bedding: (Geology) the stratification or layering of rocks. [地质](岩石的)层理 céng lǐ

Example:

Flat bedding, wavy bedding and vermiglyph were often found in the shallow lake facies.
滨浅湖相常发育水平层理、波状层理及虫迹。

weathering index (WI) ['weðəriŋ 'indeks]

Definition: 风化指数 fēng huà zhǐ shù

weathering: the various mechanical and chemical processes that cause exposed rock to decompose. 风化 fēng huà

index: a figure in a system or scale representing the average value of specified prices, shares, or other items as compared with some reference figure. 指数 zhǐ shù

Example:

According to the results of paleoclimate study and estimate of semi-quantitative weathering index, the weathering extent of sediments was low and the topographic relief between the Kuqa Depression and South Tianshan was obviously in Early-Middle Triassic.
根据风化指数半定量计算的结果，结合前人古气候的研究成果，早-中三叠世该区沉积物风化程度低，推测此时地形高差较大。

websterite [ˈwebstərait]

Definition: Aluminite is a hydrous aluminium sulfate mineral with formula. 二辉岩 èr huī yán

Origin: named after T. Webster, 19th-century Englishman; see -ite.

Extended Term:

websterite porphyry 二辉斑岩

welded tuff [weldid tʌf]

Definition: 熔结凝灰岩 róng jié níng huī yán

welded: unite (pieces of plastic or other material) by melting or softening of surfaces in contact. 焊接的 hàn jiē de

tuff: a light, porous rock formed by consolidation of volcanic ash. 凝灰岩 níng huī yán

Extended Terms:

welded tuff texture 熔结凝灰结构
welded ashflow tuff 熔结灰流凝灰岩

welded tuff texture [weldid tʌf ˈtekstʃə]

Definition: 熔结凝灰结构 róng jié níng huī jié gòu

weld: to join (metals) by applying heat, sometimes with pressure and sometimes with an intermediate or filler metal having a high melting point. 焊接 hàn jiē

tuff: a rock composed of compacted volcanic ash varying in size from fine sand to coarse gravel. Also called tufa. 凝灰岩 níng huī yán

texture: the arrangement of the particles or constituent parts of any material, as wood, metal, etc., as it affects the appearance or feel of the surface; structure, composition, grain, etc. 结构

jié gòu

Example:

the characteristics of the Welded Tuff Texture of Marine Faces at Early Triassic Period in Jiangzhou Region

江州地区早三叠世海相熔结凝灰岩结构的特征

wet chemical analysis [wet ˈkemikəl əˈnæləsis]

Definition: 化学湿法分析 huà xué shī fǎ fēn xī

wet: covered or saturated with water or another liquid. 湿的 shī de

chemical: of or relating to chemistry, or the interactions of substances as studied in chemistry. (与)化学(有关)的(yǔ) huà xué (yǒu guān) de

analysis: detailed examination of the elements or structure of something, typically as a basis for discussion or interpretation. 分析 fēn xī

Example:

Compositions of constant and trace elements in Dushan jade from Nanyang area of Henan Province were studied by PIXE and wet chemical analysis.

利用质子激发 X 荧光分析(PIXE)和湿法化学分析对河南南阳独山玉进行常量和微量元素分析。

whin [win]

Definition: hard, dark basaltic rock such as that of the Whin Sill in Northern England. 暗色岩 àn sè yán

Origin: late Middle English: probably of Scandinavian origin.

Example:

Rough as a saw-edge, and hard as whin.

跟锯齿一样粗糙，像暗色岩一样的硬。

Extended Term:

whin sill 暗色岩床

whiting [ˈwaitiŋ]

Definition: ground chalk used for purposes such as whitewashing and cleaning metal plate. 白垩粉 bái è fěn

Origin: Middle English: from Middle Dutch wijting, from wijt "white".

Example:

This object is painted with whiting.
这件物品是用白垩粉进行粉刷的。

Williston Basin [wilistən ˈbeisən]

Definition: The Williston Basin is a large intracratonic sedimentary basin in eastern Montana, western North and South Dakota, and southern Saskatchewan known for its rich deposits of petroleum and potash. 威利斯顿盆地 wēi lì sī dùn pén dì

wind ripples [wind ˈripls]

Definition: 风成波痕 fēng chéng bō hén

wind: the perceptible natural movement of the air, especially in the form of a current of air blowing from a particular direction. 风 fēng

ripple: a small wave or series of waves on the surface of water, especially as caused by a slight breeze or an object dropping into it. 波痕 bō hén

Example:

The surface rarely shows such features as wind ripples or wind granule ripples.
表面上很少显示出风成沙波纹或风成细砾波纹。

wollastonite [ˈwuləstənait]

Definition: a white or greyish mineral typically occurring in tabular masses in metamorphosed limestone. It is a silicate of calcium and is used as a source of rock wool. 硅灰石 guī huī shí

Origin: early 19th century: from the name of W. H. Wollaston (see Wollaston) + -ite.

Example:

The effects of wollastonite on the structure and properties of nylon 66 were studied.
研究了针状硅灰石粉填充对尼龙 66 的结构和性能的影响。

Extended Terms:

wollastonite ceramics 硅灰石陶瓷
wollastonite in lump 硅灰石块
wollastonite in powder 硅灰石粉
wollastonite ore 硅灰石

wormy texture [ˈwɜːmi ˈtekstʃə]

Definition: 蠕虫状结构 rú chóng zhuàng jié gòu

wormy: (of wood or a wooden object) full of holes made by woodworm. 有蛀洞的 yǒu zhù dòng de

texture: the arrangement of the particles or constituent parts of any material, as wood, metal, etc., as it affects the appearance or feel of the surface; structure, composition, grain, etc. 结构 jié gòu

X-ray CT scan [eks rei siːtiː skæn]

Definition: X 射线 CT 扫描 X shè xiàn CT sǎo miáo

x-ray: electromagnetic radiation of short wavelength produced when high-speed electrons strike a solid target. X 射线 X shè xiàn

CT scan: a medical examination that uses a computer to produce an image of the inside of somebody's body from X-ray or ultrasound pictures. 电脑断层扫描 diàn nǎo duàn céng sǎo miáo

Example:

Chest X-ray and CT scan were taken if patients showed signs of pulmonary embolism.
若患者出现肺栓塞的现象，应摄胸部 X 线片及 CT 加以确诊。

x-ray fluorescence [ˈeks rei fluəˈresns]

Definition: X 射线荧光 X shè xiàn yíng guāng

x-ray: electromagnetic radiation of short wavelength produced when high-speed electrons strike a solid target. X 射线 X shè xiàn

fluorescence: [mass noun] the visible or invisible radiation produced from certain substances as a result of incident radiation of a shorter wavelength such as X-rays or ultraviolet light. 荧光 yíng guāng

Example:

At last, the intelligent multi-channel nuclear spectrometer has been combined with X-ray fluorescence probe to analyze several mineral samples.
最后，将智能多道谱仪与 X 射线荧光探头结合起来分析了一些矿物样品。

xenocryst [ˈzenəkrist]

Definition: a crystal in an igneous rock which is not derived from the original magma. 异晶

yì jīng；捕获晶 bǔ huò jīng

Origin: late 19th century：from xeno-"foreign"+crystal.

xenolith [ˈzenəliθ]

Definition: a piece of rock within an igneous mass which is not derived from the original magma but has been introduced from elsewhere, especially the surrounding country rock. 捕虏体 bǔ lǔ tǐ；捕虏岩 bǔ lǔ yán

Origin: 1900-1905；xeno- + -lith.

Extended Terms:
accidental xenolith 外源包体
basic xenolith 基性捕虏体
congnate xenolith 同源捕掳体

yield strength [jiːld strenθ]

Definition: the stress at which a specific amount of plastic deformation is produced, usually taken as 0.2 per cent of the unstressed length. 屈服强度 qū fú qiáng dù

Example:
The rare earth additive has the tremendous influence to the compress yield strength of foam aluminum alloy, the best rare earth content is 0.4% (quality score).
稀土添加剂对泡沫铝合金的抗压屈服强度有较大的影响，最佳的稀土添加量为0.4%（质量分数）。

zeolite [ˈziːəlait]

Definition: any of a large group of minerals consisting of hydrated aluminosilicates of sodium, potassium, calcium, and barium, They can be readily dehydrated and rehydrated, and are used as cation exchangers and molecular sieves. 沸石 fèi shí

Origin: late 18th century：from Swedish and German zeolit, from Greek zein "to boil" + -lite.

Example:
Treatment and analysis of test screening equipment, has strong technical force, and can produce various specifications of species of zeolite ore slag.
筛分处理和分析测试设备，有雄厚的技术力量，能生产各种规格品种的沸石矿粉矿砂。

Extended Terms:

synthetic zeolite 合成沸石
zeolite facies 沸石相
zeolite process 离子交换法

zircon [ˈzɜːkən]

Definition: a mineral occurring as prismatic crystals, typically brown but sometimes in translucent forms of gem quality. It consists of zirconium silicate and is the chief ore of zirconium. 锆石 gào shí

Origin: late 18th century: from German Zirkon.

Example:

Tin, gold, ilmenite, zircon, monazite and kaolin, fire-resistant clay, and so on have the potential for exploration.
锡、金、钛铁矿、锆英石、独居石及高岭土、耐火黏土等具有开发潜力。

Extended Terms:

chrome-zircon red 铬锆红
zircon refractory 锆英石耐火材料
zircon sand 锆砂

zoisite zone [ˈzɒɪsaɪt zəun]

Definition: 黝帘石带 yǒu lián shí dài

zoisite: a greyish-white or greyish-green crystalline mineral consisting of a basic silicate of calcium and aluminium. 黝帘石 yǒu lián shí
zone: an area or stretch of land having a particular characteristic, purpose, or use, or subject to particular restrictions. 区域 qū yù

zonal texture [ˈzəunəl ˈtekstʃə]

Definition: 环带结构 huán dài jié gòu

zonal: of or relating to soil characterized by well-developed horizons, reflecting the type of vegetation the soil supports and the climate to which it is exposed. 带状的 dài zhuàng de
texture: the arrangement of the particles or constituent parts of any material, as wood, metal, etc., as it affects the appearance or feel of the surface; structure, composition, grain, etc. 结构 jié gòu

zone refining [zəun riˈfain]

Definition: a method of purifying a crystalline solid, typically a semiconductor or metal, by causing a narrow molten zone to travel slowly along an otherwise solid rod or bar to one end, at which impurities become concentrated. 区域熔炼 qū yù róng liàn

Example:
These metals may be purified by a technique called zone refining.
这些金属可以用所谓"区域熔炼"来提纯。

zoning [ˈzəuniŋ]

Definition: (geology) a distinctive layer or region of rock, characterized by particular fossils (zone fossils), metamorphism, structural deformity, etc. 分带 fēn dài

Origin: 1810-1820; zone + -ing.

Example:
The shallow seismic prospecting is one of indispensable important prospecting methods to civic seismic micro zoning.
浅层地震勘探是城市地震小区划中不可缺少的重要勘探手段之一。

Extended Terms:
normal zoning 正常分带作用，常态分带，正常带，顺向分带
reverse zoning 逆带现象，逆向分带逆带现象